Edward Lee Greene

Flora Franciscana

An attempt to classify and describe the vascular plants of middle California

Edward Lee Greene

Flora Franciscana
An attempt to classify and describe the vascular plants of middle California

ISBN/EAN: 9783337268596

Printed in Europe, USA, Canada, Australia, Japan

Cover: Foto ©berggeist007 / pixelio.de

More available books at **www.hansebooks.com**

FLORA FRANCISCANA.

AN ATTEMPT

TO CLASSIFY AND DESCRIBE THE VASCULAR PLANTS

OF MIDDLE CALIFORNIA.

BY

EDWARD L. GREENE,

Associate Professor of Botany, in the University of California.

SAN FRANCISCO:
CUBERY & CO., PRINTERS, 587 MISSION STREET, BELOW SECOND.
1891.

FLORA FRANCISCANA.

SERIES I.

PHANEROGAMOUS or FLOWERING PLANTS.

Vegetables having stamens and pistils, and producing seeds, of which the most essential part is a distinct embryo.

Class I. ANGIOSPERMÆ.

Seeds enclosed within a pericarp. Cotyledons two or one.

Subclass I. DICOTYLEDONOUS or EXOGENOUS PLANTS.

Embryo with two cotyledons. Leaves netted-veined. Flowers having their parts usually in fives, fours or twos.

Division I. CHORIPETALÆ.

Corolla (often wanting) of petals which are distinct, at least at base.

Order I. LEGUMINOSÆ.

Boerhaave, Hortus Academicus, ii. 22 (1729); Haller, Stirp. Helvet. 565 (1742); Juss. Gen. 345 (1789).

Herbs, shrubs or trees with alternate, stipulate, compound (in *Cercis* simple) leaves: leaflets mostly entire, a few of the upper, in some genera, converted into tendrils. Sepals more or less united and forming a 2 5-toothed or -cleft cup, the odd tooth or segment inferior. Petals 5 (sometimes by abortion fewer), more or less united above the base; the two lowest joining to form the *keel*; the two lateral enfolding this and called the *wings*; the uppermost one broader than the others, usually erect, but in the bud folded down over the others (except in *Cercis*) and called the *banner*; the corolla as a whole papilionaceous, or butterfly-shaped. Stamens usually 10, distinct or diadelphous (9 and 1), or monadelphous. Pistil 1, usually becoming a *legume*, i. e., a 2-valved 1-celled pod with 1 row of seeds: these attached to the upper suture, and containing no albumen, the large embryo filling the integuments.

Hints of the Genera.

Leaves simple,	1
" equally pinnate, mostly tendril-bearing,	2, 3
" unequally pinnate; leaflets several pairs,	4–9
" pinnately 3-foliolate,	8, 9, 11, 12
" palmately 3-foliolate,	10, 18
" palmately 5—9-foliolate,	10, 16
Unarmed shrubs; leaves variously 1–3-foliolate,	13, 15
Spinescent shrubs; leaflets 3, 1 or 0,	14, 17

1. CERCIS, *Linnæus* (RED BUD). Shrubs with simple leaves: the flowers in axillary fascicles, appearing in spring before the leaves. Calyx campanulate, with 5 broad obtuse teeth. Petals 5, the banner small, enfolded by the wings; keel-petals distinct, larger than the wings. Stamens 10, distinct. Pod thin, flat, oblong, wing-margined along the upper suture.

1. **C. occidentalis,** Torr. in Gray, Pl. Lindh. 177 (1845). Widely branching, 6–20 ft. high: leaves round-cordate, entire, obtuse or emarginate, 2 in. broad, on petioles of 1 in. or less: fl. $\frac{1}{2}$ in. long, rose-purple: pod 2 in. long, $\frac{1}{3}$ in. broad, acute at each end. In the Coast Range from near Suñol, *Behr*, northward: also in the Sierra at low altitudes. Very beautiful when in flower. Apr.

2. VICIA, *Varro* (VETCH). Weak herbs with angular stems, climbing by tendrils which terminate the pinnate leaves. Peduncles axillary, 1–∞-flowered. Calyx 5-cleft or -toothed, the upper teeth shorter. Stamens diadelphous (9 and 1). Style filiform, bent upward at apex and villous all around, under the stigma, or else on the outside only. Pod oblong, several-seeded.

* *Racemose-flowered perennials.*

1. **V. gigantea.** Hook. Fl. i. 157 (1830). Stout, 5–10 ft. high: leaflets 10–13 pairs, linear-oblong, obtuse, mucronulate, 1–2 in. long; stipules 1 in., semisagittate, toothed at base: peduncles much shorter than the leaves; the dense raceme 1-sided, 5–18-flowered; fl. dull red; pod glaucous, black when ripe. Common along streams, climbing over shrubs and small trees. Seeds as large as small peas and said to be a fair substitute for them when young. May, June.

2. **V. Americana,** Muhl. in Willd. Sp. iii. 1096 (1801). Weak, 2–5 ft. high, climbing by branched tendrils, nearly glabrous: leaflets 8–12, thin-membranaceous, vivid green above, paler beneath, closely but delicately feather-veined, elliptic-lanceolate, entire, obtuse, mucronulate, 1 in. long: peduncles shorter than the leaves, 3–8 flowered: fl. ¾ in. long, bright purple: upper calyx-teeth very short, lower well elongated: pods 1 in. long, glabrous. Var. **truncata,** Brewer: *V. truncata,* Nutt. in T. & G. Fl. i. 270 (1838). Lower and stouter than the type; leaves linear to oblong-linear, usually dentate or even serrate toward the truncate apex; fl. larger and paler. The type of this species occurs in the upper part of the Sacramento valley among the foothills of the Sierra, extending thence northward to the British boundary and eastward across the continent. The variety is also somewhat rare, and not known except from San Benito and Contra Costa counties; but there is plenty of it among the Mt. Diablo hills back of Antioch, Byron, etc. The following has been confused with this.

3. **V. linearis,** Greene. Nutt. in T. & G. Fl. i. 276 (1838), under *Lathyrus: V. Americana* var. *linearis,* Wats. Proc. Am. Acad. xi. 134 (1876). Low, decumbent or diffuse, or taller and climbing by branched tendrils: leaflets 6–10, subcoriaceous, glaucescent, the few veins prominent, confluent along the margin of the linear entire mucromate leaflet: peduncles equalling or exceeding the leaves, few-flowered: fl. 1 in. long, violet-purple: pods shorter than in the last. Common throughout middle and southern California, but in the Coast Range only. Very common also in the Rocky Mountain region, where broad-leaved smaller-flowered forms prevail; but in all these variations most easily distinguished from *V. Americana* by the color, texture, and venation of the leaflets.

4. **V. Californica.** Erect or decumbent, 6–12 in. high, villous-pubescent, scarcely climbing, the tendrils short, stiffish and not branching: leaflets 8–12, subcoriaceous, delicately feather-veined, cuneate-obovate, truncate or retuse, 5–7 lines long, more or less dentate toward the mucronulate apex: racemes exceeding the leaves, 3–5-flowered: calyx-teeth all broad and short: corolla ½–¾ in. long, deep purple. In Calaveras Co., collected by the author in June, 1889. The plant cannot

be referred to *V. Americana*, on account of its low zigzag stems, short and simple tendrils, subcoriaceous leaflets, and soft-pubescent herbage. It may be the *V. truncata*, var. *villosa*, Kell.

* * *Few-flowered annuals.*

5. **V. exigua**, Nutt. in T. & G. Fl. i. 272 (1838). Slender, 1–2 ft. high: leaflets 4 or 6, oblong-linear, obtuse: peduncles filiform, shorter than the leaves, 1–2-flowered: calyx-teeth lanceolate from a broad base: corolla white or purplish, 2 lines long: pod glabrous, 4–5-seeded.—Hillsides or plains, preferring stony or sandy soil. Mar. May.

6. **V. Hassei**, Wats. Proc. Am. Acad. xxv. 129 (1890). Taller and less delicate than the last, the leaflets ampler, more numerous, deeply notched at apex: fl. 3 lines long: pod shortly stipitate, 5–8-seeded. Of more southerly distribution than the preceding, but found at Benicia, *Bigelow*, and Santa Cruz, *Anderson*.

7. V. SATIVA, Linn. Sp. Pl. 736 (1753). Stoutish, suberect, 2–3 ft. high: leaflets 8 or 10, obovate-oblong, truncate or retuse, mucronate: fl. 1 or 2, subsessile, ½ in. long, red-purple. The Vetch or Tare, cultivated from time immemorial as a food and fodder plant, must have been brought to California a century ago by the Missionaries, and is of frequent occurrence by way-sides and in old fields in most parts of the State.

3. **LATHYRUS**, *Theophrastus* (WILD PEA). Coarser plants than *Vicia*, with broader leaves and flowers, the style villous in a line up and down the inside (next the free stamen).

* *Tendril-bearing; the racemes many-flowered.*

1. **L. polyphyllus**, Nutt. in T. & G. Fl. i. 274 (1838). Glabrous, stoutish, the stem angular, 2–3 ft. high: leaflets 5–10 pairs, oblong, obtuse, distinctly petiolulate: tendrils short: stipules as large as the leaflets, triangular, acute, with or without sharp triangular teeth at base: peduncles shorter than the leaves: calyx-teeth subulate, ciliate, the lower twice as long as the upper: corolla ¾ in. long, purple. This, common in Oregon, is attributed to Humboldt Co., *Bolander*, but is otherwise unknown within the State; but the next is very closely related to it.

2. **L. Bolanderi**, Wats. Proc. Am. Acad. xx. 363 (1885). Often shrubby below, 3–5 ft. high: leaflets 3–5 pairs, ovate, obtuse or retuse, mucronate, 1–1½ in. long, thin, on very short petiolules: tendrils ample: stipules broadly semisagittate, acute, more or less toothed: peduncles equalling the leaves: lower calyx-teeth lanceolate-acuminate, longer than the tube; upper very short, broadly triangular, all glabrous along the margin, or nearly so: corolla ¾ in. long, rose-purple, fading yellowish. Frequent on wooded slopes throughout the Bay region and northward to Oregon. June.

3. **L. sulphureus,** Brewer, in Gray, Proc. Am. Acad. vii. 399 (1867). Slightly pubescent, 3 ft. high, stem acutely angled ; leaflets 8–12, subcoriaceous, ovate- to oblong-lanceolate, acute, mucronate, less than 1 in. long, sessile ; stipules broadly semisagittate, acute or acuminate, subentire ; peduncles equalling the leaves ; lower calyx-teeth lanceolate, nearly as long as the tube, sparsely ciliolate ; corolla 1_2 in. long, yellowish, fading brownish. Lower altitudes in the mountains from Nevada Co. to Calaveras. June.

4. **L. Jepsonii,** Greene, Pittonia, ii. 158 (1890). Nearly or quite glabrous ; stem 5–8 ft. high, strongly winged along the angles and striate between them ; leaflets 8–12, linear-lanceolate, acute, 2–3 in. long, subcoriaceous, venulose ; stipules small, setaceously acuminate ; peduncles stout, about as long as the leaves ; fl. rose-purple ; calyx-teeth ovate-lanceolate, the lowest not much longer than the others ; corolla 3_4 in. long, relatively broad ; pod 2–3 in. long, sessile in the calyx, 12–16-seeded. Muddy margins of sloughs, within reach of tide-water in the Suisun marshes, *Jepson ;* also near Stockton, *Greene.*

5. **L. Californicus,** Wats. Proc. Am. Acad. xx. 363 (1885). Tall like the last, with winged stems, etc., but foliage thin and the whole herbage soft-pubescent ; leaflets 3–7 pairs, ovate-oblong or narrower, 1–2 in. long, acute or acuminate ; stipules semisagittate, acute ; peduncles equalling the leaves ; fl. large, pinkish ; calyx-teeth short ; pod 2 in. long, narrowed to a stipe. Along streams among the foot-hills on both sides of the Sacramento valley. May.

6. **L. vestitus,** Nutt. in T. & G. Fl. i. 276 (1838). Low and herbaceous, or 8–15 ft. high and shrubby at base, soft-pubescent or nearly glabrous, the stems angled ; leaflets 5–7 pairs, ovate-oblong to linear, cuspidate, subcoriaceous ; stipules broadly or narrowly semisagittate, toothed or entire ; peduncles about equalling the leaves ; fl. 3_4 in. long, broad, purplish ; lower calyx-teeth lanceolate, acuminate, as long as the tube ; ovary and pod appressed-pubescent. Very common from San Joaquin and Sonoma counties southward. Variable in size, breadth of leaflets and stipules, but with distinctive leaf-texture and pubescence. Feb.–May.

7. **L. cinctus,** Wats. Proc. Am. Acad. xxiii. 263 (1888). Stout, with angular stem and pubescent herbage ; stipules semihastate, nearly 1 in. long, the broad basal lobe coarsely toothed ; leaflets 10–15 pairs, narrow-oblong, obtuse, mucronate, 1–$1\frac{1}{2}$ in. long ; peduncles much shorter than the leaves, few-flowered ; calyx short, the lower teeth equalling the tube ; fl. $1\frac{1}{2}$ in. long ; pod $1\frac{1}{2}$ in. long, $1\frac{1}{2}$ in. broad, 3–5-seeded ; seed orbicular, $2\frac{1}{2}$ lines broad, nearly encircled by the hilum. Near Jolon, Monterey Co., *Brandegee.*

* * *Without tendrils ; peduncles 1–3-flowered.*

8. **L. Nevadensis,** Wats. Proc. Am. Acad. xi. 133 (1876). Erect, slender, 8–10 in. high, nearly glabrous: leaflets 2–3 pairs, ovate or oblong, obtuse, 1 in. long, thin: stipules narrow, acuminate at both ends: peduncles slender, at least equalling the leaves, about 2-flowered: fl. nearly 1 in. long, ochroleucous: calyx-teeth triangular, short, not very unequal, except the lowest one, which is lanceolate and much longer. At middle elevations of the Sierra Nevada; also in the Trinity Mts., *Marshall*, in open woods: strictly erect, the leaf-rachis ending in a slender recurved point.

9. **L. Torreyi,** Gray, Proc. Am. Acad. vii. 337 (1868); *L. villosus*, Torr. Pac. R. Rep. xii. 58 (1860), not of Frivalds. Erect, slender, 1–2 ft. high, the herbage thin, light green, fragrant: leaflets 4–6 pairs, with or without a reduced terminal odd one, round-ovate or oblong, $\frac{1}{2}$ in. long, mucronate: stipules narrow, acuminate, the lower lobe short or almost obsolete: fl. 1 or 2, short-peduncled, white or pinkish: calyx-teeth narrowly subulate, the upper a little shorter: pod 1 in. long, pubescent, 3–6-seeded. From Santa Clara Co., *Charles Palache*, to Napa and northward, in dry woods. Remarkable among plants of this genus as exhaling the fragrance of *Asperula odorata*. It forms a link between *Lathyrus* and the Central American genus *Cracca*. May.

* * * *Rachis dilated, ending in a rudimentary odd leaflet; peduncle 4–8-flowered.* Genus ASTROPHIA, Nutt.

10. **L. littoralis,** Endl. in Walp. Rep. i. 722 (1842); Nutt. in T. & G. Fl. i. 278 (1838), under *Astrophia*; Gray, Pac. R. Rep. xii. 58, t. 6 (1860), under *Orobus*. Stout and low, decumbent, densely silky-villous: stipules large, ovate or semihastate: leaflets 1–3 pairs, cuneate-oblong, $\frac{1}{2}$ in. long or more: peduncles exceeding the leaves: calyx-teeth nearly equal, about as long as the tube: corolla $\frac{1}{2}$–$\frac{3}{4}$ in. long, banner bright purple, wings and keel white: pod large, oblong, obtuse, villous, 3–5-seeded. Strictly maritime, in sandy or clayey soil within reach of the sea-spray: Santa Cruz, *Anderson*, San Francisco, *Andrews*, *Greene*, and far northward. The plant has as much the aspect of a *Lotus* as of a *Lathyrus*, and perhaps ought to be regarded as forming a genus; a view held by Nuttall, its discoverer.

4. **ASTRAGALUS,** *Dioscorides* (RATTLE-WEED, LOCO-WEED). Herbs either erect or decumbent, with unequally pinnate leaves, no tendrils, persistent stipules, and axillary spikes or racemes of flowers which are usually small for the size of the plant, and rather narrow. Calyx 5-toothed. Petals with slender claws, the keel obtuse. Stamens diadelphous, (9 and 1); anthers uniform. Stigma terminal, minute. Pod various, seldom or never promptly dehiscent, often coriaceous and turgid, or thin and bladdery-inflated, or thin and flat; 1-celled, or partly 2-celled by intrusion of

one or both sutures. Seeds few or many, small for the size of the pod, commonly reniform, on slender funiculi. A polymorphous genus, embracing some hundreds of species, most of them inhabiting northern Asia and North America. Most of ours have, when fresh, a heavy somewhat nauseating odor. Several are thought to be poisonous to cattle and horses.

* *Annuals.*

1. **A. didymocarpus**, H. & A. Bot. Beech. 334. t. 81. (1840). Slender, pubescent, 1 ft. high : leaflets 9–15, cuneate-oblong to linear, emarginate, 3–5 lines long : spikes long-peduncled, dense, ovate or oblong : fl. small, dull purplish : pods erect, 2 lines long and about as broad, scarcely exserted from the calyx, strongly wrinkled, 2-celled, 2-seeded. Abundant along the eastern base of Mt. Diablo Range and far southward ; apparently not in the Bay region, or near the coast, where it is replaced by the next.

2. **A. nigrescens**, Nutt. Pl. Gamb. 152 (1848). Smaller than the last, more slender, less pubescent, the less dense spikes cylindrical : pods deflexed, well exserted from the calyx, slightly wrinkled, strongly obcompressed. Common on sterile gravelly hill-sides of the Bay region ; the flowers commonly minute and dull, but on the flanks of Mt. Tamalpais and northward larger and violet. An exceedingly well marked species which eastern botanists had confused with the preceding.

3. **A. tener**, Gray, Proc. Am. Acad. vi. 206 (1864) ; *A. hypoglottis*, var. *strigosus*, Kell. Proc. Calif. Acad. ii. 115 (1863). *Phaca astragalina*, H. & A. Bot. Beech. 334 (1840). Slender, sparsely pubescent, 6–10 in. high : leaflets 9–15, linear or cuneate, acute or retuse : fl. many, capitate on a slender peduncle, purple : pod ¾ in. long, slender, incurved, 2-celled, 5–10-seeded. In moist lands, either sandy or alluvial. A handsome species ; the heads of purple and white recalling those of some kinds of clover. Apr. May.

4. **A. Breweri**, Gray, Proc. Am. Acad. vi. 207 (1864). Smaller than the last, relatively stouter, leaflets broader, heads few-flowered : pods with a short body and a very long incurved beak. Common in fields of the Sonoma valley, *Brewer*, and in Lake Co., *Mrs. Curran*. Rarely collected and perhaps somewhat local.

5. **A. Rattani**, Gray, Proc. Am. Acad. xix. 75 (1883). Strigose puberulent, 1 ft. high : leaflets 11–17, obovate-oblong, emarginate : fl. few, capitate, the peduncles exceeding the leaves : teeth of calyx shorter than the campanulate tube : corolla 5 lines long, violet : pods spreading, very slender, subulate-beaked, 2 in. long, partly 2-celled, many-seeded. Prairies of Mendocino Co., *Rattan*, and of Humboldt, *Chesnut & Drew*. Species wearing much of the aspect of a *Lotus*. June.

* * *Perennials.*

← Pods bladdery-inflated, more or less perfectly 2-celled.

6. **A. Coulteri,** Benth. Pl. Hartw. 307 (1849) ; *A. Arthu-Schottii,* Gray, Proc. Am. Acad. vi. 209 (1864). Stoutish, 1–2 ft. high, decumbent, often somewhat woody at base, whitish with a short close pubescence : leaflets 9–15, oblong or obovate, obtuse or emarginate, 1_2 in. long : spike loosely 10–20-flowered : calyx-teeth shorter than the tube : corolla $1\frac{1}{2}$ in. long, purple : pod ovate, acute, $^3_1$ in. long, chartaceous, hoary with a short pubescence. Common in the southern parts of the State, but said to have been found in the first place near Monterey, *Coulter.* I have seen it to the northward of San Luis Obispo.

7. **A. Fremonti,** Gray, Pac. R. Rep. iv. 80 (1857). Ascending, 6–10 in. high, silvery-canescent : leaflets 9–21, oval, retuse, $1\frac{1}{2}$ in. long : fl. subsessile, spreading, purple : calyx-teeth subulate, shorter than the tube : pod chartaceous, round-ovate, acuminate, 1 in. long, obscurely strigulose, mottled with purple. At Sonora Pass in the Sierra Nevada, thence southward to San Bernardino, and eastward in the desert regions. Related to *A. Coulteri,* but of different habit and with rounder pods. It has been referred to *A. lentiginosus ;* but that has much smaller pods of a very different texture.

8. **A. lentiginosus,** Dougl. in Hook. Fl. i. 151 (1830). Stoutish, diffuse, glabrous except a minute and sparse roughness upon the stems and along the margins of the 11–19 obovate or oblong leaflets : fl. spicate, or fewer and subcapitate, whitish : pod 1_2–$^3_1$ in. long including the broadly ovate body and abrupt stout beak-like acumination, the whole slightly incurved, firm-coriaceous, white or freckled.—In Sierra Co. or Plumas, if at all within our limits ; for its home is upon the plains of N. E. California, E. Oregon, etc. Readily distinguished from the foregoing by its green and glabrous aspect, as well as by its firm and hard long-pointed and incurved pods.

9. **A. platytropis,** Gray, Proc. Am. Acad. vi. 526 (1865). Dwarf, tufted on horizontal rootstocks, silvery-silky : leaflets 7–11, obovate : peduncles scapiform, equalling the leaves, capitately few-flowered : calyx-teeth subulate, shorter than the tube : pod round-ovate, scarcely 1 in. long, abruptly pointed, puberulent, white or freckled. In loose sand and gravel, near the summit of the Sierra at Sonora Pass, *Brewer.* Apparently not otherwise known within the State ; but it occurs in Nevada.

+ + *Pods bladdery, 1-celled, the dorsal suture not at all intruded.*

10. **A. Hookerianus,** Dietr. Syn. iv. 1086 (1847) ; T. & G. Fl. i. 693, under *Phaca* (1840). Less than 1 ft. high, canescently pubescent or glabrate : leaflets 13–19, rather remote, oblong to linear, 1_4–1_2 in. long : fl. short-pedicelled, whitish : calyx-teeth triangular, very short, the tube cylindrical : pod obovate-oblong, obtuse, tapering to a short scarcely exserted stipe, 2 in. long, thin, glabrous, white or mottled.—Another

northeastern species, coming within our limits in the mountain counties of Nevada, *Bolander*, and Sierra, *Lemmon*.

11. **A. Whitneyi,** Gray, Proc. Am. Acad. vi. 526 (1865). Near the last but more pubescent; fl. red-purple; pod smaller, oval, narrowed to a slender stipe which is longer, about twice the length of the calyx. -Collected at Sonora Pass, *Brewer*.

12. **A. oxyphysus,** Gray, l. c. 218 (1864). Erect, 2–3 ft. high, stoutish, canescent with a minute pubescence; leaflets 9–21, oblong, 1 in. long; peduncles exceeding the leaves, raceme elongated: calyx-teeth subulate, half as long as the oblong tube: corolla greenish-white, ⅔ in. long: pod compressed, oblique (semiobovate), acuminate at both ends, 1½ in. long, on a stipe little exceeding the calyx. Dry hills of the Mt. Diablo Range, at Arroyo del Puerto, *Brewer*, and in San Luis Obispo Co., *Lemmon*.

13. **A. curtipes,** Gray, l. c. Suffrutescent, the branches erect, 1 ft. high or more, cinereous-pubescent: leaflets 13–33, narrowly oblong, obtuse or retuse, ⅓–¾ in. long: peduncles long, raceme short: calyx-teeth setaceous-subulate, little longer than the campanulate tube: pod not compressed, semiovate, acute, 1½ in. long, pendulous on a short rigid stipe. From San Luis Obispo southward, on dry hills.

14. **A. leucophyllus,** T. & G. Fl. i. 336 (1838). Erect, tall, growing parts silvery-canescent, when older glabrate: leaflets 27–37, broadly linear, acutish, ¾ in. long: peduncles long, racemes short: calyx-teeth subulate, half as long as the oblong tube: corolla yellowish: pod obliquely oval, 1½ in. long, on a filiform pubescent stipe nearly as long. Low hills skirting the interior valley from Sacramento southward to Monterey; very common between Livermore and Niles, and probably throughout the Mt. Diablo range southward.

15. **A. leucopsis,** Torr. Bot. Mex. Bound. 56. t. 16 (1859); T. & G. Fl. i. 694 (1840) under *Phaca*. Size, habit, foliage, inflorescence, etc., as in the last, but calyx-teeth of more than half the length of the tube, which is campanulate: pod tapering at base, the stipe glabrous and only 1½ in. long. The common Rattle-Weed in San Diego Co. It is credited to Santa Barbara Co. and is very likely to be found somewhat farther northward.

16. **A. trichopodus,** Gray, Proc. Am. Acad. vi. 218 (1864); Nutt. in T. & G. Fl. i. 344 (1838), under *Phaca*. Slender, 1–3 feet high, young parts hoary, the older strigose-puberulent: leaflets very many, oblong to linear, 1½ in. long: racemes short, on peduncles exceeding the leaves: calyx-teeth much shorter than the campanulate tube: fl. yellowish: pod oval, obtuse at both ends, ½ in. long, on a short filiform stipe. Also of southern California, but perhaps in San Luis Obispo Co.

17. **A. crotalariæ,** Gray, Proc. Am. Acad. vi. 216 (1864); Benth. Pl. Hartw. 307 (1849) under *Phaca*. Stout, decumbent, glabrous, except the canescent growing parts : leaflets very many, oblong-linear to obovate, sometimes retuse, 1½ 1 in. long : stipules broadly triangular, distinct : calyx-teeth subulate, half as long as the short-campanulate tube : fl. white : pod thin, ovoid, 1 1½ in. long, sessile in the calyx. Plains and hills from San Francisco to Santa Barbara. A var. **virgatus,** Gray, is described as having narrower stipules, a looser raceme, longer and narrower calyx-teeth, etc. It may be a hybrid between this and the next.

18. **A. Menziesii,** Gray, l. c. (1864). *Phaca Nuttallii*, T. & G. Fl. i. 343 (1838) : *P. densifolia*, Smith in Rees' Cycl. (1819). Stout, erect, 2 4 ft. high, glabrous or nearly so : stipules broad, not pointed, continued around the stem, sometimes nearly meeting or even cohering opposite the base of the leaf : raceme long and dense : fl. greenish : pod thin, large as in the last. Plentiful in sandy soils along the seaboard, at Alameda, West Berkeley, San Francisco, etc. Dr. Torrey, in a Pacific Railroad Report, restored to this its first specific name given under *Phaca*, but inadmissibly ; there being a much older *Astragalus densifolius* of Lamarck.

19. **A. macrodon,** Gray, l. c. (1864) ; H. & A. Bot. Beech. 333 (1840), under *Phaca*. Erect, tall, glabrous in age, the nascent parts canescent : leaflets 23 27, oblong-lanceolate, obtuse, mucronulate : stipules small, lanceolate-acuminate : peduncles rather shorter than the leaves ; racemes long : calyx-teeth slender-subulate, equalling the campanulate tube, and almost as long as the corolla : ovary silky : pod unknown. This obscure plant was collected by Douglas only, some sixty years ago, somewhere between Monterey and Sonoma, probably near the former place. It should be carefully sought, though it may have become extinct. The long and slender calyx-teeth, according to the original description, so distinguished it from its allies, as to make its recognition easy in case it should be rediscovered.

20. **A. Douglasii,** Gray, l. c. (1864) ; T. & G. Fl. i. 346 (1838), under *Phaca*. Ascending, 1 ft. high, cinereous-puberulent when young, otherwise nearly glabrous : leaflets very many, linear or linear-oblong, 1½ -¾ in. long : spike short, dense, 10 20-flowered : calyx-teeth subulate, shorter than the campanulate tube: pod thin, obliquely ovoid, 1½ 2 in. long. In gravelly places along streams, from San Francisco to San Luis Obispo.

21. **A. Hornii,** Gray, Proc. Am. Acad. vii. 398 (1868). Slender, ascending, 2 ft. high or more, glabrous or pubescent : leaflets 21—29, narrowly oblong : peduncles much longer than the leaves, bearing a dense spike of yellowish flowers : calyx-teeth broadly subulate, scarcely as long as the campanulate tube : pods densely spicate, ovate from a broad base.

acuminate, $\frac{2}{3}$ in. long, somewhat villous. From Tulare Co. southward. Easily recognized by its cone-like spike of somewhat imbricated pods.

22. **A. Pulsiferæ**, Gray, l. c. x. 69 (1874). Slender, tufted, procumbent, white-villous ; leaflets 5—11, obovate-cuneiform, usually retuse, 3—4 lines long ; peduncles equalling the leaves, few-flowered ; calyx-teeth linear-filiform, twice the length of the campanulate tube ; corolla white with a purple tinge ; pod $\frac{1}{2}$ in. long, ovate, incurved, villous. Gravelly places in Plumas and Sierra counties. *Mrs. Ames, Mr. Lemmon.*

+ + + *Pods not bladdery, mostly of hard texture.*

23. **A. Purshii**, Dougl. in Hook. Fl. i. 152 (1830). Stems a few inches long, forming matted tufts ; herbage densely silky-villous ; leaflets 9—15, oblong ; peduncles shorter than the leaves ; fl. few, capitate, 1 in. long or less, bright purple ; calyx-teeth subulate, shorter than the tube ; pod ovate, incurved, 1 in. long, densely white-woolly, cartilaginous, obcompressed, the two sutures nearly meeting within, thus forming two incomplete compartments.—On the eastern slopes of the Sierra if at all within our limits ; but common in Nevada and northern California.

24. **A. malacus**, Gray, Proc. Am. Acad. vii. 336 (1868). Low, stoutish, densely white-hirsute ; leaflets 11—17, obovate, retuse ; peduncles exceeding the leaves, spicately few- or many-flowered ; fl. $\frac{1}{3}$ in. long, deep purple ; calyx cylindrical, dark-hairy, the slender teeth much shorter than the tube ; pods pendulous, 1 in. long, 3—4 lines wide, lunately incurved, densely hairy, 2-celled by intrusion of the dorsal suture, the cross section narrow-obcordate.—Eastern slope of the Sierra, southward.

25. **A. Andersonii**, Gray, l. c. vi. 524 (1865). Slender, canescent with a dense somewhat silky pubescence ; leaflets 13—25, oblong, oval or obovate, mucronate ; peduncles longer than the leaves ; fl. many, yellowish-white, crowded in an oblong or cylindrical spike ; calyx-teeth subulate-setaceous, nearly as long as the cylindrical whitish-villous tube ; pods pendulous, falcate, $\frac{1}{2}$—$\frac{3}{4}$ in. long, 2 lines wide, abruptly pointed, downy, 2-celled as in the last. Habitat of the preceding.

26. **A. Congdoni**, Wats. Proc. Am. Acad. xx. 360 (1885). Stems 1—2 ft. long, decumbent ; herbage soft-pubescent ; leaflets 17—21, small, orbicular, obovate or oblong, obtuse or retuse ; racemes loose, long-peduncled ; teeth of the campanulate calyx short, triangular ; corolla pale yellow ; pod chartaceous, linear, curved, 1 in. long, puberulent, 2-celled as in the above.—In Mariposa Co. along the Merced River. *Congdon.*

27. **A. Bolanderi**, Gray, Proc. Am. Acad. vii. 337 (1868). Low, suberect, slightly pubescent ; stipules scarious, united opposite the petiole ; leaflets very many, oblong-linear, $\frac{1}{2}$—$\frac{1}{3}$ in. long ; peduncles stoutish,

equalling the leaves; raceme very short; fl. ½ in. long, purplish: calyx-teeth slender-subulate, shorter than the tube: pod ovate, incurved, transversely veiny, less than 1 in. long, abruptly recurved on the ascending stipe, 2-celled. Yosemite Valley and northward to Sierra Co., in gravelly soil at 6,000 ft. altitude and upwards.

28. **A. Mortoni,** Nutt. in Gray, l. c. vi. 196 (1864). Stoutish, erect, 2 ft. high, minutely appressed-pubescent: leaflets 17—21, oblong, ½—1 in. long: fl. greenish white, in a dense oblong long-peduncled spike: pods erect, ½ in. long, oblong, minutely pubescent, 2-celled, grooved at the dorsal suture, the ventral one externally prominent. The western analogue of *A. Canadensis*; found at Mono Lake, *Brewer*, and northward along our eastern borders.

29. **A. pycnostachyus,** Gray, l. c. 527 (1865). Stout, 2 ft. high, more or less villous-hoary: leaflets about 21, oblong, ½ in. long: fl. yellowish, in dense cylindrical short-stalked spikes: pods crowded, retrorsely imbricate, ovate, acute, laterally flattened, thin-coriaceous, glabrous, coarsely reticulate, 1-celled. In moist subsaline grassy land near the entrance to Bolinas Bay, *Bolander*, 1863, *Greene*, 1888; also in a similar locality not so near the sea southwest of Mt. Tamalpais; more frequent in the southern part of the State.

30. **A. Antisellii,** Gray, in B. & W. Bot. Calif. i. 152 (1876). Rather slender, 2 ft. high, cinereous-pubescent: leaflets 21—29, crowded, linear-oblong, 2—4 lines long: raceme lax, few-flowered: calyx-teeth half the length of the campanulate tube: pod thin, linear-oblong, compressed, 1-celled, ⅔ in. long, 2 lines wide above the middle, thence tapering to the slender stipe which is thrice as long as the calyx.—From San Luis Obispo Co. southward.

31. **A. speirocarpus,** Gray, Proc. Am. Acad. vi. 225 (1864). Slender, a span high, cinereous-pubescent: leaflets 9—17, obovate or oblong, emarginate: racemes short: calyx-teeth scarcely a fourth the length of the cylindraceous tube: pod thin, flat, coiled into a ring or spiral, stipitate, 1-celled.—From Sierra Co., *Lemmon*, northward to Washington, in the var. **falciformis,** Gray, with pods curved only to the falcate, not into a ring even.

32. **A. Gibbsii,** Kellogg, Proc. Calif. Acad. ii. 161. fig. 50 (1863); *A. cyrtoides*, Gray, l. c. 201 (1864). Stoutish, erect, softly hoary: leaflets 11—21, ovate-oblong, retuse or obcordate: peduncles elongated: fl. many in a dense short raceme: calyx downy, the teeth not half as long as the long-campanulate tube: pod pubescent, 1 in. long, cartilaginous, falcate or more strongly curved, stipitate, 1-celled.—From Placer Co. to Sierra, and eastward in Nevada.

33. **A. Webberi,** Gray, in B. & W. Bot. Calif. i. 154 (1876). Low, very leafy, silvery-canescent ; leaflets 11 21, oblong or obovate, 4 7 lines long : peduncles exceeding the leaves ; spike densely 9 20-flowered : calyx-teeth subulate, half as long as the oblong-campanulate tube : corolla white or yellowish, $\frac{1}{2}$ in. long : pod glabrous, cartilaginous, 1 in. long, oblong, obtuse, arcuate or nearly straight, somewhat compressed, sessile in the calyx, 1-celled. Plumas Co., *Mrs. Ames, Mr. Lemmon.*

34. **A. Lemmoni,** Gray, Proc. Am. Acad. viii. 626 (1873). Minutely appressed-pubescent ; stems slender, diffuse, 1 ft. long or more ; leaflets 9 11, linear-oblong, mucronate, 4 5 lines long ; peduncles filiform, 1 2 in. long, having dense racemes of very small pale-purplish flowers ; calyx-teeth subulate-setaceous, as long as the short-campanulate tube ; pod chartaceous, 2 lines long, ovate-oblong, obtuse, turgid, imperfectly 2-celled, the cross section obcordate. Plumas Co. and northward, *Bolander, Lemmon, Greene.*

35. **A. Clevelandi,** Greene, Bull. Torr. Club, ix. 121 (1882). Nearly glabrous, slender, erect, 2 3 ft. high ; leaflets 15 19, less than $\frac{1}{2}$ in. long ; peduncles exceeding the leaves, the loose spicate raceme of small white flowers often 6 in. long ; pod very small. This plant, much resembling *Melilotus alba*, was first obtained by Mr. Cleveland in Indian Valley, Lake Co., afterwards by the author, on the northern slope of Mt. St. Helena, where it is common along streams in open places. June, July.

36. **A. Austinæ,** Gray, in B. & W. Bot. Calif. i. 156 (1876). Low, densely tufted, silvery-silky ; stipules scarious, mostly united into an ovate body opposite the leaf ; leaflets 9 17, oblong or oval-lanceolate, acute or mucronate, 4 5 lines long ; peduncle equalling or exceeding the leaf ; fl. capitate ; calyx-teeth filiform, longer than the campanulate tube, and nearly as long as the pale corolla, of which the banner and wings are pubescent externally ; pod chartaceous, turgid-oval, hoary-pubescent, 2 lines long and scarcely exceeding the calyx-teeth, imperfectly 2-celled. High peaks of Nevada Co., *Lemmon.*

5. **GLYCYRRHIZA,** *Dioscorides* (LICORICE). Glandular-viscid perennials with unequally pinnate leaves, and flowers in axillary peduncled spikes ; calyx 5-cleft. Stamens monadelphous or diadelphous ; the alternate anthers smaller. Pod short, compressed, prickly, indehiscent, few-seeded.

1. **G. glutinosa,** Nutt. in T. & G. Fl. i. 298 (1838) : *G. lepidota*, var. *glutinosa*, Wats. Two or three ft. high, erect or decumbent, either nearly glabrous and viscid with minute sessile resinous dots, or more decidedly glutinous by a villous or hirsute glandular pubescence, never scurfy : leaflets 13 to 19, oblong-lanceolate, 1 or 2 in. long ; stipules ovate-

acuminate to lanceolate, persistent : spikes merely oblong, 1 to 1½ in. long, on peduncles of 1 in.: pod bur-like. Common in orchards and fields about Vacaville, *Jepson*, where it is a troublesome weed ; also at Stockton, *Sanford*, and southward in the Mt. Diablo Range to Corral Hollow, *Brewer*. The species varies greatly in the degree of its hairiness and viscosity, but it is never lepidote. The absence of all scurfiness, the always short and short-stalked spikes, and above all, the thoroughly persistent stipules which, upon the lowest parts of the plant are even partly adnate to the petiole, render it impossible to merge the species in *G. lepidota* as a variety.

6. **AMORPHA,** *Linnæus*. Shrubs with unequally pinnate leaves which, with the young twigs and inflorescence, are pellucid-glandular and heavy-scented, the glands in age dark brown and opaque. Leaflets many ; stipules and stipels caducous. Flowers very small, dark purple, in long and narrow terminal spikes. Calyx obconic-campanulate, 5-toothed, persistent. Banner (the only petal present) erect, concave, unguiculate. Stamens monadelphous at the very base. Pod short, lunulate, glandular, scarcely dehiscent, 1- or 2-seeded.

1. **A. Californica,** Nutt. in T. & G. Fl. i. 306 (1838). Three to eight ft. high, the nascent parts villous-canescent : leaflets 11-15, elliptic oblong, obtuse, an inch long ; calyx-teeth acute, but broadly triangular and very short (broader than long). Santa Barbara, *Nuttall*, and doubtless farther northward, though perhaps not within our limits.

2. **A. hispidula.** Two to four ft. high, pubescent or glabrous, the glandular dots supplemented on the twigs, stalklets and leaf-rachis by acute prickle-like glands with tips more or less recurved : leaflets 17 to 25, oval to linear-oblong, an inch long, retuse or emarginate : calyx-teeth triangular-lanceolate, more than half the length of the tube : petal red-purple : pod half obcordate, very glandular, twice the length of the calyx. Frequent from Monterey Co., *Hickman*, to Marin and Napa. Confused with *A. Californica*, by Brewer & Watson, though entirely distinct. The prickle-like glands, interspersed among the depressed and sessile ones, are very characteristic ; nor are the elongated calyx-teeth less so.

7. **ROBINIA,** *Linnæus* (LOCUST-TREE). Trees or shrubs with odd-pinnate leaves and stout prickles in place of stipules ; the leaflets prickly-stipellate. Flowers showy, in pendulous racemes. Calyx slightly bilabiate, 5-toothed. Banner large, roundish, reflexed, little longer than the wings and keel. Stamens diadelphous. Pod linear, flat, several-seeded, margined along the upper suture, readily dehiscent.

1. R. PSEUDACACIA, Linn. Sp. Pl. ii. 722 (1753). Tree with large loose

racemes of very fragrant white flowers. Native of the Altantic states ; long cultivated in California for shade and ornament ; now spontaneous in many places.

8. **PSORALEA,** *Royen*. Perennials (one adventive species shrubby), punctate with dark dots and heavy-scented : leaves pinnately 3-foliolate (in No. 6 palmately 5-foliolate); stipules free from the petiole. Calyx-lobes nearly equal, the two upper sometimes connate. Keel broad, obtuse, joined to the wings. Stamens monadelphous or diadelphous ; anthers uniform. Pod ovate, indehiscent, 1-seeded.

1. **P. orbicularis,** Lindl. Bot. Reg. xxiii. t. 1971 (1837). Stem prostrate, creeping, the leaves and racemes erect, long-stalked : leaflets 2 3 in. long, the terminal one nearly orbicular, the lateral pair obovate : raceme a few inches to a foot long, the flowers subtended by large deciduous bracts: calyx villous and pedicellate-glandular, cleft almost to the base, the lowest tooth as long as the purplish corolla : stamens diadelphous ; pod ovate, acute, 3 lines long. Frequent in moist grassy places. July.

2. **P. strobilina,** H. & A. Bot. Beech. 332. t. 80 (1840). Erect, 2 3 ft. high, villous throughout, the stem and stalklets glandular ; leaflets rhombic-ovate, 2 in. long; stipules large, broadly ovate, acuminate : peduncles shorter than the leaves : spike oblong, the bracts very large, deciduous ; calyx $\frac{1}{2}$ in. long, the lower tooth much the longest, equalling the purple corolla : stamens monadelphous ; ovary pubescent. Said to inhabit the mountains of Contra Costa and Santa Cruz counties.

3. **P. macrostachya,** DC. Prodr. ii. 220 (1825). Three to twelve feet high, the pubescence variable : leaflets ovate-lanceolate ; stipules small, lanceolate ; peduncles greatly surpassing the leaves : spikes cylindrical, silky-villous ; bracts acuminate, as long as the flowers : lower calyx-tooth longest, scarcely as long as the corolla : tenth stamen almost free : pod ovate-oblong, acute, 3 or 4 lines long, compressed, villous. Very common, either on hill-sides or in low ground, but in moist places, chiefly along streams in the mountain districts ; abundant and of rank growth in the Suisun marshes ; varying from nearly glabrous to somewhat tomentose. June Oct.

4. **P. physodes,** Dougl. in Hook. Fl. i. 136 (1830). Two or three feet high, erect, nearly glabrous ; leaflets ovate, acute, 1 in. long ; stipules linear-lanceolate ; peduncles about as long as the leaves ; raceme short, dense, the bracts small : calyx covered with sessile glands and somewhat black-hairy, at length much enlarged and inflated, becoming 4 or 5 lines long, its teeth short, subequal : corolla scarcely $\frac{1}{2}$ in. long, ochroleucous, often with a deep purple tinge : stamens monadelphous ; pod rounded, compressed, 3 lines long. Common in both the Coast and Contra Costa Ranges, in open places among thickets and trees. May July.

5. P. GLANDULOSA, Linn. Sp. Pl. ed. 2. 1075 (1762). Shrubby or arborescent, with loose elongated branches; glabrous, but roughish with elevated glands: leaflets ovate-lanceolate, acuminate, 2 or 3 in. long; stipules subulate-setaceous, deciduous: racemes longer than the leaves, the bluish flowers more or less verticillate. — Native of Chile; frequent in cultivation, occasionally spontaneous.

6. P. Californica, Wats. in Gray, Proc. Am. Acad. xii. 251 (1877). Low, tufted: pubescence short, silky, appressed: leaves palmately 5-foliolate; stipules scarious, lanceolate, deciduous; leaflets broadly oblanceolate, acutish, $\frac{3}{4}$–$1\frac{1}{4}$ in. long: racemes shorter than the leaves, short-peduncled, rather loose; pedicels slender: calyx silky-villous, $\frac{1}{2}$ in. long, the linear acuminate lobes a little exceeding the petals: pod thin, villous, oblong with a lanceolate beak: seed compressed, 2 lines long or more. Head waters of the Salinas, San Luis Obispo Co., *Palmer*.

9. LOTUS, *Tournefort* (LOTUS. HOSACKIA). Herbaceous or suffrutescent, with pinnately 3–x-foliolate (in the first species often 1-foliolate) leaves; leaflets sometimes of even number but unequally distributed on the two sides of the rachis; stipules foliaceous, scarious, or more commonly reduced to dark glands. Flowers solitary, or in umbels or heads which are naked or subtended by a 1–5-foliolate bract. Calyx 5-toothed or -cleft. Corolla whitish, yellowish or purplish, changing to orange or red: petals free from the stamens; banner ovate or rounded: wings commonly meeting imperfectly and (by a twist in the claw) obliquely in front of the obtuse or acute, sometimes rostrate keel. Stamens diadelphous; the alternate filaments dilated or thickened under the anthers. Pod linear, compressed or terete, straight or arcuate, promptly or tardily dehiscent, or indehiscent, 1–x-seeded. Seeds variously rounded or elongated, sometimes quadrate, smooth, tuberculate or rugose. —A large genus, related to the clovers and of some value as forage plants. The American species, quite numerous, are mostly Californian, and have been without sufficient reason treated as constituting one or more genera distinct from *Lotus*.

* *Annuals with gland-like traces of stipules; leaflets 1–3, on a linear rachis; pods straight, readily dehiscent.*— Genus ACMISPON, Raf.

1. L. Americanus, Bisch. Hort. Heidelb. (1839); Nutt. Gen. ii. 120 (1818), under *Trigonella*: *L. sericeus*, Pursh (1814), not of DC. (1813). *Hosackia Purshiana*, Benth. Erect or decumbent, 1–2 ft. high, more or less villous: leaflets (rarely 5) ovate or oblong, acutish, $\frac{3}{4}$ in. long: peduncles slender, exceeding the leaves, the solitary salmon-colored or whitish flower subtended by a bract 3–6 lines long: calyx-tube very short, the linear teeth equalling the corolla: pod 1–$1\frac{1}{2}$ in. long: seeds oblong, smooth, dark-colored. On sunny banks, or in the dry gravelly beds of streams, or even in moist meadow lands; very widely dispersed,

and variable in size, habit, pubescence, etc. Flowering, in some localities from May until December, i. e., throughout the dry season; in this regard, as well as in general aspect, very unlike the other annual species.

* * *Stipules gland-like; leaflets 4–10, unequally distributed on opposite margins of a dilated rachis; pods readily dehiscent.* —
Genus ANISOLOTUS, Bernh.

+ *Annuals: flowers solitary, short-pedicelled, not bracted; claws of petals approximate; keel pointed.*

2. **L. Wrangelianus,** F. & M. Index Sem. Petrop. 16 (1835). *Hosackia Wrangeliana* and *subpinnata*, T. & G. Fl. i. 326 (1838), but not *Lotus subpinnatus*, Lag. (1816). Less than a foot high, ascending, much branched, densely leafy, sparsely or canescently villous: leaflets about 4, cuneate-obovate to oval or oblong, 3–6 lines long; calyx-teeth broadly subulate, equalling the tube: corolla 3 lines long, bright yellow, the broadly obovate banner erect; wings meeting above the keel, not enfolding it: pod pubescent, straight, 7–10 lines long, 5–7-seeded. Common throughout middle California, especially toward the seaboard; latterly regarded as identical with the South American *L. subpinnatus*; but that is a smaller plant with narrow leaflets, relatively long calyx-teeth, the petals all narrower, with shorter claws, etc. Apr. May.

3. **L. humistratus,** Greene, Pittonia, ii. 139 (1890); *L. brachycarpus,* Wats. Index, 225 (1878), not of Hochst. (1842). *Hosackia brachycarpa,* Benth. Pl. Hartw. 306 (1849). Low and diffuse, the branches 5–8 in. long, herbage soft-villous: fl. nearly sessile, yellow; calyx-teeth linear, much longer than the tube: pod oblong, $\frac{1}{3}$ in. long, pilose, 2–3-seeded. Clayey banks and hill-sides; as widely dispersed as the preceding, but less common. May, June.

4. **L. denticulatus,** Greene, l. c.; Drew, Bull. Torr. Club. xvi. 151, under *Hosackia* (1889). Erect, 1–2½ ft. high, fastigiately branching, pale green and glaucous, sparingly pilose: calyx-teeth longer than the tube and, with the margins of the upper leaves, somewhat denticulate: corolla 2 lines long, pale yellow or salmon-color, changing to red: pod pubescent, short, 3-seeded. A most distinct species, long confounded with *L. Wrangelianus,* but of different habitat, i. e., from Butte and Humboldt counties northward to British Columbia. It is a very common weed in grain fields of the upper Sacramento, and rank enough in its growth to be troublesome. It combines the characteristics of this group and of the next in its inflorescence: the upper axils bearing two peduncles, one short and bractless, the other elongated, bracted and sometimes 2-flowered. Apr. June.

+ + *Flowers 1 or many, on an elongated, usually bracted peduncle; claw of the banner commonly remote from the others, keel mostly obtuse.*

++ *Annuals; few-flowered.*

5. **L. micranthus,** Benth. Trans. Linn. Soc. xvii. 367 (1837). *Hosackia parviflora*, Benth. Bot. Reg. under t. 1257 (1829) : *H. microphylla*, Nutt. in T. & G. Fl. i. 326 (1838). Erect, slender, 4 10 in. high, glabrous, glaucous ; peduncle filiform, bracted, 1-flowered : fl. minute, pale salmon, turning red ; pod 1 in. long or less, compressed, constricted between the seeds ; these oval or roundish, little compressed, smooth. From Monterey northward. Apr. May.

6. **L. salsuginosus,** Greene, Pittonia, ii. 140 (1890). *Hosackia maritima*, Nutt. in T. & G. Fl. i. 326 (1838). Ascending or depressed, slightly strigose, somewhat succulent, the branches 8 18 in. long ; leaflets 4 6, obovate, obtuse ; peduncles 1 in. long, 1 4-flowered, naked or with a conspicuous 1 3-foliolate bract ; corolla yellow, 3 lines long, the banner and wings equalling the straight keel ; pod scarcely compressed, 10 12-seeded ; seeds obliquely oval, smooth. From Monterey southward, either toward the sea, or on subsaline flats of the interior. Mar. June.

7. **L. rubellus,** Greene, l. c. 141 ; Nutt. in T. & G. Fl. i. 326 (1838), under *Hosackia*. Prostrate, slender, not succulent, strigose-pubescent or nearly glabrous ; leaflets 6 10, linear-oblong, mostly acutish ; early peduncles shorter than the leaves, bractless, 1-flowered, the later longer, bracted, 2-flowered : corolla reddish, scarcely twice as long as the calyx : pod slender, straight, 7 10-seeded ; seeds quadrate, minutely granulate. Plentiful in sandy soils, San Francisco, Alameda and far southward, but apparently only along the seaboard. Apr. July.

8. **L. nudiflorus,** Greene, l. c. ; Nutt. l. c. under *Hosackia*. Near the last, but leaflets smaller and broader ; fl. thrice as large : pod broader, more flattened, slightly curved upward at apex : seeds larger, quadrate, faintly tuberculate. Eastern base of Mt. Diablo Range, near Byron, etc., on gravelly hill-tops ; thence southward throughout the State. Mar. May.

9. **L. strigosus,** Greene, l. c. Nutt. l. c. under *Hosackia*. Strigose-pubescent, decumbent or prostrate ; peduncles long, commonly 1 2-flowered and 3-foliolate-bracted : fl. 4 5 lines long, yellow ; pod pubescent, slightly curved upwards ; seeds quadrate, but somewhat cruciform, being deeply notched at each end and at the hilum, the surface closely sinuate-rugose. Same range as the last, and readily distinguished by its seeds which have something of the outline of a Maltese cross. But this and the two preceding excellent species, were confused in the " Botany of California," as elsewhere by the same eastern authors, under the name of *Hosackia strigosa*. Mar. June.

10. **L. hirtellus,** Greene, Pittonia, ii. 142 (1890). Stoutish, depressed, canescently hirsutulous, not at all strigose ; leaflets 5 7, cuneate-oblong or -obovate, obtuse ; peduncles stoutish, bracted, surpassing the leaves, 2-flowered ; pod 1 in. long, subterete, straight, 7 10-seeded ; seeds

quadrate, notched at the hilum only, faintly ruguse and coarsely granulate. Ridges above Hetch-Hetchy Valley in the Sierra Nevada, *Chesnut & Drew*; also in the Mt. Diablo Range near Livermore.

++ ++ *Perennials ; flowers capitate-umbellate.*

11. **L. leucophæus,** Greene, Pittonia, ii. 145 (1890). *Hosackia grandiflora,* var. ? *anthylloides,* Gray, Proc. Philad. Acad. 350 (1863). Low; ascending, less than a foot high, internodes short, leaves ample, herbage velvety-pubescent : leaflets 5 7, obovate, 6 9 lines long, acute ; peduncles equalling or exceeding the leaves ; umbel 1-foliolate-bracted, 5 8-flowered : fl. more than $\frac{1}{2}$ in. long, ochroleucous, becoming red-purple. Dry ridges of the inner Coast and Mt. Diablo Ranges from Colusa Co. southward throughout the State ; also in the Sierra Nevada. May.

12. **L. grandiflorus,** Greene, l. c. : Benth. Trans. Linn. Soc. xvii. 366 (1837), under *Hosackia*. Tall, slender with few leaves and long internodes, nearly glabrous : peduncles slender, elongated, small-bracted, 5 8-flowered : fl. nearly 1 in. long, deep yellow, the petals broader than in the last, turning orange. Same range as the last ; but less frequent.

13. **L. macranthus,** Greene, l. c., also in Bull. Calif. Acad. i. 81 (1885), under *Hosackia*. Stoutish, 1 ft. high, silky-puberulent : leaflets 7 9, obovate or oblong, obtuse, $\frac{1}{2}$ 1 in. long ; stipules minute, subulate, caducous, leaving a dark gland-like permanent base : peduncles 2 4 in. long, 3 7-flowered ; bract 1-foliolate : corolla 1 in. long, bright yellow, the banner $\frac{1}{2}$ in. broad : pod stout, $1\frac{1}{2}$ in. long. El Dorado Co., on Sweetwater Creek, *Mrs. Curran*. The young specimens, on account of their manifest stipules, would be referred to the next group ; the older ones, showing only the dot-like traces of them would place the species here ; but on the whole, it is a link connecting the two groups.

* * * *Perennials with true stipules ; leaflets never inequilaterally distributed ; flowers in bracted umbels ; pods long, straight, tardily dehiscent.* Types of Genus HOSACKIA, Benth.

14. **L. formosissimus,** Greene, Pittonia, i. 147 (1890). *Hosackia gracilis,* Benth. Trans. Linn. Soc. xvii. 365 (1837). Slender, glabrous, the decumbent stems several, 1 ft. long ; leaflets 5 7, from broadly obovate to obovate-oblong, obtuse, the lowest truncate or retuse ; stipules thin, ovate : umbels equalling the leaves, or shorter, the bract 3-foliolate ; calyx-teeth unequal, triangular, acute or acuminate, shorter than the campanulate tube : corolla 7 lines long, the wide-spread wings and much shorter keel rose-red, the banner yellow. Common in moist ground along the seaboard from Monterey northward. The most beautiful species. Apr. May.

15. **L. pinnatus,** Hook. Bot. Mag. t. 2913 (1829). *Hosackia bicolor,* Dougl. in Bot. Reg. t. 1257 (1829). Stoutish, glabrous, the erect stems 2 ft. high: leaflets 5-9, obovate or oblong, acutish: stipules scarious, triangular: peduncles longer than the leaves, 3-7-flowered, naked or with a small scarious 1-3-foliolate bract: calyx-teeth triangular, half as long as the tube: corolla as in the last, but keel and wings white, banner yellow. - Said to inhabit the seaboard districts from San Francisco northward; but not known to me as Californian.

16. **L. Torreyi,** Greene, l. c. 146 (1890); Gray, Proc. Am. Acad. viii. 625 (1873), under *Hosackia*. Habit of the last, but slender, more or less silky-pubescent; leaflets narrower, acute or obtuse: bract of the umbel sessile: fl. smaller, the keel and wings white, the latter not spreading. Very common along streamlets and brooks in the middle or higher Sierra and the Coast Range. June-Aug.

17. **L. oblongifolius,** Greene, l. c.: Benth. Pl. Hartw. 305 (1849), under *Hosackia*. Erect, slender, somewhat appressed-pilose: leaflets 7-11, narrowly oblong or oblanceolate, 1 in. long, acute: stipules small, acute: peduncles exceeding the leaves, 5-7-flowered: bract subsessile 1-3-foliolate: calyx-teeth subulate, about equalling the tube: corolla yellow, turning purplish or brownish: pod slender, 2 in. long: seeds turgid. From Monterey, *Coulter,* southward in the mountains.

18. **L. lathyroides,** Greene, l. c.: Dur. & Hilg. Pac. R. Rep. v. 6. t. 3 (1853). under *Hosackia*. Slender, branching and flexuous, minutely pubescent: leaflets 5-7, linear-lanceolate, acute at both ends: stipules small, scarious, ovate-acuminate; umbels 1-3-flowered, with or without a linear-lanceolate bract: fl. 5 lines long: calyx-teeth linear, acute. At Fort Miller on the San Joaquin, *Heermann,* and southward. All the specimens seen by me indicate an annual root. But the original figure and description make it perennial.

19. **L. crassifolius,** Greene, l. c. 147; Benth. Trans. Linn. Soc. xvii. 365 (1837), under *Hosackia; H. stolonifera,* Lindl. Bot. Reg. t. 1977; *H. platycarpa,* Nutt. in T. & G. Fl. i. 323 (1838). Erect, stout, 2-3 ft. high, of a dull green hue, as if glaucous, but minutely pubescent: leaflets 9-15, thickish, obovate or oblong, obtuse, mucronulate, $1\frac{1}{2}$ in. long or more: peduncles nearly equalling the leaves: umbel many-flowered, the 1-3-foliolate bract a little below it: calyx-teeth triangular, short: corolla purplish marked with green spots: pods thick, 2 in. long. From Kern Co. northward, in the mountain districts.

20. **L. stipularis,** Greene, l. c.; Benth. l. c., under *Hosackia*. Not as tall as the last, more slender, villous with spreading hairs and often somewhat glandular: leaflets 15-21, obovate-oblong, acute, mucronate. $1\frac{1}{2}$-1 in. long: stipules large, ovate: peduncles short, 4-8-flowered, the leaf-

LEGUMINOSÆ. 21

like bract near the middle, 3–9-foliolate ; calyx 2 lines long, the subulate teeth short ; corolla purple ; pod straight, 1–1½ in. long. In the Mt. Diablo Range, from Contra Costa Co. to Monterey. Seldom seen. It probably includes the *Hosackia macrophylla* and *balsamifera*, Kell. Proc. Calif. Acad. ii. 123 & 125, but this may also be doubted ; and the species ought to be carefully studied anew, in the field, by those who know where to find any of the forms referred here.

21. **L. incanus,** Greene, l. c. ; Torr. Pac. R. Rep. iv. 79. t. 4 (1857), under *Hosackia*. Low, stout, erect, densely villous ; leaflets 9–15, obovate-oblong, acute ; stipules ovate ; peduncles shorter than the leaves, 6–9-flowered ; bract above the middle, 5-foliolate ; calyx ¼ in. long, the subulate teeth half as long as the tube. Common in open pine woods about Nevada City, and for a few miles westward and northward ; but apparently of very limited distribution.

* * * * *Stipules gland-like ; leaflets few, unequally distributed ; mature calyx, with the small indehiscent usually arcuate long-pointed pods, deciduous.* Genus **SYRMATIUM,** Vogel.

+ *Perennials ; a few woody at base.*

22. **L. glaber,** Greene, l. c. 148 ; Vogel, Linnæa, x. 591 (1836), under *Syrmatium*. *Hosackia scoparia*, Nutt. in T. & G. Fl. i. 325 (1838) ; *H. glabra*, Torr. Wilkes Exp. 274 (1874). Suffrutescent, 2–8 ft. high, erect or decumbent, nearly glabrous ; leaflets mostly 3, on young shoots 4–6, oblong to linear-oblong, ¼–1½ in. long, obtuse or acute ; umbels many, sessile ; fl. 3–4 lines long, yellow, turning red ; calyx-teeth subulate, erect, rather less than half as long as the tube. Usually tufted and reedy-looking, the foliage sparse, the flowers profuse. Common about San Francisco, and from Lake Co. southward throughout the State, in the Coast Range chiefly ; flowering almost all the year round.

23. **L. Benthami,** Greene, l. c. *Hosackia cystisoides*, Benth. Trans. Linn. Soc. l. c. Resembling the last, but smaller and mostly prostrate ; umbels on peduncles which equal or exceed the leaves and are 1–3-foliolate-bracted at top ; calyx-teeth more slender, stellate-spreading in the bud and recurved in flower. Common on low hills near the sea in San Mateo Co. and southward. June, July.

24. **L. junceus,** Greene, l. c. ; Benth. l. c., under *Hosackia*. Nearly glabrous, erect, shrubby, with slender branches reedy and sparsely leafy ; leaflets obovate to oblong, 2–4 lines long ; fl. 3 lines ; calyx 2 lines long or less ; teeth very short and blunt. A more southerly species than either of the two preceding ; but said to have been found near San Francisco.

25. **L. argophyllus,** Greene, l. c. ; Gray, Mem. Am. Acad. v. 316 (1854).

under *Hosackia*. Decumbent or ascending, leafy and branching, silvery-silky throughout; leaflets 3–7, obovate and rounded, or oblong and acute, 2–7 lines long: umbels dense and capitate, on short simple-bracted peduncles; fl. 4–5 lines long; calyx half as long, its teeth filiform, silky, nearly as long as the tube. From Yosemite, *Chesnut & Drew*, and the foot-hills of the Merced, *Asa Gray*, southward. Very probably more than one species is embraced by this name and description. "*H. argentea*," Kellogg, seems distinct; but we can not at present make out characters sufficient for its restoration.

26. **L. leucophyllus,** Greene, l. c. *Hosackia sericea*, Benth. Trans. Linn. Soc. l. c. Slender, sparsely leafy, 1–2 ft. high, silvery-canescent with a close but short silky pubescence: leaflets 3, cuneate-oblong to linear, 1_2–3_1 inches long: umbels few-flowered, sessile or short-peduncled; fl. 3 lines long: calyx half as long, with slender teeth. Plant with the habit of *L. glaber*, but silky-pubescent. Evidently rare, and confined to Monterey Co.

27. **L. procumbens,** Greene, l. c. and Bull. Calif. Acad. i. 82 (1885), under *Hosackia*. Pubescence as in the last, but much more leafy and diffuse: leaflets cuneate-obovate to oblanceolate, acutish, 1_2–1 in. long: fl. sessile, 1 or 2 only in each axil: calyx-teeth subulate from a broad base, erect: pod nearly 1 in. long, straight, with an ensiform beak, 2-seeded. From Tehachapi, Kern Co., *Mrs. Curran*, southward and eastward along the mountains west of the Mohave Desert. May, June.

28. **L. Nevadensis,** Greene, Pittonia, ii. 149 (1890); Bull. Calif. Acad. ii. 148 (1886), under *Syrmatium*. *Hosackia decumbens*, var. (?) *Nevadensis*, Wats. Bot. Calif. i. 138 (1876). Diffuse, the slender branches hard and wiry, at base more or less woody: sparingly villous or somewhat tomentose: leaflets 3–5, cuneate-obovate, acute, 3–5 lines long: umbel many-flowered, short-peduncled, 1-foliolate-bracted: calyx a line long, the slender teeth 1_2 line: pod strongly arcuate, the slender beak longer than the body. Dry pine woods of the Sierra Nevada, from Sierra Co., *Lemmon*, to San Diego Co. In my first description of the species I erred in attributing to it an annual root. It is, usually at least, perennial. In habit and geographical range it is most unlike the next; but the technical characters of the two species are feeble.

29. **L. Douglasii,** Greene, l. c., *Hosackia decumbens*, Benth. Bot. Reg. under t. 1257 (1829). Larger and more leafy than the last, decumbent at base, the erect branches not wiry: pubescence more scant: fl. 4–7 lines long: calyx 1^1_2 lines; the slender teeth as long as the tube: pod longer, less arcuate. From Humboldt Co., *Chesnut & Drew*, northward.

+ + *Annuals*.

30. **L. tomentosus,** Greene, l. c.; H. & A. Bot. Beech. 137 (1836), under *Hosackia;* Vogel, Linnaea, x. 591 (1836), under *Syrmatium*. The numerous branches a foot or two long, flexuous, weak and prostrate; pubescence dense, somewhat tomentose; leaflets 5–7, obovate or cuneate-oblong, acute, 3–6 lines long; umbels short-peduncled or subsessile, bracted; fl. 3–4 lines long; calyx half as long, very villous; the filiform teeth about equalling the tube. In sandy grounds near the sea, from San Francisco southward.

31. **L. Heermanni,** Greene, l. c.; Dur. & Hilg. Pac. R. Rep. v. 6. t. 4 (1855), under *Hosackia*. Near the last, but less pubescent, neither the leaflets nor the flowers more than half as large, the leaflets broader and rounded. Same range as the last.

10. TRIFOLIUM, *Pliny* (CLOVER). Herbs with palmately (in one pinnately) 3-foliolate (in a few 5–7-foliolate) leaves and adnate stipules; the leaflets commonly denticulate. Flowers mostly very many, in roundish or ovoid or somewhat depressed capitate or umbellate clusters, on more or less elongated axillary or terminal peduncles. Calyx 5-cleft or -toothed. Corolla persistent; banner and wings commonly coherent with the stamineal tube; keel mostly obtuse and shorter than the wings. Stamens diadelphous. Pod concealed within or little exserted from the calyx, 1–6-seeded, dehiscent or indehiscent. An extensive genus of well known forage plants; the Californian species numerous; those of our district almost all annuals, flourishing between March and May.

* *Heads or spikes not involucrate.*

← *Flowers pedicellate, at length reflexed; calyx-teeth subulate, not plumose.*

1. **T. Breweri,** Wats. Proc. Am. Acad. xi. 131 (1876). Perennial, slender, diffuse, 4–12 in. high, sparsely pubescent; stipules lanceolate; leaflets mostly obcordate, toothed or serrulate, $\frac{1}{4}$–$\frac{3}{4}$ in. long on slender pedicels; calyx-teeth slender, much shorter than the whitish or pale purple corolla. —Only at considerable elevations in the Sierra; occupying open places among subalpine forests. Jul. Sept.

2. **T. gracilentum,** T. & G. Fl. i. 316 (1838); *T. denudatum,* Nutt. Pl. Gamb. 152. t. 24 (1848). Erect, slender, 1–2 ft. high, wholly glabrous; stipules ovate- or linear-lanceolate, acuminate; leaflets cuneate-obcordate, spinulose-serrulate, $\frac{1}{2}$ in. long; heads 15–25-flowered; calyx-teeth lanceolate-subulate, setaceously acuminate, thrice as long as the tube, shorter than the dull but usually deep purple corolla; pod exserted, 2-seeded; seeds obliquely oval, straw-colored, very smooth. Open plains and hillsides throughout western California. Apr. June.

3. **T. bifidum,** Gray, Proc. Am. Acad. vi. 522 (1865). Erect, very slender, 1 ft. high, pale green and glaucous, the petioles and calyx more

or less pilose-villous ; stipules ovate-lanceolate, entire, setaceously acuminate ; leaflets linear-cuneate, the sides remotely toothed, apex bifid and mucronulate ; peduncles slender, exceeding the leaves ; heads 6–15-flowered ; calyx deeply 5-parted, the teeth subulate-setaceous, about equalling the minute pale rose-colored corolla ; pod included, 1-seeded ; seed rather narrowly obovate-oblong. Var. **decipiens.** Taller and stouter, the leaflets cuneate-oblong with closely serrulate margins and only a shallow notch at apex ; heads 15–30-flowered.—The type is common from Santa Clara Co., *Torrey, Lemmon*, to Solano, *Jepson*, and Mendocino, *Bolander*, but chiefly to the east of the Coast Range. In the Bay district the variety alone is found ; and this has hitherto been confused with *T. gracilentum*, from which its pallid hue, hairy stalklets and calyces, as well as its flowers and fruits make it easily distinguishable.

4. **T. ciliolatum,** Benth. Pl. Hartw. 304 (1849) ; *T. ciliatum*, Nutt. Pl. Gamb. 152 (1848), not of Clark. Erect, 1–2 ft. high, glabrous ; stipules narrow, acuminate ; leaflets cuneate-oblong or obovate, $\frac{1}{2}$–1 in. long, obtuse or retuse, serrulate ; fl. purple, 3 lines long ; calyx-teeth lanceolate, very acute, rigidly ciliolate. Throughout the western part of the State, both seaward and in the interior. Apr.–June.

5. T. PROCUMBENS, Linn. Sp. Pl. 772 (1753). Ascending or suberect, slender, pubescent ; leaflets cuneate-oblong, emarginate, denticulate, the terminal one on a longer stalklet ; heads ovate or oblong, very dense ; fl. yellow ; upper calyx very short ; banner deflexed over the other petals in age. A small and delicate Old World clover with pinnately 3-foliolate leaves, small yellow flowers ; beginning to appear spontaneously ; frequent in lawns, having been introduced with lawn-grass seeds.

6. T. REPENS, Rivinus, Tetrap. 17 (1690). Perennial, diffuse, creeping, sending up erect long-stalked glabrous leaves and heads ; leaflets obcordate, denticulate ; heads depressed-globose, at length umbellate ; fl. white ; calyx-teeth unequal, lanceolate-subulate, shorter than the tube ; pod about 4-seeded. The common White Clover of eastern and European meadows and pastures ; a troublesome plant in lawns with us ; sparingly naturalized in our district, yet common in the wild state in Oregon, etc.

7. **T. Bolanderi,** Gray, Proc. Am. Acad. vii. 335 (1868). Perennial, cespitose, glabrous ; leaflets obovate-oblong, $\frac{1}{2}$ in. long, reticulate, serrulate ; peduncles few, elongated, exceeding the leaves ; fl. few, 3–4 lines long, purplish ; calyx-teeth lanceolate, not longer than the tube ; pod 2-seeded.—Moist ground in the neighborhood of the Yosemite, above the valley.

8. **T. Kingii,** Wats. Bot. King Exp. 59 (1871). Perennial, erect, 6–10 in. high, glabrous ; stipules lanceolate, acuminate, entire ; leaves long-

petioled: leaflets ½ -1½ in. long, ovate, oblong or lanceolate, acute, striate-veined, sharply denticulate: heads rather large, many-flowered, the rachis commonly produced above the flowers: fl. 1½ in. long or more, deep purple: calyx-teeth setaceous-subulate, equalling the tube. At Summit Station, *Bolander*, and northward, in the Sierra.

9. **T. Beckwithii,** Brewer, in Wats. Proc. Am. Acad. xi. 128 (1876). Stouter and taller than the last, often a foot high, the head of purple flowers, globose, 1 -1½ in. in diameter, the rachis not produced above the flowers: leaflets broader, often retuse or emarginate. Near Truckee, on Prosser Creek and in Sardine Valley, *Sonne;* also in Sierra Co.

10. **T. Lemmoni,** Wats. l. c. 127. Cespitose, petioles elongated, almost equalling the upright or decumbent stems, a span high or less; herbage appressed-puberulent: stipules ovate, acuminate, toothed or entire: leaflets mostly 5, obovate or oblong, obtuse, of firm texture, coarsely toothed, ½ in. long or more: heads rather few-flowered: fl. ½ in. long, deflexed in age; calyx-teeth villous, 2 lines long, exceeding the short tube. Sierra Co., *Lemmon*. Species most related to *T. Bolanderi* and *Kingii* notwithstanding its 5-foliolate leaves.

11. **T. Howellii,** Wats. Proc. Am. Acad. xxiii. 262 (1888). Stout, erect, 2 ft. high, glabrous: stipules large, ovate or lanceolate; petioles short: leaflets cuneate-oblanceolate, 1½ -3 in. long, remotely dentate; peduncles axillary, longer than the leaves: heads large, oblong: calyx-teeth narrow, about equalling the tube; corolla 4 -5 lines long, yellowish, or white. Trinity Mts., Humboldt Co., *Chesnut & Drew*. The largest, and one of the most remarkable of western clovers. July, Aug.

++ *Flowers nearly or quite sessile, not reflexed (except in No. 12); calyx-teeth-elongated, plumose, or at least hairy.*

++ *Perennials.*

12. **T. eriocephalum,** Nutt. in T. & G. Fl. i. 313 (1838). Erect or decumbent, 6 -10 in. high, soft-villous: stipules elongated, entire; leaflets oblong, 1 -1½ in. long, serrulate; fl. in dense ovate spikes, whitish, at length reflexed: calyx-teeth filiform, very villous, nearly equalling the corolla: pod hairy, 2 -4-seeded. Mendocino Co. and northward.

13. **T. pratense,** Tragus, Stirp. Hist. 586 (1552). Stoutish, ascending, 1 ft. high, pubescent: leaflets oval or obovate, often retuse, 1 in. long: heads ovate, 1 in. long, sessile: corolla elongated-tubular, rose-purple. The common *Red Clover* of eastern and Old World meadows: occasionally spontaneous with us; plentiful in a wild state northward, where the summer drought is less prolonged.

14. **T. longipes,** Nutt. in T. & G. Fl. i. 314 (1838). Erect or ascending, 3 -12-in. high, slightly pubescent: leaves and heads on long slender

stalks ; stipules foliaceous, semilanceolate, acuminate ; leaflets narrowly oblong, acute, ¾–1½ in. long, serrulate : heads round-ovate or ovate, ⅜ in. in diameter ; calyx-teeth white-villous, little shorter than the slender white (sometimes purplish) corolla. In the high Sierra, from Mono Lake northward. About Lake Tahoe occurs a dwarf form with reddish flowers. In Trinity Mts., Humboldt Co., occurs an excessively tall form approaching *T. plumosum* in size and shape of the heads.

15. **T. macrocephalum**, Poir. Suppl. v. 336 (1817) ; Pursh. Fl. ii. 479. t. 23 (1814), under *Lupinaster* : *T. megacephalum*, Nutt. Gen. ii. 105 (1818). Erect or ascending, ½–1 ft. high, stoutish, pubescent ; stipules large, foliaceous ; leaflets 5–7, oblong-cuneiform, mucronate, sharply and closely denticulate : head solitary, terminal, ovate-globose, 2 in. high ; calyx-tube short, campanulate ; the filiform teeth very long, densely plumose ; corolla ⅜ in. long, pale yellow and red : pod 2-seeded ; seed pale and smooth. A peculiar very handsome species of northern districts ; reaching our borders in Plumas Co., *Lemmon*, Apr. May.

16. **T. Andersonii**, Gray, Proc. Am. Acad. vi. 522 (1865). Dwarf, cespitose, the short depressed branching stems very stout and leafy, the whole plant densely silky-villous ; stipules lanceolate, entire : leaflets cuneate-oblong, ½ in. long, acute, almost entire : peduncles axillary ; umbels depressed-globose ; fl. ½ in. long, purplish : calyx-teeth little shorter than the petals ; pod tomentose, 2-seeded. Plumas Co., *Lemmon*, and eastward in Nevada.

++ ++ *Annuals.*

17. **T. Macræi**, H. & A. Bot. Misc. iii. 179 (1833). Much branched, decumbent or almost prostrate, the slender branches 8–18 in. long, the herbage more or less villous- or pilose-pubescent ; stipules broadly ovate, abruptly acuminate ; leaflets cuneate-oblong, obtuse, denticulate above the middle, 6–10 lines long : heads nearly or quite sessile, usually in a terminal pair, ovate, ¼–½ in. high ; calyx-teeth longer than the tube, densely plumose-hairy, nearly equalling the small purplish corolla ; pod 1-seeded. Var. **albopurpureum**. *T. albopurpureum*, T. & G. Fl. i. 313. Often 1–1½ ft. high, ascending ; heads small, ovate-conical or subcylindrical, solitary at the ends of very long slender peduncles ; calyx-teeth slender, more delicately plumose, fully equalling the white-tipped purple corolla. The true *T. Macræi* is Chilian, and is said to have lanceolate-acuminate entire stipules. No such stipules are found in our Californian plant ; and the plant above described comes nearest to it and seems to occur only on the San Francisco peninsula, where it is common on rocky or sandy slopes and hill-tops. The variety abounds almost throughout the State, running into many forms, some of which resemble the next. May, June.

18. **T. dichotomum**, H. & A. Bot. Beech. 330 (1840) ; *T. Neolagopus*,

Loja, Giorn. Bot. Ital. xv. 194 (1883). Erect or ascending, stoutish, 1–1½ ft. high, often flexuous and repeatedly dichotomous ; pubescence longer than in the last, more spreading ; stipules broadly ovate, with a short subulate point ; leaflets cuneate-obovate or oblanceolate, the upper acute, ³⁄₁ in. long, sharply denticulate ; heads long-peduncled, ovate-conical, ³⁄₁–1¼ in. high ; calyx-teeth setaceous, densely hairy, equalling the red-purple corolla ; pod with close elevated striæ. Plentiful on plains of the interior, from Vacaville, *Jepson*, to Antioch, *Greene*, and far southward. In so far as Hooker and Arnott's description goes, it accords with the plant first perfectly described by Lojacono under a new name. It is a particularly well marked species in the character of its pods.

19. **T. amœnum.** Commonly 2 ft. high, stout, simple or with few branches from the base, the heads 1–3, terminal and subterminal, herbage canescently villous ; stipules lanceolate to obliquely broad-ovate with a setaceous or triangular acumination ; leaflets broadly obovate, retuse or obtuse, erose-denticulate, 1 in. long or more, 10 lines broad ; heads globose, in age oval, 1½ in. high ; calyx-teeth linear-setaceous, plumose throughout, 3–4 lines long, much shorter than the very showy corolla ; this light rose-purple with dark centre. At Vanden Station on the Sacramento plains, *Greene* ; also at Little Oak, Solano Co., in the same general region, *Jepson*. By far the largest and handsomest of the annual clovers of middle California ; plentiful in its locality. It has been distributed for *T. dichotomum*, but wrongly. May.

20. **T. columbinum,** Greene, Pittonia, i. 4 (1887). Erect, nearly simple, 1 ft. high, somewhat silky-pubescent ; leaflets 1 in. long, cuneate-oblong, obtuse, crenulate-denticulate ; head ovate-conical, 1 in. high ; calyx-tube less than 1 line long ; the filiform segments 5 lines, soft and silky-plumose throughout, deeply concealing the minute purple corolla ; pod striate, villous at apex. Common about Vacaville ; readily known by its pale dove-colored heads altogether soft and silky, exhibiting no flowers, but seemingly made up of the long, densely plumose calyx-teeth. May.

21. **T. olivaceum,** Greene, l. c. Simple or branched from the base, 1–1½ ft. high, glabrous except an appressed pubescence on the lower face of the leaves ; petioles 1–2 in. long, with lanceolate acuminate entire stipules ; leaflets as in the last, but somewhat serrulate ; heads on long slender peduncles, hemispherical in flower, 1 in. or more broad and high ; calyx-tube 1 line long ; the linear-setaceous teeth 5–6 lines, densely plumose toward the base only, gradually less so above, nearly naked at the rather rigidly setaceous tips ; corolla deep violet-purple, very small and concealed ; pod striate, glabrous. With the preceding, but more common ; readily distinguished by its large olive-green heads.

22. T. ARVENSE, Linn. Sp. Pl. 769 (1753). Related to the last two, but

of different aspect; the numerous branches lateral, not basal; the leaves and heads short-stalked: heads oblong or cylindrical, $\frac{3}{4}$ in. long, or less: calyx-teeth silky-plumose throughout, longer than the minute whitish corolla. The *Rabbit-foot* or *Mouse-ear Clover* of Europe, naturalized on the Atlantic coast, has been reported from Alameda Co., *Kellogg*.

* * *Heads subtended by a flat or concave (sometimes nearly obsolete) involucre.*

← *Corolla not inflated in age.*

++ *Involucre flat; heads a little one-sided.*

23. **T. Wormskjoldii,** Lehm. Ind. Sem. Hort. Hamb. 17 (1825); Pugill. i. 36 (1828); Spreng. Syst. iii. 209 (1826): *T. heterodon*, T. & G. Fl. i. 318 (1838), partly. Perennial, spreading underground by slender root-stocks; stems decumbent, 3 in. 2 ft. long; herbage flaccid, glabrous: stipules lanceolate, acuminate, laciniately multifid; leaflets obovate-oblong, obtuse, pectinate-denticulate, 1 in. long or more: heads hemispherical, 1 in. broad or more; involucre $\frac{1}{2}$–$\frac{3}{4}$ in. broad, laciniate-aristate: calyx-tube scarious, 10-striate, the alternate nerves less prominent, transverse veinlets 0; teeth linear-subulate, much longer than the tube, all entire or 1 or more of them setaceously 2–3 parted: banner elliptical, deeply emarginate, pale purple; other petals darker. —Very common and variable; on hills about San Francisco only a few inches high; in springy places, or along perennial streams, coarse and fistulous, forming dense masses, the leaflets often 4 (but the fourth only half as large as the others), and the calyx-teeth more or less cut into setaceous divisions. The Mexican *T. involucratum*, Willd., to which some authors have referred our plant, has narrow acute leaflets, entire stipules, and a calyx whose tube is less diaphanous, and less prominently nerved, and of which the teeth are much shorter and less aristiform. I have taken up what is clearly the oldest name for this perennial of the Californian seaboard. And I do not think that the *T. fimbriatum*, Lindl. (1827) is a synonym of it. In Lindley's species, according to both the figure and description, the bracts forming the involucre are distinct; and the stipules are much more regularly and deeply cleft than in any Californian plant. It is very probably the northern counterpart of our middle Californian seaboard species, and it may possibly prove confluent with the next. *T. Wormskjoldii* when first published was supposed to be a native of Greenland. The author afterwards corrected this error, having learned that the seed had come from California. The handsome figure in the first "Pugillus" represents most accurately the plant of the Presidio hills, San Francisco; and no doubt it is the same. In cultivation at Berkeley, in dry soil, the species forms a sward, and continues growing and flowering throughout the spring and summer months, just as in its moist native habitat. It is a valuable forage plant.

24. **T. spinulosum,** Dougl. in Hook. Fl. i. 133 (1830); *T. atropurpureum*, Nutt. in T. & G. Fl. i. 318 (1838) under *T. heterodon*. Perennial, slender, variable as the last as to size : leaflets narrowly oblong, acute at both ends, spinulose-denticulate, ending in a stiff spinulose cusp ; stipules ovate-acuminate, spinulose-serrate : involucre deeply cleft or even divided : heads subglobose, small (less than 1 in.) : calyx-teeth narrowly subulate, stiff and pungent, nearly equalling the corolla. Moist meadows in the Sierra Nevada, from Tuolumne Co., *Chesnut & Drew*, northward. Perhaps at the far northwest confluent with *T. fimbriatum*. If so, it will take that name, as of earlier date. But according to Douglas' notes as published by Hooker, the keel and wings, both white, are shorter and more acute than in that. As compared with *T. Wormskjoldii*, it is equally tall, but always more slender, has different leaflets, smaller flowers in much smaller heads ; and its geographical distribution is entirely different. I have it from southeastern Oregon, *Mrs. Austin*, and from southwestern Nevada, *Shockley*. My own No. 880, of Siskiyou Co., Calif., 1876, distributed as *T. pauciflorum* (which is a very slender annual), is a good type of what I here have in view.

25. **T. variegatum,** Nutt. in T. & G. Fl. i. 317 (1838). Annual, glabrous, decumbent or prostrate, with very numerous slender branches : leaflets obcordate to obovate-oblong, minutely spinulose-serrate : upper stipules roundish, laciniately cleft : peduncles slender, longer than the leaves : laciniate involucre, shorter than the small (3–15-flowered) heads : calyx-tube about 15-nerved ; the teeth broadly subulate, tapering to a setaceous point, longer than the tube, shorter than the corolla : fl. dull purple or whitish. Var. **melananthum.** *T. melananthum*, H. & A. Bot. Beech. 331 (1840). *T. tridentatum*, var. *melananthum*, Wats. More rigid, spreading, the branches often a foot long or more ; heads larger : calyx-teeth more triangular and only pungently acute or acuminate, of a dark purple almost to the base ; corolla deep purple. Var. **major,** Loja. Giorn. Bot. xv. 183 (1883). *T. triste*, Nutt. in T. & G. Fl. i. 318 (1838), under *T. spinulosum*. Flaccid and procumbent, but very stout and fistulous, the branches often a yard long ; leaflets oblong-cunciform, 1 in. long or more ; heads 1 in. broad more or less : calyx-teeth dark purple ; petals purple with whitish tips.—The rather delicate, pale-flowered plant which must be the type of this species we have only from the mountains, where it is common along streamlets and in springy places. In the higher Sierra, greatly reduced and often with 1-flowered involucres, it is confused with the very different *T. monanthum*. The varieties belong to the interior plains and to the seaboard districts. These are perhaps distinct species ; but the characters on which to separate them do not appear. The var. *major*, often nearly as large every way as its perennial relative, *T. Wormskjoldii*—its branches sometimes even longer—is distinguished from that by its annual root and broader leaflets and calyx-teeth. The calyx-tube in all

the forms is more than 10-striate, the nerves numbering 15 or 20. The flaccid herbage, prostrate habit, broad leaflets and entire calyx-teeth make this easily distinguishable from *T. oliganthum* and *tridentatum*, with which it is confused in the " Botany of California."

26. **T. appendiculatum**, Loja. Giorn. Bot. Ital. xv. 181 (1883)? Glabrous, flaccid, diffuse; leaves long-petioled; leaflets cuneate-obovate or obcordate, serrulate-spinulose, mucronulate at apex ; heads hemispherical, 1 in. or less in breadth ; fl. purple ; calyx-teeth lanceolate-linear, entire, longer than the tube ; keel of the corolla rostrate-attenuate, longer than the wings.—A plant much smaller than described by Lojacono under this name, but having just the floral structure attributed to the species, certainly distinct from *T. variegatum*, which it most resembles, has been collected by Mr. V. K. Chesnut at Lake Merritt, Oakland. I also saw the same in the Torrey herbarium, from Auburn, *Bolander* (No. 4539), under the wrong name, " *T. Bolanderi*."

27. **T. oliganthum**, Stend. Nom. i. 707 (1841) : *T. pauciflorum*, Nutt. in T. & G. Fl. i. 319 (1838), not of d'Urville : *T. filipes*, Greene, Pitt. i. 66 (1887). Pale green, glabrous, erect, slender, with few ascending branches, 6–18 in. high ; upper leaflets linear, acute, 1 in. long, spinulose-serrate ; peduncles filiform, 2–3 in. long, exceeding the leaves ; head small, 7–12-flowered ; involucre reduced, laciniately divided ; fl. pale purple and white ; 2–3 lines long ; calyx-teeth ovate-acuminate, pungent, entire, equal, shorter than the 10-striate tube.—Common throughout the State, along the borders of woods and thickets ; the most slender species, yet always erect ; the herbage of a pale glaucous hue ; corolla with wings meeting in front of the keel. In all other allied species they spread away from it. Calyx laterally compressed in fruit, the segments appressed to the obovate 2-seeded pod. May.

28. **T. Watsonii**, Loja. Giorn. Ital. xv. 186 (1883). Erect, not slender, 4–12 in. high, glabrous, purplish ; leaflets 1 in. long or more, from narrowly elliptical to linear-filiform, the lower entire, upper remotely spinulose-serrate ; heads large, showy, hemispherical ; involucre small, not deeply cleft ; calyx oblong, 20-striate, the teeth $\frac{1}{4}$ as long as the tube, ovate, very abruptly contracted to a short pungent tip. Near Chico, *Mrs. Bidwell, Dr. Parry* ; related to the next, but doubtless distinct.

29. **T. tridentatum**, Lindl. Bot. Reg. under t. 1075 (1827). Erect, 8–16 in. high, glabrous, neither viscid nor clammy ; stipules setaceously laciniate, erect ; leaflets linear or lanceolate, sharply serrate ; heads 1 in. broad, the laciniate involucre much shorter than the flowers ; fl. $\frac{1}{2}$ in. long, bright purple with dark centre ; calyx with 10-nerved tube, the rigid segments broad at base, abruptly narrowed to a subulate spinulose-tipped apex which is usually subtended by a short stout tooth on each

side. Var. **scabrellum** (*T. scabrellum*, Greene, Pitt. i. 159), a slender plant with long almost filiform peduncles and broad truncate cuspidate leaflets, has a sparse scabrous pubescence upon its stalklets and growing parts. The type of this common species, sparingly leafy, the leaflets linear, the stems firmly erect and the whole herbage purplish, belongs to the seaboard, where it abounds in clayey soils, both on hills and plains. In the Sacramento valley the plant is paler, weaker and less erect, but larger and with broader leaflets ; the flowers paler and the calyx-segments mostly simple. The var. *scabrellum* is from the plains of the upper San Joaquin, and has marks enough for a species, if they were constant. But in the region of the lower San Joaquin, and in Livermore valley it appears to be confluent with *T. tridentatum*. A constant mark of the species in all its forms, as distinguished from the next two, is the 10-nerved calyx without smaller intervening striae. Mar. May.

30. **T. obtusiflorum**, Hook. Ic. Pl. iii. t. 281 and Bot. Beech. 331 (1840). Stout, erect, 1–3 ft. high, the herbage bright green, sparsely short-hairy under a lens ; the inflorescence and growing parts somewhat resinous-glandular ; stipules setaceously lacerate, broad and spreading, in age reflexed ; leaflets elliptic-lanceolate, 1–1½ in. long, spinulose-serrate : heads more than 1 in. broad, on long stoutish peduncles : calyx-tube oblong-campanulate, ¼ in. long, with 10 prominent and as many lesser nerves, these branching and forming reticulations above ; teeth subulate-spinose, entire ; corolla ⅓ in. long, lilac-purple with dark centre.- Common on clayey hill-sides and stream banks in the open country along the base of the Mt. Diablo Range, and foot-hills of the Sierra ; originally from Monterey, *Douglas* ; extending northward to Oregon, *Howell*. An exceedingly well marked species, readily known by its great size and more or less gummy heads of large flowers, 20-nerved calyx, etc. May.

31. **T. roscidum**. Erect, with ascending branches, stout, 1–2 ft. high, stems flexuous, purple, leaves deep dull green, soft-pubescent throughout and very clammy, not at all resinous : stipules spreading or reflexed, setaceously fimbriate ; leaflets (often 5) 1 in. long, linear-lanceolate, pectinately setulose : heads as in the preceding (though not glandular), calyx the same ; corolla white, with dark red-purple centre.- Plentiful in the foot-hills of the Sierra, on shaded northward slopes and along streams, near Jackson, Amador Co., and southward to San Bernardino Co., *Parish*. A remarkable species on account of its pubescence and clamminess ; the whole herbage, even at noon of the driest day, feeling as if wet with dew. But for its conspicuous pubescence it might pass in the herbarium for a form of *T. obtusiflorum* ; but it is most distinct from that and every other recognized species of clover. June.

32. **T. monanthum**, Gray, Proc. Am. Acad. vi. 523 (1865) : *T. multi-caule*, Jones, Bull. Torr. Club, ix. 31 (1882). Perennial, dwarf, 1–6 in.

high, decumbent or prostrate, branching, sparsely villous: stipules entire or slightly toothed; leaflets obcordate to oblanceolate, 1-4 lines long, mostly retuse, mucronate-dentate; peduncles equalling the leaves: involucre minute, 1-4-flowered: fl. ½ in. long, yellowish white with purple centre. Common in the higher Sierra from Kern Co., *Palmer*, to Lassen's Peak, *Lemmon*. A diminutive perennial, with which the 1-flowered state of *T. variegatum* has been confused, perhaps from the first; and Marcus Jones, in proposing *T. multicaule* may have taken the latter for the true *T. monanthum*, which Dr. Gray described as probably annual; but it is strictly perennial. The flowers are twice or thrice larger than in the annual associated with it.

++ ++ *Involucre cup-shaped ; flowers developing equally all around.*

33. **T. microcephalum,** Pursh. Fl. ii. 478 (1814). Slender, much branched, decumbent or procumbent, soft-pubescent: leaflets obovate-cuneiform or obcordate, emarginate, denticulate: stipules ovate-acuminate, nearly entire: heads subglobose, very small, x-flowered, on slender peduncles; involucre many-cleft, segments entire; calyx-teeth subulate; broad, scarious, and sometimes toothed at base: fl. minute, pinkish: pod globose, 1-seeded. Common in the Coast Range, and on plains and hillsides eastward throughout the State. May.

34. **T. microdon,** H. & A. Bot. Beech. 330, t. 79 (1840). Larger than the last, not rarely 2 ft. high, glabrous or nearly so: involucre broader, deeply cup-shaped, equalling the head, its many lobes conspicuously toothed: calyx-teeth rigid, triangular, acute, serrulate below; corolla minute, white. Abundant in many places about the Bay of San Francisco; also in the interior, and northward to Washington. May.

35. **T. cyathiferum,** Lindl. Bot. Reg. under t. 1070 (1827); Hook. Fl. i. 133 t. 50. Erect or ascending, 3-15 in. high, glabrous, pale green: stipules ovate, laciniate-toothed: leaflets obovate or oblanceolate, ½-1 in. long: heads large; involucre broad, saucer-shaped, thin, with short many toothed and nerved lobes: calyx 5-nerved, membranaceous; the nerves excurrent into branching setaceous-spinose tips which equal the small white corolla. From Highland Springs, Lake Co. *Simonds*, and near Cisco in the Sierra, *Greene*, northward. June-Aug.

+ + *Corolla more or less inflated in age.*

36. **T. barbigerum,** Torr. Pac. R. Rep. iv. 79 (1857). Branches many, stout, with short internodes, nearly prostrate, 4-10 in. long; herbage deep green, soft-pubescent: petioles elongated; leaflets broadly obovate, obtuse, denticulate, ½ in. long or less: involucre as broad as the long-peduncled heads, 4-8 lines wide, shortly lobed and setaceously toothed; calyx-tube short, thin and at length scarious: teeth setaceous-awned, plumose, sometimes 2-3-parted, usually exceeding the small purple corolla: pod 2-seeded. Frequent at Berkeley, San Francisco, etc.

37. **T. Grayi,** Loja. Giorn. Bot. Ital. xv. 189 (1883); *T. barbigerum,* var. *Andrewsii,* Gray, Proc. Am. Acad. vii. 335 (1868). Erect, stout, with long internodes, 1–2 ft. high, sparingly branched, villous with long spreading hairs: leaflets 1 in. long, cuneate-oblong or elliptic-lanceolate, obtuse or acutish, sharply serrulate: heads long-peduncled, 1 in. broad; the involucre as broad: calyx-tube scarious, villous, 10-nerved; teeth linear-subulate from a triangular base, plumose, as long as the dark red-purple corolla. First collected by *Dr. Andrews* (1856); later, at Mendocino City, *Bolander ;* also in Marin Co., *Mrs. Curran.* A fine species: not at all susceptible of being referred to *T. barbigerum* as a variety.

38. **T. fucatum,** Lindl. Bot. Reg. t. 1883 (1836). Usually stout and fistulous, the decumbent branches 1–2 ft. long; herbage light green, glabrous and somewhat succulent: stipules large, membranaceous, nearly or quite entire: leaflets 1_2–1^1_2 in. long, broadly obovate, obtuse or retuse, dentate or spinulose-denticulate; peduncles stout, far exceeding the leaves; bracts of the involucre ovate-lanceolate, acuminate, scarious-margined, connate at base: heads hemispherical, 1–2 in. broad: calyx thin, campanulate, the short teeth entire, unequal: corolla $\frac{1}{2}$–1 in. long, ochroleucous, fading with a red tinge: pod stipitate, 3–8-seeded: seed roundish, nearly 1 line broad, minutely granulate. Common along the coast and in the interior. Variable in size: often small and few-flowered.

39. **T. amplectens,** T. & G. Fl. i. 319 (1838); H. & A. Bot. Beech. 330. t. 78 (1840); *T. quercetorum,* Greene, Pitt. i. 172 (1888). Light green and glabrous like the last, but small, slender, the branches 3–10 in. long: leaflets $\frac{1}{2}$–$^3_1$ in. long, cuneate-obovate or -oblong, truncate or retuse, mucronately denticulate: peduncles slender, not surpassing the leaves: involucre half as broad as the heads, its lobes broad, scarious-margined, obtuse, sometimes cleft or toothed: calyx cleft nearly to the base, the subulate slenderly acuminate teeth very unequal, the larger rarely toothed or cleft: corolla ochroleucous, 2–3 lines long: pod membranaceous, translucent, finely reticulate with green veins, promptly dehiscent by one suture only, 4–6-seeded: seed small, transversely oval, emarginate at the hilum, coarsely tuberculate-rugose. Not common: but found at Monterey, *Douglas,* Oakland Hills, *Chesnut, Simonds,* Alameda, *Greene ;* plentiful on moist flats about Byron Springs. The type, collected by Douglas, is exactly this; and it is more allied to *T. fucatum* than to any of the following very common plants, two or more of which have long been confused with it.

40. **T. diversifolium,** Nutt. Pl. Gamb. 152 (1848); Greene, Pitt. i. 7. Diffuse, glabrous, the branches flaccid though not very slender, 1–2 ft. long: stipules ovate, entire, subulate-pointed: leaflets linear or oblong, obtuse or truncate, repandly dentate or somewhat serrulate, 1 in. long:

peduncles slender, little exceeding the leaves: heads 8–15-flowered; involucre of about 5 small ovate or oblong bracts: corolla in age oval or oblong, slightly inflated and about equally so from end to end, conspicuously striate: pod 2-seeded: seed transversely oblong, beautifully sinuous-rugose. Rather common in low moist lands along the seaboard, where it prefers the vicinity of the salt marshes; but also around ponds among the hills, and even on low subsaline plains of the lower Sacramento. A most distinct species every way, and one which, having its lowest leaves narrowest and its uppermost and later ones broadest, strangely reverses that order of leaf-widening which is otherwise universal, I believe, in Californian Clovers. Apr.–June.

41. **T. stenophyllum,** Nutt. l. c. 151; *T. amplectens,* Greene, Pitt. i. 6, not of T. & G. Near the last, the branches more slender, but wiry and upright: linear leaflets of about equal width on the lower and uppermost parts of the plant, all remotely serrate-toothed: peduncles much longer than the leaves, filiform: segments of the involucre oblong, connate at base: head small, hemispherical, the (deep purple or white) corollas in age almost obpyramidal, being gradually inflated from a narrow base to a broad, almost truncate apex: pod 2-seeded: seeds obliquely heart-shaped, strongly rugose.— One of the most common species of western California; perfectly distinct from the last by its different texture, extremely bladdery corollas and peculiar seeds. It inhabits dry plains and hill-sides, and in dry seasons is very depauperate, though always retaining all its characters. This is what Californian botanists and amateurs have always hitherto been taught to receive as the *T. amplectens* of Torrey & Gray. Hence the true *amplectens,* when first brought to the author, was published by him as a new species. Apr. May.

42. **T. depauperatum,** Desv. Journ. Bot. iv. 69, t. 32 (1814). Only a few inches high, branched from the base, flaccid, decumbent, glabrous, few-leaved: leaflets ½ in. long, cuneate-oblong, obtuse or emarginate, denticulate: head long-stalked, few-flowered: involucre greatly reduced, with truncate short lobes: corolla larger than in the last, less inflated: pod 1–2-seeded: seed little broader than long, rather angular, tuberculate-rugose.— Less common than the last, and a small, rather obscure species, seldom collected. It appears to be one of the few plants common to the western coasts of both North and South America. Mar. Apr.

43. **T. laciniatum,** Greene, Pittonia, i. 7 (1887). Slender, flaccid, glabrous, ascending, 3–6 in. high: stipules ovate, acuminate, mostly entire: lower leaflets narrowly cuneiform, denticulate, the upper broad, truncate and 3-dentate at apex, the sides laciniately toothed or pinnatifid: involucre reduced and obscure: fl. 3–5, white with purple centre, much inflated in age: pod 3–4-seeded: seed oval, with the strong corrugation running into a more or less distinctly favose coarse reticulation.

Thus far collected only by the author, on subsaline plains of the lower San Joaquin, near Byron, Bethany, etc., where it is plentiful. The leaflets, commonly laciniately and deeply pinnatifid, in occasional specimens are narrowly linear or linear-spatulate and quite entire. Mar. Apr.

11. **MELILOTUS**, *Morison* (SWEET CLOVER). Erect herbs with pinnately 3-foliolate leaves, the leaflets toothed, and small fragrant flowers in slender axillary racemes. Petals free from the diadelphous stamens, deciduous. Pod ovoid, small, scarcely dehiscent, 1- 2-seeded. Old World plants with sweet-scented herbage and very fragrant flowers. The following species are naturalized with us.

1. M. INDICA, Allioni, Fl. Ped. i. 308 (1785) ; *M. parviflora*, Desf. Fl. Atl. ii. 192 (1800); *M. occidentalis*, Nutt. in T. & G. Fl. i. 321 (1838). Annual, glabrous, 1 3 ft. high, bearing many racemes of minute yellow flowers. Common in low grounds, chiefly near the salt marshes or along rivers. A good fodder plant.

2. M. ALBA, Lam. Encycl. iv. 63 (1796). Stout, 3 6 ft. high ; fl. larger, white, very fragrant. Spontaneous in northern California ; perhaps not within our limits.

12. **MEDICAGO**, *Tournefort* (ALFALFA, BUR CLOVER). Herbs with pinnately 3-foliolate (rarely 5-foliolate) leaves and flowers 2, 3 or many on axillary peduncles. Petals free from the diadelphous stamens, deciduous. Pod 1-several-seeded, falcate-incurved or coiled into a spiral. Very valuable forage plants, natives of Asia, brought to California, by way of Mexico or South America in very early times ; some of them now naturalized almost everywhere in the State.

1. M. LUPULINA, Linn. Sp. Pl. 779 (1753). Annual, slender, procumbent, 1—2 ft. long, soft-hairy ; leaflets obovate, small ; fl. minute, in small oblong heads, yellow ; pod small, reniform or curved almost into a ring, black when ripe, 1-seeded. Foot-hills of the Sierra Nevada, at Ione and elsewhere.

2. M. DENTICULATA, Willd. Sp. iii. 1414 (1801). Annual, much branched, decumbent, glabrous ; leaflets obovate or obcordate, denticulate ; fl. 2 3, yellow ; pods coiled into 2 circles, their margins armed with hooked prickles. Common everywhere ; valuable as a forage plant, but the "burs" damaging to wool.

3. M. SATIVA, Moris. Hist. ii. 150 (1680). Perennial, erect, glabrous, 2 4 ft. high ; leaflets cuneate-oblong or oblanceolate, toothed above ; fl. x , racemose, violet ; pod spirally coiled, unarmed. This, the *Alfalfa* or *Lucerne* of the cultivators, is here and there spontaneous, but nowhere very prevalent in the wild state.

LEGUMINOSÆ.

13. CYTISUS, *Dioscorides* (BROOM). Shrubs with green very leafy or nearly leafless often angular branches, palmately or pinnately 3-foliolate leaves (leaflets entire), and solitary or racemose yellow or white flowers. Calyx with campanulate tube and bilabiate limb. Petals broad; keel obtuse. Stamens monadelphous. Pod compressed, several-seeded.— Natives of the Old World; becoming spontaneous on the Pacific coast of America.

1. C. CANARIENSIS. Linn. Sp. Pl. 709 (1753), under *Genista*. Much branched, 3–6 ft. high, soft-pubescent, the branches and branchlets very leafy: leaflets ¼–½ in. long: fl. yellow, in numerous terminal short racemes, fragrant; calyx with upper segment deeply, lower obsoletely 3-toothed at apex; banner not reflexed; keel deflexed, releasing the stamens. —Running wild in profusion on the grounds of the University at Berkeley: flowering throughout the year, but most freely in Jan. and Feb.

2. C. SCOPARIUS, Link. Enum. ii. 241 (1822); Linn. Sp. Pl. 709 (1753), under *Spartium*. Size of the last, but sparingly leafy, the branches prominently angular: leaflets glabrous, often 1 only: fl. large, bright yellow, solitary or in pairs along the branchlets, in the leaf-axils and apparently racemose: pod pilose along the margins. —Naturalized abundantly northward; more sparingly with us. Flowering in spring only.

3. C. PROLIFERUS, Linn. f. Suppl. 328 (1781). Arborescent, branches terete and, with the young leaves, etc., silky-pubescent: leaflets 3, elliptic-lanceolate, 1 in. long or more: fl. white in lateral umbellate racemes: banner reflexed: keel shorter than the wings, enclosing the stamens: pod villous. Native of Teneriffe; a valued forage shrub in some countries; escaped from cultivation at Berkeley. Jan. Feb.

14. ULEX, *Linnæus* (FURZE, GORSE). Compact very thorny shrubs with simple prickle-pointed leaf-like organs, and scattered yellow flowers. Calyx of 2 nearly or quite distinct yellowish sepals. Banner nearly as long as the other petals, not reflexed, scarcely even erect. Stamens monadelphous. Pod few-seeded, little longer than the calyx.

1. U. EUROPÆUS. Linn. Sp. Pl. 741 (1753). Three to six feet high, the numerous short branchlets villous, ending in a stout spine: lower leaves sometimes lanceolate, more commonly reduced green spines ½ in. long: fl. ½ in. long, yellow, solitary but often crowded on the branchlets; calyx villous. —Spontaneous here and there about San Francisco, where it has escaped from cultivation. Highly ornamental when in flower. Feb.–Apr.

15. SPARTIUM, *Lobelius* (SPANISH BROOM). Branches stout, terete, green and rush-like, glabrous, sparsely leafy with 1-foliolate leaves, or leafless, bearing terminal loose racemes of large yellow flowers. Calyx spathaceous, cleft to the base above, 5-toothed at apex. Banner roundish, erect: keel acuminate. Stamens monadelphous. Pod compressed.

1. S. junceum, Linn. Sp. Pl. 708 (1753). - Native of southern Europe; extremely beautiful when in flower, and, with its fascicles of long reedy branches, at no season unsightly. Spontaneous near San Francisco.

16. LUPINUS, *Catullus* (LUPINE). Leaves palmately 5 15-foliolate: leaflets entire, sessile; stipules adnate, seldom conspicuous. Flowers blue, pinkish or yellow, in terminal racemes, with bracts mostly caducous. Calyx deeply bilabiate; upper lip notched, lower usually entire, occasionally 3-toothed or -cleft. Banner roundish; wings falcate-oblong, commonly slightly united at tip in front of, and enclosing, the falcate usually slender-pointed keel. Stamens monadelphous, dimorphous, 5 with longer and basifixed anthers, the alternate 5 with shorter and versatile ones.

 * *Pods several-seeded; cotyledons distinct, petiolate.*
 + *Annuals; flowers not verticillate.*

1. **L. gracilis,** Agardh, Syn. 15. t. i. (1835). Erect, slender, 6—10 in. high, very pilose: leaflets 5—7, cuneate-obovate, ¼—1½ in. long: raceme short, lax: upper lip of calyx bifid, lower 3-toothed: petals 3 lines long, blue and white; banner shorter than the others; keel exceeding the wings: pod ½ in. long: seed 1 line broad. —From Monterey southward; seldom seen.

2. **L. concinnus,** Agardh, l. c. 6. t. i. Low, stouter than the last, canescently hirsute: leaflets 5—8, oblanceolate, 4—10 lines long, obtuse; raceme short, dense, subsessile; bracts linear-setaceous, persistent: upper calyx-lip 2-parted, lower deeply trifid: petals ⅓ in. long, violet; banner shorter than the rest: pod 4-seeded: seed orbicular.— Same range as the preceding, and more common.

3. **L. hirsutissimus,** Benth. Hort. Trans. n. ser. i. 409 (1833). Tall, erect, very hispid with viscid stinging hairs: leaflets 5—7, cuneate-obovate, retuse, obtuse or acute, mucronulate, ¾—1½ in. long: racemes loose: upper calyx-lip deeply cleft: corolla ½ in. long, reddish-purple: pod hirsute, 1 in. long. —From the Sacramento valley southward; in dry places.

4. **L. truncatus,** Nutt. in H. & A. Bot. Beech. 336 (1840). Stoutish, erect, 2 ft. high, finely pubescent, at length glabrate: leaflets 5—7, linear-cuneiform, 3-toothed or entire at the truncate apex, ¾—1½ in. long: upper calyx-lip cleft: petals deep-purple, 4—5 lines long, the banner shorter: pod 1¼ in. long. From Monterey southward; attributed to San Francisco by Brewer & Watson, I know not on what authority.

5. **L. Stiversi,** Kellogg, Proc. Calif. Acad. ii. 192. t. 58 (1863). Much branched, 1 ft. high, sparsely pubescent, rather succulent: leaflets 5—7, broadly cuneate-obovate, 1 in. long, mucronulate: racemes few-flowered, dense, pedunculate: upper calyx-lip cleft, the lobes broad, acute: fl. ½

in. long; banner short, yellow; wings rose-red, shorter than the white keel: pod 1 in. long, nearly glabrous: seeds 6 or 7, compressed, angular. At middle altitudes of the Sierra, from Yosemite northward. Our handsomest Lupine; first collected by Mr. Charles H. Stivers, an early member of the California Academy, and friend of Dr. Kellogg.

6. **L. sparsiflorus**, Benth. Pl. Hartw. 303 (1849). Branched from the base, erect, slender, 1 ft. high or more, canescently-pilose or hirsute: leaflets 5–9, linear, 1/4–1 in. long: upper calyx-lip deeply cleft: corolla violet, 5 lines long; banner shorter: pod 1½–1 in. long.—Foot-hills of the Sierra, from Sacramento Co. southward.

7. **L. leptophyllus**, Benth. Hort. Trans. n. ser. i. 409 (1833), Bot. Reg. t. 1670. Slender, erect, sparingly branched, 1–2 ft. high, villous: leaflets 8–10, narrowly linear, 1–1½ in. long, glabrous above: racemes elongated: upper calyx-lip narrow, deeply cleft: fl. ½ in. long, petals nearly equal, purple. Same range as the last.

8. **L. citrinus**, Kellogg, Proc. Calif. Acad. vii. 93 (1877). Slender, branched from the base, a span high, pubescent: leaflets 6–8, linear-spatulate, ½–¾ in. long: raceme short-peduncled, rather dense: upper calyx-lip cleft, the lobes acute: lower minutely 3-dentate: corolla yellow; wings obtuse, nearly as broad as long: pod glabrous, 4-seeded: seeds compressed, "rhomboid."—Fresno Co., *Eisen*. Perhaps related to *L. Stiversi*, though possibly belonging to the next section. Dr. Kellogg's description does not enable one to decide.

+ + *Annuals; flowers more or less verticillate.*

9. **L. micranthus**, Dougl. in Bot. Reg. t. 1251 (1829). Rather slender and weak, branched from the base, 6–18 in. high, pilose-pubescent, not at all succulent: leaflets 5–7, narrowly linear to linear-spatulate, ½–1½ in. long, on petioles twice as long: raceme peduncled, verticils 3–5, often indistinct: pedicels 1½ lines long (in fruit 3 lines): upper calyx-lip with divergent lobes; lower long, entire; corolla 2 lines long, blue, except the white and dotted middle of the erect mucronulate banner, the white spot changing to light blue; wings narrow appressed; keel woolly-ciliate toward the apex: pod 5-seeded: seed quadrate oval, whitish, with or without minute light brown dots. Common everywhere; but not at all agreeing with either the figure or description in the Botanical Register; and the type was from a very different and quite distant region of our western country, the upper Columbia River.

10. **L. polycarpus**, Greene, Pittonia, i. 171 (1888). Erect, stoutish, rather succulent, 1–2 ft. high, with firm ascending branches from midway of the stem, pubescent: leaflets 7, narrowly oblanceolate, 1 in. long; glabrous above: racemes with 4–7 very distinct verticils; pedicels 1

line long ; upper calyx-lip bifid, its ovate segments short, parallel ; lower
scarcely longer, 3-nerved, slightly notched at apex ; corolla 1½ lines long,
deep blue ; the obovate retuse banner with a white spot ; wings coherent
at tip, inflated, exposing the base of the broad short keel ; this ciliate
below the apex ; pod rigid, slightly falcate, 7—9-seeded. Very common,
preferring rich low meadow lands adjacent to the salt marshes ; also
occurring in a reduced form, on the low plains of the interior.

11. **L. trifidus**, Torr. in Wats. Proc. Am. Acad. viii. 535 (1873). Slender,
branched from the base, 6—10 in. high, pilose-canescent : racemes short
(1—3 whorls) ; upper calyx-lip deeply cleft, segments divergent ; lower
broad, deeply trifid ; corolla 2½ lines long, blue, the white spot on the
banner permanent ; keel deep, scarcely falcate, shortly and obtusely
pointed, and with a few stiffish ciliolae above the middle. In sandy land,
at San Francisco and Alameda ; in habit, pubescence, size of corolla, etc.,
very like No. 9, but most distinct.

12. **L. bicolor**, Lindl. Bot. Reg. t. 1109 (1827) ? Low, often diffuse,
stoutish, 6—10 in. high, silky-pilose : leaflets 5—7, linear-spatulate, 1 in.
long ; upper calyx-lip short, bifid ; lower twice as long, entire : corolla
4—5 lines long, blue and white, the white changing to red-purple ; banner
reflexed ; keel falcate, acute, ciliate toward the apex ; pod small, about
5-seeded. Sandy soil about San Francisco, in a slender depressed very
hairy form ; also on gravelly crests of the Oakland Hills, where it is
stouter, with ascending branches. The two may well be distinct species ;
and neither plant agrees perfectly with the original *L. bicolor* of the
Columbia River.

13. **L. pachylobus**, Greene, Pittonia, i. 65 (1887). Stout, rigid, barely
1 ft. high, with a few ascending branches from the base, hirsute throughout ; petioles slender and long ; leaflets 5—7, linear, ¾ in. long ; racemes
on stout peduncles, whorls 2—4 ; fl. 3 lines long, subsessile, deep blue ;
calyx-lips broad, the upper very short, notched ; lower entire and twice
as long ; pod large (1¼ in. long, 4—5 lines wide), very hirsute, 4—6-
seeded. Briones Hills, east of San Pablo Creek, Contra Costa Co., 15
April, 1887 ; collected only by the author. The full grown pods are very
thick and succulent just before maturity ; even bearing down to the
ground the stout branches.

14. **L. nanus**, Dougl. in Benth. Hort. Trans. n. ser. i. 409 (1833).
Commonly 1 ft. often 2 ft. high, with many decumbent branches, not
succulent, minutely and not densely villous-pubescent : leaflets oblanceolate, 1 in. long ; racemes short-peduncled, 3—7 in. long, of many rather
indistinct whorls of large deep purple fragrant flowers : upper calyx-lip
deeply cleft ; lower 3-dentate : corolla 6—7 lines long, the orbicular
retuse banner closely reflexed, the white middle part turning rose-red ;

wings lightly joined, forming an obliquely obovate inflated sac; falcate keel with a long slightly ciliate beak. Common on plains, especially in sandy soil; the most beautiful species of the group, and by no means a dwarf, except in very dry seasons.

15. **L. carnosulus,** Greene, Bull. Calif. Acad. ii. 144 (1886). Erect, 1–2 ft. high, usually simple, stout and succulent; pubescence minute, appressed: leaflets oblanceolate, 1 in. long, obtuse, but with a small recurved mucronation: raceme loose, distinctly verticillate: upper calyx-lip deeply cleft: lower entire: corolla deep blue; keel villous in the middle. Near Olema, Marin Co., *Greene*, and perhaps frequent northward; for we have it from Oregon, *Howell*. Readily distinguished from the last by its succulent herbage.

16. **L. affinis,** Agardh, Syn. 20 (1835). Very stout and succulent, irregularly branching above (not from the base), 1–2 ft. high, usually almost glabrous, the pubescence very sparse and short; stipules small, setaceous: leaflets 7, cuneate-obovate, obtuse or emarginate, 1–$1\frac{1}{2}$ in. long, on stout petioles twice or thrice as long: racemes rather short-peduncled; whorls 3–7; bracts equalling the calyx; upper calyx-lip bifid; lower entire or 3-toothed: corolla 5–6 lines long, deep bluish purple; keel broad, not strongly falcate, naked: ovary densely velvety; pod glabrate, 1–2 in. long, 5–9-seeded. Common in low, clayey soils, mostly near the sea; but also in the interior, where it is more pubescent. Although authors who have seen the types say that this plant is the real *L. affinis*, Agardh's description does not well apply to it. Doubtless No. 15, which no field botanist could confuse with the present species, has been mixed with it by various authors; possibly by Agardh himself.

17. **L. cervinus,** Kellogg, Proc. Calif. Acad. ii. 229, fig. 72 (1863). Stout, pale green, appressed-pubescent, 1 ft. high: stipules subulate, short; leaflets 7, cuneate-obovate, obtuse, mucronulate, 1–2 in. long: raceme greatly elongated; whorls distinct, many-flowered; flowers on very short pedicels: upper calyx-lip cleft; lower 3-toothed, the middle tooth slightly longer: corolla pale blue; banner orbicular, pubescent exteriorly; keel ciliate along the margin, and pubescent on the lower edge: ovary villous, 7-ovuled. Pine woods of the Santa Lucia Mts., *Lobb*. Evidently a good species, but perhaps not of this group: very possibly perennial.

+ + + *Perennials; not woody at base.*
++ *Flowers large (6–7 lines long).*

18. **L. polyphyllus,** Lindl. Bot. Reg. t. 1096 (1827). Stem solitary, nearly simple, very stout, somewhat fleshy, erect, 3–5 ft. high, pilose-pubescent, equably leafy up to the inflorescence: stipules adnate for half their length or more: petioles 6–12 in. long: leaflets 11–15, lanceolate,

acute, hirsute beneath, glabrous above, 3–6 in. long; raceme short-peduncled, dense, 1–2 ft. long; fl. subverticillate, long-pedicelled, 1_2 in. long and as broad; calyx-lips of about equal length, the upper broader, both entire; wings bluish, banner red-purple; keel falcate, acuminate, naked; pod 1–1½ in. long, ¼ in. broad, 7–9-seeded. –In open marshy ground toward the sea, from near Point Bonita light-house northward. Our most magnificent Lupine, but far from common; well marked in habit, and belonging exclusively to the seaboard. May.

19. **L. longipes.** Stems more or less clustered, erect, stoutish, not at all succulent, sparingly branched above, 2–4 ft. high, striate, glabrous or loosely hairy; leaves mostly basal, on petioles 12–18 in. long; stipules setaceous-subulate; leaflets 7–11, broadly lanceolate, acute, setaceously mucronulate, 2–4 in. long, glabrous, the margin often more or less ciliate; raceme peduncled, elongated, not dense; fl. much as in the last, but keel slightly ciliate in the middle; pod 1 in. long or more, densely hirsute, about 7-seeded; seed compressed, oval, brown with a dark diagonal line. Along streams at middle or higher elevations in the Sierra, northward to Oregon. Very distinct; and neither of the old names, *L. macrophyllus* or *grandifolius*, seems to belong to it. June–Aug.

20. **L. latifolius,** Agardh, Syn. 18 (1835); Lindl. Bot. Reg. t. 1891; *L. rivularis,* var. *latifolius,* Wats.; *L. adsurgens,* Drew. Stoutish, erect, branching, 2–4 ft. high, minutely appressed-pubescent, the stem not striate, dark green and shining, equably leafy, the basal leaves not long-stalked; stipules linear-setaceous; leaflets 5–7, broadly oblanceolate, thin, mucronulate, pilose-ciliate on the margins and the midvein beneath, 1–2½ in. long; racemes slender-peduncled, loose, the verticils often distinct; pedicels slender; calyx-teeth elongated, the upper notched slightly at the narrow apex; fl. blue, changing to a dark tawny hue; keel ciliolate below the middle; seed little compressed, the diagonal line narrow.–By streamlets and on wooded northward slopes of the Coast Range at low altitudes; common in the hills near Berkeley. May–Aug.

21. **L. cytisoides,** Agardh, l. c. *L. rivularis,* Wats., not of Dougl. Taller than the last, more rigid, the stems striate; pubescence minute, stiffish and closely appressed; stipules lanceolate-subulate; leaflets 7–9, oblanceolate, 2 in. long or more; raceme greatly elongated, but short-peduncled, dense, the flowers not whorled; calyx as in the last; corolla blue, the banner fading brownish; keel strongly falcate, densely ciliate for a short distance below the middle. Mountains of Kern Co., perhaps of Monterey, thence southward, where in Los Angeles Co. it is often six feet high, and very showy. Very unlike the last in habit, pubescence, inflorescence, etc. May–July.

22. **L. littoralis,** Dougl. in Lindl. Bot Reg. t. 1198 (1828). Stems clustered, decumbent or ascending, somewhat succulent, 1–2 ft. long.

silky, or at the nodes villous : leaflets 5–7, linear-spatulate, acute, 1 in. long, silky on both sides : fl. distinctly and rather remotely verticillate in a short-peduncled raceme ; calyx-lips subequal, entire ; banner red, shorter than the blue wings ; keel ciliate ; pod linear, hirsute ; seeds linear, brown with black spots. From near Point Reyes, *Mrs. Curran*, northward, near the sea. The coarse yellowish roots have the appearance and something of the flavor of licorice and were used as food by the aborigines.

23. **L. albicaulis,** Dougl. in Hook. Fl. i. 165 (1833). Stoutish, slightly succulent, stems decumbent, 2 ft. long or more, very leafy, appressed-puberulent ; stipules small, subulate, deciduous ; leaflets 7–9, oblanceolate, obtuse, mucronulate, 1–2 in. long, on petioles little longer ; raceme 6–10 in. long, short-peduncled ; fl. rather distinctly whorled, of a dark tawny or yellowish brown ; banner and wings shorter than the long falcate naked keel. Var. **silvestris.** *L. silvestris*, Drew, Bull. Torr. Club, xvi. 150 (1889). More slender, canescently hirsutulous throughout ; leaflets acutish ; fl. cream-color, turning brownish. The type, common from Mt. Shasta northward, may not occur within our limits. The variety, very possibly distinct, seems to differ mainly by its pubescence. It is found in the Trinity Mts., *Drew*, and something approaching it, but with pubescence less spreading, was obtained at Chico by *Dr. Parry*. June–Aug.

24. **L. formosus.** Stoutish and suberect, or more slender and decumbent, 2–3 ft. high, sparsely silky-pubescent ; stipules long, linear-setaceous, persistent ; leaflets 7–9, linear-lanceolate, very acute, 1–1$^1_3$ in. long, equalling the petiole ; raceme subsessile, more or less whorled, but rather dense ; fl. 6–7 lines long, rich violet, the banner and wings equalling, the latter entirely enfolding, the less elongated naked keel. Var. **Bridgesii.** *L. albicaulis*, var. *Bridgesii*, Wats. Stipules narrowly lanceolate, the whole plant silvery-canescent, and even villous ; raceme distinctly pedunculate, the verticils more remote and distinct. This, while like *L. albicaulis* in its glabrous keel, is most unlike it in respect to relative size of petals and their coloring. The best type of the species is of the author's own collecting on Mare Island, 1874 ; next to that, No. 857 of the State Survey. The variety will very likely prove distinct. These are our very handsomest perennial lupines : *L. albicaulis*, with its small tawny wings and banner, and long-protruding keel, one of the homeliest. Apr.–Oct.

25. **L. nemoralis.** Stem solitary, erect, slender, flexuous, leafy throughout, 1 ft. high ; pubescent throughout, the stems and stalklets somewhat appressed-pilose, the leaflets rather coarsely silky ; stipules subulate-setaceous ; leaflets 7–11, narrowly cuneate-oblong, obtuse, mucronulate, 1 in. long, on petioles of 2 in. or less ; raceme slender-peduncled, short and few-flowered, not whorled : both calyx-lips broad,

not elongate.l, entire ; corolla purple, the broad wings equalling the very falcate naked keel. In dry open pine groves at middle elevations of the mountains in Calaveras Co., collected by the author, June, 1889. Perhaps *L. Andersonii*, var. (?) *Grayi*, Wats., found by Gray in Mariposa Co., may be the same.

26. **L. sericatus**, Kellogg, Proc. Calif. Acad. vii. 92 (1877). Stoutish, decumbent, $\frac{1}{2}$–1 ft. high, very leafy, canescent with a minute closely appressed silky pubescence ; stipules setaceously acuminate from an adnate base ; leaflets 7, spatulate-oblong, obtusish, $1\frac{1}{2}$–2 in. long, on petioles as long ; raceme short-peduncled ; fl. large, in about 5 whorls, deep purple ; calyx-lips large, the upper cleft, lower obscurely 3-toothed ; keel slender-pointed, lightly ciliolate. An elegant species apparently confined to a limited area in the mountains of Lake and Sonoma counties. No. 2180 of the State Survey collection appears to be this. Dr. Kellogg described it as woody at base. We have not found it so. June Sept.

++ ++ *Flowers smaller (4–5 lines long).*

27. **L. onustus**, Wats. Proc. Am. Acad. xi. 127 (1876). Low and decumbent, sparingly silky-villous ; leaflets 5–8, oblanceolate, acutish, 1 in. long, on petioles twice as long ; fl. small, deep blue, scattered in a short, lax raceme ; keel strongly ciliate ; pod 1_2 in. broad, 1^1_2 in. long ; seeds 4 or 5, 3 lines broad. Mountains of Plumas and Sierra counties.

28. **L. confertus**, Kellogg, Proc. Calif. Acad. ii. 192, fig. 59 (1863) ; *L. Torreyi*, Gray. in Bot. King Exp. 58 (1871). Tufted, decumbent, 1–1^1_2 ft. high, densely silky-villous, most leafy at base ; leaflets 5–8, cuneate-oblong, acute, 1 in.² long or more ; raceme long-peduncled, dense ; bracts setaceous, persistent ; fl. verticillate, subsessile, purplish ; upper calyx-lip cleft ; banner narrow, keel not strongly falcate, woolly-ciliate ; pod ³⁄₄ in. long, 2–4-seeded ; seed roundish, white. Var. **Wrightii**. Taller and stouter ; leaflets obtuse, mucronulate ; raceme larger, on a longer peduncle ; fl. 5 lines long ; keel broad, scarcely falcate, densely woolly for a short space just below the apex. Yosemite Valley and northward ; the variety (No. 114 of Wright and Palmer, 1888), in the mountains of Kern Co.

29. **L. minimus**, Dougl. in Hook. Fl. i. 163 (1833) : *L. sellulus*, Kell. Proc. Calif. Acad. v. 34 (1873). Appressed silky-villous, the scapose stems 4–12 in. high ; leaflets 5–7, obovate or oblanceolate, acute, 3–8 lines long, on petioles thrice as long ; raceme dense, less distinctly verticillate ; fl. purple ; upper calyx-lip deeply cleft ; keel not falcate, woolly-ciliolate throughout ; pod 4 lines long, 2–3-seeded ; seed round-obovate, brown. Yosemite and northward. June–Oct.

30. **L. Lobbii**, Gray, in Wats. Proc. Am. Acad. viii. 533 (1873). Stems cespitose, 2–3 in. long, herbage villous ; leaflets 5–7, obovate or

oblanceolate, acute, ¾ 1 in. long: raceme dense, 2 3 in. long, the peduncle shorter than the leaves: calyx-lips deeply toothed: fl. blue; keel ciliate. Same range as the last, but little known.

31. **L. lepidus**, Dougl. in Lindl. Bot. Reg. t. 1149 (1828). Densely silky villous, a few in. to 2 ft. high, with few long-petioled leaves: stipules subulate, falcate; leaflets 5 7, oblanceolate, $\frac{3}{4}$- 1½ in. long: raceme short-peduncled; fl. scarcely whorled, deep purple, the banner with a white spot; keel abruptly falcate, woolly-margined: pod 1 in. long; seeds white. Attributed to the eastern slope of the Sierra in the "Botany of California"; but no plant answering to the description of *L. lepidus* is known to us as Californian.

32. **L. parviflorus**, Nutt. in H. & A. Bot. Beech. 336 (1840). Stem solitary, erect, tall, branching: herbage somewhat hirsute, at length glabrate, pale and glaucescent: stipules minute, subulate; leaflets 5 7, oblanceolate: raceme slender, 1½ -1 ft. long; fl. subverticillate, small, blue: upper calyx-lip notched, lower entire: pod ¾ in. long, 2 4-seeded: seed light-colored.—Yosemite, *McLean*, thence northward and eastward.

33. **L. calcaratus**, Kellogg, Proc. Calif. Acad. ii. 195, fig. 60 (1863). Stems clustered, erect, stoutish, 1 2 ft. high: pubescence minute, silky: leaflets about 9, oblanceolate, acute, 1 2 in. long; raceme short-peduncled, dense: fl. scattered, small, purplish or white; calyx-tube produced above into a spur: upper lip notched, and with the spur colored like the corolla; lower longer, entire: banner silky-pubescent on the back; keel broad, abruptly apiculate, the margin ciliolate: pod 1 in. long: seed light-colored.— Eastern base of the Sierra, on the borders of Nevada.

34. **L. caudatus**, Kellogg, l. c. 198, fig. 61. Near the last, but the herbage canescent with a dense silky-velvety pubescence: racemes sub-sessile; fl. subverticillate, deep violet; spur of calyx shorter; keel narrower and cymbiform, its margin densely woolly-ciliate. Same habitat as the last; obtained near Boca, *Sonne*, in 1888.

35. **L. meionanthus**, Gray, Proc. Am. Acad. vi. 522 (1868). Resembling the last, but dense pubescence shorter; racemes short, subsessile; fl. smaller (only 2 lines long); calyx not spurred: banner glabrous; keel merely ciliate. In the Donner Lake district, near the highest summits, thence eastward and southward in Nevada.

36. **L. holosericeus**, Nutt. in T. & G. Fl. i. 380 (1840). Slender, 12 18 in. high; pubescence silvery, closely appressed: leaflets 6—8, narrowly oblanceolate, acute, arcuate, ¾—1½ in. long: racemes subsessile, 3 - 6 in. long: fl. whorled, small; calyx slightly spurred; lips broad, nearly equal, the upper notched: banner broad, pubescent on the back; keel

ciliate ; pod 1 in. long ; seeds large. Along the eastern base of the Sierra, and northward in Oregon.

+ + + + *Suffrutescent or shrubby species.*
++ *Small-flowered dwarfs of the higher mountains.*

37. **L. Danaus,** Gray, Proc. Am. Acad. vii. 335 (1868). Matted branches only 2–3 in. long, few-leaved ; pubescence strigose-hirsute : leaflets 4–5, oblanceolate, 2–4 lines long, on long petioles : raceme dense, oblong and 1 in. long, or shorter and capitate : upper calyx-lip bifid ; lower tridentate : corolla purple and white ; keel straight, ciliate. A rare species, inhabiting the summit of Mt. Dana, and apparently only once collected.

38. **L. Breweri,** Gray, l. c. 334. Larger than the last, more decidedly suffrutescent, the branches prostrate or only decumbent ; pubescence dense, silvery-silky and closely appressed : leaflets 7–10, spatulate or cuneate, obtuse, 3–4 lines long : raceme 1 in. long, dense : upper lip of calyx cleft : corolla violet ; keel scarcely ciliate. Apparently common in the high Sierra from Kern, *Rothrock*, to Sierra Co., *Lemmon*.

++ ++ *Large-flowered species.*

39. **L. Grayi,** Wats. Proc. Am. Acad. xi. 126 (1876). Stems stoutish, 1 ft. long, decumbent from a woody branching caudex ; hoary-tomentose throughout : leaflets 5–9, cuneate-oblong, $1\frac{1}{2}$–$1\frac{1}{2}$ in. long, on petioles twice as long : racemes short-peduncled ; fl. verticillate, 6–7 lines long, deep blue : calyx-lips elongated, entire : petals subequal ; keel ciliate : pod 1 in. long or more, 4–6 seeded.—In dry pine woods of the middle Sierra. June.

40. **L. Ludovicianus,** Greene, Bull. Calif. Acad. i. 184 (1885). Decidedly shrubby and the stout branches erect from the base, 1–2 ft. long ; densely white-tomentose and with a coarser hirsute pubescence on the branches and stalklets : leaflets 7–9, broadly oblanceolate, obtuse, 1 in. long : fl. of middle size, subverticillate in a short dense short-peduncled raceme : calyx-lips broad, entire, subequal : keel strongly falcate, surpassing the other petals, somewhat woolly-ciliolate : pod 1 in. long, tomentose, 5-seeded.—Mountains above San Luis Obispo, *Mrs. Curran*.

41. **L. ornatus,** Dougl. Bot. Reg. t. 1216 (1828). Woody branches stout, short, numerous, horizontal ; flowering branches erect, 1–3 ft. high, leafy below ; pubescence short, silky, appressed : leaflets 5–7, oblanceolate, acute, 1–2 in. long : raceme long-peduncled, rather lax : calyx-lips subequal, the upper toothed or bifid : petals subequal, blue, the banner paler ; keel falcate, ciliate toward the apex : pod $1\frac{1}{4}$ in. long : seed white, almost orbicular, compressed, $2\frac{1}{2}$ lines long.—Species of the dry northeastern plains ; common north of Mt. Shasta, and occurring in Plumas Co., *Mrs. Austin* ; always pronouncedly shrubby, though the woody branches are short and depressed.

42. **L. albifrons,** Benth. in Lindl. Bot. Reg. t. 1642 (1833); *L. Douglasii,* Agh. Syn. 34 (1835). Arborescent, the distinct trunk-like woody stem 1–3 ft. high, parted into numerous short leafy and flowering branches, these ending in a rather long-peduncled loose raceme; leaflets 7–9, oblanceolate, 1 in. long or more, silvery-silky on both sides: fl. verticillate, large, deep blue; upper calyx-lip broad, cleft to the middle, or less deeply; lower entire; petals subequal, the broad banner with a whitish spot which soon changes to rose-purple; keel ciliate: pod 2 in. long, 5–9-seeded; seed oval, 2 lines long, brownish, encircled marginally by a dark line. Var. **collinus.** Smaller in all its parts and with no trunk-like stem, the branches decumbent from a short caudex. Very common along the seaboard, though not near the shore, but on clayey slopes and along ravines; the variety on rocky summits about the Presidio, San Francisco, and on the islands in the Bay. Very distinct from the next, although foreign authors having mixed them in their herbaria, seem to have found the confusion hopeless. Feb. Apr.

43. **L. Chamissonis,** Esch. Mem. Acad. St. Petersb. 288 (1826). Commonly 3 ft. high, but never arborescent: the suffrutescent branches forming a more or less dense tuft and leafy throughout; foliage much as in the last, but petioles shorter; raceme more elongated and dense, but scarcely peduncled: fl. less distinctly whorled, paler blue, or of a lavender shade; banner with a permanent yellowish spot. Apparently confined to the sand dunes of the San Francisco peninsula and Point Reyes. Probably not in the southern part of the State; never away from the sea. The "var (?) *longebracteatus*" of the "Botany of California" must needs be the very type of the species. Much later in its flowering than the last. Apr. July.

44. **L. variicolor,** Steud. Nom. Bot. ii. 78 (1841); (Greene, Pitt. i. 216; *L. versicolor,* Lindl. (1837), not of Sweet; *L. littoralis,* B. & W. Bot. Calif. i. 118 partly, not of Dougl. Woody basal branches short, slender, very tough, the decumbent, or often assurgent annual ones very leafy, 1 ft. long or less; pubescence of the leaves scant, appressed, the stems often sparingly hirsute: leaflets 7–9, narrow, acute: raceme short, the whorls often 3, 2 or 1 only: fl. large; banner white or pale blue; wings blue; keel ciliate throughout its length: pods large. Frequent on grassy northward slopes at the Presidio, San Francisco, and southward, in San Mateo Co.; most nearly related to the next, the flowers occasionally yellowish, and there are manifest hybrids between the two. Apr. June.

45. **L. arboreus,** Sims, Bot. Mag. t. 682 (1803); Bot. Reg. xxiv. t. 32; *L. rivularis,* Dougl. in Lindl. Bot. Reg. t. 1595 (1833). From arborescent and 6–10 ft. high to suffrutescent and bushy; slightly silky-pubescent: leaflets about 9, narrowly lanceolate, $\frac{3}{4}$–$1\frac{3}{4}$ in. long, acute, glabrate above: raceme often 1 ft. long; fl. whorled, sulphur-yellow, or the wings

and keel bluish, or all the petals blue ; keel ciliate ; pod 1½ 3 in. long, 8 12-seeded ; seed oblong, little compressed, dark-colored. Very common in the Bay region, especially on hills not far from the sea ; usually yellow-flowered ; but about Tomales and Bolinas Bays more blue than yellow, just as figured by Lindley under the name *L. rivularis* as the type of that supposed species. Southward, at Santa Barbara, etc., the flowers are wholly of a rather intense blue. Apr. Aug.

* * *Pods 2-seeded ; cotyledons connate. Annuals with whorled flowers and persistent bracts.* Subgenus PLATYCARPUS, Wats.

46. **L. microcarpus**, Sims. Bot. Mag. t. 2413 (1823). Branched from the base, or near it, 1 ft. high or less, somewhat succulent, villous throughout ; leaflets 9, cuneate-oblong, 1 in. long or more ; racemes short-peduncled ; bracts subulate-setaceous, equalling the calyx or shorter ; fl. short-pedicelled, purplish or flesh-color ; calyx densely hirsute ; upper lip short, subscarious, emarginate or cleft ; lower obscurely 2 3-toothed. Throughout the State, apparently in the interior only. The plant is of rather doubtful identity with the Chilian species so named.

47. **L. densiflorus**, Benth. Trans. Hort. Soc. n. ser. i. 409 (1833) ; Bot. Reg. t. 1689. Stem stout, simple below, parted in the middle into numerous wide-spread branches, 2 ft. high, succulent, sparsely villous ; racemes 6–10 in. long, long-peduncled ; bracts setaceous from a broad base ; fl. white or rose-color, sometimes yellow, the banner greenish-dotted ; calyx sparingly villous ; upper lip scarious, deeply cleft ; lower long, toothed. Very common, both on the seaboard and plains of the interior. The yellow-flowered plant, possibly distinct (*L. Menziesii*, Agh.) occurs in Napa Valley and near Antioch, and has sometimes been confused with the next. Apr. May.

48. **L. luteolus**, Kellogg, Proc. Calif. Acad. v. 38 (Apr. 1873) ; *L. Bridgesii*, Gray, Proc. Am. Acad. viii. 538 (Nov. 1873). More slender, simple below, loosely branching above, 2 ft. high or more, rigid, not succulent ; racemes shorter and more dense ; bracts linear-setaceous ; fl. rather small for the group (6 lines long), pale yellow, subsessile ; upper lip of calyx ovate-lanceolate, entire ; lower 3-toothed. A mountain species, from Sonoma Co. and Mendocino northward. June, July.

17. **PICKERINGIA**, *Nuttall*. A rigid much branched spinescent shrub, with small nearly sessile 1 3-foliolate exstipulate leaves, and large solitary almost sessile purple flowers. Calyx campanulate, repandly 4-toothed. Petals equal ; banner orbicular, the sides reflexed ; keel-petals oblong, obtuse, distinct. Stamens distinct. Pod linear, compressed, straight, several-seeded.

1. **P. montana**, Nutt. in T. & G. Fl. i. 389 (1840) ; Torr. Bot. Mex. Bound. 51. t. 14. Shrub 3–6 ft. high, the branches spreading widely ;

leaves crowded; leaflets 1-3 in. long, oblanceolate, acute, entire, somewhat silky when young: fl. near the ends of the stiff spinescent branchlets, on short 2-bracteolate peduncles, from pale rose- to deep red-purple, about ¾ in. long. At middle elevations in the Coast Range; often forming impenetrable thickets on hill-sides and summits; the flowers very beautiful. Apr. June.

18. **THERMOPSIS**, *Robert Brown* (FALSE LUPINE). Stout erect perennial herbs with palmately 3-foliolate leaves, foliaceous stipules, and a terminal raceme of yellow flowers; the pedicels subtended by persistent bracts. Calyx campanulate, cleft to the middle, the two upper teeth often united. Banner roundish, shorter than the wings, the sides reflexed; keel nearly straight, obtuse, equalling the wings. Stamens distinct. Pod long, linear, flat, several-seeded. An Asian and West American genus analogous to *Baptisia* of the Atlantic slope, but very distinct.

1. **T. macrophylla**, H. & A. Bot. Beech. 329 (1840). Stipules ovate, 1 in. long or more; leaflets elliptic-oblong, acute at each end, 4 in. long, villous-tomentose beneath, nearly glabrous above; pod straight, nearly erect, 4-5-seeded. An obscure and little known plant (unless the next be specifically the same), collected by Douglas probably near Monterey, and at a comparatively recent date (1876) by Mr. Joseph Clarke in Mendocino Co.

2. **T. Californica**, Wats. Proc. Am. Acad. xi. 126 (1876). Stipules broadly lanceolate, less than 1 in. long; leaflets obovate or oblanceolate. 1—2 in. long, silky-tomentose on both faces; pod 6-8-seeded. Common on low hills in Marin, Sonoma and Napa counties; also southward. May.

ORDER II. **DRUPACEÆ.**

De Candolle, Flore Française, iv. 479 (1805).

Shrubs or trees with bark exuding gum; bark, leaves and seeds more or less keenly bitter (containing hydrocyanic acid). Leaves alternate, simple, with small caducous stipules. Flowers perfect (except in *Nuttallia*), regular. Calyx tubular or campanulate, free from the ovary, the tube lined with a disk, deciduous; limb 5-lobed, imbricate in æstivation. Petals 5, perigynous. Stamens about 20, inserted within the petals on the disk of the calyx-tube. Pistil 1 (in *Nuttallia* 5); style simple; ovary 1-celled, 2-ovuled, becoming a drupe. Seed pendulous; cotyledons large, thick, fleshy; albumen 0. A small order, commonly appended to Rosaceæ as a suborder; nearly allied to Pomaceæ, less intimately to Rosaceæ proper; of economic importance, on account of the fruits plums, cherries, almonds, etc.

1. **AMYGDALUS**, *Theophrastus* (ALMOND-TREE). Leaves conduplicate in the bud. Flowers solitary or in pairs, from lateral buds, appearing

before or with the leaves. Drupe velvety-pubescent; sarcocarp more or less fibrous, often thin and in maturity dehiscent, falling away from the putamen; this osseous or suberous, smooth or rugose.

1. **A. Andersonii,** Greene. Gray, Proc. Am. Acad. vii. 337 (1868), under *Prunus*. Shrub 3–6 ft. high, the branches short, rigid, somewhat spinescent; leaves fascicled, spatulate or oblong, obtuse or acute, somewhat serrulate, 1_2 1 in. long; fl. 1–3 from each bud, appearing with the leaves, 1_2 in. broad, rose-colored, the petals orbicular; drupe 1_2 in. long, compressed, acute; stone acutely margined on one edge, furrowed upon the other, acute at each end, faintly rugose. Eastern borders of the State, from Sierra Co., *Lemmon*, southward. A handsome bush when in flower; a true almond in its affinities, not a plum or cherry.

2. **A. fasciculata,** Greene. Torr. Pl. Frem. 10, t. 5 (1854), under *Emplectocladus*. Shrub 2–3 ft. high, very rigid, divaricately branched; leaves fascicled, narrowly spatulate, entire, 1_3 in. long, nearly sessile; fl. very small, sessile; petals linear, recurved; stamens only 10 or 15; drupe subglobose, 5–6 lines long, hirsute-tomentose; sarcocarp thin; stone subglobose, smooth, obtuse on both margins, acute at each end.— Eastern slope of the Sierra, in Mono Co. or Inyo, if within our limits. The linear and plane petals would be exceptional in any of the old genera of this order; and Dr. Torrey's *Emplectocladus* may perhaps be valid as a genus.

3. A. COMMUNIS, Linn. Sp. Pl. 473 (1753). The Almond Tree, native of Asia Minor, and perfectly at home in California, where it is one of the most valued of orchard trees, is already spontaneous here and there, and will inevitably become naturalized in course of time. The same may be said of

4. A. PERSICA, Linn. l. c. 472, the Peach Tree, a native of Persia, more hardy than the Almond, and more generally cultivated in America.

2. PRUNUS, *Varro* (PLUM-TREE. PRUNE). Leaves convolute in the bud (in our species). Flowers in umbellate clusters from lateral buds, appearing before or with the leaves. Drupe ovoid, glabrous, glaucous; the thick sarcocarp pulpy, sweet or pleasantly acidulous, and with the distinctive flavor of plums; putamen bony, smooth, compressed, acutely edged on one margin, grooved on the other.

1. **P. subcordata,** Benth. Pl. Hartw. 108 (1849). Arborescent, 3–10 ft. high, much branched, more or less spinescent; nascent leaves and twigs finely pubescent, in age glabrate; leaves ovate, cuneate or obcordate at base, obtuse or acute, sharply serrulate, about 1 in. long, short-petioled; umbels 2–4-flowered; pedicels 1_4–1_2 in. long, fl. white, 1_2 in. broad; drupe 3_4 in. long, red, the pulp rather hard and unpalatable. Var

Kelloggii, Lemmon, Pitt. ii. 67 is a larger shrub with yellow drupe larger, more pulpy and sweeter. The variety in Sierra Co. and northward; the type common in the middle and southerly parts of the State. Fl. Apr., fr. Sept.

3. **CERASUS,** *Theophrastus* (CHERRY-TREE. CHOKE-CHERRY. [SLAY). Leaves conduplicate in the bud. Flowers corymbose or racemose from lateral buds which are often leaf-bearing. Drupe globose, glabrous, destitute of bloom; the sarcocarp sweet rather than acidulous (in our species), often keenly bitter, sometimes sour and astringent; putamen osseous or ligneous, smooth, mostly globose, not prominently margined.

* *Flowers corymbose, from lateral buds; drupe small, with bony putamen.* — CERASUS proper.

1. **C. emarginata,** Dougl. in Hook. Fl. i. 169 (1830); *C. glandulosa,* Kell. Proc. Calif. Acad. i. 59 (1855). *Prunus emarginata,* Walp. Rep. ii. 9 (1843). Arborescent, 10–30 ft. high, the twigs reddish but dull, often pubescent: leaves obovate or oblanceolate, obtuse or acute, rarely emarginate, 1½–3 in. long, finely serrulate, distinctly biglandular at summit of the short petiole: corymbs much shorter than the leaves, few-flowered: drupe oval, dark red, bitter and astringent. My only Californian specimens properly referable to this species were collected in Humboldt Co., *Chesnut & Drew;* but Dr. Kellogg had it from Placerville, and, under the impression that our Coast Range cherry-bush was the true *C. emarginata,* gave to the old species a new name.

2. **C. Californica.** *Prunus emarginata,* B. & W. Bot. Calif. i. 167, excl. syn. & var. *mollis,* not of Walpers. Shrub 3–8 ft. high, branched from the base and clothed throughout with a smooth shining bark: leaves obovate, oblong or oblanceolate, obtuse, retuse or emarginate, on sterile twigs acutish, ¾–1½ in. long, finely crenate-serrulate, mostly uniglandular, and that on the lower part of the blade, well above the junction with the petiole: fl. few, in a short corymb: fruit bright red, intensely bitter. Hills of the Coast Range, from Humboldt Co., *Marshall,* to Mt. Tamalpais, *Bolander,* and frequent in the Oakland Hills; also, in a narrow-leaved form from Donner Lake in the Sierra, *Rev. Dr. Bouté,* northward to Siskiyou Co., *Greene;* along streamlets or on drier ground, but always a mere shrub, in aspect as well as in character quite distinct from *C. emarginata.*

* * *Flowers racemose, from axillary leafless buds; drupe large, with ligneous putamen.* Old genus LAUROCERASUS.

3. **C. ilicifolia,** Nutt. in H. & A. Bot. Beech. 340. t. 83 (1840). *Prunus ilicifolia,* Walp. l. c. 10 (ISLAY). Shrubby or arborescent, evergreen, often 12–18 ft. high, with well rounded head and trunk clothed with a dark rough bark: leaves ovate or ovate-lanceolate, obtuse or acute, truncate or rounded at base, coarsely spinose-toothed, coriaceous, glossy above,

glabrous throughout, 1–2 in. long, short-petioled : racemes 1–2 in. long, leafless ; fl. small ; drupe ½ in. thick or more, slightly obcompressed ; putamen thin, scarcely ligneous ; sarcocarp thin, sweetish, scarcely astringent when ripe. Oakland Hills, thence southward throughout the State ; a mere shrub in the Bay region, but attaining the size and proportions of a small but shapely tree in southern Monterey Co.

* * * *Flowers racemose at the ends of leafy branchlets ; drupe small, astringent.* Old genus PADUS.

4. **C. demissa,** Nutt. in T. & G. Fl. i. 411 (1840). *Prunus demissa,* Walpers, l. c. (CHOKE-CHERRY). Shrub deciduous, 3–12 ft. high ; leaves ovate or oblong-ovate, acute or acuminate, rounded or cordate at base, sharply serrate, more or less pubescent beneath, 2–4 in. long, with 1 or 2 glands on the petiole just below its summit ; racemes 3–4 in. long, many-flowered ; drupe globose, red or dark purple, astringent ; putamen ligneous, globose. Hills behind North Berkeley, but more frequent back from the seaboard throughout the State. Fl. Apr. fr. Sept.

4. **NUTTALLIA,** *Torrey & Gray* (OSO BERRY). Shrub with the habit of Amelanchier, but flowers diœcious, in pendulous racemes terminating short leafy branchlets. Calyx turbinate-campanulate, 5-lobed. Petals 5, broadly spatulate, erect in the pistillate flowers, spreading in the staminate. Stamens 15, in two rows, 10 inserted with the petals, 5 lower down within the calyx-tube ; filaments slender, short. Pistils 5 ; styles short, lateral, jointed at base. Drupes 1–4, ovoid, with thin pulp and osseous putamen. Seed solitary ; cotyledons convolute.

1. **N. cerasiformis,** T. & G. in H. & A. Bot. Beech. 336. t. 82 (1840) ; Greene, in Garden and Forest, ii. 219. Shrub with clustered stems 2–6 ft. high, the bark dark brown ; leaves broadly oblanceolate, entire, obtuse or acutish, mucronulate, 2–3 in. long, short-petioled ; racemes shorter than the leaves ; bracts conspicuous ; fl. white, very fragrant ; drupes 6–8 lines long, slightly compressed, blue-black ; pulp bitter. Common in the Coast Range hills, near the sea ; also in the Sierra northward. The flowers exhale a rich odor of almonds. The fruit though bitter is eaten greedily by birds and mammals. Jan.-Apr.

ORDER III. **POMACEÆ.**

Loiseleur-Deslongchamps, Manuel des Plantes Usuelles, i. 211 (1819) ; S. F. Gray, Nat. Arr. ii. 532 (1821) ; Lindl. Trans. Linn. Soc. xiii. 93 (1821) ; Bartl. Ord. Nat. 399 (1830) ; Spach. Phanerog. ii. 49 (1834) ; M. J. Rœmer. Syn. iii. 97 (1847).

Trees and shrubs with astringent but neither bitter nor poisonous properties ; not gummiferous. Leaves alternate, simple or unequally pinnate, with caducous stipules. Flowers perfect, regular, racemosely or corymbosely clustered, white or reddish. Calyx-tube urceolate or campanulate, more or less coherent with the ovary, the usually short free portion lined with an annular or laminar staminiferous disk; limb 5-lobed, imbricate in aestivation. Petals 5, perigynous. Stamens mostly 20, inserted on the disk. Pistil compound; ovary of 2, 3 or 5 carpels, becoming a pome; styles as many as the carpels. Seeds usually 2 in each cell, collateral, ascending; cotyledons fleshy; albumen 0. Like the preceding a small order, important as yielding such fruits as the apple, pear, quince, etc.

1. **AMELANCHIER,** *Lobelius* (SERVICE-BERRY. JUNE-BERRY). Shrubs with deciduous oblong or rounded serrate or subentire leaves, and bracted racemose white flowers appearing with them in early spring; the bracts caducous. Calyx-tube broadly turbinate; segments as long as the tube, erect or reflexed in flower. Petals from linear-oblong and plane to obovate and concave. Stamens 20, much shorter than the petals. Styles 3–5, coalescent at base or distinct; carpels as many, incompletely 2-celled, but only 1-seeded. Fruit small, berry-like, crowned with the persistent calyx-lobes, of a dark purple and more or less glaucous, the pulp sweet and edible. Seeds small, with a thin black testa.

1. **A. alnifolia,** Nutt. Journ. Philad. Acad. vii. 22 (1834): Gen. i. 306 (1818), under *Aronia*; *A. florida*, Lindl. Bot. Reg. t. 1589 (1833). Arborescent, but seldom 10 ft. high: leaves nearly full grown at flowering time, but thin, dark green, oval or oblong-ovate, obtuse at both ends, coarsely serrate toward the apex, otherwise entire, woolly-pubescent beneath, even in age: racemes x-flowered; bracts setaceous, long-woolly: calyx densely tomentose, the triangular lanceolate teeth closely reflexed: petals spatulate-linear, $\frac{3}{4}$ in. long, plane: stamens very short, not equalling the calyx-teeth. Along streams in the mountain districts, from Lake and Mendocino counties northward; perhaps also southward. What is probably a dwarf condition of this species, with leaves almost entire, is found on northward slopes of the Oakland Hills, and in similar situations among the Mission Hills, San Francisco. Fl. Apr. fr. June.

2. **A. glabra.** Stout, divaricately branched: leaves at flowering time very thin, deep green, $\frac{1}{2}$ in. long, broadly obovate to orbicular, truncate or retuse, the margin sharply and rather deeply serrate all around except at the very base, glabrous throughout: racemes few-flowered: bracts glabrous: calyx glabrous: the broadly triangular sharply acuminate lobes erect: petals cuneate-oblong, less than $\frac{1}{2}$ in. long; stamens a little exceeding the calyx. -In the Donner Lake region of the Sierra Nevada, *Rev. Dr. Bouté*, June, 1888: the specimens in flower only. The total absence of all pubescence from the floral organs and even the bracts of the raceme, together with the shape and attitude of the calyx-teeth, mark this as a very distinct species.

3. **A. pallida.** Stems clustered and bushy, 3–5 ft. high, rigid and with an ashy bark: leaves even at flowering time almost coriaceous, pale green, 3_4 in. long, from oblanceolate and obovate to oblong, obtuse or retuse at apex, cuspidate, entire or sparingly serrate-toothed at apex: racemes ∞-flowered, but short and somewhat corymbose, the lower pedicels elongated: calyx more or less tomentose; teeth triangular, acute, erect: petals obovate or obovate-oblong, slightly concave, 3–5 lines long. A most distinct species, as well marked in floral character as in its very pale glaucous-looking coriaceous foliage. Common on dry hills of the northern and northeastern parts of the State; perhaps not within our limits. Fl. May, fr. July.

2. **CRATÆGUS,** *Tournefort* (THORN). Thorny shrubs or small trees, with simple toothed or lobed leaves and corymbose heavy-scented white flowers. Calyx-tube urceolate; the limb 5-lobed. Petals rounded, concave. Stamens 5–20. Styles 2–5. Pome more or less berry-like, red or purple, crowned with the calyx-teeth and containing 2–5 bony 1-seeded carpels.

1. **C. rivularis,** Nutt. in T. & G. Fl. i. 464 (1840). Tree 10–15 ft. high, nearly or quite glabrous: thorns short, stout: leaves oblong-ovate or ovate, obtuse or acute, cuneate at base, incisely serrate, occasionally somewhat lobed, 1–2 in. long: corymbs small: fl. 4–5 lines broad: calyx-lobes short, obtuse, often pubescent on the margin: pome small. Sierra and Plumas counties, and far to the northward and eastward.

2. **C. Douglasii,** Lindl. Bot. Reg. t. 1810 (1835). Tree 10–25 ft. high. leaves and young shoots villous-pubescent: thorns 1 in. long: leaves $1\frac{1}{2}$–3 in., broadly ovate, more or less incisely cleft or lobed and finely serrate: corymb many-flowered: fl. 5–8 lines broad; calyx-teeth lanceolate, nearly as long as the tube, pubescent: fruit dark red or purple, sweetish but rather insipid. In the northern counties; perhaps not within our limits.

3. **HETEROMELES,** *M. J. Rœmer* (TOLLON. CALIFORNIA HOLLY. CHRISTMAS BERRY). A small evergreen tree with simple coriaceous serrate leaves, and numerous small white flowers in terminal corymbose panicles. Calyx turbinate; limb 5-parted, the lobes at length inflexed over the carpels and becoming fleshy. Petals 5, rounded, concave. Stamens 10; filaments dilated at base and slightly connate. Ovary 2–3-celled, 4–6 ovuled; styles and stigmas 2–3. Fruit ovoid, red, berry-like with dry mealy pulp of acid and astringent taste; carpels free from the fleshy calyx-tube above the middle. Seeds 1–2 in each cell, erect: testa thin-cartilaginous.

1. **H. arbutifolia.** M. J. Rœmer, Syn. Monog. iii. 105 (1847); Ait. f. Kew. iii. 202 (1811), under *Cratægus*; Lindl. Bot. Reg. t. 491 (1820), under

Photinia. Usually 10–25 ft. high ; nascent parts tomentulose : leaves dark green and shining, narrowly oblong to oblong-lanceolate, acute at both ends, sharply but not very closely serrate or dentate, 2–4 in. long : pome 3 lines long : seed one-half as long. –Very common along streams and on northward slopes, in the Coast Range. Fl. July, fr. Dec.

4. **SORBUS,** *Theophrastus* (MOUNTAIN ASH). Unarmed deciduous shrubs or small trees with few and coarse branchlets and large winter buds. Leaves large, unequally pinnate, the leaflets serrate. Flowers numerous, small, white, in terminal compound cymes. Stamens 20. Styles distinct and as many as the cells of the ovary (3–5). Pome very small, globose or pyriform, the coriaceous cells (2-ovuled in the ovary) 1-seeded by abortion. Seeds brownish ; testa rather thin.

1. **S. occidentalis,** Greene. Wats. Proc. Am. Acad. xxiii. 264 (1888), under *Pyrus; P. sambucifolia,* B. & W. Bot. Calif. 189, not of Cham. & Schl. Bushy and only 2–6 ft. high, glabrous ; leaflets 3–5 pairs, oblong or elliptical, obtuse, serrate only from about the middle, 1–2 in. long : cyme small, few-flowered : calyx glabrous : pome somewhat pyriform, coral-red. –In the Sierra from near the Yosemite northward.

5. **MALUS,** *Tournefort* (APPLE-TREE. CRAB-APPLE). Small deciduous trees with more slender branchlets and small winter buds. Leaves simple, more or less serrate. Flowers rather large, reddish or white, corymbose at the ends of short lateral branchlets. Stamens 20. Styles 5, more or less united at base. Carpels 5, wholly covered by the adnate calyx-tube, chartaceous in fruit, 2-seeded. Pome large, globose, depressed at each end, the flesh of acidulous rather than saccharine taste and destitute of grit-cells.

1. **M. rivularis,** M. J. Roemer, Syn. Monog. iii. 215 (1847) ; Dougl. in Hook. Fl. i. 203. t. 68 (1833) under *Pyrus.* Tree 15–25 ft. high : leaves ovate-lanceolate, acute or acuminate, 1–3 in. long, often slightly 3-lobed, sharply serrulate, more or less pubescent when young : corymb somewhat racemose ; pedicels slender, 1 in. long : petals orbicular, 3–4 lines broad, white : pome red or yellow, short-cylindrical, $\frac{1}{2}$ in. long or more. –The *Oregon Crab-apple* has been found as far southward in the State as Sonoma Co., *Bigelow,* and may perhaps be expected in Marin.

2. M. COMMUNIS, DC., the common apple of the orchards already of frequent occurrence by waysides, is destined to become naturalized in California, as in many parts of the world where it has been long cultivated.

6. **PERAPHYLLUM,** *Nuttall.* Shrub low, diffuse, unarmed, with rigid, lanceolate, deciduous leaves, and the inflorescence, flowers and fruits of *Malus,* save that the cells of the carpels are divided each into two 1-seeded apartments by an incomplete partition.

1. **P. ramosissimum**, Nutt. in T. & G. Fl. i. 474 (1840). Only 2—3 ft. high, but the long recurved and more or less tortuous slender branches spreading widely : leaves narrowly oblanceolate, short-petioled, acute, sparingly denticulate, 1—2 in. long : fl. rose-color, 3_4 in. broad ; petals strongly concave : pome globose, 5—7 lines in diameter, watery and acidulous, with the flavor of apples. This peculiar wild apple of our western deserts, entering California along the eastern base of the Sierra, seems hardly distinct from *Malus*; though herbarium botanists misled by a technicality of the carpellary structure have wished to reduce it to *Amelanchier*, to which it is not, in our judgment, at all closely allied ; although *A. pallida*, of similar habitat, resembles it in habit and in foliage.

Order IV. ROSACEÆ.

Jussieu, Genera Plantarum, 334 (1789), partly ; Endl. Gen. 1240 (1840).

Herbs or shrubs often prickly, with alternate frequently compound leaves and mostly foliaceous commonly adnate stipules. Flowers perfect or unisexual, solitary, cymose, corymbose, or paniculate. Calyx free from the ovary, 4—5-cleft, the segments valvate (rarely imbricate) in æstivation. Petals perigynous, as many as the calyx-lobes and alternate with them, or 0. Stamens 5—∞, perigynous (in *Aruncus* hypogynous). Pistils 1—∞ : ovary usually 1-celled and with 1 ovule, sometimes many-ovuled; ovules pendulous or ascending. Styles as many as the ovaries, inserted terminally or laterally, persistent or deciduous. Fruit an achene or an aggregation of drupelets, sometimes follicular, dehiscent by the ventral suture. Seeds with little or no albumen. A large order, of the temperate or boreal regions of the northern hemisphere chiefly ; furnishing some choice fruits (raspberry, blackberry, strawberry) and flowers (rose, spiræa, Kerria), and medicinal plants of astringent properties.

Hints of the Genera.

Unarmed shrubs ;
 Leaves simple,
   Pistils 3—5 ; fr. dry, dehiscent, 1—2-valved, - - 1, 2, 6
   " 1—5 ; " an achene, - - - - - - 6—9
   " ∞ ; " fleshy (mass of drupelets), - - - 17
  Leaves minutely 2—3-pinnately dissected, - - - - - 4, 5
   " small, pinnately 5-foliolate, - - - - - - 15
Prickly shrubs, - - - - - - - - - - - 17, 18
Herbs ; leaves ample, 3-pinnate, - - - - - - - - - 3
  " palmately 3-nate or 5-nate, - - - - - - 15—17
  " unequally pinnate,
    Calyx prickly, - - - - - - 11, 12
    " unarmed, - - - - - - 10, 13—16

1. **NEILLIA,** *Don* (NINE-BARK). Shrubs unarmed with surculose shreddy-barked stems and simple more or less lobed and toothed decidu-

ous leaves ; stipules free and deciduous. Flowers white, in corymbs terminating lateral leafy branchlets. Calyx 5-lobed with campanulate tube. Petals 5, rounded. Stamens ∞, in several rows. Pistils 1-5, becoming as many inflated 2-valved several-seeded capsules which are alternate with the calyx-lobes when of the same number, slightly coherent toward the base. Seeds several, obovoid, with a shining crustaceous testa, and copious albumen.

1. **N. capitata**, Greene, Pittonia, ii. 28 (1889) ; Pursh, Fl. 1, 342 (1814) ; under *Spiræa*. Surculiform stems 10-20 ft. long, more or less tortuous and reclining or interlacing among the branches of small trees ; leaves short-petioled, ovate, acute, more or less distinctly 3-lobed and coarsely toothed, 2-3 in. long, glabrous or stellate-pubescent ; fl. in hemispherical corymbs, or these on vigorous shoots racemosely elongated ; pedicels and calyx more or less tomentose ; calyx-lobes shorter than the petals ; follicles usually 4, exceeding the calyx, 3-4 lines long, ultimately splitting into 2 valves. Common along streams in the Coast Range from the Bay region northward. Apr. May.

2. **SPIRÆA,** *Tournefort* (HARDHACK. STEEPLE-BUSH). Erect unarmed shrubs with simple leaves and no stipules. Flowers numerous, small, white or rose-purple, crowded in a terminal corymb or thyrsoid panicle. Calyx 5-lobed. Petals 5, rounded, imbricate in bud. Stamens 20 or more, inserted with the petals. Pistils about 5, becoming several-seeded follicles which are alternate with the calyx-lobes when of the same number, cartilaginous, not inflated. Seeds linear ; testa thin ; albumen 0.

1. **S. betulæfolia**, Pallas, Fl. Ross. t. 16 (1784). Slender, widely branching, 1-2 ft. high, with reddish bark ; leaves ovate-oblong, 1 in. long, subsessile or short-petioled, rounded at base, obtuse, sharply serrate except toward the base ; fl. rose-purple in small compound fastigiate corymbs ; carpels 1 line long ; ovules 5-8. Moist places among rocks in the higher Sierra. July, Sept.

2. **S. Douglasii**, Hook. Fl. 172 (1830). Stouter, 3-5 ft. high ; growing parts and lower face of leaves tomentose ; leaves oblong, 1-3 in. long, serrate toward the apex ; fl. rose-colored, crowded in a long thyrsoid panicle. Humboldt Co., *Chesnut & Drew*, and northward toward the coast or in the mountains. July.

3. **ARUNCUS,** *Linnæus* (GOATS' BEARD). Herbaceous perennial with ample tripinnate leaves and no stipules. Flowers diœcious, small, white, in numerous slender panicled spikes at summit of the tall stem. Petals 5, spatulate, convolute in bud. Stamens hypogynous. Pistils 3-5, becoming several-seeded follicles which are alternate with the calyx-lobes, otherwise as in *Spiræa*, but seeds pendulous.

1. **A. vulgaris,** Raf. Sylv. Tell. 152 (1838); *A. silvester*, Kost. (1844). *Spiræa Aruncus*, Linn. Sp. Pl. 490 (1753). Glabrous, branching, 3–6 ft. high : leaves 12–18 in. long : leaflets 2–5 in. long, ovate or lanceolate, acuminate, often with a pair of lobes at base, sharply and doubly serrate-toothed, short-petiolulate, thin, sometimes pubescent beneath ; panicle large, compound, pubescent : fl. a line broad, nearly sessile ; filaments elongated : carpels 3–5, glabrous.— In woods of the Coast Range northward.

4. **BASILIMA,** *Rafinesque*. Unarmed shrub with stout branches, smooth dark-colored bark, coriaceous bipinnately dissected stipulate leaves, and terminal panicles of middle-sized white flowers. Calyx 5-lobed. Petals 5, rounded, imbricate in bud. Stamens ∝, perigynous. Pistils 5, becoming coriaceous several-seeded carpels which are opposite the calyx-lobes connate at base, but ultimately 2-valved by a tardy second dehiscence along the dorsal suture. Seeds with distinct albumen.

1. **B. Millefolium,** Greene. Torr. Pac. R. Rep. iv. 83 (. 5 (1857), under *Spiræa ; Chamæbatiaria Millefolium,* Maxim. ; *Sorbaria Millefolium,* Focke in Engl. & Prantl. Rigidly erect and widely branching, 3–8 ft. high : leaves 2 in. long or less, with crowded pinnæ and minute pinnules ; stipules small, free, entire : the growing foliage, inflorescence, and even the carpels more or less densely stellate-tomentose : fl. 1_2 in. broad ; fruit small, little exserted. Eastern slope of the Sierra Nevada and eastward, in dry rocky places. Of special interest as representing in America the otherwise exclusively Asiatic genus, *Basilima* (more recently named *Schizonotus* Lindl., and still later *Sorbaria* Maxim.), of which it has all the characters, though scarcely the habit. The foliage is resiniferous and fragrant as in the next genus.

5. **CHAMÆBATIA,** *Bentham* (TAR-BUSH). A low unarmed evergreen shrub with smooth bark, coriaceous tripinnately dissected stipulate leaves, and terminal few-flowered cymes of rather large white flowers. Calyx with turbinate tube and deeply 5-lobed limb. Petals 5, obovate, imbricate in bud. Stamens ∝, perigynous. Pistil 1, simple, becoming a coriaceous obovoid large achene. Seed with scant albumen.

1. **C. foliolosa,** Benth. Pl. Hartw. 308 (1849) ; Torr. Pl. Frem. 11. t. 6 ; Hook. Bot. Mag. t. 5171. Dwarf, rather slender, with wide-spread branches, 1–2 ft. high, tomentose-pubescent and resiniferous, the calyx only somewhat glandular-hispid : leaves obovate-oblong, 2 in. long ; ultimate pinnules minute, crowded ; stipules small, linear, entire ; petals 3–4 lines long ; achene nearly filling the persistent calyx, abruptly acute.—Western slope of the Sierra from Nevada Co. southward to Mariposa, in pine woods at an elevation of 3,000 to 6,000 ft. Very resinous and heavy-scented ; commonly called *Tar-weed*.

6. SCHIZONOTUS, *Rafinesque.* Unarmed deciduous shrubs with simple toothed or lobed extipulate leaves and terminal panicles of numerous small white flowers. Calyx deeply 5-cleft, nearly rotate. Petals 5, rounded, imbricate in bud. Stamens 20, inserted on an annular peryginous disk. Pistils 5, wholly distinct, becoming 1-seeded hairy carpels, alternate with the calyx-lobes, very tardily dehiscent by the dorsal suture only, or indehiscent.

1. **S. discolor,** Raf. Sylv. Tell. 152 (1838); Pursh, Fl. i. 342 (1814), under *Spiræa*; Maxim. Adn. Spir. 150 (1879), under *Holodiscus*. Shrub 2–6 ft. high, the branches short, rigid, clothed with a gray more or less broken and shreddy bark : leaves ovate, cuneately narrowed to a short winged petiole, above the middle pinnately toothed or lobed, the lobes when present entire, deep green and nearly glabrous above, whitish-tomentose beneath ; panicles erect on short erect or ascending branches : carpels more or less densely hirsute throughout.—On the eastward slope of the Sierra, on dry rocky slopes and summits ; perhaps including the *Spiræa dumosa*, Nutt., of the Rocky Mountains. July.

2. **S. ariæfolius,** Greene. Smith in Rees' Cycl. xxxiii. (1819), under *Spiræa*, also Lindl. Bot. Reg. t. 1365 (1830). Commonly 8–18 ft. high with long spreading or recurved slender branches, these and the stem clothed with a smooth unbroken dark brown bark : leaves short-petioled, deltoid-ovate, 2–3 in. long, two-thirds as broad at the almost truncate base, pinnately shallow-lobed from base to apex, the lobes entire or toothed, green and glabrate above, slightly paler beneath with sparse villous-appressed pubescence : panicle ample, 6–10 in. long, drooping in fl., erect in fr. : carpels compressed, hirsute along both sutures, the sides glabrous and covered with sessile globular resin-dots.—Woods of the Coast Range, near the sea-level or at low elevations among the hills. One of the most beautiful of shrubs when laden in June with its creamy-white half-pendulous panicles. In Marin Co. it often attains the height of 20 feet or more.

7. CERCOCARPUS, *Humboldt, Bonpland & Kunth* (MOUNTAIN MAHOGANY). Unarmed evergreen shrubs or trees with simple leaves, small stipules, and axillary solitary or fascicled apetalous flowers. Calyx salverform, the 5-lobed limb deciduous. Stamens ∞, in 2 or 3 rows on the limb of the calyx. Pistil 1 ; style terminal ; stigma terminal ; ovule solitary, ascending. Fruit a terete villous achene surmounted by a long villous twisted style. Seed linear ; albumen 0.

1. **C. ledifolius,** Nutt. in Hook. Ic. t. 324, and T. & G. Fl. i. 427 (1840). Arborescent, 10–30 ft. high : leaves thick-coriaceous, narrowly lanceolate, entire, more or less revolute, the midvein alone prominent, glabrous above, pubescent beneath, 1–1½ in. long, very short-petioled : persistent

calyx-tube becoming 3–5 lines long : tail of achene 2 in. long. Eastern slope of the Sierra, in the Mono Lake region, etc.

2. **C. betulaefolius**, Hook. Ic. t. 322 (1840) : Greene, Bull. Calif. Acad. ii. 396 (1887) : *C. betuloides*, Nutt. in T. &. G. Fl. i. 427 (1840) ; Greene, in Garden and Forest, ii. 470 (1889) : *C. parvifolius*, var. *glaber*, Wats. Bot. Calif. i. 175 (1876). Shrubby or arborescent, 6–15 ft. high, the stem with a gray thin flaky bark ; branches spreading or recurved : leaves somewhat coriaceous, broadly obovate with more or less cuneate entire base, but coarsely serrate-toothed above the middle, conspicuously feather-veined, glabrous above, pubescent beneath, 1½–2½ in. long : calyx-tube at length ½ in. long : tail of achene often 3 in. long. In the Coast Range throughout the State.

3. **C. parvifolius**, Nutt. in H. & A. Bot. Beech. 337 (1840). Distinguished from the last by the dark thick fissured and persistent bark of the trunks, always bushy habit, and strictly cuneiform foliage.—A Rocky Mountain species, probably reaching the eastern slope of the Sierra in middle California.

8. **PURSHIA**, *De Candolle*. Unarmed shrubs with numerous stout branches and branchlets. Leaves small, crowded and fascicled, cuneiform and 3-dentate, with minute 3-angular stipules. Flowers subsessile, solitary or few at the ends of the short lateral leafy branchlets. Calyx tubular-funnelform, persistent, 5-cleft at summit. Petals 5, obovate, unguiculate. Stamens 25, perigynous in a single series. Pistil 1 (rarely 2) ; style terminal ; stigma lateral, *i. e.*, decurrent down the side of the style : ovule 1, ascending. Fruit a pubescent oblong or obovate achene. Seed without albumen, but with a layer of bitter resinous matter between the two integuments.

1. **P. tridentata**, DC. Trans. Linn. Soc. xii. 158 (1818) ; Pursh, Fl. i. 333. t. 15 (1814), under *Tigarea* ; Lindl. Bot. Reg. t. 1446 ; Hook. Fl. i. 170. t. 58. Branches strict and virgate : leaves cuneiform, ¾ in. long, plane, acutely or obtusely 3-dentate at apex, white-tomentose beneath : calyx glandular : petals pale yellow, exceeding the obtuse calyx-lobes : achene oblong, attenuate at each end, well exserted from the calyx. Eastern slope of the Sierra northward.

2. **P. glandulosa**, Curran, Bull. Calif. Acad. i. 153 (1885). Of widely spreading habit, the branches divaricate : leaves ½ in. long or less, of cuneate-obovate outline, more abruptly narrowed to a distinct petiole : blade 3-lobed almost to the base, the lobes obtuse or emarginate at apex, their margins closely revolute : achene obovate, scarcely exserted from the calyx. –On the Mohave slope of the Kern Co. mountains, *Mrs. Curran*, and eastward in Nevada.

ROSACEÆ.

9. ADENOSTOMA, *Hooker & Arnott* (CHAMISO). Unarmed evergreen shrubs with rigid linear entire sessile fascicled stipulate leaves, and small white flowers in closely panicled terminal racemes. Calyx obconical, 5-toothed, 10-striate, the orifice bearing 5 oblong glands. Petals 5, orbicular. Stamens 10—15, inserted in bundles alternate with the petals. Pistil 1, simple; style laterally inserted and flexuous toward the base; ovary 1-celled, 1- or 2-ovuled, becoming an achene covered by the hardened persistent calyx-tube.

1. **A. fasciculatum,** H. & A. Bot. Beech. 139 (1840). Shrub 2—20 ft. high, with virgate branches covered with leaf-fascicles: leaves linear-subulate, 2—5 lines long, pungently acute, glabrous, often resinous; stipules small, acute; fl. crowded, sessile; calyx 1 line long, bracted at base, the teeth much shorter than the small petals: ovary obliquely truncate. One of the commonest and the most characteristic bushes of the summits and elevated slopes of the Coast Range from Lake Co. southward; also in the Sierra Nevada, but less common there. June.

10. SANGUISORBA, *Fuchs* (BURNET). Herbs with unequally pinnate leaves, coarsely toothed leaflets, foliaceous adnate stipules, and small bibracteolate polygamous or diœcious flowers in crowded spikes on naked peduncles. Calyx-tube turbinate, contracted at the throat, becoming 3—4-angled or winged and persistent; limb 4-parted, imbricate in the bud, deciduous. Petals 0. Stamens 2—∞. Carpels 1—3, free from the calyx; style terminal, filiform; stigma tufted; ovule 1, suspended. Achene membranaceous, closely invested by the hardened and angular smooth or rugose calyx-tube.

1. **S. officinalis,** Linn. Sp. Pl. i. 116 (1753); Brew. & Wats. Bot. Calif. i. 186 (1876), under *Poterium*. Perennial, glabrous, 2—4 ft. high: leaflets about 9, ovate or oblong, cordate at base, 1_2—2 in. long: fl. polygamous, deep purple or red, in oblong spikes 1_2—1 in. long: fr. 1 line long.— Reported from Mendocino Co., *Bolander*. An Old World plant, perhaps not native with us.

2. **S. annua,** Nutt. in T. & G. Fl. i. 429 (1840), but in Hook. Fl. i. 198 (1833), under *Poterium*. Annual, branching, glabrous, 1—2 ft. high: leaflets 9—13, oval, 1_2 in. long, deeply pinnatifid, segments linear: fl. perfect, greenish, in ovoid or cylindrical spikes 1_4—1 in. long: stamens 2 or 4, short: fr. rugose between the 4 angles.—From Monterey and Kern counties northward.

11. ACÆNA, *Mutis*. Perennial herbs, or the stems somewhat woody at the decumbent or creeping base. Leaves unequally pinnate; leaflets incised or pinnatifid. Flowers in terminal more or less spicate clusters. Calyx-tube oblong, contracted at the throat, persistent, at length armed with retrorsely barbed prickles; limb 3—7-parted, valvate, deciduous,

Petals 0. Stamens 1 10. Pistils 1 or 2: ovary free from the calyx; style terminal; stigma capitate, multifid; ovule 1, suspended. Achene enclosed in the hardened calyx-tube.

1. **A. trifida**, Ruiz & Pavon, Fl. Peruv. i. 67 t. 104 (1798). Stems 1 ft. high, leafy mostly at the creeping and woody base; herbage silky-villous: leaflets 9-13, oblong-ovate, 3-5 lines long, pinnately cleft into 3-7 segments: fl. small, greenish purple in an interrupted spike; filaments exserted: fr. ovate. 2 lines long, 3-4-angled; angles with 2-4 stout prickles, the intervals with shorter ones. Grassy summits or northward slopes of the hills along the sea coast; common at the Presidio and elsewhere near San Francisco; also Chilian.

12. **AGRIMONIA**, *Brunfels* (AGRIMONY). Tall perennials with odd-pinnate leaves and long slender terminal racemes of small yellow flowers. Calyx-tube urceolate; throat encircled by a border of hooked prickles; limb 5-lobed, at length conniveut. Petals 5. Stamens 5-15, in 1 row. Pistils 2, distinct, free from the calyx; styles terminal; stigma dilated, 2-lobed; ovule pendulous. Achenes 1 or 2, enclosed in the bur-like persistent calyx.

1. **A. Eupatoria**, Linn. Sp. Pl. i. 448 (1753). Hirsute or glabrate, 2-4 ft. high, sparingly branched above: leaflets 5-7, usually 2-3 in. long with very small ones intervening, oblong-obovate, coarsely toothed, acute at each end: stipules large, semicordate, toothed or lobed: calyx in fruit ¼-⅓ in. long, the tube 10-sulcate above: achene 1, subglobose, 1 line thick. Apparently widely disseminated in California, but seldom seen; common on the Atlantic slope and in Europe.

13. **GEUM**, *Gesner*. Perennials with mostly radical lyrate or pinnate leaves, adnate stipules and large solitary or corymbose flowers. Calyx persistent, concave, the upper part of the tube bearing 5 bracteoles alternately with the valvate segments of the limb. Petals 5. Stamens ∞. Pistils ∞, on a conical or clavate receptacle: style terminal, straight or geniculate; ovule 1, ascending. Achenes compressed, caudate with the persistent elongated naked or plumose styles. Seed erect.

* *Styles jointed near the middle, the upper part deciduous, the lower naked and hooked; calyx-lobes reflexed.*—GEUM proper.

1. **G. macrophyllum**, Willd. Enum. i. 557 (1809). Hispid, stoutish, 1 ft. high: radical leaves lyrately and interruptedly pinnate, the terminal leaflet very large (3-5 in. broad), somewhat 3-lobed, subcordate at base: fl. scattered, yellow, ½ in. broad: bractlets of calyx small: achenes hispid.—In the Sierra Nevada, *Bolander*, *Lemmon*, and in Humboldt Co., *Chesnut & Drew*, but more common far northward.

* * *Styles straight, wholly persistent, in age elongated and plumose; calyx-lobes erect.*—Genus SIEVERSIA, Willd.

2. **G. triflorum,** Pursh, Fl. ii. 736 (1814). Soft-villous, 1 ft. high : radical leaves interruptedly pinnate with numerous crowded cuneate-oblong incised leaflets : fl. large, few on long peduncles : calyx reddish : the linear bractlets 4 -9 lines long, equalling the lobes and the erect petals : plumose tails of the achenes 2 3 in. long. In the higher Sierra, and northward and eastward to Arctic America.

14. **ALCHEMILLA,** *Tragus.* Herbs of various habit ; ours small annuals with leafy stems, and minute green flowers fascicled in the axils of the palmately lobed leaves. Calyx-tube urceolate : limb 4- 5-cleft, with or without as many minute bractlets or intervening teeth. Stamens 1 or 2, minute. Pistils 1 or 2 ; style basal or lateral ; ovule 1, ascending. Achene ovate, compressed.

1. **A. arvensis,** Scop. Fl. Carn. ed. 2, i. 115 (1772); Linn. Sp. Pl. i. 123 (1753), under *Aphanes*: *A. occidentalis,* Nutt., probably. Slender, simple or much branched from the base, 1 4 in. high, leafy, floriferous and hirsute-pubescent throughout. the calyx-tube densely hirsute : leaves 3-parted, the segments 2 3-cleft : calyx-tube much contracted under the 4-parted limb, bractlets minute. Var. **glabra.** Glabrous, even to the calyx-tube. which is broader than in the type, less constricted at the orifice, with relatively larger bractlets.- Common along streams, borders of thickets, or on open plains ; the variety in the valley of the Sacramento.

2. **A. cuneifolia,** Nutt. in T. & G. Fl. i. 432 (1840). Differs from the preceding in having leaves longer than broad, cleft at the summit only, and a calyx-limb 5-cleft, without intervening bractlets.- Santa Barbara, *Nuttall,* and possibly within our limits.

15. **POTENTILLA,** *Brunfels* (CINQUEFOIL. FIVE-FINGER). Herbs (one species shrubby) with pinnately or palmately compound leaves, the leaflets usually toothed or cleft, and adnate stipules. Flowers axillary and solitary or in terminal cymes. Calyx from flat to campanulate, 5-cleft, valvate, with 5 alternating bractlets. Petals 5, rounded or elongated, yellow, red or white. Stamens 5 ∞ ; filaments filiform or dilated. Pistils 1 ∞ : styles more or less lateral. deciduous. Achenes on a glabrous or hairy dry (in one spongy-fleshy) receptacle.

* *Stamens 5 ∞, uniform ; filaments filiform, or dilated at the base only.*

+ *Petals minute, linear-oblong ; stamens 5.* Genus SIBBALDIA, Linn.

1. **P. procumbens,** Clairv. Man. (1811); Linn. Sp. Pl. i. 284 (1753), under *Sibbaldia.* Perennial, dwarf, creeping. the stems leafy and flowering at the ends : leaflets 3, cuneate, 3-toothed at the truncate apex, $\frac{1}{4}$—1 in. long : peduncles shorter than the leaves : fl. cymose, yellow ; petals acute : achenes 5 10, raised on short hairy stipes. High summits of the Sierra.

+ + *Petals spatulate, dark purple : stamens and pistils ∞ ; ripe receptacle enlarged, spongy-fleshy.* Genus COMARUM, Linn.

2. **P. palustris,** Scop. Fl. Carn. ed. 2. i. 359 (1772); Linn. Sp. Pl. i. 502 (1753), under *Comarum*; Crantz, Stirp. Austr. 73 (1769), under *Fragaria*. A stout, subaquatic perennial, with creeping stems and large leaves pinnately 5–7-foliolate: leaflets obovate-oblong, 2 in. long or more, coarsely and sharply serrate: fl. 1 in. broad, the inside of the calyx dark purple like the petals; these much shorter than the calyx-lobes; the juicy strawberry-like receptacle oval, 1_2 in. thick.— A common plant of northern and subarctic swamps, both in Europe and America. Sierra Co., *Lemmon*, and Butte, *Mrs. Bidwell*.

+ + + *Shrubby; petals broad, rounded, yellow; achenes villous.*— Genus DASIPHORA, Raf.; PICROPOGON, Bunge.

3. **P. fruticosa,** Linn. Sp. Pl. i. 495 (1753). *Dasiphora floribunda*, Raf. Ant. Bot. 167 (1838). Much branched, 1–4 ft. high, more or less villous; stipules scarious; leaflets 5–7, oblong-lanceolate, entire, approximate, 1_f 1 in. long; fl. solitary or cymose; petals exceeding the calyx, 1_f 1_2 in. long; stamens 30; achenes 20, very villous. Higher Sierra, and far northward around the northern hemisphere.

+ + + + *Herbs of various habit; petals rounded; receptacle small, dry.*—
POTENTILLA proper.
++ *Perennial; flowers axillary, solitary.*

4. **P. Anserina,** Linn. l. c. Leaves odd-pinnate, often 1 ft. long; leaflets 7–21, with smaller ones interposed, oblong, sharply serrate, white-tomentose beneath, silky or glabrate above; stems prostrate, with long internodes, rooting at each joint and producing at each a tuft of leaves and one or more long peduncled large yellow flowers; petals 1_f 1_2 in. long, exceeding the calyx; stamens 20–25; achenes 20–40; receptacle villous.—Along stream-banks, margins of ponds, or in springy places both along the seaboard and in the mountains.

++ ++ *Perennials; flowers terminal, cymose.*
Leaves ternate.

5. **P. gelida,** C. A. Meyer, Ind. Pl. Cauc. 167 (1831). Leaves mostly radical; leaflets broadly cuneiform, 1_2–2_4 in. long, rounded at apex and coarsely toothed, the terminal one short petiolulate, the lateral sessile: fl. few; bractlets and calyx-lobes nearly equal; petals 2–3 lines broad, exceeding the calyx; achenes x. High Sierra, from the Donner Lake district, *Bolander, Sonne*, northward. July, Aug.

6. **P. Grayi,** Wats. Proc. Am. Acad. viii. 560 (1873); *P. Clarkeana*, Kell. Proc. Calif. Acad. vii. 94 (1877). Small and slender, deep green, scantily pubescent; leaflets obovate, 1_2 in. long, toothed at the rounded or truncate apex, the terminal one long-petiolulate; bractlets half as long as the calyx-lobes; petals exceeding the calyx; achenes 15–20. Yosemite and southward, in the higher mountains. June, July.

Leaves palmate or pinnate; leaflets 5 or more.

7. **P. Wheeleri,** Wats. l. c. xi. 148 (1876). Decumbent, silky-villous, 2—3 in. high, leafy and flowering from the base: leaves palmate; leaflets 3—5, cuneiform, 3—5-toothed at the rounded summit, $\frac{1}{2}$ in. long; stipules entire; fl. opposite the leaves: calyx 3 lines long; bractlets obtusish, smaller than the lobes: petals obcordate, little exceeding the calyx: achenes 20. Head-waters of Kern River, at 8,200 ft., *Rothrock*.

8. **P. dissecta,** Pursh, Fl. i. 355 (1814). Stems 1 ft. high, more or less, ascending from a decumbent base; pubescence usually almost none; herbage of a deep rich green; leaves usually palmate, or closely pinnate; leaflets 5—7, the lowest much reduced in size, uppermost 1 in. long, all pinnately cleft into narrow segments: fl. few, large and showy, in an open cyme: petals exceeding the lanceolate calyx-lobes: achenes 10—15. In the higher Sierra; frequent. July, Aug.

9. **P. Nuttallii,** Lehm. Ind. Sem. Hamb. 12 (1852); Revis. 89. t. 33: *P. rigida*, Nutt. in T. & G. Fl. i. 440 (1840), not of Wallich (1828); *P. gracilis*, var. *rigida*, Wats. Stout, 1—2 ft. high, decumbent at the very base, sparingly hirsute with short appressed hairs, no part tomentose or even canescent, but somewhat glandular: leaves palmate: leaflets 5—7, the lowest pair much smaller than the rest, obovate- or oblong-cuneiform, coarsely pinnate-toothed or cleft: cyme rather contracted: petals broadly obcordate, larger than the ovate-lanceolate calyx-lobes: achenes 40 or more, smooth, slightly margined. Frequent in the higher Sierra, but more common northeastward in the Rocky Mountain region.

10. **P. gracilis,** Dougl. in Hook. Bot. Mag. t. 2984 (1830): *P. Blaschkeana*, Turcz. in Lehm. Revis. 107. t. 64 (1856). Taller than the last, often 3 ft. high, villous throughout, not glandular, lower face of leaves densely white-tomentose: leaflets 7 or more, often pinnate, more or less deeply pinnatifid or only coarsely serrate-toothed: cyme loose and ample: fl. and fr. as in the last. —Frequent in the higher mountains both west and east in the State.

11. **P. Plattensis,** Nutt. in T. & G. Fl. i. 439 (1840). Decumbent or depressed, the slender stems 5—10 in. long; herbage light green, sparingly silky or glabrous: stipules large; leaflets 7—15, approximate, nearly alike in size, $\frac{1}{4}$—$\frac{3}{4}$ in. long, cut into 3—5 linear segments: pedicels slender: cyme open: petals obcordate, 3 lines long, exceeding the lanceolate calyx-lobes: achenes 25—40, very thick, smooth, marginless. From Sierra Valley northward and far eastward, in mountain meadows, or on elevated plains.

12. **P. Breweri,** Wats. Proc. Am. Acad. viii. 555 (1873). Suberect, rather rigid, 3—10 in. high, densely white-tomentose: stipules broad; leaflets 7—13, rather crowded, cuneate-obovate deeply incised, $\frac{1}{4}$—$\frac{1}{2}$ in. long: fl. as in the last, though smaller: achenes 20—25. —An alpine species prevailing from the Donner Lake district of the Sierra southward.

13. **P. glandulosa,** Lindl. Bot. Reg. t. 1583 (1833); *P. Wrangeliana*, Lehm. Revis. 49. t. 19 (1856). Erect, 1–2 ft. high, glandular-pubescent and ill-scented: leaves pinnate; leaflets 5–9, ovate or rhombic-ovate, coarsely and doubly serrate: cyme lax, leafy-bracted: fl. small; the pale yellow obovoid petals scarcely equalling the calyx: stamens 25, in 1 row, on the margin of the thickened disk: styles attached below the middle of the ovary. Var. **Nevadensis,** Wats. Slender, scentless and scarcely glandular: leaflets small, simply incised: floral bracts inconspicuous; petals exceeding the calyx, light yellow. Var. **reflexa.** Cyme very lax, few-flowered: petals deep yellow, small, but equalling the calyx-lobes and with them somewhat reflexed when fully expanded. Var. **lactea.** Leaflets cuneate-obovate, simply and deeply toothed: corolla large, exceeding the calyx, white. In the mountains everywhere, from near the sea-level to the higher wooded parts of the Sierra. Most probably comprising several species. The common type, which is of the Coast Range, is more accurately represented in Lehmann's than in Lindley's figure. The first variety belongs to the Sierra Nevada at almost subalpine elevations. Var. *reflexa* is of the foot-hills, in dry ground, usually under cover of shrubbery or in groves of pine. Var. *lactea* is of higher elevations in Fresno and Kern counties.

++ ++ ++ *Annuals or biennials; flowers inconspicuous; achenes minute, very numerous.*

14. **P. millegrana,** Engelm. in. Lehm. Ind. Hort. Hamb. (1849); *P. rivalis,* var. *millegrana,* Wats. Tall, flaccid, soft-pubescent, leafy up to the inflorescence: leaves 3-nate, the radical on long slender petioles; leaflets cuneate-obovate, obtusely serrate at apex only; stipules ovate-lanceolate, entire: cymes diffuse; fl. very numerous; petals yellow; stamens about 10: achenes whitish. Eastern slope of the Sierra, and eastward; but also on the lower San Joaquin, on muddy banks of the river.

15. **P. biennis.** Biennial, branched from the base, erect and rather stout, 1 ft. high or more, the stems purple, leafy, the whole herbage pubescent and minutely glandular: stipules oblong-lanceolate, obtuse or acute, the lowest entire, the upper more or less toothed or lobed: leaflets 3 (rarely), cuneate-flabelliform, irregularly incised, the broad teeth or lobes mucronulate: cymes mostly contracted and dense: petals small, yellow, spatulate-oblong, scarcely equalling the calyx: stamens about 10: achenes minute, whitish. In moist places in the mountains, from Butte Co. to Kern and San Luis Obispo. In habit and aspect very distinct from those mostly extra-Californian plants which have been referred to *P. rivalis*. They are all annuals; this certainly biennial.

* * *Perennials; petals obovate to linear; stamens 10 (20 in our species), alternately long and short, the filaments more or less dilated throughout.* Genus **Horkelia,** Ch. & Schl.

+ *Cymes lax, dichotomous; bractlets large, often exceeding the calyx-lobes.*

16. **P. frondosa,** Greene, Pittonia, i. 300 (1889). Erect or decumbent, 1½—3 ft. high, leafy throughout, viscidly hirsute and heavy-scented : radical leaves with 7 9, cauline with 5 7 leaflets ; these 1 2 in. long, oval or oblong, doubly incised, thin and finely rugose ; stipules ovate-lanceolate, coarsely incised : cyme widely spreading, loose and leafy : calyx short-campanulate, the large spreading bractlets exceeding the segments, tripid at apex ; stamens very unequal ; petals ligulate, erect or little spreading, white. A very well marked species, perhaps somewhat local, but plentiful near Martinez, *Frank Swett*; also at Santa Cruz, *Parry.* May, June.

17. **P. Californica,** Greene, l. c. 100 (1887); Ch. & Schl. in Linnæa, ii. 26 (1827). Size and habit of the last, but stem less leafy, leaves mostly radical : herbage glandular-pubescent, very fragrant : leaflets 11 21, the uppermost more or less confluent, the lower distinct but approximate : leaflets ¹⁄₁ -⁵⁄₁ in. long, broadly cuneiform, toothed or deeply incised at the rounded apex : cymose-dichotomous inflorescence lax : calyx ½ in. high, short-campanulate ; bractlets exceeding the calyx-lobes and 3-toothed at the broad apex, the middle tooth longest : petals white, spatulate, spreading or suberect. Var. **elata.** *P. elata,* Greene, l. c. More slender than the type, equally fragrant ; leaflets deeply and incisely once or twice cleft ; bractlets of the calyx like the segments triangular-lanceolate, entire. The type is common on wooded slopes about San Francisco and Oakland. The variety, first described, as a species, from fruiting specimens which were misleading as to the filaments, is of Napa and Humboldt counties, *Greene, Chesnut & Drew.*

18. **P. multijuga,** Lehm. Revis. 29. t. 7. ? (1856); *P. Lindleyi,* Greene, Pitt. i. 101 (1887). *Horkelia cuneata,* Lindl. Bot. Reg. t. 1997 (1837). Erect, 1 ft. high, leafy at base only, glandular-villous and fragrant : leaflets 19 25, rounded and incised above the middle, cuneate at base, 1½ in. long : inflorescence distinctly terminal, though ample, at summit of the slender nearly leafless stem : calyx cyathiform ; bractlets entire, smaller than the lobes : petals narrowly oblong, white, spreading: filaments subulate-dilated, the alternate ones little shorter. Santa Cruz, *Pringle, Parry,* and southward along the coast hills. Most distinct from *P. Californica* in habit, inflorescence, calyx, etc., though confused with it in the "Botany of California," as was the next.

19. **P. Kelloggii,** Greene, Pittonia, i. 101 (1887), also Bull. Calif. Acad. ii. 416, under *Horkelia.* *H. Californica,* var. *sericea,* Gray, Proc. Am. Acad. vi. 529 (1868). Stems stout, ascending, or almost prostrate, 1 2 ft. long ; herbage glandless, scentless, canescent with a short dense silky pubescence : leaflets 11 15, obovate, coarsely toothed, 1½- ³⁄₁ in. long : calyx-tube cupulate ; lobe, lanceolate, ¹⁄₁ in. long, equalled by the oblong entire bractlets : petals pure white, spatulate oblong, ¹⁄₁ in. long. In

ROSACEÆ.

sandy soil, near the bay at Alameda; also above Lake Merced in San Francisco Co., and near Santa Cruz, *Parry*.

+ + *Cymes usually more condensed ; bractlets smaller than the calyx-lobes.*
++ *Leaflets in many pairs, deeply incised or lobed.*

20. **P. Parryi,** Greene, l. c. Slender stems 6–10 in. high from a tufted leafy caudex : herbage dark green with a sparse villous pubescence and some glands about the inflorescence : leaflets cuneate-obovate, cleft scarcely to the middle : cyme lax : calyx rotate (no proper tube) ; bractlets narrow, half the length of the lanceolate lobes : petals cuneate-oblong, $1\frac{1}{4}$ in. long, white : achenes gray, oblong-reniform, minutely reticulate. Near Ione. *H. Edwards, Mrs. Curran, Dr. Parry.*

21. **P. laxiflora,** Drew, Bull. Torr. Club, xvi. 151 (1889). Stems slender, ascending, 1 ft. high : herbage sparingly long-villous : leaflets approximate in 10–12 pairs, divided into 2 or 3 segments, these linear and entire or deeply cleft into 2 or more linear lobes : cyme lax : calyx campanulate : bractlets much narrower and shorter than the ovate-lanceolate acute segments : petals spatulate-oblong, entire, exceeding the calyx, white : filaments all petaloid-dilated : achenes only 2 or 3, light-brown, smooth and shining. In pine woods of the Hy-Am-Pum Valley, Humboldt Co., *Chesnut & Drew.*

22. **P. Bolanderi,** Greene, Pittonia. i. 103 (1887) ; Gray, Proc. Am. Acad. viii. 338 (1868), under *Horkelia*. Size and habit of the last, but cyme condensed and herbage densely hoary-pubescent, leaflets $1\frac{1}{4}$–$1\frac{1}{2}$ in. long, cuneate-obovate, with 3–5 oblong rounded lobes : petals oblong-spatulate, white, equalling the calyx-lobes ; achenes dark, ovate-reniform, minutely granular. In Lake and Colusa counties, *Bolander, Mrs. Curran.*

23. **P. tenuiloba,** Greene, l. c. 105 (1887) ; Gray, Proc. Am. Acad. vi. 529 (1865), under *Horkelia ; H. congesta*, var. *tenuiloba*, Torr. Pac. R. Rep. iv. 84 (1857). Stems 1 ft. high ; herbage canescently villous : leaflets $1\frac{1}{4}$–$1\frac{1}{2}$ in. long, cuneate-obovate, deeply parted into 4–8 linear lobes, or the uppermost narrower, few-lobed or linear and entire : cymes compact : calyx 2 lines long ; lobes linear, surpassed by the oblong-spatulate white petals. Sonoma Co., *Bigelow*, and southward to San Luis Obispo, *Mrs. Curran*. A somewhat rare or local species.

24. **P. Douglasii,** Greene, l. c. *Horkelia fusca*, Lindl. Bot. Reg. t. 1997 (1837). Erect, $1\frac{1}{2}$–$1\frac{1}{2}$ ft. high : herbage purplish or dark green, glandular-villous : leaflets ovate-cuneiform or cuneate-oblong, $1\frac{1}{2}$–1 in. long, deeply cleft : calyx $1\frac{1}{4}$ in. long, more than equalled by the cuneate-oblong emarginate white petals. Var. **tenella** (Wats.) is distinguished as slender and small, with more deeply cleft leaflets, smaller fl., etc. In the Sierra from Yosemite northward ; the variety in Sierra Co.

25. **P. ciliata,** Greene, l. c. Habit and aspect of the last, but herbage not purplish, villous: radical leaves with more than 20 crowded and imbricated pairs of leaflets, these divided into 3 oblong linear entire segments: cymes capitate-congested: calyx-segments lanceolate, exceeding the turbinate tube and, with the narrower and shorter bractlets, villous-ciliate: petals with linear-oblong blade and slender claw of equal length: filaments scarcely dilated: pistils very few (1–3). Owens' Valley, Inyo Co., *Dr. Kellogg.* Plant with the aspect of the preceding species, but in floral character an "*Ivesia.*"

26. **P. purpurascens,** Greene, l. c.; Watson, Proc. Am. Acad. xi. 148 (1876), under *Horkelia.* Short-pubescent and somewhat pilose, 6 in. high: leaflets 2–4-parted into oblong or obovate segments: fl. few, in a rather open cyme: calyx ⅓ in. long, purplish; bractlets small, narrow: petals broadly cuneate-oblong, nearly equalling the calyx-lobes, rose-colored: stamens 20 in 2 rows, the filaments in one row filiform. Head of Kern River, at 9,000 ft., *Rothrock.*

++ ++ *Leaflets in few pairs, 3-dentate at apex.*

27. **P. Tilingi,** Greene, l. c.; Regel, Gart. Fl. 1872. t. 711, under *Horkelia: H. tridentata,* Torr. Pac. R. Rep. iv. 84. t. 6 (1857): *Ivesia tridentata,* Gray, Proc. Am. Acad. vii. 338 (1868). Stems slender decumbent, 6–12 in. high: herbage canescent with an appressed silky pubescence: leaves mainly radical: leaflets cuneate-obovate to narrowly oblong, ½–1 in. long, 3-dentate or entire: fl. small, in a much branched dense cyme: petals white, spatulate to linear, little exceeding the calyx: filaments varying from subulate-dilated to filiform: achenes more or less tuberculate.—Middle elevations in the Sierra. This and the preceding completely nullify both *Horkelia* and *Ivesia* as supposed genera.

* * * *Perennials; leaflets ∞, crowded or even imbricated on the rachis (except in No. 31); stamens 5—20, uniform, filaments not dilated (except in No. 32); pistils very few, or 1 only.* Genus
IVESIA, Torr. & Gray.

+ *Stems leafy; flowers cymose-panicled.*

28. **P. Pickeringii,** Greene, Pittonia i. 105 (1887); Torr. in Bot. Wilkes Exp. 288. t. 4 (1862), under *Ivesia.* Densely white-villous: leaflets at first closely imbricated, in later development only approximate, oblong, entire or 2–5-lobed, 1–4 lines long: cymes dense, but arranged in an open panicle: calyx 2 lines long: bractlets linear: petals spatulate, equalling the calyx, yellowish: achenes 4—6.—A northeastern species, reaching Sierra Valley. *Lemmon.*

29. **P. unguiculata,** Greene, l. c.; Gray, Proc. Am. Acad. vii. 339 (1868). Resembling the last, but less densely villous: cymes less crowded: calyx 1¼ in. long, the lobes and bractlets acuminate; petals white, with

narrow claw and rounded limb; stamens about 15; achenes 5–8.— Eastern foot-hills of the Sierra northward; also in the Yosemite Valley.

30. **P. santolinoides**, Greene, l. c. 106 ; Gray, l. c. vi. 531 (1868), under *Ivesia*. Stems slender, $\frac{1}{2}$–$1\frac{1}{2}$ ft. high: leaves densely white-villous, 2–4 in. long, terete with the many small imbricated leaflets : fl. scattered in a very ample cymose-dichotomous panicle, on pedicels slender and at length elongated ; calyx 1 line long, villous or glabrate ; bractlets short ; petals spatulate or obovate, exceeding the calyx, white ; stamens 15 ; filaments long, slender ; anthers purple ; achene 1. In the high Sierra, from Lake Tahoe southward to perhaps Kern Co.

+ + *Alpine species with scapiform almost leafless stems, and a terminal more or less compact cymose inflorescence.*

31. **P. Webberi**, Greene, l. c. ; Gray, l. c. x. 71 (1874). Low, loosely villous ; leaflets 9–13, only approximate, 3–5 lines long, 2–5-parted into linear segments ; fl. mostly long-pedicelled in congested cymes ; calyx 2–3 lines long ; lobes lanceolate ; bractlets small ; petals narrowly oblong, yellow ; stamens 5–10 ; achenes 3 or 4, large, ovate. Sierra and Indian valleys in the Sierra northeastward, *Webber*, *Lemmon*.

32. **P. decipiens**, Greene, l. c. 106. *Ivesia pygmaea*, Gray, l. c. vi. 531 (1868). Dwarf, 1–4 in. high : viscid glandular and sparsely villous : leaflets much crowded, 1–6 lines long ; inflorescence in age open-cymose, in fl. congested ; calyx nearly rotate : petals oblong-obovate, retuse or emarginate, yellow ; stamens 5–10 ; filaments subulate ; achenes 2–4. Alpine summits of the Sierra.

33. **P. Gordoni**, Greene, l. c. ; Hook. in Kew Journ. Bot. v. 341 t. 12 (1853), under *Horkelia* ; T. & G. Pac. R. Rep. vi. 72 (1855), under *Ivesia*. Taller than the last, sometimes 1 ft. high : leaves similar but more ample ; cyme dense, terminating an erect scapiform stem ; calyx campanulate ; segments erect or little spreading ; petals yellow ; stamens 5 ; filaments filiform ; achenes 2 or 3. Var. **lycopodioides** (Wats.). *Ivesia lycopodioides*, Gray, Proc. Am. Acad. vi. 530 (1868). Dwarf, nearly glabrous ; leaves terete or nearly so by imbrication of the minute rounded leaflets, otherwise apparently as in the type. The type at subalpine elevations of the Sierra. The variety may be a reduced alpine state of it or it may prove a distinct species.

34. **P. Muirii**, Greene, l. c. ; Gray, l. c. viii. 627 (1873), under *Ivesia*. Dwarf, densely silky-villous ; stems 1 in. high from a thick caudex : leaves terete by imbrication of the minute silky leaflets ; fl. small, in a close cyme ; calyx purplish, 1 line long ; segments exceeding the narrow spatulate yellow petals ; stamens 5 ; filaments short ; achenes 2.—On Mt. Hoffmann in the Sierra Nevada, at 9,000 ft. altitude. *John Muir*.

16. **FRAGARIA,** *Braufels* (STRAWBERRY). Perennial stoloniferous herbs with 3-foliolate leaves; the leaflets coarsely toothed: scapes cymosely ∞-flowered. Flowers as in *Potentilla,* but the numerous achenes borne on an enlarged pulpy edible receptacle. Petals in all ours white.

　　* *Leaves light-green, of thin texture; achenes superficial.*

1. **F. Californica,** Ch. & Schl. in Linnæa, ii. 20 (1827). Often 10 in. high, commonly smaller: leaflets cuneate-obovate, rounded, sparingly villous on both sides: scapes and petioles slender: fl. 1_2 in. broad; calyx-teeth and often the petals also more or less toothed: fr. small, globose. Common along the seaboard; preferring wooded or bushy moist slopes among the more elevated hills.

2. **F. vesca,** Linn. Sp. Pl. i. 494 (1753), partly. Smaller than the preceding: leaflets thinner, more conspicuously veiny: calyx-teeth and petals smaller and entire; fr. ovoid.—Frequent at middle elevations in the Sierra Nevada only.

　　* * *Leaves deep or dark green, of firmer texture; each achene inserted in a small depression of the receptacle.*

3. **F. Chilensis,** Ehrh. Beitr. vii. 26 (1792). Diœcious: scapes and petioles short, the dark-green coriaceous leaves commonly depressed; leaflets cuneate-obovate, nearly glabrous and somewhat shining above, villous beneath: fl. 1 in. broad.—Sandy banks and grassy slopes near the sea from the vicinity of San Francisco to Alaska.

4. **F. Virginiana,** Ehrh. l. c. 24. Smaller than the preceding: the deep-green but rather dull subcoriaceous foliage pubescent on both sides: fl. 1_2 in. broad: fr. as in the last.—Sierra Nevada in Tuolumne Co., *Chesnut & Drew.*

17. **RUBUS,** *Virgil.* Shrubs or almost herbaceous undershrubs with stems unarmed or prickly, erect, reclining or prostrate. Leaves simple and lobed, or compound; stipules adnate. Flowers white or red, solitary, corymbose or panicled. Calyx persistent, 5-lobed, without bractlets. Petals (5) and stamens (∞) perigynous. Pistils 2–∞, crowded on an elevated receptacle, ripening into a coherent body of small drupes, so forming the aggregate fruit called a raspberry or blackberry.

　　* *Fruit hemispherical or conical, concave beneath, parting freely from the receptacle* (Raspberry).

　　　← *Unarmed; leaves ample, palmately lobed.*

1. **R. parviflorus,** Nutt. Gen. i. 308 (1818): *R. Nutkanus,* Moç. in DC. Prodr. ii. 566 (1825). Erect, 3–8 ft. high, the bark of the main stem becoming brown and shreddy; branchlets and pedicels hirsute and more or less glandular-hispid: leaves membranous, 4–12 in. broad, irregularly serrate, the 3–5 lobes acute or acuminate: fl. few, in loose terminal

clusters, white or pinkish, 1 2 in. broad: carpels ∞, tomentose: fr. hemispherical, scarlet when ripe, "sweet and pleasantly flavored." Var. **velutinus,** (Greene, in Bull. Torr. Bot. Club, xvii. 14 (1890). *R. velutinus,* H. & A. Bot. Beech. 140 (1840). Leaves smaller, of much firmer texture, densely velvety-pubescent, evenly serrate : fr. dry, insipid. The type is found only in the mountains of the interior or easterly parts of the State. The variety, very possibly a good species, belongs to the seaboard, where it is common along the banks of streams. Fl. Mar., fr. June.

+ + *Stems prickly ; leaves 3-foliolate.*

2. **R. spectabilis,** Pursh, Fl. i. 348. t. 16 (1814) ; Lindl. Bot. Reg. t. 1424. Stoutish, 5 10 ft. high, sparingly armed with stout straight prickles : leaves occasionally simple ; leaflets ovate, acute or acuminate, doubly serrate, often more or less lobed, the veins beneath and the stalks and stalklets sparingly villous : fl. 1 3, large, red : fr. large, ovoid, red or yellow, glabrous. Var. **Menziesii.** Wats. *R. Menziesii,* Hook. Fl. i. 141 (1833). Foliage somewhat tomentose and silky. Mendocino Co., *Bolander,* northward, in moist woods. The variety is of the San Francisco district, growing on wooded banks of streams, mostly near the sea. Apr. June.

3. **R. leucodermis,** Dougl. in Hook. Fl. i. 178 (1833) : *R. occidentalis* var. Hook. l. c. : *R. glaucifolius,* Kell. Proc. Calif. Acad. i. 67 ? (1855). Stems 3 5 ft. long, surculose, recurved or trailing, the epidermis and lower face of leaves very glaucous : prickles abundant, short, straight or recurved : leaves 3 5-foliolate : leaflets ovate or lanceolate, acuminate, doubly serrate ; stipules setaceous : fl. few : sepals long-acuminate, exceeding the white petals : fr. hemispherical, glaucous, black and sweet, or red and acidulous. The Black Raspberry of Oregon and Washington has a truly black fruit : and it is the type of *R. leucodermis.* If the common accounts of the fruit in the Californian shrub be true, ours should be a distinct species ; and Dr. Kellogg's name for it would be restored.

+ + + *A prostrate, unarmed, nearly herbaceous undershrub.*

4. **R. pedatus,** Smith, Ic. Rar. t. 63 (1793) ; Hook. Fl. i. 181. t. 63. Stem slender, pubescent : leaves glabrous or sparsely villous ; leaflets 3, but the lateral commonly parted to the base, 1 in. long, incisely toothed ; stipules ovate-oblong : fl. 1, on a long slender peduncle, white, $1\frac{1}{2}-\frac{3}{4}$ in. broad : fr. of 2 or more large elongated (oblong-pyriform) red acidulous drupelets. One of the prettiest ornaments of mountain woods from N. California near the coast, to British Columbia and Alaska. The large drupelets, usually 2 only, lie on the ground ; being too heavy for their almost filiform peduncle. The plant is attributed to "woods near the coast above San Francisco," *Newberry.* It should be sought among the redwoods in Marin Co. and Mendocino.

* * *Fruit oblong or cylindrical, the drupelets persistent upon their elongated receptacle* (Blackberry).

5. **R. vitifolius,** Ch. & Schl. in Linnæa, ii. 10 (1827), also *R. ursinus*, l. c. 11. Stems woody, very prickly and glaucous, weak and trailing or suberect, 5 20 ft. long: leaves simple (on young plants; also often on flowering branchlets), or pinnately 3 5 foliolate; leaflets ovate to oblong, coarsely toothed, glabrous or more or less pubescent or tomentose; stipules oblanceolate to linear: fl. imperfect; staminate large, with elongated petals; pistillate small, with petals short and relatively broad: fr. oblong, black and sweet.- Very common on banks of streams throughout the Coast Range and in the interior: variable and perhaps embracing more species than one. Fl. Jan. Apr.; fr. May, June.

18. **ROSA,** *Varro* (WILD ROSE). Prickly shrubs with unequally pinnate leaves, adnate stipules and solitary or corymbose large flowers. Calyx-tube globose or urceolate; limb 5-parted; bractlets 0. Petals 5, rounded, spreading. Stamens ∞, on a thickened margin of the silky disk which lines the calyx-tube. Pistils ∞; ovaries free and distinct; styles subterminal; ovules solitary, pendulous. Fruit of few or many osseous large achenes enclosed in the fleshy-enlarged red berry-like calyx-tube.

* *Calyx-lobes deciduous from the fruit.*

1. **R. gymnocarpa,** Nutt. in T. & G. Fl. i. 461 (1840). Slender, 1 4 ft. high, armed with scattered slender and weak straight prickles: leaflets 5 9, rather remote, glabrous, oval, sharply doubly serrate, $1\frac{1}{2}$- 1 in. long: fl. 1, 2 or 3, barely 1 in. broad: calyx-lobes ovate, with few or no appendages: fr. 3 5 lines long, oval or oblong, nearly or quite closed at summit: seeds few, smooth. Var. **pubescens,** Wats. Finely pubescent. Common in shady places, near streams and on bushy northward slopes of the Coast Range; the variety in the Sierra Nevada. Mar.- May.

* * *Calyx-lobes persistent.*

2. **R. Sonomensis.** Slender, 1 ft. high, with many flexuous very leafy branches well armed with straight prickles: stipules short, almost truncate, narrow, the margin closely glandular-ciliolate, at length revolute: leaflets 5, remote, broadly ovate or nearly orbicular, truncate or somewhat cordate at the slightly inaequilateral base, $\frac{1}{4}$ $1\frac{1}{2}$ in. long, the margin evenly and coarsely serrate, the serratures minutely glandular-denticulate, both surfaces glabrous: fl. many, small, in dense terminal corymbs: calyx-tube round-pyriform, glandular-hispid; lobes ovate-lanceolate, acuminate, without foliaceous tip or appendages, erect in fruit.- At the Petrified Forest in Sonoma Co., collected by the author late in August, 1888, and distributed as *R. spithamæa*, to which it is allied, but from which the characters of leaflet and stipule abundantly distinguish it.

3. **R. spithamæa,** Wats. Bot. Calif. ii. 444 (1880). Glabrous and sparingly prickly, low and slender 4--12 in. high sparingly branched and

sparsely leafy ; stipules narrow, acuminate, glandular-ciliate, not revolute ; leaflets 3–7, obovate or elliptical, cuneate at base, 1_2–1 in. long, serrate and glandular-serrulate ; fl. solitary or few in a corymb ; calyx-tube globose-oblong, densely glandular-hirsute ; lobes broader than in the last, with a longer and more attenuate acumination. Very common in pine woods at middle elevations of the Sierra from Yuba Co. southward at least to the Soda Springs west of Donner Lake, but apparently seldom collected ; and Mr. Rattan obtained it first on Trinity River. A well marked species, confluent with no other, and of very special habitat.

4. **R. gratissima.** Erect, much branched, 4–6 ft. high, well armed with long straight rather weak prickles of which, on vigorous growing shoots only, two very long ones are infrastipular ; foliage thinnish, bright green, glandular, very fragrant, the rachis decidedly prickly beneath and, with the stalklets, stipules and calyx-lobes, very minutely velvety-tomentose : stipules not glandular, those of the flowering branchlets entire, of the growing shoots deeply and closely serrate-incised : leaflets 5–7, ovate, acute, 1_2–3_4 in. long, regularly simply and rather deeply serrate, the teeth somewhat falcate : fl. 3 or more in a corymb, 1–1^1_2 in. broad : calyx-tube globose ; lobes with foliaceous tips.—Borders of wet meadows, and about springy places in the mountains of Kern Co. A shrub with the habit of *R. Californica*, but strikingly unlike any forms of that species in that the almost glabrous thin foliage is of a bright sweetbriar green, with much of the glandular indument and fragrance of that species. June, July.

5. **R. Californica,** Ch. & Schl. in Linnæa, ii. 35 (1827). Erect, branching 3–8 ft. high ; prickles few, stout, usually recurved, mostly infrastipular in pairs ; foliage deep green, of firm texture, more or less glandular and tomentose ; stipules entire ; leaflets 5–7, ovate or oblong, acute or obtuse, the serratures mostly simple, spreading rather than falcate-incurved ; corymb few- or many-flowered ; pedicels pubescent and glandular ; calyx-lobes foliaceous-tipped : fruit globose, 4–6 lines thick, the persistent lobes erect.—The common wild rose of middle and southerly parts of the State ; most frequent along the seaboard and on banks of rivers in the interior. On the lower San Joaquin, among trees, we have seen it fifteen feet high, and showing a tendency to climb, the specific characters remaining the same.

Order V. CALYCANTHEÆ.

Lindley in Botanical Register, under t. 404 (1819).

A small order, placed here on account of the analogy subsisting between it and some Rosaceæ in point of floral structure ; but probably in no wise related to that order. It is represented in our district by one species of the genus *Calycanthus*.

CALYCANTHUS, *Linnæus* (SWEET-SCENTED SHRUB). Fragrant shrubs with opposite entire exstipulate leaves, and solitary terminal large red or purple flowers. Sepals ∞, in many ranks, inserted on a persistent obconical tube; the outer successively shorter and bract-like, the inner longer and colored like the petals; all deciduous. Petals ∞, on the mouth of the tube, the inner shorter. Stamens ∞, inserted on the upper part of the tube within, the inmost without anthers; filaments short, persistent. Pistils ∞, distinct, inserted on the base and sides of the calyx-tube; styles terminal. Achenes enclosed in the dry thin fibro-ligneous calyx-tube. Seed erect; albumen 0; cotyledons foliaceous, convolute.

1. **C. occidentalis,** H. & A. Bot. Beech. 340. t. 84 (1840). Shrub 6 12 ft. high : leaves dark-green, ovate to oblong-lanceolate, scabrous, 3- 6 in. long : peduncles 1-3 in. long; petals and larger sepals linear-spatulate, 1 in. long or more; inner petals incurved: sterile filaments linear-subulate, densely villous: fruiting calyx ovate, $1\frac{1}{4}$ in. long: achenes villous, 4 lines long.- Common along streams in the lower mountains. Flowers of a dull dark red. May—Aug.

Order VI. JUGLANDEÆ.

De Candolle, Théorie Elementaire, 215 (1813).

Represented by a single species of the genus

JUGLANS, *Pliny* (WALNUT TREE). Trees with hard wood, alternate exstipulate unequally pinnate somewhat resinous-aromatic leaves, and unisexual flowers. Staminate flowers in long aments, 12—40 stamens to each of the 3-lobed green perianths; the pistillate solitary, or few and spicate, their calyx adherent to the ovary, 4-toothed and bearing 4 small petals. Pistil 1; style short; stigmas 2, linear or clavate, fringed. Pericarp large, fleshy, indehiscent, enclosing a rugose nut which in germination parts into two valves. Seed without albumen; cotyledons fleshy, 2-lobed, rugose.

1. **J. Californica,** Wats. Proc. Am. Acad. x. 349 (1875); *J. rupestris,* var. *major,* Torr. Sitgr. Rep. 171. t. 16 (1854). Tree 40-60 ft. high, the trunk 2 4 ft. thick : leaflets 5 8 pairs, oblong-lanceolate, acute, 2 $2\frac{1}{2}$ in. long: aments loose, 4 8 in. long: fruit globose, little compressed, 1 in. thick: nut shallow-sulcate. -Frequent along streams, chiefly back from the seaboard.

Order VII. RUTACEÆ.

De Candolle, Prodromus, i. 709 (1824). RUTÆ, Juss. Gen. 296 (1789).

Represented by a single species of the genus *Ptelea.*

PTELEA, *Linnæus* (HOP-TREE). Shrubs or small trees with alternate 3-foliolate aromatic pellucid-dotted leaves, and corymbose regular flowers.

Sepals, petals and stamens each 4 or 5, the latter inserted outside of a disk encircling the ovary. Ovary 2-celled, surmounted by a short style, and becoming an orbicular broadly winged 2-seeded samara.

1. **P. crenulata,** Greene, Pittonia, i. 216 (1888). Tree 10 - 25 ft. high, strongly aromatic when fresh ; glabrous except the tomentulose flowers, and a sparse pubescence on the lower face of the leaves and on the fruit: leaflets cuneate-obovate, obtuse or acute, 1 3 in. long, crenulate or crenate-serrate ; filaments villous near the base ; samara $^3{}_1$ in. long and as broad, truncate or emarginate at both ends, often triquetrous and 3-seeded. In the Coast Range, from Lake Co. southward through Contra Costa, etc. May.

Order VIII. SAPINDACEÆ.

Jussieu, Annales du Museum, xviii. 476 (1811).

Trees or shrubs with opposite compound, or at least deeply lobed leaves, without stipules. Inflorescence compound, usually racemose or thyrsoid. Sepals 5, nearly distinct, or joined into a tubular calyx. Petals 4 or 5, distinct, and, with the few and definite stamens, inserted hypogynously, or around a hypogynous disk. Fruit a 3-celled capsule, or a double samara. Seeds large ; without albumen.

1. **STAPHYLEA,** *Linnæus* (Bladder-Nut). Shrub with opposite stipulate, pinnately 3 5-foliolate leaves, the leaflets stipellate, and flowers in pendulous racemose panicles. Calyx deeply 5-parted, the base bearing a thick disk, the oblong lobes whitish. Petals 5, alternate with the sepals. Stamens 5, alternate with the petals. Ovary 2 3-lobed, becoming a bladdery 2 3-celled capsule dehiscent at the apex. Seeds roundish, bony ; cotyledons fleshy.

1. **S. Bolanderi,** Gray, Proc. Am. Acad. x. 69 (1874). Leaflets 3, glabrous, broadly oval, 1 2 in. long obtuse, cuspidate or abruptly acute, serrulate ; sepals ¼ in. long ; petals slightly longer, somewhat spatulate, white ; stamens and style well exserted : fruit linear-oblong, 2½ in. long. From Shasta Co. to Fresno, at considerable elevations in the middle ranges of the Sierra.

2. **ACER,** *Pliny* (Maple. Box-Elder). Trees or shrubs with opposite palmately lobed or pinnately compound leaves without stipules. Flowers small, greenish or reddish, in terminal racemes, umbel-like corymbs, or fascicles, perfect or unisexual. Calyx usually 5-lobed. Petals 5 or 0. Stamens usually 8 (3 12), in the perfect flowers inserted with the petals upon a lobed disk. Ovary 2-lobed, 2-celled ; styles 2.

elongated. Fruit a double samara, the two 1-seeded parts separating at maturity, each long-winged. Cotyledons large and thin.

* *Leaves simple ; trees not diœcious.*- ACER proper.

1. **A. macrophyllum,** Pursh, Fl. i. 267 (1814); Hook. Fl. i. 112. t. 38 ; Nutt. Sylva, ii. 77. t. 67. Tree 50 90 ft. high, 2–3 ft. in diameter : leaves 1_2- 1 ft. broad, deeply 5-lobed, the sinuses rounded, the segments often 3-lobed, coarsely toothed : fl. large, in large crowded pendulous racemes which appear with the unfolding leaves, greenish yellow or reddish : stamens 9 or 10 ; filaments hairy : fruit densely hirsute or almost hispid, the glabrous wings 1 in. long or more, divergent. Along mountain streams or on hillsides throughout the Coast Range ; also in the middle Sierra, where it is taller, more slender, and with a harder wood than in the Coast Range. A good quality of maple sugar has been made from the mountain form. The wood is white and susceptible of a fine polish. The young twigs when cut exude a milky juice.

2. **A. glabrum,** Torr. Ann. Lyc. N. Y. ii. 172 (1828) ; *A. tripartitum*, Nutt. in T. & G. Fl. i. 247 (1838). Shrub 8 15 ft. high, glabrous, slender : leaves 2 4 in. broad, round cordate in outline, laciniately 3 5-lobed, or sometimes completely 3-foliolate ; the lobes or leaflets doubly serrate, the teeth very acute : fl. corymbose on short 2-leaved branchlets : sepals and petals greenish, linear, 2 3 lines long : filaments glabrous : fruit with slightly spreading wings 1 in. long or less. From Yosemite northward in the Sierra. Only a bush in California, but often a small tree in more northerly regions.

* * *Leaves unequally pinnate ; tree diœcious.* Genus NEGUNDO, Mœnch.

3. **A. Californicum,** Greene. T. & G. Fl. i. 250 & 684 (1838), under *Negundo* ; H. & A. Bot. Beech. 327. t. 77 ; Nutt. Sylv. ii. 90. t. 72. Tree 30 70 ft. high, the young twigs and partly developed leaves villous-canescent : leaflets 3, ovate, or the lateral ones oblong, acute, 3 4 in. long, the terminal largest and 3 5 lobed, or coarsely serrate : fl. of sterile tree umbellately clustered, the pedicels long and capillary, those of the fertile in drooping racemes : fruit pubescent 1- $1\frac{1}{2}$ in. long, including the nearly erect wings.--In the Coast Ranges from San Luis Obispo northward. Often planted for shade, along with the very distinct *A. Negundo* of the Atlantic slope.

3. **ÆSCULUS,** *Linnæus* (BUCKEYE. HORSE-CHESTNUT). Trees with opposite palmately compound exstipulate leaves, and a large thyrsoid inflorescence, the flowers on jointed pedicels. Flowers polygamous. Calyx tubular, unequally 5-toothed. Petals 4 or 5, unguiculate. Stamens 5 8, exserted, often unequal. Ovary 3-celled : ovules 2 in each cell, 1 abortive. Fruit a large coriaceous 3-valved capsule. Seed very large :

testa chestnut-brown, showing a large white hilum. Cotyledons large, fleshy, somewhat coherent.

1. **Æ. Californica,** Nutt. in T. & G. Fl. i. 251 (1838); Spach, Phaner. iii. 35 (1834), under *Calothyrsus*. A low spreading tree, glabrous, except the petiolules and inflorescence which are minutely pubescent: leaflets 5, on distinct stalklets, oblong or elliptic-oblong, mostly rounded at base, acute or acuminate at apex, serrulate, 3–5 in. long; thyrsus cylindrical, often 1 ft. long; calyx 2-lobed, the lobes scarcely toothed; corolla white with a faint tinge of rose, 1_2 in. long; stamens 5–7, long-exserted; fruit smooth, usually 1-seeded; seed 1 in. thick. Tree often 25 or 30 feet high, the rounded or depressed head of still greater breadth; very common throughout middle California, but much smaller in the foothills of the Sierra than along the coast. Admirable specimens are seen at Shell Mound, and on Point Isabel. Fl. May; fr. Nov.

Order IX. ANACARDIACEÆ.
Lindley, Introduction to Nat. Syst. 2 ed. 166 (1836).

Shrubs or trees with resinous and often acrid juice, alternate exstipulate leaves, and small variously clustered regular flowers. Stamens definite in number, as many or twice as many as the petals. Pistil 1; ovary free from the calyx. Fruit drupaceous.— Represented here by two species of the genus

RHUS, *Theophrastus*. Ours deciduous shrubs with trifoliolate leaves and small perfect or unisexual flowers in axillary bracted panicles or spikes. Sepals and petals usually 5. Stamens inserted under the edge of a disk lining the base of the calyx. Pistil 1; styles 3, distinct or united. Fruit a small compressed drupe with thin flesh and ligneous putamen. Seed erect; albumen 0.

* *Flowers greenish, in small axillary panicles, appearing with the leaves; drupe white; putamen striate.*- Genus Toxicodendron, Tourn.

1. **R. diversiloba,** T. & G. Fl. i. 218 (1838); Lindl. Bot. Reg. xxi. t. 38; Hook. Fl. i. 127. t. 46. Erect and 3–6 ft. high, or ascending trees by aerial roots to the height of 15 ft. or more; leaflets ovate, obovate or elliptical, 1–4 in. long, variously lobed or toothed, the indentations obtuse, or the leaflet rarely entire; panicles short-peduncled, more or less pendulous; fl. $1\frac{1}{2}$ lines long; fr. 2–3 lines broad. Copious in the Coast Range hills, preferring cool northward slopes and the banks of streams; absent from the more elevated portions of the Sierra; the terror of many excursionists and of some botanists, and commonly called *Poison Oak*.

* * *Flowers yellow, in small dense spikes, appearing before the leaves; drupe red, hairy; putamen smooth.* Genus Lobadium, Raf.

2. **R. trilobata,** Nutt. in T. & G. Fl. i. 219 (1838). Diffusely branching, 2—5 ft. high, aromatic-scented, more or less pubescent when young: terminal leaflet thrice as large as the lateral, cuneate-obovate, 1—2 in. long, 3-lobed and coarsely toothed above the middle; lateral pair round-obovate, scarcely lobed, but coarsely crenate: spikes ½—¾ in. long, short-pedicelled: fr. viscidly hirsute, the thin pulp keenly and pleasantly acid. At middle elevations of the mountains, but not very common. The very tough and flexible branches were employed by the Indian women in their finest basket-work, and southeastward the shrub is locally known as *Squaw Bush*. The species is clearly enough distinct from the Atlantic *R. Canadensis*, in which the leaflets are of about equal size, alike in form, and none lobed. These shrubs are scarcely congeneric with *Poison Oak*; and probably both *Toxicodendron* and *Lobadium* are defensible as distinct from *Rhus*.

Order X. CELASTRINEÆ.

Robert Brown, in Flinder's Voyage, 22 (1814).

Shrubs with simple exstipulate leaves, and small perfect regular flowers. Sepals and petals 4 or 5, imbricate in bud. Stamens as many as the petals, inserted alternately with them on or under the edge of a perigynous disk. Ovary free from the calyx, but immersed in the disk or encircled by it, 3- or 4-celled; cells 1- or several-ovuled. Fruit capsular, loculicidal; seed without albumen.

1. EUONYMUS, *Theophrastus* (BURNING BUSH). Deciduous shrub with 4-angular green branches, opposite leaves, and flowers in loose axillary cymes. Sepals and petals 4 or 5, widely spreading. Stamens very short, on a broad angled disk. Ovary immersed in the disk, 3—5-celled; style short or 0. Capsule coriaceous, 3—5-lobed and -valved. Seeds 1—4 in each cell, covered with a fleshy red aril.

1. **E. occidentalis,** Nutt. in Torr. Pac. R. Rep. iv. 74 (1857). Erect, slender, 7—15 ft. high: leaves ovate or oblong-lanceolate, acuminate, serrulate, short-petioled, 2—4 in. long: peduncles slender, 2—4-flowered: fl. 5-merous, dark brown-purple, 4—6 lines wide: fr. smooth, deeply lobed. Apparently one of our rarest shrubs; found in Santa Cruz Co., *Anderson*, San Mateo, *Behr*, and Marin, *Bigelow*, *Bolander*. Fl. Apr.

2. PACHYSTIMA, *Rafinesque*. Low evergreen shrubs with opposite leaves and 1 or more greenish small axillary 4-merous flowers. Sepals joined at base into a short obconical tube. Stamens inserted at the edge of the disk lining the calyx-tube. Ovary free, 2-celled; style very short; stigma conspicuous. Capsule small, oblong, coriaceous, 2-valved, 1—2-seeded. Seed with a white many-cleft membranaceous aril.

1. **P. Myrsinites,** Raf. Am. Monthly Mag. ii. 176 (1818); Pursh. Fl. i. 119 (1814), under *Ilex*; *Myginda myrtifolia*, Nutt. Gen. i. 109 (1818); *Oreophila myrtifolia*, Nutt. in T. & G. Fl. i. 259.   Branching and leafy, 1–2 ft. high: leaves ovate or oblong, obtuse or acute, cuneate at base, serrate or serrulate, ½–1½ in. long: fl. 1 line wide, on pedicels 1–2 lines long: fr. 2 lines long. Woods of the middle Sierra from Yuba Co. northward.

Order XI. RHAMNEÆ.

De Candolle, Prodromus, ii. 19 (1825). *Rhamni*, Juss. (1789).

Shrubs with simple leaves; stipules minute, mostly caducous. Flowers 4–5-merous, small, perfect or unisexual regular. Calyx 4–5-cleft, valvate in æstivation. Petals cucullate or convolute, sometimes 0. Stamens as many as the calyx-lobes and alternate with them, *i. e.*, opposite the petals. Ovary more or less free, surrounded by a fleshy disk, 2–3- or 4-celled; ovules solitary, erect. Fruit baccate or capsular. Seeds erect; albumen fleshy or 0.

1. **RHAMNUS,** *Nicander* (BUCKTHORN). Shrubs evergreen or deciduous with alternate leaves and axillary clusters of small greenish 4–5-merous flowers. Disk thin, lining the tube of the calyx. Fruit sub-globose, the juicy pulp enclosing 2 or 3 large nut-like seeds.

* *Seeds sulcate and somewhat concave on the back; cotyledons foliaceous, with recurved margins.*—RHAMNUS proper.

1. **R. aluifolia,** L'Her. Sert. Angl. 5 (1788); Hook. Fl. i. 122. t. 42. Erect, unarmed, 4–6 ft. high, deciduous: leaves oval, acuminate, 2–3 in. long, crenately serrate: fl. solitary or clustered, 5-merous, apetalous: fr. obovate, 3-seeded, ¼ in. long, black. Eastern base of the Sierra from near Truckee, *Sonne*, northward.

* * *Seeds convex on the back, sulcate ventrally if at all; cotyledons fleshy, flat.* Old genus FRANGULA.

+— *Evergreen species.*

2. **R. crocea,** Nutt. in T. & G. Fl. i. 261 (1838). Low, intricately branched, slender and spinescent, 2–5 ft. high: leaves rigidly coriaceous, ½ in. long, bright-green above, yellow beneath, roundish ovate in outline, glandular-denticulate: fl. 4-merous, apetalous, often unisexual, short-pedicelled, solitary or few in a fascicle: fr. small, obovoid, scarlet, 1- or 2-seeded.   Var. **ilicifolia.** *R. ilicifolia*, Kell. Proc. Calif. Acad. ii. 37 (1863).   Arborescent, 12–15 ft. high, with ample leafy branches not at all spinescent: fl. 5-merous.—The low spinescent type, found by Nuttall at Monterey, occurs on Angel Island also, but is more rare than that arborescent variety which may very likely prove distinct. This is

frequent from Clear Lake down along the Mt. Diablo Range to the southern part of the State.

3. **R. Californica,** Esch. Mem. Acad. Petrop. x. 281 (1826): *R. oleifolia*, Hook. Fl. i. 123. t. 44 (1830): *R. laurifolia*, Nutt. in T. & G. Fl. i. 260 (1838). Bushy or arborescent, 4–20 ft. high, nascent parts pubescent, otherwise glabrous: leaves thin-coriaceous, elliptic-oblong, acute or obtuse, denticulate or entire, 1–4 in. long: fl. subumbellate, 5-merous: petals small, ovate, emarginate: filament long: anther exserted from the cucullate petal: fr. globose, $\frac{1}{4}$–$\frac{1}{2}$ in. in diameter, copiously pulpy, black: seeds usually 2, hemispherical, as broad at base as at summit. Along the seaboard, on sandy plains near the shore, where it is a low compact bush, or along streams among the lower and middle Coast mountains in arborescent form; not in the interior, nor in the Sierra, except at low elevations northward and beyond our limits. Well known to druggists under the Spanish name of *Cascara sagrada*; sometimes called *Wild Coffee*, from the likeness which the seeds bear to coffee-grains; these said to have been sometimes used as a substitute for that article. Fl. Mar. Apr.; fr. Sept.

4. **R. tomentella,** Benth. Pl. Hartw. 303 (1849). Near preceding, of similar habit, but never either low-bushy or arborescent: leaves 2 in. long, narrowly oblong or elliptical, abruptly acute or acuminate, entire, the margin narrowly revolute, glabrate above, minutely and very densely silvery- or yellowish-tomentose beneath: fl. and fr. as in the last. Foothills of the Sierra only, and from Butte Co. southward to Lower California. Pubescence peculiar.

++ *Deciduous species.*

5. **R. rubra,** Greene, Pittonia, i. 68 & 160 (1887). Shrub 3–6 ft. high, diffusely branched, the branches glabrous, with a thin reddish smooth and shining, or dull and slightly pubescent bark: leaves thin, short-petioled, obovate to elliptic-oblong, obtuse or acute, closely serrulate, glabrous on both faces, or puberulent beneath: flowers few in umbellate clusters, 5-merous: petals concealing the anthers, the filaments short and deltoid: fr. broadly obovoid, $\frac{1}{4}$ in. in diameter, deep red or purple, mostly 3-seeded: seeds narrowed at base. Higher Sierra, from Lake Tahoe, *Parry*, and Truckee, *Sonne*, southward; chiefly on the eastern slope, but also on the western, from Calaveras Co. southward; perhaps extending into Arizona.

6. **R. Purshiana,** DC. Prodr. ii. 25 (1825). Arborescent, sparingly branched, 6–20 ft. high; growing parts tomentose-pubescent: leaves thin, obovate- or elliptic-oblong, often ample, 2–8 in. long, 1–$3\frac{1}{2}$ broad, obtuse at base, acute at apex, margin often repand, always finely penticulate: fl. in umbellate cymes, rather few and large, 5-merous:

petals minute, cucullate, bifid: fr. round-obovate, 1_3 – 1_2 in. thick, black, 3-seeded; seed obovate. — A northern species, credited to Mendocino Co. in the "Botany of California;" found near Arcata, Humboldt Co., *Chesnut & Drew.*

2. **CEANOTHUS,** *Linnæus* (CALIFORNIA LILAC). Arborescent, shrubby or suffrutescent, unarmed or spinescent, with petioled leaves and mostly thyrsoidly arranged, caducous-bracted fascicles or cymes of small perfect blue or white flowers. Calyx campanulate, 5-cleft, the lobes acute, connivent; disk thick, adnate to the calyx and base of the ovary. Petals 5, cucullate and arched, on long claws. Stamens 5; filaments filiform, longexserted. Ovary 3-lobed; style short, 3-cleft. Fruit 3-lobed and capsular, though coated with a thin layer of bitter resinous pulp; ultimately separating into 3 unilocular 1-seeded carpels which are elastically dehiscent by the ventral suture. Seeds obovate without a furrow.

* *Leaves alternate, membranous or thin-coriaceous, glandular-toothed or entire; fruit unappendaged or slightly crested.* CEANOTHUS proper.

    ← *Branches flexible, not spinescent.*
    ++ *Leaves thin, plane, entire.*

1. **C. Andersonii,** Parry, Proc. Davenp. Acad. v. 172 (1889). A slender graceful glabrous shrub 10–15 ft. high, the young branches and twigs terete, the bark smooth, light-green: leaves narrowly oblong, mostly obtuse, 1_2–$1_{1/2}$ in. long, delicately pinnate-nerved, pale beneath, on short slender petioles: thyrse elongated, lax, on a long leafy peduncle: fl. white, on filiform pedicels: fr. small, smooth. Santa Cruz Mountains, near Ben Lomond, *Anderson, Parry.* One of the most beautiful species, and only recently discovered, in a region where men had ceased to expect new shrubs.

2. **C. parvifolius,** Trel. Proc. Calif. Acad. 2d ser. i. 110 (1888); *C. integerrimus,* var. *? parviflorus,* Wats. Proc. Am. Acad. x. 334 (1875). Erect, branched from the base, slender, very leafy, 2–3 ft. high: branches and branchlets terete, pale-green and glabrous: leaves oblong, obtuse or nearly truncate, 3–9 lines long, obscurely 3-nerved from the base and faintly reticulate, wholly glabrous: thyrse shorter but more lax than in the next, the peduncle leafless: fl. sky-blue or paler. In open woods of the Sierra Nevada from near the Calaveras Big Trees northward, at a higher altitude than the next, with which it has been confounded, but with which it does not seem to be confluent. In some points it is more like the preceding species. June.

3. **C. integerrimus,** H. & A. Bot. Beech. 329 (1840); *C. Nevadensis,* Kell. Proc. Calif. Acad. ii. 152. fig. 45 (1863). Tall, loosely branching and sometimes arborescent, 5–12 ft. high, the branchlets green, more or less

angular when young, and warty in age: leaves ovate, 1-3 in. long, prominently triple-veined, pubescent or glabrate, entire or very slightly glandular serrate: thyrse long and dense, terminating leafy branchlets: fl. from deep-blue to white. One of the most common species of the higher Coast Range hills and foot-hills of the Sierra, and very ornamental.

++ ++ *Leaves thin or subcoriaceous, plane, glandular-toothed.*

4. **C. diversifolius,** Kellogg, Proc. Calif. Acad. i. 58 & 65 (1855); *C. decumbens*, Wats. Proc. Am. Acad. x. 335 (1875). Semi-herbaceous, decumbent and somewhat creeping, the branches a few inches to a foot long or more, hirsutely pubescent: leaves thin, $\frac{1}{2}$ — $1\frac{1}{2}$ in. long, elliptic-oblong, obtuse or acutish, glandular-denticulate, the glands stipitate: thyrse short, the umbels sessile: fl. sky-blue, few and long-pedicelled: capsule with narrow wing-like crests. At middle elevations on the western slope of the Sierra, in pine woods, where it carpets the ground almost for miles in certain districts.

5. **C. Lemmoni,** Parry, Proc. Davenp. Acad. v. 192 (1889). Branches erect or ascending, rather rigid, never even decumbent; bark of a light gray, only the growing shoots and the foliage pubescent: leaves oblong or elliptical, 1 in. long or less, glandular-serrate, often whitish tomentulose beneath: fl. as in the preceding: fr. conspicuously crested at summit. Foot-hills of the Sierra along the upper Sacramento, *Mrs. Gates, Mr. Lemmon*. Very unlike the last in vegetative character, habit, fruit, etc., and of another habitat. First indicated to Dr. Parry as an undescribed species, in specimens collected by Mrs. Gates at Rose Springs in 1874, and preserved in the Herbarium of the University at Berkeley.

6. **C. foliosus,** Parry, Proc. Davenp. Acad. v. 172 (1889). Low, slender, the erect stems 2—3 ft. high, with many ascending very leafy branches; nascent parts pubescent: leaves subcoriaceous, often fascicled, glaucous beneath, deep but dull green above, 2—5 lines long, obovate or oval, obtuse, short-petioled, closely denticulate, the mucronate teeth having very large rather deciduous resin-glands: fl. few, light blue, in a simple usually capitate raceme on a slender more or less leafy-bracted peduncle: capsule sharply crested at summit.— Wooded hills of Napa, Sonoma and Lake counties. Collected by the author, near the Geysers, in 1874; ten years later, near St. Helena by Mr. Rivers, whose specimens were shown to Dr. Parry, and the specific characters indicated.

7. **C. tomentosus,** Parry, l. c. 190. Erect, 4—8 ft. high, with rather few slender spreading branches; bark glabrous and brownish on old branches, rusty-tomentose on the growing ones: leaves coriaceous, short-petioled, $\frac{1}{2}$—1 in. long, ovate, obtuse, serrate, dark green and tomentulose above, densely white-tomentose beneath: thyrsus short and short-peduncled: umbels pedicellate, few-flowered: fl. deep blue: fr.

small, crested. Foot-hills of the Sierra in Amador Co., back of Ione.
Parry, Greene. It is in the State Survey collection under No. 4558. A
fine species, in habit different from any of those Coast Range shrubs
which resemble it in foliage, and with which it was long confounded.

8. **C. velutinus,** Dougl. in Hook. Fl. i. 125. t. 45 (1830). Stout,
diffusely branching, 2–4 ft. high: leaves subcoriaceous, broadly oval,
1½–3 in. long, shining and thick-glutinous above, more or less velvety-
pubescent and strongly 3-ribbed beneath; petioles stout, ½ in. long:
thyrse compound, loose and broad, rather short-peduncled: fl. white.
Higher parts of the Coast Range from Mt. St. Helena northward, and in
the Sierra from near Donner Lake. June.

9. **C. thyrsiflorus,** Esch. Mem. Acad. Petrop. x. 285 (1826); Lindl.
Bot. Reg. xxx. t. 38; Nutt. Sylv. ii. 44. t. 57. Arborescent, 6–15 ft. high,
glabrous or nearly so, branches angular, foliage firm-membranous,
bright and shining: leaves 1–2 in. long, short-petioled, ovate-oblong,
strongly 3-ribbed: thyrse dense, sometimes broader than long, on short
leafy peduncles: fl. deep blue: fr. small, smooth. From Monterey
northward, preferring northward slopes and cool ravines; very showy,
and said to have been cultivated in early days, but now seldom seen
except in its native wilds, and these wilds are now almost obsolete in the
vicinity of San Francisco. There are fine specimens on Angel Island.
In the Berkeley hills it is associated with *C. sorediatus*, with which it
also hybridizes so freely that the undiscerning may regard the two as
confluent; but the last named is of the rigid and spinescent group.

++ ++ ++ *Leaves pinnate-veined; margins glandular-toothed, undulate or
revolute; surface mostly papillose or rugose.*

10. **C. Parryi,** Trel. Proc. Calif. Acad. 2d ser. i. 109 (1888). Arborescent, 6–10 ft. high; branches sparingly villous or glabrate, angular,
more or less papillose: leaves subcoriaceous, oblong, obtuse, ½–1½ in.
long, the pinnate veins supplemented by a pair of laterals which run
near the more or less strongly revolute margin; surface of leaf glabrate,
lower face more or less tomentose-canescent: thyrse narrowly oblong,
umbels subsessile: fl. blue: fr. small, smooth. In the hill-country
between Napa and Sonoma counties, *Mr. Rivers, Dr. Parry*, northward to
the interior of Humboldt Co., *Marshall*; very closely allied to the next,
but probably distinct. The geographical ranges of the two are entirely
different. May, June.

11. **C. papillosus,** T. & G. Fl. i. 268 (1838); Hook. Ic. Pl. t. 272, and
Bot. Mag. t. 4815. Stouter than the last, less arboreous, 4–6 ft. high;
branchlets and stalklets hirsute-pubescent: leaves narrowly oblong,
1–2 in. long, glandular-serrate, the surface rugose and glandular-papillose: fl. blue, in short, mostly simple and short-stalked racemes: fr.
small, smooth. Hills along the seaboard, from Monterey to San Francisco.

12. **C. impressus**, Trel. Proc. Calif. Acad. 2d ser. i. 112 (1888). Near the last, but the leaves much broader, not papillose, though strongly rugose, the midvein much depressed, the margin strongly revolute: inflorescence dense, subglobose.— A little known species of some station southward in the mountains toward Santa Barbara.

13. **C. dentatus**, T. & G. Fl. i. 268 (1838). Low and much branched, very leafy, the branchlets and veins of the leaf beneath rusty-tomentose: leaves crowded and fascicled, 1/2 in. long or more, oblong-cuneiform, truncate or retuse, the margins undulate and revolute: fl. deep blue, in nearly simple slender-peduncled racemes, these very numerous and clustered at the ends of lateral branchlets: fr. small, smooth. –Monterey and southward. Apr. May.

+ + *Branches spinescent; flowers in simple clusters.*

++ *Leaves entire, triple-veined (except in the first).*

14. **C. spinosus**, Nutt. in T. &. G. Fl. i. 267 (1838). Arborescent, 20–30 ft. high: branches lax, spreading, leafy and glabrous: leaves coriaceous, shining, 3/4–1 1/4 in. long, oblong, obtuse or retuse, entire; petioles slender: fl. deep blue, very fragrant, in a thyrse or simple raceme: fr. smooth, resinous, 1/4 in. thick.—Mountains towards Santa Barbara and southward. The most arboreous of our mainland species, though often merely shrubby.

15. **C. divaricatus**, Nutt. l. c. 266. Rigidly and diffusely branched, the branches spinescent and divaricate, nearly glabrous: leaves ovate to oblong, 1/3–1 1/4 in. long, rounded at base, acute or obtuse at summit, not tomentose beneath, entire or minutely glandular-serrulate: racemes rather lax, often leafy: fl. blue or white: fr. of middle size, very resinous.—In the Coast Range; very common.

16. **C. incanus**, T. & G. Fl. i. 265 (1838). Spinescent branches thick and stout minutely canescent, the foliage also cinereous-velvety and pale: leaves coriaceous, tomentose beneath, broadly ovate or elliptical obtuse, subcordate at base or somewhat cuneate, 3/4–2 in. long: fl. white, in short racemes from thick spurs or axillary branchlets: fr. 2 lines in diameter, resinous and warty.—In the Coast Range from Santa Cruz Co. to Lake and Humboldt.

17. **C. cordulatus**, Kellogg, Proc. Calif. Acad. ii. 124. fig. 39 (1863); *C. eglandulosus*, Trel. l. c. 110 (1888). Low and widely straggling, pale-cinereous and more or less hirsute: leaves oval or rounded, 1/2–1 1/2 in. long, usually more or less cordate at base, entire or at apex serrulate: fl. white in short clusters emanating from axillary rigid branchlets: fr. small, not resinous-warty.—In the higher Sierra, and eastward in Nevada.

++ ++ *Leaves triple-veined, glandular-serrate.*

18. **C. oliganthus**, Nutt. in T. & G. 1. c. 266. "Stem and branches villous: leaves elliptic-ovate, nearly glabrous above, villous beneath, glandularly serrulate, rather obtuse: panicles lateral and terminal, very short, few-flowered, naked, or leafy toward the base: disk pentagonal; ovary with 3 protuberances at the angles nearly as large as itself." Mountains near Santa Barbara.

19 ? **C. hirsutus**, Nutt. l. c. "Somewhat spiny and almost hirsute, particularly the young branches: leaves cordate-ovate, glandularly serrulate, nearly sessile, rather obtuse: panicle terminal, elongated, leafy: disk obscurely pentagonal: protuberances of the ovary small." Habit of the preceding. We give Nuttall's own descriptions of his two species which later writers of less than his experience in these shrubs have probably unadvisedly combined. Dr. Parry was of opinion that more than one species was embraced by the "*C. hirsutus*" of later writers. We note that if both Nuttall's names apply to the same shrub, *oliganthus* has the precedence over *hirsutus*.

20. **C. sorediatus**, H. & A. Bot. Beech. 328 (1840); *C. azureus*, Kell. Proc. Calif. Acad. i. 55 (1855); *C. intricatus*, Parry, Proc. Davenp. Acad. v. 168 (1889). Shrubby or arborescent, 5–10 ft. high, nearly glabrous: branches spreading or recurved, and with short stiff branchlets: leaves subcoriaceous, glossy above, glabrous or somewhat tomentose beneath, but silky along the rims, oblong-ovate, ½–1½ in. long, rounded or subcordate at base: racemes of deep blue ½–2 in. long, usually not longer than broad. Plentiful on Mt. Tamalpais, on the northern slope; common in the Berkeley Hills, and far southward. Mar.–May.

* * *Evergreen shrubs; branches mostly short, rigid and small-leaved, with warty stipules; leaves mostly opposite, hard-coriaceous, closely pinnate-nerved, spinose-toothed or entire; fruit with 3 horns.*– Subgenus CERASTES, Wats.

+ *Shrubs prostrate or trailing.*

21. **C. prostratus**, Benth. Pl. Hartw. 302 (1849). Prostrate, glabrous, the branches rooting, repeatedly subdivided, the whole forming a close mat a yard or two in diameter: leaves obovate or oblong-cuneiform, ⅓–1 in. long, obtuse or truncate, with 2 or 3 pairs of coarse spinose teeth above the middle: fl. few, bright blue, on short stout peduncles: fr. ½ in. thick, horns erect.—One of the characteristic undershrubs of pine woods in the middle Sierra from Mariposa Co. northward; also in the higher Coast Range northward; rejoicing in the alliterative common name of *Mahala-Mats*; very beautiful when in flower; fruit the largest in the genus. June, July.

22. **C. connivens**, Greene, Pittonia, ii. 16 (1889). Diffuse but not prostrate or rooting, the branches mostly simple, pliable rather than

rigid, 3 ft. long or more : leaves cuneate-obovate or oblanceolate, $\frac{1}{2}$-1 in. long, entire except at the truncate or retuse mostly 3-toothed apex, glabrous and rugulose above, beneath white-tomentulose between the veins : fl. white : fr. in umbellate clusters, small ; horns narrow and elongated, closely appressed, imbricately overlapping at the summit of the capsule. Habitat of the preceding nearly, but apparently at a somewhat lower altitude and not common.

+ + *Shrubs erect, with short rigid branchlets.*

23. **C. cuneatus,** Nutt. in T. & G. Fl. i. 267 (1838) ; Hook. Fl. i. 124 (1830), under *Rhamnus.* Stems clustered, covered with a smooth gray bark, 6-12 ft. high, the branchlets short and remote, glabrous or nearly so : leaves cuneate-obovate or oblong, obtuse or retuse, entire, $1\frac{1}{2}$ in. long or less, exceeded by the profuse simple subsessile umbellate clusters of rather large dull-white heavy-scented flowers : fr. rather large : horns short, erect. Var. **ramulosus.** Smaller, the branchlets more numerous and more leafy : leaves narrower and longer, more tomentose beneath : fl. half as large, scentless, deep blue : fr. smaller and more elongated. The type abundant at middle elevations throughout our whole district, extending northward to the Columbia : the variety in the Coast Range only, and from Santa Cruz Mts., *Greene*, to Marin and Napa counties, *Mrs. Curran, Dr. Parry.* Feb.—Apr.

24. **C. crassifolius,** Torr. Pac. R. Rep. iv. 75 (1857). Taller than the last, the stems less clustered, more arboreous, and with darker bark : young branchlets whitish-tomentose : leaves ovate or ovate-oblong, $1\frac{1}{2}$-1 in. long, obtuse, entire, or remotely spinose-denticulate, glabrous above, white-tomentose beneath : umbels often leafy-peduncled, larger than in the last, the fl. on longer pedicels : fr. rather small ; horns inconspicuous. A species of the southern parts of the State mainly ; but said to occur in the Coast Range as far north as Mendocino Co. This, however, only on the authority of the "Botany of California," whose authors probably had the next in view as a part of their " *C. crassifolius.*"

25. **C. divergens,** Parry, Proc. Davenp. Acad. v. 173 (1889). Low, much branched, the branchlets stout and divaricate, hoary when young : leaves 1 in. long very rigidly coriaceous, cuneate and entire below, above bearing 2 or 3 pairs of opposite coarse spinescent serrate teeth, the truncate apex with or without a similar tooth : umbels peduncled or subsessile : fl. large, rose-purple : fr. large, elongated, with 3 prominent horns and as many alternating crests.—Not common ; apparently first collected in Lake Co., *Dr. Torrey*, later in Napa Co., *Dr. Parry ;* also on hills near San Jeronimo in Marin Co., *Dunn.* Apr. May.

26. **C. rigidus,** Nutt. in T. & G. Fl. i. 268 (1838) ; Hook. Bot. Mag. t. 4664 ; Torr. Bot. Mex. Bound. t. 9. Erect, 6 ft. high, the branchlets

short, crowded, pubescent and very leafy: leaves coriaceous, broadly obovate, truncate, retuse or obcordate, $1\frac{1}{2}$ in. long, often nearly as broad, sharply but rather minutely spinose-dentate or -serrate, glabrous above, more or less tomentulose beneath: umbels many, few-flowered, sessile: fl. small, bright blue: fr. $\frac{1}{4}$ in. thick, with short erect horns.—From Monterey southward along the coast. Apr. May.

27. **C. vestitus,** Greene, Pittonia, ii. 101 (1890). Stems clustered, 6 ft. high, widely branching, the rigid branchlets cinereous-tomentose: leaves $\frac{1}{2}$ in. long or less, hard coriaceous, subsessile, somewhat concave above, round-obovate, obtuse or retuse, (on young shoots acute) sharply spinose-toothed all around, cinereous-puberulent on both faces but more so beneath: fl. small, white, in numerous subsessile corymbs; pedicels $\frac{1}{2}$ in. long, rather stout: fr. (immature) small, the short saliently spreading horns inserted at about the middle. Borders of pine forests on mountains south of Tehachapi, Kern Co. Collected by the author in 1889. *C. cuneatus,* occurring in the same region, and to which it is related, was in mature fruit while the present species was but just passing out of flower late in June.

28. **C. megacarpus,** Nutt. Sylv. ii. 46 (1848); *C. macrocarpus,* Nutt. in T. & G. (1838), not of Cavanilles (1794). Greene, Bull. Calif. Acad. i. 80. Arborescent, often 8–12 ft. high, with dark-colored rough bark and rounded head of not very rigid branches; branchlets rusty-pubescent and marked by large warty stipular glands: leaves alternate, thick coriaceous, obovate-cuneiform, entire, emarginate, whitish-tomentose beneath: fl. in pedunculate usually simple umbels, snow-white: fr. $\frac{1}{2}$ in. long, with 3 horns at summit. Summits of Santa Ynez Mountains in Santa Barbara Co., *Nuttall, Greene, Parry,* and to be sought in similar localities on the unexplored higher mountains to the northward. A most distinct species; the only one of its group, except *C. verrucosus* of the San Diego mesas, with alternate leaves. Very showy, and much like a wild plum-tree in aspect, when in full bloom. Feb.

Obscure species, and hybrids.

29. **C. Veitchianus,** Hook. Bot. Mag. t. 5127 (1859). Branches stalklets and surface of leaves glabrous: stipules obvious, membranous or subscarious; blade of the leaf strongly pinnate-veined, obovate-cuneiform, obtuse, the young acutely, the older obtusely glandular-serrate, $\frac{3}{4}$ in. long, smooth and shining above, tomentose between the stout veins beneath: fl. bright blue, in numerous dense corymbs at the ends of all the branches. Unknown in the living state except as cultivated in England from seeds collected by Thomas Bridges in California. Judging from the figure in the Botanical Magazine, the shrub could not be referred to any species known, even as a hybrid. The leaf-outline and

indentation much more the strongly and exclusively pinnate venation — must exclude it altogether from *C. thyrsiflorus*, with which Dr. Parry placed it as a probable hybrid. We think it must be a good species, perhaps local and awaiting rediscovery.

30. **C. floribundus,** Hook. l. c. t. 4806 (1851). Pilose-scabrous throughout : leaves small, remotely pinnate-veined, less than $1\frac{1}{2}$ in. long, oblong, acute, undulate and glandular-denticulate : fl. blue, in very many globose sessile corymbs at the ends of the branchlets.—Also known to us by Hooker's figure and description only, and apparently a species. The figure shows to our eye, no mark which we should construe as indicative of a hybrid genesis.

31. **C. Lobbianus,** Hook. l. c. t. 4810 (1854). Of more lax habit than either of the preceding : leaves 1 in. long, obovate, distinctly 3-nerved, glandular-dentate : thyrse dense, oval or roundish, on peduncles which equal the leaves. Like the last, grown in England from Californian seed. Very possibly a cross between *C. thyrsiflorus* and *sorediatus*. Shrubs much like it are often seen in the Berkeley Hills.

32. **C. rugosus.** Stems stout but pliable, prostrate, glabrous in age, the growing parts canescently puberulent : leaves coriaceous, 1_2–$1\frac{1}{2}$ in. long including the short petiole, obovate- or elliptic-oblong, acute at both ends, closely and saliently spinulose-serrate, 3-nerved, finely rugose on both faces, tomentulose beneath : fl. pale blue or white, in a short nearly simple raceme, the peduncle equalling the leaf.- Top of a high hill near Truckee, June, 1890, *Sonne*. Doubtless a hybrid, of which *C. cuneatus* is one of the parents. Mr. Sonne suggests that *C. velutinus* may be the other. The young leaves are subtended by triangular-subulate stipules a line long.

Order XII. **TITHYMALOIDEÆ.**

Ventenat, Tabl. du Reg. Veget. iii. 483 (1799). Euphorbiaceæ, Robert Brown in Flinder's Voyage, Gen. Rem. 23 (1814).

Herbs shrubs or trees, often with milky acrid juice, the leaves simple, stipulate. Flowers axillary or terminal, bracted, imperfect, monœcious or diœcious, in all ours apetalous. Stamens 1–∞. Pistil 1 ; ovary superior, 1–3-celled. Fruit a 1–3-celled capsule with as many lobes as cells ; the lobes in maturity separating from a central axis as a 1-celled 1-seeded carpel ; this elastically dehiscent by two sutures and exposing or ejecting the usually arilled or strophiolate seed. Ovules and seeds pendulous. Embryo embedded in fleshy albumen ; cotyledons flat.—A large family mainly tropical, often possessing poisonous qualities ; sometimes harmless or even wholesome. Very analogous to Rhamneæ in fruit-structure ; as nearly related to Malvaceæ. Feebly represented in middle California.

1. **CROTON,** *Linnæus*. Pale scurfy or stellate-hairy plants with alternate exstipulate entire leaves, and racemose, cymose or solitary unisexual apetalous flowers. Staminate calyx 4 6-parted, slightly imbricate in bud. Stamens 5–7, on a hairy receptacle ; anthers inflexed in bud. Pistillate calyx when present 5-parted. Ovary simple and 1-celled, or 2–3-lobed with as many cells ; styles as many as the ovary-cells, simple or once or twice forked. Seed grayish, smooth and shining.

* *Fruit 3-lobed ; styles forked.*

1. **C. Californicus,** Müll. Arg. in DC. Prodr. xv². 691 (1862). *Hendecandra procumbens,* Esch. Mem. Acad. Petrop. x (1826) ; H. & A. Bot. Beech. 389. t. 91. Suffrutescent, the woody basal part of the stem decumbent or prostrate ; leafy branches erect, 1 ft. high ; these and the foliage silvery-canescent with a fine scurf and a minute stellate pubescence : leaves narrowly oblong or elliptical, obtuse at each end, 1–2 in. long, on slender petioles half as long : staminate flowers greenish, in short subsessile racemes ; calyx-lobes about 1 line long ; filaments hairy : pistillate fl. mostly solitary, on short pedicels ; styles twice forked : capsule deeply 3-lobed, 1/4 in. thick ; seed 2 1/2 lines long, with a small appressed caruncle. Plentiful among the sand-hills about San Francisco and southward.

* * *Fruit of a single 1-seeded carpel ; style simple.*- Genus EREMOCARPUS, Benth.

2. **C. setigerus,** Hook. Fl. ii. 141 (1840) ; Benth. Bot. Sulph. 53. t. 26 (1844) under *Eremocarpus*. A stout low annual with short but widespread leafy branches, the heavy-scented herbage with a spreading hispid and an appressed stellate pubescence : leaves ovoid or rhomboid, 1/2– 2 in. long, on slender petioles, the upper crowded and appearing opposite or whorled : staminate fl. few in a corymb, long-pedicelled ; calyx with oblong obtuse segments a line long : pistillate fl. 1, 2 or 3 in an axil ; ovary and style densely pubescent : capsule and seed 2 lines long. Plant often a foot or two broad and only a few inches high, yet all the branches clear of the ground ; regarded as a troublesome weed, though the seeds are greedily devoured by wild fowl, whence the common name of *Turkey Mullein* has been derived, the nutritive character of the seeds being taken in conjunction with the mullein-like herbage. It is far more prevalent in the interior valleys and foot-hills than along the seaboard, in middle California. July–Nov.

2. **EUPHORBIA,** *Pliny* (SPURGE). Herbs with milky juice, alternate or opposite toothed or entire leaves, and inflorescence either terminally clustered, or solitary in the forks of the many branches. Both staminate and pistillate flowers within the same involucre ; this cup-shaped and like a calyx, the 4 or 5 lobes minute, usually alternating with as many

glands which have often a colored margin resembling a petal. Staminate flowers many, of a single naked stamen jointed upon a short pedicel which has often a minute bract at base. Pistillate flower 1, in the center of the involucre, pedicellate and soon exserted from it, consisting of a single 3-celled ovary, 3 forked styles and 6 stigmas each 2-lobed. Capsule 3-seeded. Seeds smooth, reticulate, rugose or pitted, with or without a caruncle.

* *Stems erect ; stipules 0 ; involucres in forked or umbellate terminal cymes ; glands flattened or convex ; seed carunculate.*

Genus TITHYMALUS, Scopoli.

+— *Glands crescent-shaped or 2-horned.*

1. **E. LATHYRIS**, Linn. Sp. Pl. 2d ed. i. 655 (1762); 1st ed. i. 457 (1753), as *E. Lathyrus*. Annual or biennial, erect, stout, 1- -3 ft. high, glabrous throughout: leaves opposite, 4-ranked, linear-lanceolate, sessile, entire, obtuse, cuspidate, 3–4 in. long: inflorescence bracted, the branches twice or thrice dichotomous, the leaf-like bracts oblong-ovate: glands crescent-shaped, with broad obtuse horns: capsule 1_3 in. thick, the lobes rounded, in age wrinkled: seeds reticulate rugose.—Native of the Mediterranean region; spontaneous in middle California, as an escape from gardens.

2. **E. EXIGUA**, Linn. Sp. Pl. i. 456 (1753), partly; Amœn. Acad. iii. 118 (1756). Annual, slender, glabrous, 3—10 in. high: leaves alternate, linear, entire, acute or obtuse, the floral dilated at base and subcordate: inflorescence lax, repeatedly dichotomous: glands semilunate, the horns divergent: capsule smooth, scarcely a line wide: seed ovate-quadrangular, whitish, minutely tuberculate.—A weed of the grain fields in Europe; reported as occurring at Santa Clara, *B. F. Leeds*.

3. **E. leptocera**, Engelm. in Torr. Pac. R. Rep. iv. 135 (1857): *E. crenulata*, Engelm. in Bot. Mex. Bound. 192 (1859). Annual or biennial, erect and simple, or with decumbent basal branches, 1 ft. high: leaves alternate, obovate-spatulate, obtuse, $\frac{1}{2}$ –$1\frac{1}{2}$ in. long, entire or erose-denticulate; the floral opposite or ternate, broadly rhombic-ovate, sometimes connate, acute, $\frac{1}{4}$—$\frac{3}{4}$ in. broad: involucre turbinate, the oblong lobes nearly entire; glands large, crescent-shaped, the slender horns entire or cleft: styles long, bifid: capsule 2 lines broad: seeds ash-colored, oblong-ovate, dark-pitted, about $1\frac{1}{4}$ lines long, prominently carunculate. Common in bushy places either in sandy or clayey soil. Mar.- Sept.

+— +— *Glands discoid, entire.*

4. **E. dictyosperma**, F. & M. Ind. Sem. Petrop. 37 (1835): *E. Arkansana*, E. & G. Pl. Lindh. 53 (1845). Annual, erect, $\frac{1}{2}$—$1\frac{1}{2}$ ft. high, glabrous,

stem simple below, or branched from the base : cauline leaves alternate, oblong- to obovate-spatulate, obtuse or retuse, obtusely serrulate, 1_2 $1\frac{1}{2}$ in. long ; floral opposite, round-ovate, subcordate, mucronate, 2- 6 lines long: involucres and glands small : style deeply bifid : capsule rough with small warty protuberances : seeds subglobose, dark-colored. delicately net-veined, the caruncle thin and flat.—Of wide dissemination in the State, but less common than the last. Mar.—June.

* * *Stems diffusely branched, nearly or quite prostrate ; leaves all opposite. unequal at base, stipulate ; involucres solitary, the glands with petaloid appendages (except in No. 5): seeds ash-colored.* Genus ANISOPHYLLUM, Haworth.

5. **E. ocellata**, Dur. & Hilg. Pac. R. Rep. v. 15. t. 18 (1855). Annual, prostrate, glabrous, the branches 4—10 in. long : leaves thick, deltoid to ovate-oblong, often cordate at base, acute, entire, 2 5 lines long : stipules setaceous, entire or cleft : involucres campanulate, less than a line long ; lobes fringed ; glands 2 4, yellowish or purplish, short-stipitate, circular and discoid, with or without a narrow margin : capsules a line broad : seeds round-ovate, obtusely angled, smooth or obscurely rugose. –Plains of the San Joaquin and southward.

6. **E. albomarginata**, T. & G. Pac. R. Rep. ii. 174 (1855) : *E. Hartwegiana*, Boiss. in DC. Prodr. xv^2. 31 (1862). Perennial, prostrate, slender, glabrous and pallid, the branches 3 6 in. long : leaves nearly orbicular, entire, obtuse above, somewhat cordate at base, 2 4 lines broad ; stipules joined into a triangular entire or lacerate white scale : involucres campanulate or turbinate, less than a line long ; the 4 brownish glands with a conspicuous white or rose-colored petaloid appendage : capsule nearly a line long : seeds oblong, 4-angled.- Southern like the last, but in the Kern Co. mountains.

7. **E. serpyllifolia**, Pers. Syn. ii. 14 (1807), var. *consanguinea*, Boiss. in DC. Prodr. xv^2. 43 (1862) ; Millsp. in Pitt. ii. 84 : *E. sanguinea*, Greene, Bull. Calif. Acad. ii. 56. Diffuse annual, with ascending or horizontal but seldom prostrate slender branches: herbage glabrous, deep green, reddening in age : leaves obovate- to spatulate-oblong, 1 3 lines long, obscurely pinnate-veined, sharply serrate above the middle ; stipules setaceous, lacerate or subentire : glands of involucre minute, transversely oblong, reddish and with narrow 2 3-lobed or entire white or rose-colored appendages : seed quadrangular, the length scarcely twice the breadth, the sides more or less rugose-pitted, the angles somewhat prominent, hardly " sharp." —Common on plains and in the lower foothills of the Sierra. Perhaps wholly distinct from the Mexican type on which *E. serpyllifolia* was founded.

8. **E. occidentalis**, Drew, Bull. Torr. Club, 152 (1889) : Millsp. l. c. 89.

Habit of the last, but the glabrous herbage of a dull rather yellowish green : leaves oval or broadly oblong, only slightly unequal, very obtuse at each end, serrate above the middle or quite entire, mucronulate, 2 4 lines long ; stipules setaceous-lacerate : appendages of involucre crenate-lobed : seed ½ line long, whitish, the faces more or less distinctly sinuate-rugose between the rather prominent angles.—Humboldt Co., *Chesnut & Drew*, and on Mt. St. Helena, *Greene;* by streamlets and in moist situations. More nearly closely allied to *E. serpyllifolia* than is the following.

9. **E. rugulosa,** Greene. *E. serpyllifolia,* var. *rugulosa,* Engelm. Millsp. l. c.: *E. serpyllifolia,* Greene, Bull. Calif. Acad. ii. 57, not of Persoon. Wholly prostrate and very closely depressed, rather succulent (very brittle when dry), much branched and in age forming a very close mat a foot broad or more : herbage glabrous, pallid and glaucescent : leaves veinless, sharply serrate or almost entire : stipules, involucre, etc. as in the preceding : seeds whitish, finely and transversely rugose between the scarcely prominent angles.—Native of the southern extremity of the State, but well established along the railroads in our district, even at Dwight Way Station, Berkeley, where it has appeared annually since 1886. Totally unlike the preceding species in aspect, the very earliest branches lying flat upon the ground, and taking a peculiar zigzag course in their growth ; the stem lacking fibrous tissue, very excessively milky-juicy, the herbage peculiarly pallid. June– Oct.

10. **E. humistrata,** Engelm. in Gray Man. 3d ed. 386 (1859). Annual, prostrate, slender, the branches hirsute-pubescent : herbage dark green or purplish : leaves obovate or elliptical, very oblique at base, serrulate above the middle, 3—4 lines long, marked with a brown spot above : stipules lanceolate, fimbriate : involucre cleft on the back, its red or white appendages truncate or crenate : pods sharply angled, puberulent: seed 1½ line long, ovate, obtusely angled, minutely rugose-roughened.— Near Ione, collected only by the author ; common eastward, in the valley of the Mississippi.

11. **E. Preslii,** Gussone, Prodr. Fl. Sicul. i. 539 (1827) : *E. hypericifolia,* Engelm. not of Linn. Erect or ascending, branched from the base, 1- 2 ft. high, glabrous or sparsely pubescent : leaves oblong-linear, often more or less falcate, serrate, ½- 1½ in. long, often with a dark spot ; stipules triangular : peduncles longer than the petioles, collected in leafy cymes toward and at the summits of the branches : appendages of involucre white or reddish, entire : seed ⅔ line long, obtusely angled, wrinkled and tuberculate, dark-colored.- In the upper Sacramento valley, by roadsides : not frequent, perhaps recently arrived from the eastern states where it is a common weed.

Order XIII. POLYGALEÆ.

A. L. de Jussieu, in Annales du Museum, xiv. 386 (1809).

Herbs or shrubs often with milky juice. Leaves simple, entire, exstipulate. Flowers, except as to the pistil, simulating the papilionaceous; but the affinities apparently with certain allies of *Euphorbia*. We have but two species; both of the genus

POLYGALA, *Dioscorides*. Ours low undershrubs with alternate leaves and few irregular flowers in terminal cymes. Sepals 5, two larger than the others, lateral and petal-like. Petals 3, joined to each other and to the stamen-tube, the middle one hooded above and beaked or crested. Stamens 6–8, unequal, monadelphous, forming a sheath, this open on one side, adnate to the base of the petals; anthers 1-celled, opening at top. Ovary short, 2-celled; ovules solitary, pendulous; style long, curved, dilated. Capsule membranaceous, flattened contrary to the narrow partition, rounded and notched at summit, dehiscent at the margin. Seed carunculate; embryo large, in a thin albumen.

1. **P. Californica,** Nutt. in T. & G. Fl. i. 671 (1840): *P. cucullata*, Benth. Pl. Hartw. 299 (1849); *P. Nutkana*, Torr. Bot. Mex. Bound. 49. t. 12 (1859), not of Moç. Stems many, slender, 2–8 in. high, from a woody base, mostly simple: leaves glabrous or slightly pubescent, oblong-lanceolate or ovate-elliptical, acute or obtuse, ½–1 in. long: fl. rose-purple, on bractless pedicels 1–3 lines: outer sepals 2½ lines long, rounded saccate at base: inner ones broadly spatulate, ½ in. long or less: lateral petals linear-lanceolate, somewhat ciliate, equalling the broad obtuse somewhat curved beak of the rounded hood: fr. mostly from apetalous fl. near the root; capsule glabrous, broadly ovate, ¼ in. long, subsessile, retuse, narrowly margined: seed 2 lines long, pubescent; caruncle wrinkled and bladdery, calyptriform, half the length of the seed. In the Coast Range only, from Mendocino Co. southward. Nuttall's description of his *P. Californica*, and Bentham's account of *P. cucullata* are fully concordant, and the fact is certain that it was the Coast Range species which each had in view.

2. **P. cornuta,** Kellogg, Proc. Calif. Acad. i. 62 (1855): *P. Californica*, B. & W. Bot. Calif. i. 59, not of Nuttall. More woody than the last, stouter and more freely branching, ½–1 ft. high: fl. greenish white, all fruiting: sepals densely tomentose: lateral petals only equalling the hood, this with a narrow straight beak: capsule ovate, ⅓ in. long, emarginate or 2-toothed at summit, narrowly winged: seed ¼ in. long, densely hairy; caruncle terete, its thin lateral wing partially covering the body of the seed.— In the Sierra Nevada, at middle altitudes from El Dorado Co. northward; not reported from the Coast Range; hence not collected by Nuttall, nor known to Bentham when, unwittingly, he republished Nuttall's species under the new name of *P. cucullata*.

ORDER XIV. **LINEÆ.**
De Candolle, Theorie Elementaire 217 (1819).

A small order, comprising little besides the one genus

LINUM, *Vergil* (FLAX). Herbs with tough-fibrous bark, alternate (opposite in No. 4) entire leaves without stipules or with glandular organs in the place of them, and cymose-panicled very regular and symmetrical 5-merous flowers. Sepals imbricate, persistent. Petals convolute, fugacious. Stamens monadelphous at the very base. Styles 2, 3 or 5, often united below. Ovary of as many carpels as styles, each more or less divided into 2 cells by a partition proceeding from the dorsal suture. Fruit capsular, septicidally dehiscent. Seeds 1 in each half-cell, ovate, compressed, mucilaginous when moistened; embryo large; albumen thin; cotyledons broad, flat.

* *Flowers ½—1 in. broad, blue; sepals not glandular-margined.—*
LINUM proper.

1. L. USITATISSIMUM, Linn. Sp. Pl. i. 277 (1753), partly. Annual, glabrous, glaucous, 1—2 ft. high, simple up to the ample inflorescence: sepals oval, short-acuminate, 3-carinate-nerved at base, the inner scarious-margined and ciliate: petals broad-cuneiform, blue, with deeper veins, ½ in. long: capsule round-ovoid, equalling the calyx, tardily dehiscent, incompletely 10-celled, the septa not ciliate. One of the cultivated flaxes; occasionally spontaneous.

2. L. HUMILE, Mill. Dict. ed. 8 (1768); *L. usitatissimum*, Linn. in part. Much like the last, but lower and more branching: capsule more elongated, promptly dehiscent, the septa ciliate. Another of the long-cultivated flaxes of the Old World; sometimes found wild by waysides.

3. **L. Lewisii,** Pursh, Fl. i. 210 (1814); Trel. Trans. St. Louis Acad. v. 12; *L. perenne*, B. & W. Bot. Calif. i. 89, not Linn.; *L. decurrens*, Kell. Proc. Calif. Acad. iii. 44. fig. 11. Perennial, glabrous, glaucous, 1—2½ ft. high, densely leafy below, lax-corymbose above: sepals broadly ovate, not ciliate, 3—7-carinate-nerved: petals large, deep blue: capsule broadly ovate, obtuse, 3—4 lines long, twice as long as the sepals, the 10 valves dehiscing widely, the septa ciliate.—While Messrs. Brewer and Watson make this beautiful plant "Common on dry soils nearly throughout the State," Prof. Trelease limits its westerly range to the "Great Plains," thus excluding it from California altogether. It is frequent in our middle and higher mountains northward; less common in the Coast Range, but found on the Salinas, *Brewer*, and at Colma and Millbrae, *Behr*. It is absent from the interior and drier sections of the State.

* * *Annuals; leaves often with stipular glands; fl. small, white, rose-purple or yellow; sepals usually glandular-ciliate; petals commonly with lateral teeth and ventral appendages, pistils only 2 or 3.—*
Subgenus HESPEROLINON, Gray.

+ *Petals yellow.*

4. **L. digynum,** Gray, Proc. Am. Acad. vii. 334 (1868). Glabrous, 6 in. high : leaves opposite, oblong, acutish, ¼ – 1½ in. long ; stipular glands 0 ; sepals lanceolate, acuminate, ciliate-denticulate, 1 line long : petals not appendaged : styles and carpels only 2 or 3.—Near Yosemite Valley, *Bolander*, and northward in the Sierra Nevada.

5. **L. Breweri,** Gray, l. c. vi. 521 (1865). Slender, 3–12 in. high, glabrous, glaucous, few-flowered : leaves linear-setaceous, 6–8 lines long : stipular glands conspicuous : sepals 1½ lines long, ovate, acute, glandular on the margin : petals spatulate, emarginate, ¼ in. long, 3-appendaged at base : capsule ovoid, acute, about equalling the calyx.— Common on the Mt. Diablo foot-hills ; also found on Lone Mountain, San Francisco, *Palmer*.

6. **L. Clevelandi,** Greene, Bull. Torr. Club, ix. 121 (1882). A foot high or less, diffusely and loosely paniculate : leaves oblong, obtuse or acute, without stipular glands ; fl. minute : sepals narrow, acute, sparingly glandular ciliate ; petals obovate-oblong, constricted toward the base, retuse or emarginate, scarcely exceeding the calyx, the median appendage oblong, glabrous : capsule ovoid, acute, exceeding the calyx, the false septa complete to about the middle, then abruptly narrowed. Lake Co., *Cleveland*.

7. **L. adenophyllum,** Gray, Proc. Am. Acad. viii. 624 (1873). A foot high or less, loosely paniculate above, villous-pubescent or glabrate : leaves broadly linear, subcordate at base, margined with stipitate glands: stipular glands 0 : sepals 1 line long, lanceolate, acute, glandular-denticulate ; petals obovate-spatulate, mostly emarginate, 2 lines long, 3-appendiculate and hairy at base, the median appendage obovate : filaments abruptly dilated and obtusely bidentate at base : capsule ovoid, acute, equalling the calyx, false septa narrow. -Lake Co., *Bolander, Kellogg, Mrs. Curran*.

+ + *Petals white or pale purplish*.

8. **L. drymarioides,** Curran, Bull. Calif. Acad. i. 152 (1885). Sparingly villous, 4 –10 in. high, loosely dichotomous, with long slender internodes : leaves opposite or subverticillate, broadly ovate, acute or acuminate, with crowded marginal glands : fl. rose-colored, very small and remote, on short pedicels ; sepals lanceolate, acute or mucronulate, serrulate and sometimes glanduliferous ; petals ovate, emarginate, 2-toothed and 3-appendaged at base : capsule ovoid, acute, equalling the calyx, 6-valved, the false septa incomplete, narrow, widening gradually to the base. - Lake Co., near Epperson's, *Mrs. Curran*.

9. **L. micranthum,** Gray, l. c. vii. 333 (1868). Puberulent and somewhat glaucous, 6 - 15 in. high, very loosely dichotomous-paniculate, the minute white flowers on almost capillary pedicels ; leaves spatulate-

oblong, obtuse or acutish, with or without stipular glands : sepals ovate-oblong, acutish, the inner slightly glandular-ciliate ; petals white, obovate, a line long, not toothed, destitute of lateral appendages : filaments round-toothed at base and slightly hairy : capsule ovoid, equalling the calyx. - In the Sierra Nevada from middle California to Oregon.

10. **L. spergulinum,** Gray, l. c. Rather larger than the last, glabrous or with scattered hairs : leaves linear, obtuse, little narrowed at base, with or without stipular glands : pedicels slender nodding : sepals ovate, glandular-ciliate ; petals white or rose-colored, obovate, 2 - 3 lines long, 3-appendaged at base : capsule ovoid, acute, exceeding the calyx. - Dry woods of the Coast Range only ; common in Marin and Sonoma counties.

11. **L. Californicum,** Benth. Pl. Hartw. 299 (1849). Glaucous, glabrate or puberulent, 5 - 15 in. high, with angular branchlets : leaves remote, linear, the stipular glands prominent : pedicels short, erect, not exceeding the rose-colored flowers, these clustered at the ends of the branchlets; sepals ovate-lanceolate, acute, carinate below, sparingly glandular-ciliate; petals obovate, $1\frac{1}{3}$ in. long, twice the length of the calyx, dilated and 3-appendiculate below : filaments not toothed : capsule ovoid, acute, little shorter than the calyx, the false partitions broad, gradually narrowed upwards. Var. **confertum,** Gray, in Trel. l. c. Low, densely leafy, the inflorescence condensed ; median appendage of petals obovate. Eastern slope of Mt. Diablo Range, both northward and southward ; also about San Francisco ; the variety on Mare Island.

12. **L. congestum,** Gray, Proc. Am. Acad. vi. 521 (1865). Size of the last, glabrous except the calyx, the branches short and crowded : stipular glands small : fl. rose-purple, in close terminal clusters ; sepals pubescent, lanceolate, acuminate, not glandular ; petals $\frac{1}{4}$ in. long, 2-toothed, 3-appendiculate, the median appendage long and hairy : capsule subglobose, shorter than the calyx. A rare species, to be sought among the hills at the northern base of Mt. Tamalpais.

Order XV. GERANIACEÆ.

De Candolle, Flore Française, iv. 828 (1805). Gerania, Juss. (1789).

Ours soft-herbaceous plants with acidulous, pungent or aromatic properties, and perfect mostly 5-merous flowers. Sepals and petals distinct, the latter deciduous, their insertion, like that of the 5 - 15 stamens, hypogynous. Filaments distinct or slightly connate at base ; anthers versatile, 2-celled, dehiscing lengthwise. Carpels as many as the sepals and alternate with them (or fewer), united around a central column, becoming distinct and 1-seeded in maturity, or else forming an elastically dehiscent 5 - 10-valved many-seeded capsule.

1. GERANIUM, *Dioscorides* (CRANESBILL). Stems with enlarged joints. Leaves mostly opposite, palmately lobed ; stipules scarious. Peduncles umbellately few-flowered, or 1-flowered. Flowers regular ; sepals and petals imbricate in bud. Fertile stamens 10. Carpels 5, 2-ovuled, 1-seeded ; styles persistent, coherent with the central column until the carpel is ripe, then with it splitting away from it elastically from below upwards, each forming a coil, not bearded within.

* *Annuals ; flowers less than a half-inch broad.*

1. **G. Carolinianum,** Linn. Sp. Pl. ii. 682 (1753). Erect, much branched from the base, 1 ft. high, the pubescent herbage light-colored : leaves 5-parted, the divisions cleft into many oblong-linear lobes : sepals awn-pointed, as long as the pale flesh-colored emarginate petals : carpels pubescent : seeds ovoid-oblong, blackish, minutely reticulate.— Common in the Bay region ; flowering in the early spring, but disappearing with the beginning of the dry season.

2. G. DISSECTUM, Linn. Amœn. Acad. iv. 282 (1759). Taller than the last, the herbage of a darker green ; leaves cut into narrower and more acute segments : fl. larger, bright red-purple, the petals more deeply emarginate : seed roundish, more strongly reticulate.— Rather common : preferring moist and partially shaded situations ; continuing in flower until the end of June.

3. G. MOLLE, Linn. Sp. Pl. ii. 682 (1753). Low, slender, diffuse, the branches a few inches to 1 ft. long, the herbage softly and somewhat clammily villous : leaves 1 in. broad or more, cleft into oblong obtusish lobes : sepals ovate-oblong, not awn-pointed : petals very small, rose-color : carpels glabrous, transversely rugose : seed minutely striate.— Plentiful northward, from northwestern California to British Columbia ; well established about the U. S. Marine Hospital, San Francisco.

4. G. RETRORSUM, L'Her. in DC. Prodr. i. 645 (1824). Stouter than any of the foregoing, light green, glabrous except a short stiffish retrorsely appressed pubescence on the stems and growing parts : leaves 2 in. broad, 5-parted, the segments obtusely and not deeply 3-lobed : petals 2 lines long, obtuse, purple, equalling the aristate sepals : carpels slightly hairy : seeds oblong, minutely striate-reticulate.— Collected by the author, on moist sandy soil near the salt marsh not far from Mastic Station, Alameda, May, 1887. Native of New Zealand. Possibly perennial.

* * *Perennials ; flowers about an inch broad.*

5. **G. incisum,** Nutt. in T. & G. Fl. i. 206 (1838) : *G. erianthum*, Lindl. Bot. Reg. xxviii. t. 52. excl. syn. Erect, 1- 2 ft. high, villous and glandular-pubescent : leaves 2—5 in. broad, 3—5-parted and laciniately cleft : sepals villous and glandular ; petals red-purple, these (within) and the

filaments white-villous: beak of fruit glandular, 1½ in. long, short-pointed. —In dry open places of the Sierra, from Fresno Co. northward.

6. **G. Richardsonii**, F. & M. Ind. Sem. Petr. iv. 37 (1837): *G. albiflorum*, Hook. Fl. i. 116. t. 40 (1830), not Ledeb. (1829). More slender, somewhat retrorsely pubescent, the stalklets more or less villous and glandular: uppermost leaves lanceolate, serrate but not lobed: sepals canescent and glandular: petals clear white, villous on the inside. Range of the preceding, but at higher altitudes and only in moist soils. Flowers invariably white.

2. **ERODIUM**, *L'Heritier* (STORKSBILL). Vegetative characters of *Geranium*, but leaves often pinnate. Flowers and fruit almost the same; but fertile stamens 5 only, as many scale-like sterile filaments alternating with them. Beak of carpel when ripe silvery-bearded within and spirally twisted.

* *Naturalized species; leaves pinnate.*

1. E. CICUTARIUM, L'Her. in Ait. Kew, ii. 414 (1789); Linn. Sp. Pl. ii. 680 (1753), under *Geranium*. Leaves chiefly radical, in a depressed rosulate tuft, usually 6–10 in. long, the many leaflets laciniately pinnatifid with narrow acute lobes; cauline leaves reduced; peduncles exceeding them and bearing an umbel of 4–8 small bright purple flowers: beak of carpels 1–2 in. long.—Frequent in the Bay region; perhaps more common in the interior and southward. This is one of the pasture plants commonly called *Pin-clover* and *Alfilerilla*; but it is not the important one. The herbage is rather strongly aromatic for a good fodder plant.

2. E. MOSCHATUM, L'Her. l. c.; Rivinus, Pentap. t. 110 (1699), under *Geranium*. Coarser and larger, the radical leaves ascending, 1 ft. long or more; cauline more ample; leaflets unequally and doubly serrate: corolla pale and rather dull purple or rose-color: herbage with a delicate marshy odor. This is the prevalent *Pin-clover* of middle California, where it is a hundred fold more abundant than any other species, of ranker growth, its foliage not depressed, and a most valued plant for pasturage. *E. cicutarium* is less common and of little relative value. Both species are annual, and begin their growth with the coming of the first autumnal rains, and are in flower through all the later winter and early spring months. Though well established in California from a very early period, neither of them is with reason believed to be indigenous. The New World type of *Erodium* is simple-leaved.

3. E. BOTRYS, Bertoloni, Amœn. Ital. 35 (1819). Radical leaves rosulate, closely depressed, shining above, of oblong obtuse outline, the segments coarsely dentate: stems short: sepals 4 lines long; pale

purple or lilac petals longer: beak of carpels 2 3 in. long. Common toward the foot-hills skirting the eastern borders of the interior plains, near Sacramento, Ione etc.; also in Marin Co. in many places.

* * *Native species; leaves simple, rounded.*

4. **E. macrophyllum,** H. & A. Bot. Beech. 327 (1840). Subacaulescent. 4 10 in. high, soft-pubescent and with some gland-tipped pilose hairs: leaves 1 3 in. broad, reniform-cordate with a broad open sinus, crenate-serrate: peduncles exceeding the leaves: sepals oblong, accrescent, at length 1_2 in. long; petals equalling them, pure white: carpel clavate, 1_3 in. long (excluding the 1 in. beak), densely velvety-pubescent: seed oblong linear, ¼ in. long, dull, smooth. Plains of the interior; also toward the seaboard in Marin Co., northward to Oregon. Mar. Apr.

5. **E. Californicum.** Caulescent, the stem exceeding the rather few radical leaves, 6 12 in. high; herbage without soft pubescence, but upper part of stem and growing parts with abundant spreading hairs tipped with purple glands: leaves broadly cordate-ovate with closed sinus, slightly 5-lobed, rather coarsely crenate, the teeth obtuse, mucronulate: fl. much as in the preceding but petals deep rose-red: fruit unknown. Berkeley Hills and eastward in the Mt. Diablo Range. Sufficiently unlike the preceding, though little known and seeming rare. The herbage has a reddish tinge, and the leaf bears a deep red-purple spot or zone near the base. The stem-leaves are more deeply and sharply lobed than the radical, and the species is in some points more like *E. Texanum*, yet very distinct from that also. Apr.

3. **TROPÆOLUM,** *Linnæus* (NASTURTIUM). Tall leafy climbing plants, the succulent herbage with a pungent juice. Leaves alternate, simple, exstipulate. Flowers large, axillary, solitary, irregular. Sepals not quite distinct; the 3 upper somewhat conjointly produced at base into a long spur. Petals 5, unequal; the 3 lower often shorter. Stamens 8, distinct from the very base. Carpels 3, becoming large corky sulcate achenes.

1. T. MAJUS, Linn. Sp. Pl. i. 345 (1753); Curt. Bot. Mag. t. 23. Leaves orbicular, peltate, repandly lobed: petals usually orange-red, 1 2 in. long, broad and obtuse, unguiculate, the 3 lower fimbriate lacerate at the base of the blade: achenes ⅓ ½ in. in diameter.—Native of Peru; escaped from cultivation in many places in California, especially southward, at Santa Barbara, etc.; also near Belmont, San Mateo Co.

4. **FLŒRKEA,** *Willdenow.* Low annuals, slightly succulent, the juice pungent. Leaves alternate, pinnately cleft, exstipulate. Flowers axillary, solitary, regular, 3 5-merous (all ours 5-merous, or by exception 4-merous). Sepals valvate in bud. Petals convolute, as many hypogynous

glands alternating with them. Stamens 10, distinct. Style 5-cleft: carpels distinct, subglobose, fleshy when young, becoming soft variously roughened achenes separating from their short axis.

1. **F. Douglasii,** Baillon, Hist. v. 20 (1874) : R. Br. in Lond. & Edinb. Phil. Mag. ii. 70 (1833), under *Limnanthes*, also Lindl. Bot. Reg. t. 1673. Glabrous throughout, 6–18 in. high : leaflets narrowly cuneiform, incisely lobed or parted : peduncles 2–4 in. long : sepals lanceolate, $\frac{1}{4}$–$\frac{1}{3}$ in. long : petals yellow, $\frac{3}{4}$ in. long, obovate, emarginate : achenes obovate-pyriform, more or less tuberculate. Coast Range and foot-hills of the Sierra, in very very wet places. Apr. May.

2. **F. rosea,** Greene. Hartw. in Benth. Pl. Hartw. 301 (1849), under *Limnanthes*. Very near the preceding, rather stouter ; leaf-lobes almost linear ; petals broader, white, fading with a tinge of rose : achenes broader, more coarsely and more sharply roughened. Common in the interior, along the lower Sacramento and San Joaquin, on moist subsaline lowlands. Mar., Apr.

3. **F. alba,** Greene. Hartw. l. c., under *Limnanthes*. Smaller, relatively stouter, but the stems only 2–5 in. long : leaf-segments short, broad, 3-lobed : young parts and calyx rather densely long-woolly : petals white, little exceeding the calyx. A very distinct species, belonging to the upper Sacramento, thence ranging northward.

5. **OXALIS,** *Linnæus* (WOOD-SORREL). Herbs with sour juice (containing oxalic acid), alternate palmately 3-foliolate leaves and cymose or umbellate regular 5-merous flowers. Sepals imbricate, distinct or slightly coherent at base, persistent. Petals convolute, deciduous. Stamens 10, more or less monadelphous, those opposite the petals longer than the others. Ovary of 5 united carpels ; styles distinct. Fruit an ovoid or columnar loculicidally dehiscent capsule ; the valves remaining attached to the central axis ; cells 2–several-seeded. Seeds pendulous, the testa aril-like, at length splitting and becoming recurved.

1. **O. Oregana,** Nutt. in T. & G. Fl. i. 211 (1838) : *O. Acetosella*, var. *Oregana*, Trel. Mem. Bost. Soc. iv. 90 (1888). Acaulescent, perennial by simple or sparingly branched scaly rootstocks ; herbage rusty-pubescent : leaves 1 ft. high : leaflets broadly obcordate, ciliate, 1 in. long, $1\frac{1}{2}$ in. broad : scapes 1-flowered, shorter than the leaves, bibracteolate above the middle : petals oblong-obovate, emarginate, white with purple veins : capsule ovoid. Shaded slopes in the Coast Range.

2. **O. corniculata,** Linn. Sp. Pl. i. 435 (1753). Perennial, erect or decumbent, 3–10 in. high, branching, pubescent : leaflets broadly obcordate : peduncles mostly 2-flowered : fl. small, yellow : capsule columnar, $\frac{3}{4}$ in. long, densely pubescent, many-seeded.—Not common in

California. There are several named varieties and reputed species, and our forms deserve more thorough study in the field; particularly as to the nature of the stipules, if present. A mere variety with brown-purple foliage is in the gardens as a border plant, and here and there spontaneous.

Order XVI. MALVACEÆ.

Jussieu, Genera, 271 (1789). Malv.e, Adanson, Fam. ii. 390 (1763).

Herbs or shrubs, with mucilaginous juice, tough-fibrous inner bark, alternate stipulate leaves and a more or less stellate pubescence. Flowers usually perfect, complete and regular; the 5-cleft valvate (rarely imbricate) and persistent calyx often subtended by a supplementary whorl of bracts and thus appearing double. Petals 5, hypogynous, at base commonly joined to each other and to the base of the tube of the monadelphous stamens, convolute in bud. Stamens $5-\infty$, more or less completely monadelphous and sheathing the styles; anthers usually reniform, 1-celled. Ovaries distinct, forming a ring around a central columnar elevation of the receptacle and becoming achenes, or joined into one 5-10-celled organ and becoming more or less capsular. Seeds usually roundish, with little or no albumen.

Hints of the Genera.

Stamens ∞; anthers reniform, 1-celled,
    Calyx with cup-like involucre at base, - - - - - - - 1
    " with or without 1–3 bracts at base,
        Fruit a whorl of 1-seeded carpels, - - - - 2, 3, 5–7
        " " " 2-9-seeded carpels, - - - - 4, 7, 8
        " a 5-celled capsule, - - - - - - - - - 9
Stamens 5; anthers elongated, 2-celled, - - - - - - - - 10

1. **LAVATERA,** *Tournefort.* Stout shrubs with coarse flexible branches, ample palmately lobed leaves, and axillary showy flowers. Involucel 3-lobed. Stamineal tube divided at summit into numerous filaments. Style-branches stigmatose lengthwise, on the inside. Fruit a depressed whorl of 5-8 crowded achenes surrounding the angular column of the receptacle which scarcely exceeds them, and covered by the persistent calyx.— Genus artificially separated from *Malva* on account of the gamophyllous involucel mainly; referred to *Althæa* by Baillon.

1. **L. assurgentiflora,** Kellogg, Proc. Calif. Acad. i. 14 (1854); Greene, Pitt. i. 77; Baker, Journ. Bot. xxviii. 240. Coarse, stout, soft-woody, flexuous-branched, 6-15 ft. high, the young branches, pedicels and calyx, rarely the leaves also, stellate-hairy or -tomentose: leaves long-petioled, 3-6 in. broad, angularly 5-7-lobed, the lobes coarsely toothed: fl. solitary, on a long deflexed and curved pedicel: petals 1-1½ in. long, cuneate-obovate, truncate or retuse, abruptly reflexed from near the

base, rose-red with crimson veins: stamineal column glabrous: styles exserted: fr. 1/2 in. broad; carpels not beaked, equalling the summit of the axis.—Native of the islands off Santa Barbara and San Pedro: long cultivated about San Francisco, where it is become spontaneous both in the sand-dunes and along the seashore. Jan. - May.

2. **MALVA,** *Pliny* (MALLOW). Herbs with broad angular or rounded leaves, and axillary solitary or glomerate flowers. Involucel 3-leaved. Stamens and pistils as in *Lavatera*. Column of receptacle short, seeming depressed below the whorl of achenes.

1. M. PARVIFLORA, Linn. Amœn. Acad. iii. 416 (1756); Greene, in W. Am. Scientist, 155; Baker, Journ. Bot. xxviii. 341; *M. obtusa*, T. & G. Fl. i. 225 (1838). Simple or branching, the branches depressed and only a few inches long, or the main stem erect and 2–6 ft. high: herbage more or less pilose-hairy: leaves long-petioled, obsoletely 5 7-lobed, round-cordate, crenate, 1–3 in. broad: fl. glomerate, small, the pale blue corolla little exceeding the calyx: bractlets linear: calyx accrescent, the broad-lobed limb rotately spreading away from the mature fruit: achenes glabrous or pubescent, transversely and sharply rugose on the back, the acutely winged margins distinctly toothed. A homely weed, extremely common, often small and depressed when growing in the streets or along country waysides, but in good soil erect with ascending branches and sometimes (in southern parts of the State) ten feet high.

2. M. BOREALIS, Wallm. in Liljebl. Sv. Fl. 2d ed. 218 (1798). Habit, aspect and foliage of the last, but herbage more conspicuously pilose and often a little stellate-hairy: bractlets lanceolate: calyx-lobes deep, closed over the mature fruit: corolla pale blue, 1/2 in. long, surpassing the calyx: achenes reticulate-rugose, the acute margins entire.—Rather common about Berkeley; easily distinguished from the foregoing by the larger flowers, connivent calyx-lobes, entire-margined and irregularly rugose achenes, etc. The *M. borealis* of the "Botany of California" is the preceding. I formerly took the present plant for *M. Nicæensis*, All., and specimens from middle Europe so named, but not authenticated, misled me. Both our species of this genus are naturalized from Europe. *M. rotundifolia*, so common on the Atlantic slope, has not appeared in California.

3. **SIDALCEA,** *A. Gray*. Herbs with rounded and commonly lobed or parted leaves; occasionally diœcious. Flowers in terminal racemes or spikes, rose-purple or white. Involucel 0. Stamineal column double; filaments of the outer series united into about 5 sets; of the inner distinct. Style-branches stigmatic lengthwise, as in *Malva*; fruit the same, except that the achene is sometimes beaked.

* *Annuals.*

1. **S. diploscypha,** Gray, Pl. Fendl. 19 (1849) & Gen. Ill. ii. 58. t. 120; T. & G. Fl. i. 234 (1838), under *Sida*. Erect, 1-2 ft. high, paniculately branching, pilose-hirsute with long spreading hairs: leaves long-petioled, rounded, the radical deeply crenate; cauline 7-parted with 2-3-cleft oblong segments: inflorescence umbellate, the umbels many, at the ends of the branchlets, 3-5-flowered: fruiting calyx $\frac{3}{4}$ in. long, deeply cleft, the segments lanceolate, acuminate: corolla 1 in. long, pale rose-color: achenes cochleate and nearly orbicular, scarcely a line in diameter, reticulate-rugulose on the back. Very common on hills and in fields along the Mt. Diablo Range, both in the hills and upon the plains adjacent. Mar.—May.

2. **S. secundiflora.** Pubescence and foliage as in the last, but plant less branching, the flowers in terminal rather lax spicate racemes: petals oblique, purple, with a very dark spot at base: achenes nearly 2 lines long, semiobcordate, strongly favose-reticulate. Only less common than the last; often associated with it, and confused with it in the herbaria, but never confluent with it; and the specific characters appear quite sufficient. It may possibly be the same as *S. diploscypha,* var. *minor* (Gray, Pl. Fendl.); but there is nothing in the description of that variety to indicate it.

3. **S. Hartwegi,** Gray in Benth. Pl. Hartw. 300 (1849); Pl. Fendl. 20: *S. tenella*, Greene, Bull. Calif. Acad. i. 7 (1884). Erect, with slender ascending branches, 1-2 ft. high; branchlets, pedicels and calyx short-pubescent: lower leaves 5-parted, the lobes linear-cuneiform, entire or trifid, of the upper linear, entire: fl. $\frac{1}{2}$ in. long or more, rose-purple: outer phalanges of stamens narrow, closely approximate to the inner: achenes a line long, strongly incurved, the favose reticulation elongated.— Common in the foot-hills from Butte Co. to Calaveras, perhaps farther southward; also on the plains adjacent. Mixed with the next in the " Botany of California," and the present writer, mistaught as to the type of *S. Hartwegi*, in separating the two, appears to have made a synonym.

4. **S. hirsuta,** Gray, Pl. Wright i. 16 (1852); Proc. Am. Acad. xxii. 286 (1887): *S. delphinifolia*, Gray, Pl. Fendl. 19 & Gen. Ill. ii. 58. t. 120 (1849), not *Sida delphinifolia*, Nutt. Stout, erect, simple or almost fastigiately much branched, 2-4 ft. high; the rather densely spicate inflorescence and the growing parts densely hirsute: lower leaves round-cordate, slightly crenate-lobed; cauline completely divided into 7-9 narrowly linear entire segments or leaflets: calyx $\frac{1}{2}$ in. long, the lobes deep, acuminate: corolla rose-purple, 1 in. long: achenes rugose-reticulate, tipped with a long rather soft but hispid erect beak.—Valleys among the Coast Range hills northward, in Mendocino Co. etc.; also in the interior of the State, on the lower Sacramento and San Joaquin.

5. **S. calycosa**, Marcus Jones, Am. Nat. xvii. 875 (1883) : *S. sulcata,* Curran, Bull. Calif. Acad. i. 79 (1884). Stout, rather widely branching, 2 ft. high, glabrous below, sparingly hirsute above : inflorescence loosely spicate : calyx-lobes ovate-lanceolate, abruptly acuminate : corolla 1 in. long, deep or pale purple : achenes more or less perfectly sulcate on the back, by obliteration of the usual transverse ridges.—In Marin Co., toward the coast, *Jones ;* also at the eastern base of the Sierra Nevada, near Folsom, *Mrs. Curran.*

* * *Perennials.*

+— *Erect species, with branched inflorescence.*

6. **S. Hickmani**, Greene, Pittonia, i. 139 (1887). Stems tufted from a woody-fibrous tap-root, 2–3 ft. high, leafy throughout, rough with a stellate-hispidulous pubescence : lowest leaves orbicular, small, with slight crenate lobes ; cauline larger, round-flabelliform, coarsely and irregularly toothed around the semicircular margin, the petiole longer than the blade : racemes numerous, axillary and terminal, few-flowered : pedicels short, subtended by 3 filiform bractlets 1½ in. long : calyx-lobes oblong-ovate, acuminate : corolla purple, 1 in. long : achenes nearly orbicular, 1 line long, marked on the back by scattered transverse short and sharp ridges.—A remarkable species, in aspect quite unlike any other known, but in character a true *Sidalcea.* Cañons of the Salinas valley, *Hickman.*

7. **S. oregana**, Gray, Pl. Fendl. 20 (1849) ; Greene, Bull. Calif. Acad. i. 77 ; Nutt. in T. & G. Fl. i. 234 (1838), under *Sida. Sida malvæflora,* Lindl. Bot. Reg. t. 1036 (1826), not of DC. Stems solitary or few from the root, 2–6 ft. high, naked and paniculately branched above, leafy below ; inflorescence stellate-tomentose, peduncles and lower part of stem sparingly hirsute, the plant otherwise glabrous : lower leaves orbicular, 7–9-lobed, the cuneate-obovate lobes 3-cleft at summit ; upper 7–9-parted, narrowly and deeply cleft : spicate racemes usually dense but elongated : calyx-lobes broadly ovate, acute, not longer than the tube : corolla 1½–1 in. long : achenes small, straight (semiorbicular), slightly beaked, smooth and glabrous, 1 line long.—In the northeastern counties of the State, doubtless within our limits, but more common in Oregon. It is also to be sought in the Coast Range ; for a fruiting spike only, evidently of this species, was collected by Mr. V. K. Chesnut in Sonoma Co., near Santa Rosa, in 1887.

8. **S. spicata**, Greene, Bull. Calif. Acad. i. 76 (1885) ; Gray, Proc. Am. Acad. xxii. 288 ; Regel. Gart. Fl. 291. t. 737 (1872), under *Callirhöe.* Smaller and more slender than the last, occasionally with a simple spike, but always erect ; herbage of a light green, pilose-hispid throughout and with a rough stellate or fascicled pubescence on the calyx, lower face of

leaves, etc.; racemes short, spicate-crowded or more open and elongated: calyx deeply cleft, the ovate acute or acuminate lobes twice as long as the tube, very hairy : petals ½ in. long, deeply emarginate, red-purple or paler : achenes small, depressed, pubescent but not reticulate. In moist ground in open woods, or along fences and borders of thickets in the higher Sierra ; the densely spicate typical form at Cisco and Donner Lake ; a taller, more branching and slender-spiked state occurring far northward, *Mrs. Austin.* Everywhere and in all its forms readily distinguishable from *S. Oregana* by the long and harsh pubescence of the whole herbage, and its vivid light green hue ; the more deeply cleft calyx, with more attenuate as well as relatively much longer lobes, shorter and incurved achenes, etc. June- Sept.

9. **S. malvæflora** (Moç. & Sesse), Gray, Pl. Wright. i. 16 (1852), not of Bot. Calif., or of later papers : *S. Neo-Mexicana*, Gray, Pl. Fendl. 23 (1849); Proc. Am. Acad. xxii. 287 ; Hemsl. Biol. Centr. Am. i. 99. *Sida malvæflora*, Moç. & Sesse, in DC. Prodr. i. 474 (1824). Stems slender at least at the base, clustered and strictly erect, from a thick more or less grumose or tuberous root, 2-8 ft. high, occasionally simple, commonly with a few paniculate branches ; glabrous, except a slight scarcely stellate or even fascicled pubescence on the calyx and pedicels, and a few scattered solitary hirsute hairs on the stem above and below : leaves small, the lowest truncate at base and incised-crenate, the upper 5-cleft or -divided, the segments entire and narrow, or broader and with a few pinnate lobes : calyx small, deeply cleft, the segments deltoid-ovate, acute or acuminate : achenes less than a line long, nearly as broad, minutely apiculate, glabrous, smooth or more or less rugose-reticulate.- This Arizono-Mexican species, common in mountain meadows of New Mexico and Arizona, has been found in San Bernardino Co., *Parish*, and is almost certain to occur on the eastern side of the Sierra within our limits, in Kern, Inyo and Mono counties. It is wholly distinct from each and all of the five or six Californian species collectively designated *S. malvæfolia* in the State Survey volumes.

++ *Stems decumbent at base, simply racemose above.*

10. **S. delphinifolia** (Nutt.), Greene. not of Gray, Pl. Fendl. & Gen. Ill. (1849). *Sida delphinifolia*, Nutt. in T. & G. Fl. i. 235 (1838). *Sidalcea malvæflora*, Gray, in later writings, not of Pl. Wright. (1852). Stems clustered from a tuberous-enlarged or somewhat ligneous crown, stout, erect from a decumbent base, or ascending, 1-3 ft. high : whole herbage rather stiffly hirsute, the hairs often fascicled in threes, spreading or somewhat retrorse, lower face of leaves with a sparse stellate pubescence beneath the hirsute : radical leaves crenate-incised, round, with open sinus ; cauline 7-parted, the segments deeply trifid : fl. often unisexual : calyx in fr. sometimes ½ in. long, very deeply cleft, the lobes broadly

lanceolate, acute ; rose-red corolla often 1½ in. long ; achenes large, semiorbicular rugose-reticulate. Var. **humilis**, *S. humilis*, Gray, Pl. Fendl. 20. Often only 6 10 in. long and much depressed : hirsute pubescence scant or none ; achenes more cochleate (almost orbicular). The hairy type is southern, but is found in San Mateo Co. The variety, or perhaps a species, is the common early flowering perennial of the whole seaboard region of middle California. It is usually diœcious, the pistillate flowers being of less than half the size of the staminate.

11. **S. Californica,** Gray, Pl. Fendl. 19 (1849) ; Proc. Am. Acad. xxii. 286 ; Nutt. in T. & G. Fl. i. 233 (1838), under *Sida*. Size and habit of the last, but not in the least hirsute, the whole herbage cinereous with a short and soft pubescence of radiate-clustered hairs. Of the Santa Ynez Mts., but to be expected in San Luis Obispo Co. if not in Monterey, on the higher mountains.

12. **S. asprella,** Greene, Bull. Calif. Acad. i. 78 (1885) ; Gray, l. c. (1887). More slender, 1- 2 ft. high, without hirsute or hispid hairs, but roughish with a minute dense stellular pubescence : leaves of lower and upper parts of plant alike in form, all 5-lobed, the lobes with about 5-rounded teeth : achenes rugose-reticulate throughout, dorsally somewhat concave, the margin angled. In the foothills of the Sierra only, and from Butte Co. to Calaveras. May July.

13. **S. glaucescens,** Greene, l. c. ; Gray, l. c. Slender. 1 2 ft. high, glabrous, or with a minute and obscure pubescence on the altogether pallid and seemingly glaucous herbage : leaves all palmately divided, the cuneate divisions 3 –5-lobed or -toothed, or of the uppermost leaves linear, entire : raceme lax, few-flowered : corolla deep purple, the petals obtuse or truncate ; achenes with elongated reticulation. In the higher Sierra only, from Mt. Shasta to Kern Co.; common there, and very beautiful. June- Sept.

* * * *Anomalous species ; perhaps sui generis, and if so, to be called* HESPERALCEA.

14. **S. malachroides,** Gray, Proc. Am. Acad. vii. 332 (1868) ; H. & A. Bot. Beech. 326 (1840), under *Malva*. Root perennial : stems clustered, erect, 2 6 ft. high : herbage rough-hispidulous throughout with a stellate-clustered short pubescence : leaves ample, (2 5 in. wide) cordate, 3 7-angled, the lobes coarsely toothed : inflorescence diœcious : fl. small, white, in short dense spikes at the ends of the numerous paniculate leafy branches : calyx-lobes broad, acute : some outer stamens joined in pairs beyond the tube, the others distinct : carpels somewhat stellate-pubescent or glabrous, the surface even, apparently dehiscent by a dorsal suture. In swampy lands of the Coast Range from Santa Cruz Co. to Humboldt : not common. June Aug.

4. MODIOLA, *Mœnch.* Prostrate and more or less creeping herbs with palmately divided leaves, and small flowers on long axillary peduncles. Involucel 3-bracted. Calyx 5-cleft. Stamineal tube simple. Stigmas capitate. Carpels numerous, 2-valved, partly 2-celled by the intrusion of a horizontal valve-like process between the 2 seeds.

1. M. CAROLINIANA, Don, Dict. i. 466 (1831); Gray, Gen. Ill. ii. 72. t. 128; Linn. Sp. Pl. 688 (1753), under *Malva*. *Modiola multifida*, Mœnch. Meth. 620 (1794). Stems several feet long, more or less hirsute, leafy and flowering throughout: leaves of broad-ovate outline, truncate at base, palmately and deeply 5–7-cleft, the segments subdivided or coarsely toothed: peduncles mostly solitary, 1–2 in. long, about equalling the petioles: corolla $\frac{1}{2}$ in. broad, purple-red: carpels lunate, much flattened, hispid along the upper edge. — Naturalized at Auburn, *Miss Harrison.* Native of the southern U. S. and West Indies.

5. SIDA. Herbs with undivided leaves. Involucel 0 (except in ours where it is 3-bracteate as in the preceding). Calyx 5-cleft. Stamineal tube simple. Stigmas capitate. Carpels 1-celled, 1-seeded, dehiscent or indehiscent, forming a short-conical fruit. Seed pendulous.

1. **S. hederacea,** Torr. in Gray, Pl. Fendl. 23 (1849); Dougl. in Hook. Fl. i. 107 (1830), under *Malva*; *S. obliqua*, T. & G. Fl. i. 233 (1838). Perennial, stoutish, erect-spreading or prostrate, very leafy, $\frac{1}{2}$–1 ft. high, hoary- or yellowish-tomentose throughout: leaves short-petioled, about 1 in. long, reniform, very oblique at base, plicate, serrate or crenate: fl. axillary, solitary or several arranged paniculately, the pedicels slender, at length deflexed: calyx subtended by 1 or 2 slender bractlets: lobes acuminate: corolla $\frac{3}{8}$ in. long, cream-color: fr. short-conical, smooth, glabrous; carpels 6–10, triangular, $1\frac{1}{2}$ lines long, attached by a straight ventral edge to the slender axis. — A depressed hoary weed, very common in low and subsaline clayey soils, throughout the interior of the State, and along the seaboard near salt marshes southward; apparently easily propagated by its roots or rootstocks, springing up on railway embankments remote from its native soil. The fruit is very seldom seen. The author's many years' search for it has been rewarded with but a single whorl of full grown carpels. The cause of the plant's sterility should be enquired into. It is morphologically exceptional in either genus, *Malva* or *Sida*, in which men have placed it.

6. MALVASTRUM, *A. Gray.* Herbaceous or shrubby (ours mostly hoary-tomentose shrubs), with usually angular foliage, and solitary or racemose-panicled flowers. Calyx with an involucel of 1–3 bractlets, or none. Stamineal tube simple; free filaments terminal and distinct. Styles 5 or more; stigmas capitate. Carpels 1-seeded, bivalvate-dehiscent or indehiscent. Seed ascending. An artificial genus; some species taken out of *Malva* solely on account of the capitate stigmas; others easily referable to *Sphæralcea;* while the ascending ovule alone distinguishes others from *Sida.*

* *Annuals.*

1. **M. exile,** Gray, Ives Exp. 8 (1860). Stems slender, diffuse or ascending, 3–12 in. high : herbage green but minutely and sparsely stellate-pubescent : leaves round-ovate, cordate or truncate at base, 5-lobed, sparingly toothed, 1 in. long, on slender petioles as long : fl. axillary, solitary or few, on slender pedicels ½—1 in. long : involucel of 3 linear bracts : calyx deeply 5-parted ; segments triangular, acuminate : corolla 3–4 lines long, white : achenes 12–15, cochleate-rounded, less than a line wide, sharply rugose transversely. —A small species, of the south-eastern deserts ; but said to occur in Merced Co.

2. **M. Parryi.** Prostrate or ascending, the purplish and often rough-hairy branches 1—2 ft. long : herbage cinereous or hoary with a rough stellate pubescence : leaves 1—2 in. long, deeply 5-parted, the segments coarsely toothed or lobed : fl. mostly solitary, on slender peduncles 1½—4 in. long : involucel of 3 linear-setaceous persistent bracts ½ in. long : calyx-lobes deltoid-ovate, long-acuminate : petals deep-purple, ½—¾ in. long : carpels 15—20, strongly cochleate, a line wide, sharply rugose transversely.—A very common early flowering annual of the plains and valleys of Monterey Co., and perhaps San Luis Obispo, *Parry, Lemmon ;* also near Tulare, *Parry.* There is a State Survey specimen (No. 542), obtained on the Nacimiento River in 1861, which Dr. Gray had called "*Sidalcea diploscypha ;*" and, although the species bears much general resemblance to the annual Sidalceas—more than to any *Malvastrum*—it has probably been referred latterly to *M. exile.* Mar.—May.

* * *Shrubby or suffrutescent species.*

3. **M. fasciculatum,** Greene. Nutt. in T. & G. Fl. i. 225 (1838), under *Malva.* *Malvastrum Thurberi,* Gray, Pl. Thurb. 307 (1854) : *M. splendidum,* Kell. Proc. Calif. Acad. i. 65 (1855). Usually 6–8 ft. high, often larger and arborescent, the main stem a few inches thick ; bark smooth, gray ; branches long, wand-like, slender, racemose or amply racemose-paniculate above, these and the lower face of the leaves canescently short-tomentose : leaves angularly 5-lobed and coarsely toothed, 1½–3 in. long, and almost as broad : calyx-lobes triangular, as broad as long, acute : corolla rose-purple, ¾ in. long : carpels smooth, tomentose above, promptly dehiscent : seed with a stellular-hairy minute reticulation.—A very handsome shrub or small tree, and the most common species, occurring from Mt. Diablo, *Rattan,* and Monterey Co., *Abbott,* southward throughout the State.

4. **M. Palmeri,** Wats. Proc. Am. Acad. xii. 250 (1877). Stouter than the last, the branches shorter, flexuous, the inflorescence terminal,

subsessile : stellate tomentum coarser, yellowish or brownish : leaves
3 –5-lobed, the lobes and teeth obtuse ; stipules lanceolate, conspicuous :
bractlets of the few large calyces linear-lanceolate, nearly equalling the
acuminate calyx-lobes : petals 1 in. long, dull pinkish or yellowish
white : carpels and seed as in the last. –Plentiful on bushy hills in San
Luis Obispo Co., *Palmer*, *Michael*, also in Monterey and San Benito,
Hickman. A homely species, stouter than *M. fasciculatum*, equally
shrubby though never as tall.

5. **M. marrubioides**, Dur. & Hilg. Pac. R. Rep. v. 6. t. 2 (1855).
Suffrutescent, 2 ft. high, densely tomentose : leaves thick, short-petioled,
½ in. long, ovate, subcordate, obscurely 3-lobed, sharply toothed : fl.
subsessile in glomerate clusters in the axils of the upper leaves, or
running out into a naked interrupted spike : calyx-lobes long-acuminate:
petals rose-color, ½ –¾ in. long : carpels rounded or oblong, smooth and
glabrous. –A rare or local species, as to the type, which is from Millerton,
on the San Joaquin ; but a shrub six or eight feet high, with less condensed inflorescence, but apparently much the same, is found along
stream-banks back of Belmont, San Mateo Co.

6. **M. orbiculatum**. Suffrutescent, the stout erect and simple
branches 2–3 ft. high ; whole plant densely tomentose : leaves short-
petioled, 1–2 in. long and as broad, the lower and smaller round-
reniform, the upper orbicular, not even obscurely lobed but coarsely
crenate, very obtuse or slightly retuse : fl. many, nearly sessile and densely
glomerate in the axils of the upper-leaves and at almost leafless
subterminal nodes : bractlets setaceous, much shorter than the lanceolate
acuminate deep calyx-lobes : corolla deep rose-color, ½ in. long or more :
fr. unknown.—Collected by the author, in the mountains south of
Tehachapi, Kern Co., June, 1889 ; well marked in leaf-outline as distinct
from the preceding and the following.

7. **M. Fremonti**, Torr. in Gray, Pl. Fendl. 21 (1849). *Sphæralcea
Lindheimeri*, Bot. Calif. i. 85, by mistake. Suffrutescent, very stout,
2–3 ft. high, densely white-tomentose : leaves very thick, short-petioled,
1–3 in. long, broadly ovate, cordate at base, slightly 3–5-lobed and
crenate : fl. in short axillary pedunculate racemose clusters : calyx
ovate, ½ in. long, only the setaceous tips of its lobes visible amid the
deep and dense white tomentum, almost equalled by the 3 linear
setaceous involucral bractlets : corolla ¾ in. long, rose-color : carpels
thin, smooth, promptly dehiscent. –From Mt. Diablo, *Rattan*, southward
in the same range of hills ; also in Calaveras Co., according to Gray.

8. **M. multiflorum**. Shrubby, slender, canescently stellate throughout : leaves ½ in. long or more, thin, ovate, slightly lobed and finely
toothed : fl. numerous, small, on axillary leafy paniculate rather crowded

branchlets: calyx 3 lines long, subtended by 3 setaceous bractlets; segments ovate-lanceolate, acuminate: corolla 1_2 in. long or less: carpels suborbicular, 1_2 line long, wholly indehiscent, very conspicuously favoso-reticulate both dorsally and laterally; the minute reniform seed smooth, glabrous. Known only in a single fragmentary specimen, deposited in the herbarium of the University long ago, with a few other specimens, all said to have been gathered near Monterey; the collector's name not known. There is a blunt incurved very short vacant apex to the achene, suggesting that the shrub may perhaps as well be a *Sphæralcea*.

9. **M. Munroanum**, Gray, Pl. Fendl. 21 (1849); Dougl. in Bot. Reg. t. 1306 (1830), under *Malva*. Branching from a mere woody base, 1-2½ ft. high, hoary with a scattered stellate-pubescence: leaves broadly ovate, cordate, 3-5-lobed and acutely or crenately toothed, 1-2 in. long, exceeding the petiole: fl. in short and rather dense terminal and subterminal racemes: calyx-lobes acute or acuminate, 2—4 lines long: corolla scarlet, $1_2 - 3_4$ in. long: carpels 2 lines long, oblong, rounded or short-beaked above and pubescent, reticulate on the sides toward the base. A fine showy species of the Interior Basin, reaching our borders along the eastern base of the Sierra.

7. **SPHÆRALCEA,** *A. St. Hilaire.* In all respects like *Malvastrum* except that the fruit is conical rather than depressed, the carpels being longer and 2-ovuled: the lower seed ascending, the upper pendulous (when not wanting by abortion of the ovule).

1. **S. incana,** Torr. in Gray, Pl. Wright. i. 21 (1852), Pl. Fendl. 23 (1849), name only: *S. Emoryi,* Torr. in Gray ll. cc. Rather slender, 1-2 ft. high, softly canescent with minute pubescence: leaves ovate-cordate, slightly 3-5-lobed, crenate: calyx and scarlet corolla as in *Malvastrum Munroanum,* which the plant much resembles, differing chiefly in the longer 2-seeded carpels forming a truncate-conical fruit.—The publication of this dates from the Plantæ Wrightianæ, only the names having been given in the Plantæ Fendlerianæ. In both places the name *incana* precedes *Emoryi*. The plant comes barely within our limits along the eastern base of the Sierra, but is common in Arizona, New Mexico, etc.

8. **ABUTILON,** *Camerarius.* Herbs or shrubs, usually soft-tomentose, with axillary solitary mostly yellow flowers. Involucel 0. Staminal tube simple, antheriferous at summit. Styles 5 or more, with capitate stigmas. Fruit truncate-globose or -conical; carpels dehiscent, several-seeded.

1. A. Avicennæ, Gærtn. Fr. et Sem. ii. 251 (1802). *Sida Abutilon*, Linn. Sp. Pl. ii. 685 (1753). A stout erect branching annual, 2-6 ft. high, the herbage green but velvety-pubescent and almost oily to the

touch ; leaves round-cordate, acuminate, crenate-dentate, 3 6 in. long,
on petioles of 2 5 in. ; peduncles axillary, erect, shorter than the
petiole ; fl. small, orange-yellow ; carpels about 15, inflated, obliquely
birostrate, pubescent, 3-seeded. A common weed in cultivated grounds
at the East ; reported as established about Santa Rosa, Sonoma Co.

9. **HIBISCUS,** *Dioscorides.* Stout herbs, with large and showy
axillary and solitary flowers. Involucel of many bractlets. Stamineal
column antheriferous below the summit ; above naked and truncate or
5-toothed. Styles united ; stigmas 5, capitate. Carpels united into a
5-celled loculicidal capsule ; cells several-seeded.

1. **H. Californicus,** Kellogg. Proc. Calif. Acad. iv. 292 (1873) ; *H.
Moscheutos*, var. *occidentalis*, Torr. Bot. Wilkes Exp. 256 (1874) ; *H. lasiocarpus*, var. *occidentalis*, Gray, Proc. Am. Acad. xxii. 303 ; Wats. in
Garden and Forest, i. 425. fig. 68. Perennial, stout, erect, branching,
5 -7 ft. high, velvety-pubescent : leaves cordate-ovate, acuminate, coarsely
but not deeply toothed, 3 5 in. long, exceeding the petioles : peduncle
jointed above the middle, 2 3 in. long, 1-flowered : calyx 1 in. long,
cleft to the middle, the lobes acute : corolla 3—4 in. long, yellowish or
cream-color, with dark purple center : capsule 1 in. long, acute, velvety-
pubescent : seeds a line in diameter, globose, striate and tuberculate-
roughened. In moist or swampy places along the rivers of the interior,
from the lower San Joaquin to Butte Co. It is possible that more than
one species is included in the above synonymy.

10. **FREMONTIA,** *Torrey.* A stellate pubescent shrub or small tree
with alternate and rather small lobed leaves ; stipules small, caducous.
Flowers axillary, on short pedicels. Involucel of 3- 5 small bracts.
Calyx 5-cleft almost to the base, imbricate in bud, yellowish and petaloid,
pitted at base, persistent. Corolla 0. Stamens 5, united to the middle
of the filaments ; anthers linear, adnate, 2-celled, curved. Ovary 5-celled,
many ovuled ; style acute, stigmatic at the apex. Capsule loculicidally-
dehiscent ; cells 2 - 3-seeded. Seeds ovate ; embryo small, in copious
fleshy albumen ; cotyledons ovate. Genus monotypical, as far as known ;
unless our shrub be, as some very learned authorities assert, only a
second species of the Mexican *Cheiranthodendron*.

1. **F. Californica,** Torr. Pl. Frem. 6. t. 2 (1850) ; Hook. f. Bot. Mag. t.
5591 ; Greene, in Garden and Forest, ii. 470 ; Sarg. Silv. i. 47. t. 23 ;
Baill. Hist. iv. 70 (1873), under *Cheiranthodendron*. Arborescent and
12 -20 ft. high, or small and bushy ; branches stout ; bark dark gray :
leaves thick, rusty-tomentose beneath, $\frac{1}{2} - 2\frac{1}{2}$ in. long, broadly ovate-
cordate, entire or 3-lobed, the lobes obtuse, mucronate ; petiole short :
flowers many and almost crowded on the branches ; the corolla-like
calyx 1 3 in. wide, yellow within, partly of a rusty-red without : capsule

ovate, 1 in. long, densely hairy; cells villous within: seeds ovate, 2 lines long, pubescent.—Frequent in the lower mountains southward. Occasional, in a small bushy form, in Shasta and Nevada counties. A shrub of singular aspect, highly ornamental when in flower, especially in its southern and more tree-like development. It is not impossible that the northern and shrubby one, the very existence of which is virtually denied in the new Silva of North America, may prove a marked variety or second species. Fremont's type must have been the southern form. The mucilaginous properties of the bark have led to its employment as a substitute for that of the *Slippery Elm* of the East; and that name has been applied to our tree.

Order XVII. HYPERICEÆ.

J. St. Hilaire, Exposition des Familles Naturelles, ii. 23 (1805). Hyperica, Juss. (1789).

A small family, analogous to *Malvaceæ*, but scarcely allied to them; here represented by a few species of the one principal genus of the order.

HYPERICUM, *Dioscorides* (St. John's-wort). Glabrous perennials; the bright green herbage punctate with pellucid or dark-colored dots. Leaves opposite, simple, entire, exstipulate. Inflorescence cymose; flowers yellow. Sepals 5, imbricate in bud. Petals 5, convolute in bud, rotate in expansion. Stamens ∞, usually connate at base, into 3–8 clusters. Styles 2–5, nearly or quite distinct; ovary 1-celled with 3 parietal placentæ, or 3-celled by union of the placentæ with the axis. Capsule with many minute seeds.

1. **H. concinnum,** Benth. Pl. Hartw. 300 (1849); *H. bracteatum*, Kell. Proc. Calif. Acad. i. 65 (1855). Erect, slender, wiry, very leafy, suffrutescent at base, $\frac{1}{2}$–1 ft. high: leaves thickish and somewhat conduplicate, linear or linear-oblong, acute: cyme few-flowered: fl. 1 in. broad: sepals ovate, acuminate: stamens ∞, in 3 fascicles.—Common on dry bushy hillsides in hard clayey soil, at middle elevations of the Coast Range, from San Mateo Co., *Behr*, northward. May, June.

2. **H. Scouleri,** Hook. Fl. i. 111 (1830); *H. formosum*, var. *Scouleri*, Coult. Bot. Gaz. (1886), p. 108. Erect, slender, simple or branched above, altogether herbaceous from running rootstocks, 1–2 ft. high: leaves thin, shorter than the internodes, 1 in. long or less, oblong, obtuse, sessile, clasping: fl. large, in more or less panicled cymes: sepals oval or oblong, obtuse, 2 lines long or less: petals $\frac{1}{2}$ in. long: stamens ∞, in 3 fascicles: capsule 3-celled. In wet grassy places throughout the mountain districts at middle altitudes. If this be specifically identical with the Mexican *H. formosum*, it will stand almost alone among plants not of alpine or even subalpine habitat enjoying a range of almost three

thousand miles north and south. Doubtless other characters than those of the form of the sepals are to be found by which the two species may be more satisfactorily distinguished.

3. **H. anagalloides**, Ch. & Schl. Linnæa, iii. 127 (1828). Diffusely branching, very slender, prostrate or assurgent, stoloniferous, forming a mat a foot or more in breadth : leaves oval or elliptical, $\frac{1}{4}$ $\frac{1}{2}$ in. long, obtuse, clasping, only half as long as the internodes : inflorescence leafy-paniculate-cymose ; fl. scarcely 2 lines long, the obovate- or linear-oblong sepals exceeding the petals ; stamens 15–20, nearly or quite distinct. Var. **Nevadense**. Erect from a merely decumbent and scarcely stoloniferous base, only a few inches high : leaves equalling or exceeding the internodes : cyme rather ample, strictly terminal, on a short naked peduncle : fl. 3 lines long. The type frequent in springy places along the seaboard and among the coast hills, from Santa Cruz northward. The variety, of pronouncedly different aspect and inflorescence, is perhaps limited to the foot-hills of the Sierra. It may prove a species.

Order XVIII. ELATINEÆ.

Cambessèdes, in Mémoires du Muséum, xviii. 225 (1829).

Low annuals with opposite leaves, membranous stipules, and axillary regular symmetrical 2–5-merous flowers. Sepals, petals and stamens all distinct, hypogynous. Styles distinct ; stigmas capitate ; ovary 2–5-celled, becoming a 2–5-celled capsule with central placenta and a septicidal or septifragal dehiscence. Seeds straight or curved.

1. **ELATINE**, *Linnæus* (WATER-WORT). Glabrous dwarf and rather succulent plants of wet places, sometimes aquatic and floating. Flowers axillary. Sepals 2–4, nerveless, obtuse, persistent. Petals 2–4. Stamens as many or twice as many as the petals. Styles, or sessile stigmas, 2–4. Pod thin, globose, 2–4-celled, several- or many-seeded. Seeds cylindrical, straight or curved, striate-pitted.

1. **E. brachysperma**, Gray, Proc. Am. Acad. xiii. 361 (1878). Commonly terrestrial : leaves oblong or oval, attenuate at base, sometimes lanceolate, $\frac{1}{4}$ in. long or less : fl. sessile, mostly dimerous ; stamens 2 or 3 : seed oval, nearly straight, $\frac{1}{4}$ line long, coarsely pitted in 6 or 7 lines of 10–12 pits. Plains of the interior of the State, in very wet places ; also near the coast southward.

2. **E. Californica**, Gray, l. c. Floating : leaves obovate, narrowed at base, the lowest with petiole as long as the blade : fl. short-pedicellate : sepals and petals 3 or 4 each ; stamens twice as many : seeds circinate-incurved, $\frac{1}{3}$ line long, minutely pitted in 10–12 lines of about 25 pits. In Sierra Valley, *Lemmon*.

2. **BERGIA**, *Linnæus*. Coarser annuals, not succulent, pubescent.

Flowers pedicellate and often fascicled, 5-merous. Sepals with strong midrib, acute. Capsule crustaceous, more or less of the partitions remaining with the axis.

1. **B. Texana**, Seubert, in Walp. Rep. i. 285 (1842); Hook. Ic. t. 278 (1839), under *Merimea*; T. & G. Fl. i. 678, under *Elatine*. Diffusely branched, the branches a foot long more or less; herbage glandular-pubescent: leaves oblanceolate, acute, serrulate, $\frac{1}{2}$ — $1\frac{1}{2}$ in. long, narrowed to a short petiole: fl. fascicled, pedicellate; sepals carinate, $1\frac{1}{2}$ lines long, exceeding the petals and stamens: capsule globose: seeds smooth and shining. Moist or very wet places along rivers and ditches, from the middle Sacramento valley, southward to Merced. Although first detected in California by the author, in 1874, the plant is no rarity in the interior of the State. June- Oct.

Order XIX. FRANKENIACEÆ.
A. St. Hilaire, Bull. de la Soc. Philom. 22 (1815).

An order embracing scarcely more than the genus

FRANKENIA, *Linnæus*. Herbs or undershrubs with opposite entire small exstipulate leaves usually sessile and even united at base by a slight membranous continuation of the blade. Fl. small, solitary and sessile in the axils of the very numerous branches and branchlets, usually 5-merous and complete. Calyx tubular, furrowed; the lobes valvate and induplicate in bud. Petals hypogynous, narrowed to a claw which bears an appendage on its inner face. Stamens hypogynous. Style cleft into 2—4 filiform divisions; ovary 1-celled. Capsule invested by the persistent calyx; the few or several seeds attached to the margins of the 2—4 valves.

1. **F. grandifolia**, Ch. & Schl. Linnæa, i. 35 (1826); Torr. Bot. Mex. Bound. 36. t. 5. Somewhat woody at base, erect, much branched and slender, $\frac{1}{2}$—1 ft. high, glabrous or soft-pubescent, very leafy: leaves obovate to narrowly oblanceolate, revolute, $\frac{1}{4}$—$\frac{1}{2}$ in. long, of a dull green: calyx linear, $\frac{1}{4}$ in. long, strongly furrowed, the lobes short, acute: petals small, red, the blade 1 line long or more, erose at summit, the appendage of the claw bifid: stamens 4—7: style 3-cleft: capsule shorter than the calyx, linear, angular: seeds numerous. A homely plant of the salt marshes along the seaboard, and subsaline moist plains of the interior; glistening with a briny dew in a moist atmosphere, more or less incrusted with salt when the air is dry. Flowering all summer.

Order XX. CARYOPHYLLEÆ.
Linnæus, Philosophia Botanica, 31 (1751).

Herbs or suffrutescent plants with inert watery juice, mostly opposite leaves and swollen nodes. Inflorescence usually dichotomous. Flowers mostly 5-merous, complete and regular. Sepals united or distinct, imbricate in bud, persistent. Petals imbricate or convolute, often bifid, sometimes wanting. Stamens usually 10, occasionally 5, distinct, hypogynous around a ring-like disk, or perigynous by cohesion of disk with calyx-tube. Styles 2–5, mostly distinct and with decurrent stigmas. Fruit a capsule opening by valves or teeth. A large order, chiefly of northern or temperate regions; of small economic importance, though the Old World genus *Dianthus* furnishes the carnations, picotees and other pinks of the florists; and some other genera, such as *Lychnis* and *Silene* are more or less cultivated as ornamental plants.

Hints of the Genera.

Calyx gamophyllous and tubular, at least below; toothed or cleft at summit,   1–4
  Sepals distinct, or nearly so;
    Stipules wanting;
      Petals bifid, or at least emarginate, - - 5, 6
      " entire, or at most only retuse, - - 7, 8
    Stipules present (scarious or setiform);
      Petals, if present, conspicuous,   9, 10
      " minute, - - - - 11, 12

1. VACCARIA, *Dodonæus*. A glabrous glaucous annual much branched above. Calyx gamosepalous, pyramidal with 5 prominent angles. Petals 5, unguiculate, not appendaged. Stamens 10. Styles 2. Capsule ovate, 1-celled, but with rudimentary partitions at base, 4-toothed at apex.

1. V. VULGARIS, Host, Fl. Austr. i. 518 (1827). *Saponaria Vaccaria*, Linn. (1753). *Lychnis Vaccaria*, Scop. (1772). Erect, 1–2 ft. high, simple below, cymose-paniculate above: leaves cordate-ovate, acute, entire, sessile: petals red; blade obcordate; claw linear: styles short: seeds dark-colored.—An Old World weed of grain-fields, becoming frequent in our region.

2. AGROSTEMMA, *Linnæus* (CORN-COCKLE). Tall annual, sparingly branched above; pubescent, not viscous. Calyx gamosepalous, tubular, coriaceous, 10-ribbed, 5-toothed. Petals 5, unguiculate. Capsule coriaceous, 1-celled, 5-toothed.

1. A. GITHAGO, Linn. Sp. Pl. i. 435 (1753); Lam. Encycl. iii. 643 (1789). under *Lychnis*. Erect, 2–4 ft. high, soft-hirsute: leaves linear-lanceolate, connate at base: fl. solitary on long upright peduncles: calyx $1\frac{1}{2}$ in. long, the linear teeth as long as the tube, deciduous from the mature fruit: petals purple, not equalling the calyx-teeth; limb broad, obtuse, entire; claw unappendaged. A weed of the grain-fields, more pernicious than *Vaccaria*, but not yet common in California. It has been found at Berkeley.

3. LYCHNIS, *Theophrastus*. Herbs usually pubescent and more or less viscid. Calyx gamosepalous, membranaceous, striate, 5-toothed. Petals commonly with cleft limb and appendaged claw. Stamens 10. Styles 5 or 4. Pod 1-celled, opening by as many or twice as many teeth as there are styles.

1. **L. Californica,** Wats. Proc. Am. Acad. xii. 248 (1877). Perennial, cespitose, 2–4 in. high, glandular-puberulent above; leaves linear to linear-oblanceolate : fl. 1–3, on slender pedicels ; calyx ovate-campanulate, ½ in. long or less ; blade of petal obovate, bifid, each segment lobed at the side ; capsule short-stipitate.—A somewhat rare plant of the high Sierra.

4. **SILENE,** *Lobelius* (CATCHFLY). Habit of *Lychnis*, and flowers about the same, often vespertine. Styles 3. Pod sometimes 3-celled at base, at summit dehiscent by 3 or 6 teeth.

* *Annuals.*

1. **S. antirrhina,** Linn. Sp. Pl. i. 419 (1753). Erect, slender, loosely paniculate throughout, or more commonly simple below, glabrous, glandless except a viscid belt of an inch, more or less, in the middle of each internode of the branches ; leaves lanceolate, acute, 1 in. long ; pedicels erect ; mature calyx oval, 3 lines long, the teeth short ; petals red, the blade emarginate a line long ; crown inconspicuous ; seeds minutely papillose. Common enough, in sandy soil, both along the seaboard and in the interior. The petals are very seldom seen, though perhaps never wanting. Mar. Apr.

2. S. GALLICA, Linn. l. c. 417. Slender, 1 ft. high, sparingly branched or nearly simple, hirsute ; leaves spatulate, 1–1½ in. long ; fl. racemose on very short pedicels, rose-color ; petals with obovate entire blade and small appendages. One of the commonest weeds of fields and waysides ; the small flowers usually forming a one-sided spike or raceme ; the petals not withering so early in the day as in other species of the group. Mar. Jun.

3. S. RACEMOSA, Otto, in DC. Prodr. i. 384 (1824). Stoutish, rather roughly pubescent, 1½–2 ft. high, dichotomously racemose from near the base ; leaves lanceolate ; fl. white, fragrant, ½ in. broad, unilateral ; blade of petal cuneate-obovate, deeply bifid. Occasional in fields about Berkeley ; flowers pure white, very fragrant, strictly vespertine ; plant about twice as large as *S. Gallica* and quite showy.

4. S. NOCTIFLORA, Linn. Sp. Pl. i. 419 (1753). Stoutish, loosely dichotomous, 1–3 ft. high, viscid-pubescent ; lower leaves spatulate 3–4 in. long, upper lanceolate ; fl. few, peduncled ; calyx more than 1 in. long, the teeth subulate ; corolla 1 in. broad ; petals bifid and appendaged. Uncommon, but occasionally met with along railroads and by waysides

in the mountain districts. Flowers of a rather dull greenish white and very strictly vespertine.

* * *Perennials ; leafy throughout, usually low.*

5. **S. Menziesii,** Hook. Fl. i. 90. t. 30 (1830) : *S. Dorrii,* Kell. Proc. Calif. Acad. iii. 44. fig. 12 (1863). Stems numerous, slender, decumbent, dichotomous, branching freely, $\frac{1}{2}$ 1 ft. high; herbage glandular-puberulent : leaves ovate-lanceolate or -oblong, acute at each end, 1 2 in. long : peduncles lateral and terminal, equalling the leaves, 1-flowered : petals white, 3 4 lines long, equalling the ovate calyx, bifid, unappendaged : capsule ovate-oblong : seeds nearly black, tuberculate.—Dry woods of the Sierra Nevada, from Mono Lake northward. A small-flowered species, somewhat resembling an *Arenaria*, and the corolla not vespertine.

6. **S. Hookeri,** Nutt. in T. & G. Fl. i. 193 (1838) ; Hook. f. Bot. Mag. t. 6051 : *S. Bolanderi,* Gray, Proc. Am. Acad. vii. 330 (1865). Hoary-pubescent, 3 10 in. high, very slender : leaves 1 -2 in. long, spatulate, acute : fl. few, large, erect, on pedicels $1\frac{1}{2}$ in. long : calyx $\frac{3}{4}$ in.: petals twice as long, pale pink or white, the broad claw ciliate below, the blade cut into 4 6 lanceolate or linear entire or bifid segments ; appendages lanceolate, decurrent upon the claw. From Mendocino and Plumas counties northward, on hillsides in the wooded country.

7. **S. Californica,** Durand, Pl. Pratt. 83 (1855). Puberulent and more or less glandular, 4 in. to 4 ft. high, simple or sparingly branched above : leaves ovate to oblanceolate, $1\frac{1}{2}$ 4 in. long, acute or acuminate : fl. large, on short pedicels : calyx 7 10 lines long : petals scarlet, deeply parted, the segments bifid, their lobes 2 3-toothed or entire, often with a linear lateral tooth; appendages oblong-lanceolate : capsule $\frac{1}{2}$ in. long, ovate, short-stipitate.—From Placer and Mendocino counties southward in the mountains perhaps throughout the State.

8. **S. campanulata,** Wats. Proc. Am. Acad. x. 341 (1875). Glandular-puberulent, $\frac{1}{2}$ -$1\frac{1}{2}$ ft. high, dichotomously branched above : leaves lanceolate, acute, 1 $1\frac{1}{2}$ in. long : fl. on short deflexed pedicels : calyx inflated, campanulate, $\frac{1}{2}$ in. long, rather deeply lobed, the lobes broad, obtuse or acutish, finely net-veined : petals $\frac{3}{4}$ in. long, pale flesh color or greenish ; claw pubescent ; limb 4-parted nearly to the base, the lobes bifid to the middle or the lateral ones entire or merely notched ; appendages oblong, entire, fleshy : filaments pubescent, exserted : ovary sub-globose, short-stipitate. In the Coast Range from Mendocino Co. northward.

9. **S. Lyallii,** Wats. l. c. Stems slender, ascending ; herbage glabrous, the inflorescence puberulent : leaves narrowly oblanceolate, 1 2 in. long : fl. few, in a dichotomous cyme, the slender pedicels erect : calyx

inflated, campanulate, 1/3 in. long, net-veined above; teeth broad, obtuse : petals brownish purple, 7 lines long; claw naked; limb shortly bifid; appendages oblong, entire : anthers not exserted : ovary narrowly oblong.—Near Gold Lake and in Sierra Valley, Sierra Co.

* * *Perennials ; erect, leafy only below the panicled or racemose inflorescence ; calyx oblong or clavate, never inflated.*

+ *Petals 4-parted or -cleft.*

10. **S. laciniata,** Cav. Icon. vi. 44. t. 564 (1801); Lindl. Bot. Reg. t. 1444. *Lychnis pulchra*, Ch. & Schl. Linnæa, v. 234 (1830). Pubescent and viscid, erect or ascending, 1—2 ft. high : leaves narrowly oblanceolate to linear, 2—3 in. long: fl. few or many, on elongated branches, 1 in. broad, bright scarlet, the pedicels (1/2—3 in. long) not reflexed in fruit : petals deeply 4-cleft, the lobes linear, acute; appendages ovate : capsule oblong : seed strongly tuberculate on the back. From the vicinity of Sacramento southward, but not common within our limits ; more prevalent in the southern counties.

11. **S. Lemmoni,** Wats. Proc. Am. Acad. x. 342 (1875). Only the inflorescence viscid-puberulent : slender stems erect, from a decumbent perennial branching base, 8—12 in. high, branched : leaves mostly on the young shoots, oblong-lanceolate to spatulate, acute : fl. in an open panicle, the slender pedicels erect or at length deflexed, 1/2 — 3/4 in. long : calyx ovate-cylindric, 1/3 in. long ; teeth acutely triangular : petals rose-color, 6—8 lines long ; blade broad, 4-cleft nearly to the base, the lobes linear, entire or notched : appendages lanceolate, entire ; claw villous : ovary oblong.—Webber Lake Valley in the Sierra Nevada, *Lemmon*.

12. **S. occidentalis,** Wats. l. c. 343. Often somewhat tomentose below, glandular-puberulent above : stems stoutish, erect, 1 1/2—2 ft. high, simple or branching : leaves oblanceolate, acute, more or less ciliate at base, 2—4 in. long : fl. in an open panicle, erect or nodding on slender pedicels 1/2—1 1/2 in. long : calyx cylindrical, 1/2 in. long or more ; teeth ovate, obtuse : petals larger, deep purple, 4-cleft into almost equal lobes, or the lateral ones shorter ; appendages linear, entire, half as long as the limb ; claw slightly villous, without auricles : filaments slightly exserted : ovary oblong, 1/4 in. long, on a stipe as long.—Plumas Co.

13. **S. montana,** Wats. l. c. Puberulent : stems slender, from a decumbent branching perennial base, mostly simple, 1 ft. high : leaves narrowly oblanceolate, acuminate, 1 1/2—2 in. long : panicle narrow ; fl. erect, on short pedicels : calyx cylindrical, 3/4 in. long ; teeth oblong, acutish : petals little exceeding the calyx, rose-color ; blade deeply 4-cleft into linear entire equal segments ; claw with basal auricles and terminal appendages somewhat lacerate : capsule oblong, its stipe 2 lines long. Sierra Co., and eastward in Nevada.

14. **S. Bernardina,** Wats. Proc. Am. Acad. xxiv. 82 (1889). Glandular-puberulent: stems slender, from slender rootstocks, 1 ft. high, few-flowered: leaves narrowly linear-oblanceolate, 1–2 in. long: peduncles slender, 1–3-flowered: calyx $1\frac{1}{2}$ in. long, cylindrical, with oblong-ovate teeth: petals greenish, 8 lines long; limb cleft to below the middle into 4 equal narrow lobes; appendage nearly half the length of the limb. 2-parted, the outer segment linear, entire, the inner oblong and lacerate; claw naked, broadly auricled: stamens unequal, included: capsule oblong, short-stipitate. Tulare Co., *Palmer* (n. 185), June, 1888.

15. **S. Shockleyi,** Wats. l. c. xxv. 127 (1890). Puberulent, slender, 3–8 in. high: leaves linear-oblanceolate, 1–2 in. long: fl. few or solitary: calyx viscid-pubescent, cylindrical, 6–8 lines long, the acute lobes $1\frac{1}{2}$ lines: petals rose-color or greenish; limb equally 4-cleft to below the middle; claws auricled; appendages broad and more or less laciniate: stamens and styles equalling the petals: capsule oblong, long-stipitate: seeds tuberculate on the back. At high altitudes on the White Mountains, Mono Co., *Shockley*: collected in August, 1888.

+ + *Petals bifid.*

16. **S. pectinata,** Wats. l. c. x. 344. Viscid-pubescent, erect, stout, 1–$1\frac{1}{2}$ in. high: leaves lanceolate, acuminate, $1\frac{1}{2}$–$2\frac{1}{2}$ in. long, the radical long-petioled: panicle narrow, strict or spreading; fl. erect, on pedicels $\frac{1}{2}$–1 in. long: calyx oblong, $1\frac{1}{2}$–$\frac{3}{4}$ in. long, cleft almost to the middle into narrow acute teeth: petals rose-color or purple, 1 in. long; claw naked, not auricled; limb broadly oblong, deeply bifid, the segments obtuse; appendages lanceolate, entire: ovary oblong, subsessile: seeds finely tuberculate.—In Plumas and Sierra counties, *Lemmon*, *Mrs. Ames*.

17. **S. incompta,** Gray, Proc. Am. Acad. vii. 330 (1868). Viscid-puberulent: stems cespitose and leafy at base, 1 ft. high, simple or loosely branching above: leaves broadly lanceolate, acute, $1\frac{1}{2}$–$2\frac{1}{2}$ in. long: fl. on slender and short pedicels: calyx oblong-cylindric, $\frac{1}{2}$ in. long, the oblong teeth acute: petals $\frac{3}{4}$ in. long, light rose-color; limb cleft into 2 ovate-oblong toothed lobes; claw naked, narrowly auricled; appendages short, toothed: capsule ovate, short-stipitate: seed small, not tuberculate.—Neighborhood of Yosemite.

18. **S. Douglasii,** Hook. Fl. i. 88 (1830): *S. multicaulis*, Nutt. in T. & G. Fl. i. 192 (1838). Finely puberulent, rarely somewhat glandular above: stems erect, slender, $\frac{1}{2}$–$1\frac{1}{2}$ ft. high, few-flowered: leaves narrowly oblanceolate or linear, 1–2 in. long: fl. erect, on slender pedicels: calyx oblong-cylindric, often somewhat inflated, 5–7 lines long; teeth broad, acutish: petals rose-purplish or white, 8–10 lines long, with broad obtuse lobes, auricled claw and narrow appendages:

capsule oblong-ovate, long-stipitate ; seeds strongly tuberculate on the back. — In the Sierra, from Donner Lake northward.

19. **S. Bridgesii**, Rohrb. Ind. Sem. Berol. (1867) & Monogr. Sil. 204 (1868). Finely pubescent below, viscid above : stems erect, slender, simple, 1 ft. high or taller ; leaves narrowly oblanceolate, acute or acuminate, 1–2 in. long ; fl. racemose, on slender spreading pedicels 3–6 lines long ; calyx oblong-cylindric, 4–5 lines long ; teeth narrow, acute : the white narrow bifid petals $\frac{2}{3}$ in. long ; lobes narrowly linear ; appendages very small : styles long : capsule ovate. Yosemite.

20. **S. verecunda**, Wats. Proc. Am. Acad. x. 344 (1875) : *S. Engelmannii*, var. *Behrii*, Rohrb. in Linnæa, xxxvi. 264 (1869). Pubescent and viscid throughout : stems clustered, $\frac{1}{2}$–$1\frac{1}{2}$ ft. high, erect or decumbent : leaves oblanceolate, acute, 1–2 in. long : fl. few, erect, on stoutish pedicels $\frac{1}{2}$–1 in. long : calyx oblong-cylindric, $\frac{1}{2}$ in. long ; teeth triangular, acutish : petals $\frac{3}{4}$ in. long, rose-color : limb bifid to the middle ; lobes linear, the inner entire, outer commonly with a tooth near the base : appendages notched at apex ; claw narrowly auricled : capsule oblong-ovate : seeds strongly tubercled on the back.— Common on the San Francisco peninsula from near the Presidio, and the Mission Hills, to Point San Pedro, San Mateo Co. Mar. June.

21. **S. Ludoviciana**, Wats. l. c. xxiii. 261 (1888). Glandular-pubescent throughout, 1 ft. high : leaves narrowly linear, 2 in. long or less : peduncles 1–2-flowered, equalling the floral leaves : calyx narrow-cylindrical, $\frac{1}{2}$ in. long ; teeth oblong-ovate, scarious-margined and ciliate : petals $\frac{3}{4}$ in. long, the oblong limb bifid to the middle, with or without lateral teeth ; claw narrowly auriculate ; narrowly oblong appendages acute, more or less lacerately toothed : capsule subcylindric, on a stipe $1\frac{1}{2}$ lines long : seeds flattened, tuberculate.— Monterey and San Luis Obispo counties.

5. **CERASTIUM**, *Dillenius* (MOUSE-EAR CHICKWEED). Soft-pubescent and slightly clammy low herbs, with white flowers in leafy- or scarious-bracted dichotomous cymes. Sepals 5, neither carinate nor 3-nerved. Petals 5, bifid or emarginate. Stamens 10. Styles 5, rarely 4 or 3. Capsule cylindric, often incurved, thin and translucent, 1-celled, ∞-seeded, dehiscent at apex by about 10 teeth. Seeds roundish-reniform, scarcely compressed, commonly granulate.

1. **C. viscosum**, Linn. Sp. Pl. i. 437 (1753). Annual, soft-pubescent and somewhat clammy, the branches erect or ascending from a decumbent base, $\frac{1}{2}$–1 ft. high : leaves ovate, obovate, or oblong-spatulate, $\frac{1}{2}$–1 in. long ; bracts of the inflorescence herbaceous throughout : cymes in early state rather dense : pedicels even in fruit only 2 lines long ; the calyx as long, the sepals acute : petals shorter than the calyx : capsule

nearly straight, much longer than the calyx. One of the most common weeds of early spring ; the corolla expanding only in sunshine or at mid-day. Native of Europe. Feb. Apr.

2. **C. vulgatum**, Linn. Sp. Pl. 2d ed. i. 627 (1762). Much like the last, but root perennial ; stems cespitosely branched at base : leaves oblong ; bracts scarious tipped ; fl. more densely clustered and larger ; fruiting pedicels much longer ; sepals obtuse ; petals equalling the calyx. Common enough in Oregon and Washington, this has been but once noticed in California ; Plumas Co., *Mrs. Austin*.

3. **C. arvense**, Linn. Sp. Pl. i. 438 (1753). Perennial, cespitose, downy with reflexed hairs, the inflorescence somewhat viscid : branches 4–8 in. high: leaves linear-lanceolate, 4–10 lines long, acutish: cyme contracted, bearing about 3 flowers (sometimes 5 ; as often 1 only), the branches ascending, often little exceeding the pedicel of the first flower ; bracts ovate, obtuse, suberect : sepals ovate-oblong, obtuse, scarious-margined, $1\frac{1}{2}$–2 lines long ; the obcordate petals twice as long : capsule nearly straight, little exceeding the calyx. Var. **maximum**, Holl. & Britt. Bull. Torr. Club. xiv. 47. t. 64 (1887). Often $1\frac{1}{2}$ ft. high : leaves linear to lanceolate, often 2 in. long ; cyme not only repeatedly dichotomous (12–20-flowered), but the branches almost divaricate ; floral bracts lanceolate and spreading : capsule more than twice the length of the calyx. The type, quite like the European plant in all respects, is common on rocky and bushy hills about San Francisco, crossing the straits into Marin Co., but not reported east of the Bay. The variety, of which the most pronounced type is from Humboldt Co., *Marshall*, is in the State Survey collection as from Mendocino Co., *Bolander*.

4. **C. pilosum**, Ledeb. Mem. Acad. St. Petersb. v. 359 (1815): *C. oblongifolium*, Pac. R. Rep. iv. 70. Perennial, erect, stout, more or less densely pilose, the inflorescence glandular-viscid : leaves oblong-lanceolate, $\frac{1}{2}$–1 in. long, 1–6 lines broad, acute, almost sheathing at base : fl. few, large, in a terminal leafless cyme : sepals 3–4 lines long, obtuse ; petals longer ; capsule 6–10 lines long, the slender teeth at length circinate-revolute. A remarkable Siberian and Alaskan species, said to have been found long ago, on Point Reyes, *Bigelow*.

6. **STELLARIA**, *Linnæus* (CHICKWEED). Low herbs with mostly quadrangular stems, no stipules, and small axillary and solitary, or terminal and cymose white flowers. Flowers as in *Cerastium*, but styles usually 3 only, sometimes 2 or 4. Capsule globose or oblong, cleft below the middle into twice as many valves as there are styles.

* *Leaves ovate, petiolate ; root annual.*

1. S. media, With. Bot. Arr. 418 (1776); Vill. Dauph. iii. 615 (1789):

Smith, Fl. Brit. ii. 473 (1800). *Alsine media*, Camerarius, Hort. Med. 11 (1558); Bauh. Pin. 250 (1623); Linn. Sp. Pl. i. 272 (1753). Weak, flaccid, procumbent, rooting at the lower joints; stems marked by a pubescent line: leaves ovate, $\frac{1}{4}$ – $\frac{3}{4}$ in. long, on slender petioles, or the upper sessile: floral bracts foliaceous; pedicels slender, deflexed in fruit: calyx pubescent: stamens 3–10: capsule oblong-ovate, 2–3 lines long, equalling or exceeding the calyx.—A very common weed of shady places, or sometimes in open ground. Dec.–June.

* * *Leaves lanceolate, sessile; perennials, except n. 2.*

2. **S. nitens**, Nutt. in T. & G. Fl. i. 185 (1838). Annual, the stems almost capillary, diffuse, sparingly leafy, 3–6 in. high, the whole plant very glabrous and shining, or with a slight pubescence below: leaves $\frac{1}{4}$–$\frac{1}{2}$ in. long, acute, the lower short-petiolate: fl. erect, on short pedicels, in a very lax bractless cyme: sepals 3-nerved, narrow, acuminate, 2 lines long: petals deeply bifid, only half as long, sometimes 0: capsule oblong, shorter than the calyx. Of the Coast Range, plains of the interior, and Sierra foot-hills; very common, yet so delicate and inconspicuous as to be easily overlooked. Mar.–May.

3. **S. umbellata**, Turcz. Cat. Baikal. n. 245 (1838). Glabrous, very slender, ascending, from slender creeping scaly rootstocks, 3–5 in. high, sparingly leafy, umbellate-cymose: leaves spreading, elliptic- or oblong-lanceolate, acute at each end, 4–8 lines long: pedicels elongated: sepals ovate-lanceolate, 1–1$\frac{1}{2}$ lines long, 1-nerved: petals 0: capsule about 2–3 lines long.—High Sierra, in the Yosemite region. July.

4. **S. longipes**, Goldie, in Edinb. Phil. Journ. vi. 185 (1822). Glabrous and of a bright shining green, or glaucescent, erect or ascending, 2 in. to 1$\frac{1}{2}$ ft. high: leaves linear or linear-lanceolate, $\frac{1}{2}$–1$\frac{1}{2}$ in. long, acute, rather rigid, usually ascending: fl. few, on long and slender pedicels, these scarious-bracted or bractless: sepals 1$\frac{1}{2}$–2$\frac{1}{2}$ lines long, scarcely nerved: petals as long or longer: capsule ovate-oblong, exserted and dark-colored at maturity: seed smooth.—Yosemite and northward.

5. **S. borealis**, Bigel. Fl. Bost. 2d ed. 182 (1824). Glabrous and bright green, the slender stems usually weak and decumbent, leafy up to the leafy-bracted cyme, $\frac{1}{2}$–1$\frac{1}{2}$ ft. high: leaves ovate-oblong to linear-lanceolate, $\frac{1}{2}$–2 in. long, acute, usually spreading, sometimes ascending: pedicels 1–2 in. long, spreading or deflexed: sepals ovate to lanceolate, 1–2 lines long: petals not longer, 2-parted or 0: capsule ovate, 1$\frac{1}{2}$–2 lines long: seed smooth.—In marshes of Mendocino Co., *Bolander*; a very large form.

6. **S. crispa**, Ch. & Schl. in Linnæa, i. 51 (1826). Near the last, but smaller, more leafy, less distinctly cymose: leaves ovate, elliptical or lanceolate, acute or acuminate, $\frac{1}{4}$–$\frac{3}{4}$ in. long, the margin more or less

undulate-crisped ; fl. solitary in the axils, or few in a cyme : sepals ovate-lanceolate ; the petals not equalling them : capsule ovate-oblong, little exserted : seeds smoothish.—From the Calaveras Big Trees, *Hooker & Gray*, to Auburn, *Miss Harrison*, and northward to Alaska.

7. **S. Jamesii**, Torr. in Ann. Lyc. N. Y. ii. 169 (1827). Erect and rather stout, ½–1 ft. high ; herbage deep dull green, pubescent and very viscid : leaves lanceolate, acuminate, 1–3 in. long : cyme leafy, rather contracted, the branches short and divaricate : sepals oblong, acute, 2–3 lines long, shorter than the 2-parted petals : capsule ovate, shorter than the calyx : seed smooth. —Dry pine woods of the middle Sierra, from Kern Co., *Greene*, northward.

8. **S. littoralis**, Torr. Pac. R. Rep. iv. 69 (1857). Pubescent, ascending, stoutish, 1 ft. high : leaves 1 in. long, ovate, acute, rounded at base, rather thick : fl. in a terminal compound cyme : sepals lanceolate, acute, 3 lines long, obscurely 3-nerved, shorter than the 2-parted petals : capsule included within the calyx.—At Point Reyes, near the seashore, *Bigelow*.

7. **ARENARIA**, *Chabræus* (SANDWORT). Mostly low tufted herbs with sessile usually subulate often rigid leaves and no stipules ; flowers white, cymose-panicled or capitate-clustered. Sepals 5 or 4. Petals as many, entire, emarginate, or 0. Styles 3, opposite as many sepals. Capsule globose or ovoid, dehiscent into as many entire, 2-cleft, or 2-parted valves as there are styles. Seed reniform-globose, or laterally compressed.

* *Cæspitose perennials with scarious-bracted inflorescence; valves of the capsule 3, cleft or parted ; seeds not carunculate.*—ARENARIA proper.

1. **A. congesta**, Nutt. in T. & G. Fl. i. 178 (1838). *Brewerina suffrutescens*, Gray, Proc. Am. Acad. viii. 620 (1873). Glabrous, 6–10 in. high, from a cespitose and suffrutescent base : leaves linear-subulate, scabrous on the margin, rigid and somewhat pungent, those of the low sterile shoots 1–2½ in. long, of the stem ½–1 in. long, spreading or suberect : fl. subsessile in several capitate-congested terminal fascicles, or pedicelled and subumbellate : sepals ovate-oblong, scarious-margined, obscurely 3-nerved, 1½–2½ lines long, acute : petals narrowly oblong, 3–4 lines : stigmas capitellate : capsule coriaceous, about equalling the calyx : seeds small, angular.—In the higher Sierra from near Donner Lake northward ; the flowers less crowded than in Nuttall's type from the northern Rocky Mt. region, the herbage less glaucous, etc. The capitellate character of the stigmas is exceptional in this family ; and on this Dr. Gray at one time held the plant to be of a distinct generic type ; a view afterwards relinquished.

2. **A. capillaris**, Poir. Encycl. vi. 380 (1804) : *A. nardifolia*, Ledeb.

in Hook. Fl. i. 98. t. 32 & Fl. Alt. ii. 166 (1830). Habit and foliage of the preceding, but leaves shorter : stems 6–10 in. high, very viscid above, especially the pedicels of the umbellate or more open and dichotomous cyme ; bracts small, lanceolate : calyx with a broad almost truncate base ; sepals ovate, acute, 3-nerved, membranously margined, $1\frac{1}{2}$–2 lines long ; petals longer : capsule somewhat exceeding the calyx.—High Sierra, but less common than the preceding ; Donner Pass, *Torrey*; Webber Lake, *C. T. Blake*.

3. **A. pungens**, Nutt. in T. & G. Fl. i. 179 (1838). Rather compactly cespitose but not woody ; the short leafy and flowering branches erect ; herbage glandular-pubescent : leaves linear-subulate, pungent, 3–5 lines long, crowded : fl. few in an open leafy-bracted cyme : sepals lanceolate acuminate, pungent, $1\frac{1}{2}$–3 lines long, obscurely 3-nerved : petals scarcely equalling the calyx : capsule shorter, few-seeded : seeds smooth. In the higher Sierra, from Tuolumne Co. northward.

* * *Low annuals ; cymes foliaceous-bracted ; valves of capsule 3, entire ; seeds not carunculate.*—Old genus ALSINE.

4. **A. Douglasii**, T. & G. Fl. i. 674 (1840) ; Fenzl. (1833), under *Alsine*. Sparsely pubescent with spreading gland-tipped hairs, or glabrous, slender, branching, 3–12 in. high : leaves filiform, $1\frac{1}{4}$–$1\frac{1}{2}$ in. long, ascending or spreading, slightly connate at base : fl. large, on long slender pedicels : sepals oblong-ovate, acute, $1\frac{1}{2}$ lines long, 1–3-nerved : petals obovate, 2 lines long or more : capsule globose, equalling the calyx ; seeds large, smooth, compressed and acutely angled. On stony hill-tops and sandy or gravelly plains throughout the State. Mar., Apr.

5. **A. Californica**, Brewer, in Boland. Catal. 6 (1870) ; *A. brevifolia* var. (?) *Californica*, Gray, Proc. Calif. Acad. iii. 101 (1864). Glabrous, very slender, 2–3 in. high : leaves lanceolate, obtusish, 1–2 lines long : fl. small, on slender pedicels : sepals oblong-ovate, acute, 3-nerved, 1–$1\frac{1}{2}$ lines long ; petals spatulate, 2 lines : capsule oblong, as long as the calyx : seeds small, sharply muriculate. Less common than the preceding, and confined to middle parts of the State. Apr., May.

6. **A. palustris**, Wats. Bot. Calif. i. 70 (1876) : Kell. Proc. Calif. Acad. iii. 61 (1863), under *Alsine*. Glabrous, flaccid, decumbent, leafy throughout, $\frac{1}{2}$–2 ft. high : leaves linear-lanceolate, acute, $\frac{1}{2}$–1 in. long : fl. few, large, long-pedicelled : sepals elliptic, obtuse or acutish, nerveless, herbaceous, but with a narrow scarious margin, $1\frac{1}{2}$–2 lines long : petals oblong, twice longer : capsule oblong, shorter than the calyx : seeds numerous. Dr. Kellogg, writing of this from San Francisco almost thirty years ago says: "A plant very abundant in swamps in this vicinity, known to us for the last ten years." It does not appear to have been seen by any one in recent years. Dr. Behr, who knew the original

locality, says it is extinct there. Mr. Parish has discovered it, or a plant very like it, in a marsh near San Bernardino.

* * * *Flowers sometimes 4-merous; valves of capsule bifid; seeds with a small caruncle at the hilum.* Genus MŒHRINGIA, Linn.

7. **A. macrophylla**, Hook. Fl. i. 102. t. 37 (1830). Stems low, ascending, from running rootstocks, mostly simple, leafy, puberulent above: leaves in 3 or 4 pairs, lanceolate, acute at each end, 1—2 in. long, thin and flaccid: fl. few, on slender pedicels: sepals ovate-oblong, acuminate, 1½—2½ lines long, 1-nerved, longer than the obtuse petals: capsule ovoid, nearly equalling the calyx: seeds few, large, smooth.—In shady places, from Marin and Sierra counties northward; rather rare. In specimens from Humboldt Co., *Chesnut & Drew*, the leaves are broader than in the type, and scarcely at all acuminate.

8. **ALSINELLA**, *Dillenius* (PEARLWORT). Diminutive herbs with subulate or filiform exstipulate leaves, and minute long-pedicelled often apetalous flowers. Sepals 4 or 5, commonly rotate-spreading in fruit. Petals when present as many, entire or emarginate. Styles 4 or 5. Capsule 1-celled, ∞-seeded, dehiscent to the base into as many entire valves as there are styles; the valves alternate with the sepals.

1. **A. occidentalis**, Greene. Wats. Proc. Am. Acad. x. 345 (1875), under *Sagina*. Annual, glabrous or nearly so, almost capillary, decumbent at base or ascending, 1—6 in. high: leaves in pairs (none fascicled), slightly connate, acute, ¼—½ in. long: fl. 5-merous, on long pedicels, these erect in fruit: sepals 1 line long · petals nearly as long: stamens 10: capsule exceeding the calyx. Very common, in almost every variety of soil, throughout our whole district from San Francisco northward, but often a minute and obscure plant. Mar.—May.

2. **A. saginoides**, Greene. Linn. Sp. Pl. i. 441 (1753), under *Spergula*. *Sagina Linnæi*, Presl. Rel. Hænk. ii. 14 (1835). Biennial or perennial, cespitose, glabrous and somewhat succulent, 1—2 in. high: leaves ¼—½ in. long, somewhat fascicled: pedicels elongated, nodding after flowering: sepals a line long, obtuse, exceeding the petals: stamens 10: capsule twice the length of the calyx.—A plant of the far north; found, however, at Webber Lake, *Lemmon*.

3. **A. crassicaulis**, Greene. Wats. Proc. Am. Acad. xviii. 191 (1883), under *Sagina*. Perennial, stoutish and succulent, decumbent: leaves broadly linear, acute, 2—6 lines long, scarious and connate at base: pedicels 4—8 lines long; fl. erect or nodding, large, the sepals more than a line long; petals smaller: styles very short: capsule ovate, scarcely exserted from the closed fruiting calyx. —A little known apparently maritime species found at Dillon's Beach, Marin Co., 1880, *J. W. Congdon*.

4. **A. ciliata.** Annual, very slender and diffuse, 1–3 in. high, roughish with short gland-tipped hairs: leaves subulate-linear, 1/4 in. long, their scariously dilated and connate bases sparsely ciliolate: fl. very numerous, short-pedicelled, erect, 4-merous: sepals ovate-oblong, very obtuse, surpassed by the capsule: seeds dull, muriculate. Vicinity of Ione; a plant of compact habit, very different from the other species.

9. **SPERGULA,** *Dodonæus* (CORN-SPURREY). Herbs with linear and apparently whorled leaves; the opposite pair (subtended by a pair of scarious stipular scales) being augmented by several crowded and spreading fascicled ones of nearly their own size which along with them seem to form a verticil. Flowers 5-merous, perfectly symmetrical (stamens 10 or 5); the 5 styles alternate with the sepals, the 5 valves of the capsule opposite the sepals. Petals entire. Seeds laterally compressed, acutely margined or winged. Embryo spiral.

1. S. ARVENSIS, Linn. Sp. Pl. i. 440 (1753). Glabrous or pilose-pubescent and slightly clammy, 1–2 ft. high, simple or with many decumbent basal branches: leaves almost filiform, 1–2 in. long: cyme terminal, ample, dichotomous, the long pedicels nodding after flowering, but erect in flower and again when the capsule is mature: sepals oblong or ovate, 2–3 lines long, the white petals rather long, unfolding only in sunshine: capsule ovoid: seeds acutely margined. Naturalized in fields and by waysides everywhere in the Bay region. Jan.–Sept.

10. **TISSA,** *Adanson* (SAND SPURREY). More or less succulent herbs of maritime districts or subsaline plains inland. Leaves linear or subulate, with scarious stipules. Flowers arranged dichotomously or unilaterally. Sepals 5. Petals 5, entire (sometimes fewer than 5 or even 0). Stamens 2–10. Styles 3, rarely 5. Capsule 3-valved. Seeds winged or wingless. Embryo annular.

* *Perennials with fusiform fleshy roots.*
  ← *Internodes not short (about 1 in.); fascicled leaves few.*

1. **T. macrotheca,** Britt. Bull. Torr. Club, xvi. 129 (1889) partly; Hornem. in Ch. & Schl. Linnæa, i. 53 (1826), under *Arenaria;* F. & M. in Kindb. Monogr. Lep. 16 (1863), under *Lepigonum.* Stems ascending or depressed, stoutish, terete, often 1 ft. high; whole herbage deep green and rather densely viscid-pubescent: leaves semiterete, linear-subulate, acute, often longer than the internodes (1–2 in.); stipules ovate-triangular, 2 lines long: pedicels 1/3 in. long or more, subtended by leafy bracts often nearly as long: sepals 1/4 in. long, with narrow scarious margins: petals as long, lilac: capsule ovoid, about equalling the calyx: seeds triquetrous-obovate, smooth, dark brown, with a very narrow or sometimes obsolete scarious wing.—Maritime only, and common from the Bay region southward, in sandy soil along the borders of salt marshes.

2. **T. leucantha,** Greene, Pittonia, i. 301 (1889). Habit of the preceding, but glabrous except a glandular pubescence on the more ample and loosely dichotomous inflorescence; branches more or less distinctly quadrangular : leaves linear, acute, little exceeding the internodes (1 in. or more); stipules deltoid-ovate, acuminate, 2–3 lines long : pedicels 1 in. long or more, at length abruptly deflexed, subtended by reduced and linear-subulate bracts : sepals 2–3 lines long, with broad scarious margins : corolla ½ in. broad or more : filaments broadly subulate and almost petaloid : apex of capsule exserted, distinctly triquetrous : seed brown, smooth, of round-obovate outline and with a broad scarious wing. Confined to clayey subsaline or alkaline plains of the interior; plentiful at several points on the lower San Joaquin, also on the eastern side of the Livermore Valley, and southward perhaps throughout the State; for Mr. Parish collects it near San Bernardino. It is the only showy species. Mar.–May.

+ + *Internodes short; axillary leaf-fascicles conspicuous.*

3. **T. pallida,** Greene in Britt. l. c.; also *T. macrotheca,* var. *scariosa*, Britt. l. c. Prostrate, diffusely branching and densely cespitose, the geniculate stems stoutish below, often naked and appearing suffrutescent; herbage pale, densely pubescent and very viscid : primary leaves oblong-linear, very acute, ½ in. long or more ; those of the fascicles shorter and relatively broader ; stipules ovate-acuminate, often 4–5 lines long : fl. either scattered singly on short branchlets or in reduced terminal cymes: pedicels ½ in. long : calyx ⅙ in. long : petals lilac : capsule as long : seeds obliquely orbicular, light brown, very smooth, broadly margined.—On high and dry clayey bluffs overhanging the ocean in San Francisco Co., also across the channel in Marin. In floral character quite like *T. macrotheca* with which it has been long confounded, though in vegetative characters very distinct. It appears to have been collected only by Dr. Torrey (n. 41), Mr. Meehan (at Monterey ?) and the present writer. May–July.

4. **T. Clevelandi.** Prostrate, slender, very diffuse, forming deep green mats ½–1½ ft. broad; herbage pubescent but only slightly viscid : leaves narrowly linear, the fascicled ones subulate, all equalling or exceeding the internodes : fl. in terminal cymes only, small (⅓ in. broad), pure white.—The plant to which I here give a provisional name as a probable new species was formerly abundant on rather sandy uplands about San Diego, and occurs sparingly on gravelly knolls at the Presidio, San Francisco. It is a part of the *T. villosa*, Britt., and may possibly be proven identical with the South American plant. But the name is so inapplicable to ours, that I believe the two will be found distinct. According to Dr. Britton, it has been found also at San Jose. Its close-matted habit, profusion of fascicled leaves, and clear white corollas, as

* * *Annuals; flowers usually lilac or lavender-color.*

5. **T. rubra**, Britt. Bull. Torr. Club, xvi. 126 (1889); Linn. Sp. Pl. i. 423 (1753), under *Arenaria;* Fries, Fl. Hall. 76 (1817), under *Lepigonum*. Stems slender, terete, prostrate, a few inches long, glabrous below, pubescent and more or less glandular above: leaves narrowly linear or subulate, acute or mucronate, $\frac{1}{4}$–$\frac{1}{2}$ in. long; stipules lanceolate, acuminate, 1–2 lines long: pedicels slender, 2–3 lines long: sepals oblong, obtuse, scarious-margined: petals reddish, about equalling the sepals: capsule ovate, obtuse, not exserted: seeds brownish, tuberculate, wingless, triquetrous-obovate, with a marginal elevation. Roadsides above Petaluma, collected by the author in September, 1888. Doubtless introduced from Europe, where it is common in many districts, chiefly at some distance from the sea. It was also found in the interior of Oregon, in the summer of 1890.

6. ? **T. diandra**, Britt. l. c. 128; Guss. Prodr. Fl. Sicul. i. 515 (1827), under *Arenaria*. *A. salsuginea*, Burge. in Ledeb. Fl. Alt. ii. 163 (1829). Near *T. rubra*, but more slender; stipules very short; cyme leafless and widely divaricate, more profusely flowering; capsules minute, subglobose; seeds very small, black. A plant doubtfully referred to this Old World species by Dr. Britton, was found in Sierra Valley by Mr. Lemmon.

7. **T. marina**, Britt. l. c. 126. *Arenaria rubra*, var. *marina*, Linn. Sp. Pl. i. 423 (1753). *Spergula marina* of pre-Linnæan authors. *Lepigonum marinum*, Wahlb. Root thickish, not much branched, perhaps sometimes perennial: stems ascending, 3–8 in. high, somewhat compressed or angular, glabrous or somewhat glandular-pubescent: leaves semiterete, narrowly linear, acute, light green, glabrate, seldom exceeding the internodes; stipules broadly ovate, abruptly acuminate: cymes scarcely leafy: pedicels about twice as long as the capsules: sepals acute or acuminate, with a broad or narrow scarious margin: petals broadly ovate, obtuse, scarcely equalling the sepals, whitish or pale rose-color: capsule ovate, obtuse, nearly twice the length of the calyx: seeds orbicular, with an elevated margin, reddish-brown, smooth, winged or wingless. Common and variable, occurring mostly near the sea; perhaps also on subsaline plains of the interior.

8. **T. salina**, Britt. l. c. 127; Presl. Fl. Cech. 93 (1819), under *Spergularia*. Roots slender and tufted, simple or much branched: stems 6 in. high, much branched, usually ascending, rarely divaricate and prostrate: herbage glabrous or pubescent: leaves flat, linear-filiform, obtuse or acutish, glabrous, light or livid green, seldom longer than the internodes; stipules broadly ovate, short-acuminate, not shining: pedicels

leafy-bracted, or the upper bractless, none of them longer than the capsules; sepals oblong or oblong-ovate, obtuse, scarious-margined; capsule acute, much longer than the calyx ; seeds round-obovate, tuberculate or muriculate, the marginal elevation distinct ; hyaline wing narrow or wanting. Var. **sordida.** Stems ascending ; herbage very viscid and hairy ; fl. in unilateral leafless racemes ; seeds nearly black, sharply muriculate, wingless. Var. **Sanfordi.** Stems erect repeatedly dichotomous ; herbage scarcely viscid and only slightly pubescent ; inflorescence partly dichotomous, only the ultimate branchlets unilaterally racemose ; seeds dark brown, nearly smooth, wingless. Nothing quite like the type of this, as defined by Kindberg, is known to me as Californian ; but a plant very near it is common on the seaboard. The first variety is very abundant in low rich soil above the salt marshes on the "Island," near Alameda. The second belongs to the plains about Stockton, Lathrop and elsewhere in the interior. Mar. May.

10. **T. tenuis,** Greene, in Britt. l. c.; Pittonia, i. 63 (1887), under *Lepigonum.* Slender, prostrate, very diffuse, the whole plant 1 ft. broad, glabrous, or the inflorescence sparsely glandular-pubescent ; leaves linear-filiform, 1 in. long, equalling the internodes ; stipules broader than long, acute, but small and inconspicuous ; fl. very numerous, crowded and often subsessile on the countless dichotomous-cymose branchlets, apetalous ; stamens 2 ; capsule triquetrous, acute, more than twice the length of the oblong obtuse scarious-margined sepals ; seeds reddish-brown, obliquely obovate, compressed, smooth, margined, wingless.- On hard clayey soil not far from Mastic Station, Alameda ; collected by the author in May, 1887 ; and the plant has not reappeared.

11. **POLYCARPON,** *Loefling.* Low annuals, diffusely dichotomous, with flat leaves, small scarious stipules and minute cymose flowers. Sepals 5, carinate-concave. Petals 5, minute, hyaline. Stamens 3 5. Ovary 1-celled ; style short, 3-cleft. Capsule 3-valved, several-seeded.

1. **P. depressum,** Nutt. in T. & G. Fl. i. 174 (1838). Very slender, prostrate, the many branches 1 2 in. long ; leaves opposite, spatulate, glabrous ; stipules small, narrow ; fl. minute ; the pedicels with small bracts ; petals almost filiform, shorter than the sepals, entire ; capsule globose, 6 12-seeded. Said to have been found as far north as Santa Cruz Co.; otherwise known only from about San Diego and on the peninsula of Lower California.

12. **LOEFLINGIA,** *Linnaeus.* Low much branched rather rigid and pungent-leaved annuals ; the leaves with adnate and connate setaceous stipules. Flowers small, sessile in the axils of the leaves and branches. Sepals 5, rigid, carinate. Petals minute or 0. Stamens 3 5. Ovary 1-celled ; style very short or 0. Capsule 3-valved, several-seeded. Genus

closely connecting this order with Polygoneæ, through *Lastarriæa*, and with Amarantaceæ through *Amarantus*.

1. **L. squarrosa,** Nutt. in T. & G. Fl. i. 174 (1838); Gray, Gen. Ill. ii. 24. t. 106; Brandg. in Zoe, i. 219. The numerous prostrate or erect-spreading branches 2–6 in. long; herbage glandular-pubescent: leaves and sepals subulate-setaceous, rigid and recurved, the leaves 2–3 lines long, the sepals somewhat shorter: capsule elongated, triquetrous, at length exserted, ∞-seeded. Plains of the lower San Joaquin, near Lathrop, *Greene;* also in the Sacramento valley, according to Brandegee; more plentiful farther south.

2. **L. pusilla,** Curran, Bull. Calif. Acad. i. 152 (1885); Brandg. l. c. 220. Habit of the preceding, but smaller; stems 2–3 in. long: sepals narrowly lanceolate, abruptly acute, neither pungent nor recurved: stamens 5: style 0: capsule triquetrous, not exserted, ∞-seeded. Mountains near Tehachapi, Kern Co., *Mrs. Curran.*

Order XXI. ILLECEBREÆ.

Robert Brown, Prodromus Floræ Novæ Hollandiæ, 413 (1810).

Diffuse or tufted herbs or suffrutescent plants with mostly opposite entire stipulate leaves, and clustered or open-cymose apetalous flowers. Sepals 5, 4 or 3, seldom distinct. Stamens as many as the calyx-lobes and opposite them, or fewer, perigynous (except in n. 1). Ovary superior, 1–3-celled, 1–several-ovuled; styles 2 or 3, often united below. Fruit 1-seeded and utricular, or a several-seeded capsule.

1. **MOLLUGO,** *Linnæus* (CARPET-WEED). Low much branched annuals with exstipulate whorled leaves, and axillary flowers on slender pedicels. Sepals 5, white within, thus resembling petals when expanded. Stamens 5, hypogynous, alternate with the sepals, or 3 and alternate with the cells of the ovary. Stigmas 3. Capsule 3-celled, 3-valved, loculicidal, the partitions breaking away from the many-seeded axis.

1. M. VERTICILLATA, Linn. Sp. Pl. i. 89 (1753). Prostrate, much branched, forming a mat 1–2 ft. broad; branches rather slender, the glabrous linear-spatulate leaves 3–8 in an imperfect whorl, unequal, the largest less than 1 in. long, scarcely equalling the internodes: pedicels about as many as the leaves, 1–2 lines long: capsule oblong, obtuse, little exserted from the calyx: seeds minute, round-reniform, black and shining, somewhat striate. By waysides in the upper part of the Sacramento valley; not common in California, and doubtless introduced; native of Africa and perhaps of tropical America.

2. **PARONYCHIA,** *Clusius.* Herbs with opposite entire leaves and a pair of scarious stipules at each node; flowers (in ours) clustered in the

axils. Sepals 5, imbricate, somewhat cucullate under the apex and aristate or mucronate at the very tip. Stamens 5 or fewer, inserted on the base of the sepals, these often slightly united. Petals represented by 5 small setiform organs alternating with the stamens. Ovary 1-celled, 1-ovuled; ovule attached by a slender basal funiculus, ascending or subpendulous. Utricle enclosed in the persistent calyx, at length bursting longitudinally. Seed smooth. Embryo annular.

1. **P. Chilensis,** DC. Prodr. iii. 370 (1828); Gay, Fl. Chil. ii. 521. Greene, in W. Am. Sc. iii. 156. Perennial, diffuse, cespitose, the tough and pliable short-jointed stems suffrutescent; leaves oblong-linear, $1\frac{1}{2}$ 3 lines long, membranaceous, pungent at tip, minutely appressed pubescent; stipules thin-hyaline, ovate-lanceolate, 1 2 lines long; fl. few in the axils, very shortly pedicelled; calyx scarcely $\frac{3}{4}$ line long, purplish; sepals spinulose-tipped and only slightly cucullate; seed reddish-brown. Frequent on grassy hillsides and summits at the Presidio; evidently indigenous; otherwise known only as South American.

2. **P. pusilla,** Greene, Pittonia, i. 302 (1889). Annual, slender, parted from the base into a few ascending branches, these with many short distichous branchlets; herbage canescent with setulose straight or uncinate-tipped hairs; leaves oblong-lanceolate, acute, sessile, 1 2 lines long; stipules hyaline, minute, broadly ovate; fl. crowded, sessile, minute, the sepals $\frac{1}{2}$ line long, scarcely cucullate, the terminal bristle little longer than those scattered up and down the back; seed black, smooth and lustrous. On an isolated outcropping of rock, in the mouth of a cañon opening to the plains, at the eastern base of Mt. Diablo, near Bethany; collected only by the author, 30 Apr., 1889.

3. **PENTACÆNA,** *Bartling.* Perennials of cespitose habit, with alternate subulate rigid and pungent leaves, silvery hyaline stipules, and sessile flowers clustered in the axils. Sepals 5, united at base, very unequal, cucullate, the 3 outer large and with a stout divergent terminal spine, the 2 inner much smaller and with but a short awn. Petals minute, scale-like. Stamens 3 - 5; staminodia 0. Style very short, bifid. Utricle enclosed in the rigid persistent calyx. Embryo curved.

1. **P. ramosissima,** H. & A. in Bot. Misc. iii. 338 (1833); *P. polycnemoides,* Bartl. in Presl. Rel. Haenk. ii. 5, t. 49 (1835). *Lœflingia ramosissima,* Weinm. (1820); DC. (1829), under *Paronychia.* Stems prostrate, forming mats 6 in. to 2 ft. broad, woolly-pubescent; leaves 3 5 lines long, squarrose when old; stipules lanceolate, shorter than the leaves, 1-nerved; calyx-tube nearly a line long, the divergent outer lobes 2 lines; utricle apiculate.— On sandy plains and dry gravelly hilltops toward the sea throughout the State; also South American.

Order XXII. POLYGONEÆ.

Jussieu, Genera, 82 (1789). PERSICARIÆ, Adans. Fam. ii. 273 (1763).

Herbs, or rarely shrubs, with alternate or whorled leaves of revolute vernation; stipules when present cohering around the stem and forming a sheath. Inflorescence various, but commonly racemose and terminal. Calyx of 4–9 nearly or quite distinct sepals, often colored and petaloid, persistent. Stamens as many as the sepals, or fewer, perigynous. Styles 2–4, distinct or somewhat connate, opposite the angles of the lenticular or triquetrous 1-ovuled ovary. Fruit a compressed or triquetrous achene. Seed erect; embryo straight, in the midst of a farinaceous albumen, or curved around it.—An extensive family, containing many homely and weedy plants; but the tender leaves of the sorrels, and petioles of rhubarb, abounding in oxalic acid, are valued substitutes for fruit in northern latitudes. The roots of the docks abound in tannin. Buckwheat is a well-known cereal. There are the closest connections between this order and the Caryophylleæ on the one hand, and the Amarantaceæ on the other.

Hints of the Genera.

Leaves alternate, stipulate;
    Sepals 4–6, equal, appressed to the achene, - - - - - 1
    " " the outer smaller and spreading, - - - - 2
    Perianth tubular below, 6-lobed above, - - - - - - 3
    " campanulate, 6-cleft, - - - - - - - - 4
Leaves alternate or verticillate, exstipulate;
    . Involucre wanting;
        Flowers capitate, each with an herbaceous bract, - - 5
        Perianth tubular, cuspidately 6-toothed, - - - 9
    Involucre tubular or campanulate,
        with 4–8 obtuse or merely acute teeth, - - - - - 6
        with 3–5 usually awned lobes, - - - - - - 7
        with 3–6 cuspidate often hooked teeth, - - - - 8
    Involucre 2-lobed, 1-flowered, enlarged in fruit, - - - - 10

1. POLYGONUM, *Columna.* Herbs or undershrubs with alternate entire leaves and sheathing stipules. Flowers small, in axillary fascicles or terminal spikes or racemes. Perianth of 5 or 6 nearly distinct often colored and petaloid sepals. Petals 0. Stamens 4–9, commonly in 2 sets or circles. Styles 2 or 3, distinct, or connate below, often very short; stigmas capitate. Fruit a triangular or lenticular achene, usually closely invested by the persistent perianth. Embryo lateral, half immersed in the hard albumen, curved: cotyledons narrow.—A vast genus as now received by most botanists, but probably embracing several quite natural genera.

* *Leaves jointed upon a short petiole adnate to the 2-lobed or lacerate sheath; flowers axillary to leaves or bracts; filaments of the 3 inner stamens broad at base; achenes triquetrous; cotyledons incumbent.—*
                POLYGONUM proper.

    ⊢ *Glabrous and suffrutescent; sheaths conspicuous; sepals colored.*

1. **P. Paronychia,** Ch. & Schl. Linnæa, iii. 51 (1828); Meisn. DC. Prodr. xiv. 89. Stems stoutish, tough and pliable, ascending or prostrate, 1–2 ft. long, leafy above, below clothed with the scarious sheaths, these $\frac{1}{2}$ in. long, brownish and 5-nerved below, lacerate above : leaves subcoriaceous, 1 in. long, linear-lanceolate, revolute : fl. densely crowded at the ends of the branches, the spikes more or less leafy-bracted : perianth white or rose-color veined with green or brown, $\frac{1}{4}$ in. long : sepals oblong-obovate : stamens 8 : styles as long as the ovary : achene 2 lines long, smooth and shining. In sandy soil near the sea, from Santa Cruz northward, flowering almost all the year through.

2. **P. Shastense,** Brew. in Gray, Proc. Am. Acad. viii. 400 (1872). Stems stout, rigid, ascending, 6–8 in. long, sparingly leafy : sheaths with herbaceous base, and a 2-lobed scarious usually deciduous lamina: leaves oblanceolate, acute, not revolute, 4–6 lines long : fl. 1–3 in each of the lower axils, white or rose-color with darker veins, $1\frac{1}{2}$–$2\frac{1}{2}$ lines long, attenuate to a naked pedicel : sepals round-obovate : stamens 8 : styles much shorter than the ovary, persistent : achene $2\frac{1}{2}$ lines long, smooth and shining. Common on alpine slopes of the Sierra Nevada.

3. **P. Bolanderi,** Brew. l. c. Stems slender, woody and very brittle, tufted and strictly erect, $\frac{1}{2}$–2 ft. high : sheaths much shorter than the nodes, herbaceous below, scarious and lacerate above, persistent : leaves narrowly linear or subulate, acute or cuspidate, $\frac{1}{4}$–$\frac{1}{2}$ in. long, not revolute : fl. solitary or few in the axils of short leafy branchlets, each involucrate with a sheath-like scarious bract on the joint of the short pedicel : sepals oblong-ovate, $1\frac{1}{2}$ lines long, rose-color or white, slightly spreading : stamens 8 or 9 : styles half as long as the ovary.- Plentiful near the Soda Springs above Napa ; also on the eastern base of the same range of mountains bordering the valley of the Sacramento, but evidently somewhat local. Aug.–Oct.

+ + *Annuals, with striate stems and less conspicuous sheaths; sepals mostly green with whitish margins.*

++ *Branches leafy to the summit, floriferous throughout.*

4. P. AVICULARE, Linn. Sp. Pl. i. 362 (1753). Stoutish, much branched, prostrate, the branches 1–3 ft. long ; herbage glabrous, bluish-green : leaves oblong or lanceolate, acutish, $\frac{1}{2}$–$2\frac{1}{2}$ in. long : fl. on very short pedicels : sepals 1 line long, green, with white or rose-colored margin : achene broadly ovate, 1 line long or less, dull black and minutely granular.- A very prevalent weed in summer fields and vineyards ; native of Europe. Apr.–Oct.

5. **P. minimum,** Wats. Bot. King Exp. 315 (1871). Low and slender, simple or with a few branches, 1–6 in. high : stems nearly terete, reddish, more or less scabrous-puberulent : leaves broadly ovate or ovate-oblong, $\frac{1}{2}$ in. long, acute : fl. less than a line long, erect on slender

short pedicels ; margin of sepals often rose-tinted ; stamens 5–8 ; achene smooth and shining, exceeding the sepals. In the higher Sierra, from near Yosemite northward. July–Oct.

6. **P. Kelloggii.** Erect, slender, simple or with a few widely divergent branches from the base, 1–3 in. high, the internodes very short, scarcely equalling the lobes of the sheaths, the whole plant glabrous ; leaves linear, acute, $\frac{1}{4}$–$\frac{1}{2}$ in. long, almost divaricately spreading, never imbricated, not much smaller at the summits of the branches than below, all the axils floriferous ; fl. several at each node, subsessile, $\frac{3}{4}$ line long, greenish ; achene $\frac{1}{2}$ line long, light brown, smooth or obscurely striate, the face rhombic-ovate.—Common in the Donner Lake region of the Sierra ; near *P. imbricatum*, but with spreading and equal leaves, no distinct inflorescence, the achenes light chestnut-brown. It is in the State Survey collection under n. 6005. Aug.–Oct.

++ ++ *Upper nodes approximated and more floriferous, their leaves reduced and bract-like.*

7. **P. ramosissimum**, Michx. Fl. i. 237 (1803). Erect or ascending, 2–4 ft. high, branching from the middle ; herbage glabrous and of a yellowish green ; leaves lanceolate or linear, 1–$2\frac{1}{2}$ in. long, acute, attenuate to a slender base ; sheaths scarious and loose, becoming lacerate to the base ; fl. $1\frac{1}{2}$ lines long, yellowish, drooping on the slender pedicel ; achene smooth and shining. Said to occur in the lower Sierra ; not known near the seaboard.

8. **P. Douglasii**, Greene, Bull. Calif. Acad. i. 125 (1885) ; *P. tenue*, Wats. not Michx. Erect, slender, sparingly branched above the base, 1–$1\frac{1}{2}$ ft. high, glabrous, or somewhat scabrous at the nodes ; leaves thinnish, oblong to lanceolate, 1-nerved, with or without a few veinlets diverging from the midvein, the margin smooth, often somewhat revolute ; stipules hyaline throughout, the sheathing portion short or wanting, the limb more or less lacerate ; fl. 1 or more in each upper axil, on short deflexed pedicels ; sepals $1\frac{1}{2}$–2 lines long, tipped with white or rose-color ; achene black and shining. Var. **latifolium**, Greene, l. c. Lower, often flexuous, the leaves broader and shorter ; fl. commonly more spicate-congested at the ends of the few branches ; achenes broader. Common in the mountains at middle elevations, almost throughout the State ; long confounded with the eastern *P. tenue*. July–Oct.

9. **P. coarctatum**, Dougl. in Hook. Fl. ii. 133 (1840). Much like the last, but more freely branching, often diffuse, the herbage more or less scabro-puberulent throughout ; leaves firmer in texture, acute ; fl. more spicate-crowded and on erect pedicels ; sepals larger, rose-color or white with only a broad midvein of green ; achenes very minutely punctate toward the apex.—Of more northerly distribution than the last, but plentiful as far south as the Petrified Forest in Sonoma Co. July–Sept.

10. **P. imbricatum,** Nutt. in Wats. Am. Nat. vii. 665 (1873). Stems 1–8 in. high, slender, angular, the branches few or many, all from above the base, ascending; herbage glabrous or at the nodes a little scabrous: leaves $\frac{1}{2}$–1 in. long, linear, acute, 1-nerved; sheaths rather large, bifid or lacerate above the short scarious base : fl. rather densely spicate, the bracts loosely imbricate, 2–4 lines long, often with a narrow scarious margin : fl. subsessile, 1 line long or less, whitish or rose-tinted : stamens 3 or 5 : styles as long as the ovary : achene $\frac{3}{4}$ line long, black, minutely tuberculate-striate. A subalpine species common from near Donner Lake northward in the Sierra. July–Oct.

* * *Leaves not jointed with the petiole, striately 3-nerved (except in n. 14); sheaths 2-lobed or fimbriate; stamens 8, the inner 3 scarcely dilated.* **Subgenus DURAVIA,** Wats.

+ *Stems hard and rigid; flowers spicate, solitary in the axils of bracts; styles persistent.*

11. **P. Californicum,** Meisn. in DC. Prodr. xiv. 100 (1856). Erect, slender, 3–6 in. high, panicled-spicate, the stem and branches glabrous, dark brown : leaves rigid, linear or filiform, $\frac{1}{2}$–1$\frac{1}{4}$ in. long, pungently acute: spikes very slender, elongated, the subulate bracts 1–2 lines long; sheaths 1 line long, deeply lacerate-fringed, nearly equalling the pale rose-colored flowers: achene narrow, slightly exposed; styles slightly divergent. Valleys and dry hills of the interior from near Napa and Sacramento northward. July–Sept.

12. **P. Greenei,** Wats. Proc. Am. Acad. xiv. 294 (1879); Bot. Calif. ii. 480. Resembling the last, but with denser stouter spikes; bracts and finely fimbriate sheaths 2 lines long : achene oblong-ovate, with very short and stout almost erect styles. Upper part of the Sacramento valley, *Mrs. Bidwell*, and northward. July–Sept.

13. **P. Bidwelliæ,** Wats. ll. cc. Smaller than the preceding, the branches fewer, more divergent: spikes short and dense : stipules very conspicuous, large, scarious and white-chaffy, 2 lines long or more, equalling or surpassing the bracts, entire or only slightly lacerate at the 2-lobed summit : fl. a line long. pale rose-color : achene oblong-ovate, less than a line long including the very divergent styles. Known only from near Chico, *Mrs. Bidwell*, *Dr. Parry*, but a very well marked species.

+ + *Stems leafy and floriferous throughout; flowers 2 or 3 in each axil; styles deciduous.*

14. **P. Parryi,** Greene, Bull. Torr. Club, viii. 99 (1881). Usually with several upright branches 1–3 in. high, short-jointed, very leafy, rather sharply angled, glabrous : leaves $\frac{1}{2}$–1 in. long, linear, acute, 1-nerved, the upper scarcely diminished in size, all somewhat spreading or merely ascending; sheaths 2 lines long, broad and nearly distinct from the

short petiole, cleft to the middle, or more deeply, into a lacerate and somewhat curled fringe : fl. several, sessile, less than a line long : achene chestnut-brown, smooth and shining. Subalpine in the vicinity of the Yosemite, *Parry;* but it is represented in numbers 6355 and 6451 of the State Survey, and may have been referred to *P. imbricatum*. It is a link connecting *Duravia* with true *Polygonum*.

* * * *Leaves not jointed, more ample, pinnately veined; sheaths cylindrical, oblique or truncate; fl. in dense spikes or loose cymelets; stamens 4–8, all the filaments filiform; styles deciduous; cotyledons accumbent.*

← *Stems usually branching, leafy, the spikes often panicled; styles short, often only 2 and the achene lenticular.*—Old genus PERSICARIA.

++ *Weedy annuals of fields and gardens.*

15. **P. nodosum,** Pers. Syn. i. 440 (1805). Stoutish, erect or ascending, 1–4 ft. high, freely branching, glabrous except the rough glandular peduncles, and scabrous leaf margins and veins beneath ; stem often purple-dotted throughout : leaves lanceolate, 2–5 in. long, acuminate, short-petioled ; sheaths naked in age, glandular-ciliolate when young : spikes linear, usually drooping, 1 in. long or more : fl. white or pale rose, 1 line long : stamens 6 : styles 2 : achene lenticular, ovate. Very common in cultivated lands, preferring moist places. July–Oct.

16. **P. Persicaria,** Linn. Sp. Pl. i. 361 (1753). *Persicaria maculosa*, S. F. Gray, Nat. Arr. ii. 269 (1821). Much like the last but the sheaths and bracts conspicuously ciliate : leaves less acuminate, subsessile : spikes shorter and erect : fl. rose-color : achenes often triquetrous.— Apparently uncommon in California, but reported in the Botany of Beechey's Voyage as here, and found in Humboldt Co. more recently, *Rattan, Chesnut & Drew*. Both this and the last are natives of Europe and Asia. It is impossible to say whether with us they are indigenous or introduced weeds. July–Oct.

++ ++ *Perennials, either aquatic or of wet places.*

17. **P. acre,** HBK. Nov. Gen. ii. 179 (1817). Decumbent, rooting at the lower joints, 2–5 ft. high ; herbage light green, pellucid-punctate and acrid, glabrous or a little scabrous : leaves lanceolate, acuminate, short-petioled ; sheaths bristly-ciliate : spikes narrow and lax, 1–3 in. long, erect : sepals greenish and glandular-dotted, 1 line long : stamens 8 : achene commonly triquetrous.—Very common in marshy places, along streamlets, etc. June–Nov.

18. **P. Hartwrightii,** Gray, Proc. Am. Acad. viii. 294 (1870). Stems stout and simple, rooting at the decumbent base, above equably leafy to the summit ; herbage more or less strigose-hirsute : leaves broadly lanceolate, acute, 2–7 in. long, on very short petioles ; stipules with an

abruptly spreading foliaceous border ; fl. rose-red, in a dense ovate or oblong terminal spike ; stamens 5 ; style 2-cleft ; achene lenticular. Not common in California, but occurring between Berkeley and Temescal, in low ground ; also in the upper valley of the Sacramento.

19. **P. Muhlenbergii**, Wats. Bot. Calif. ii. 43 (1880) ; *P. amphibium*, var. (?) *Muhlenbergii*, Meisn. in DC. Prodr. xiv. 116 (1856). Stoutish, erect, 2–3 ft. high, leafy throughout, scabrous with short appressed or glandular hairs, with more or less of a softer pubescence ; leaves broadly lanceolate, narrowly acuminate, 4–7 in. long, on petioles of nearly 1 in.; sheaths with no spreading margin ; spikes 1 or 2, elongated and narrow, 1–3 in. long ; fl. and fr. as in the last. Perhaps as rare as the preceding; but found by the author along the shore of the lakelet in front of the U. S. Marine Hospital, San Francisco, in July, 1888.

20. **P. amphibium**, Linn. Sp. Pl. i. 361 (1753) ; S. F. Gray, Nat. Arr. ii. 268 (1821), under *Persicaria*. Aquatic with floating leaves, or geniculate and rooting in the mud along the shores of ponds and lakes ; herbage glabrous, or nearly so ; leaves elliptical or oblong, obtuse or acutish, very smooth and shining above, 2–5 in. long, on petioles half as long ; spike mostly solitary, dense, ovate or oblong, 1–1½ in. long ; fl. rose-color ; fr. lenticular. Common in mountain lakes, both in the Coast Range and the Sierra ; also in the sloughs about Stockton.

+ + *Stems from stout creeping rootstocks, simple, scape-like, leafy at base chiefly ; spike terminal, solitary ; styles 3, elongated ; achenes triquetrous.* — Old genus BISTORTA.

21. **P. Bistorta**, Linn. l. c. 360. *Bistorta major*, Ray, Gerarde, S. F. Gray, Nat. Arr. ii. 267 (1821). Glabrous and somewhat glaucous ; the ample oblong-lanceolate radical leaves often 6–8 in. long, acute ; cauline reduced and sessile on the long thin sheath, commonly revolute ; stem 1–3 ft. high ; spike-like raceme oblong, 1½–1½ in. long, white or rose-tinted ; bracts ovate, acuminate ; stamens and styles exserted. Frequent in subalpine meadows of both ranges of mountains ; the inflorescence always shorter and broader, and the flowers paler, in the American than in the European plant. July– Oct.

+ + + *Branching and leafy perennials from running rootstocks ; fl. in more or less racemose or panicled cymelets ; styles short ; achenes triquetrous.* — Subgenus ACONOGONON, Meisn.

22. **P. polymorphum**, Ledeb. Fl. Ross. iii. 524 (1849). Stout, erect, 3–7 ft. high, nearly glabrous, the leaf-margins scabrous ; leaves short-peduncled, ovate- to oblong-lanceolate, acuminate, connate or rounded at base, decurrent on the petiole, 3–7 in. long ; panicles loose, many-flowered, scarcely leafy ; sepals greenish white, 1–2 lines long, exceeding the pedicels, shorter than the achene and not very closely appressed. Subalpine in the Sierra, from Yosemite northward. July– Oct.

23. **P. Davisiæ,** Brew. in Gray Proc. Am. Acad. viii. 399 (1872). Somewhat decumbent, 1 ft. high, stout, leafy throughout, flexuous and branching, pubescent with short spreading hairs, or nearly glabrous, the leaf-margins scabrous-ciliate: leaves 1-2 in. long, ovate or oblong, acute or obtuse, cuneate or rounded at base, subsessile: fl. in small axillary and terminal cymose clusters, much shorter than the leaves: sepals yellowish or purplish green, 1½-2 lines long, narrow at base, shorter than the achene.—On barren slopes near the highest summits of the Sierra, from Alpine Co. northward. Aug. Oct.

* * * * *Twining or climbing annuals with broad leaves and flowers in loose axillary panicles or racemes; sepals green with whitish margins, enlarging in fruit; stigmas 3, subsessile; achenes triquetrous.—* Genus BILDERDYKIA, Dumort.

24. P. CONVOLVULUS, Linn. Sp. Pl. i. 364 (1753). *Fagopyrum carinatum,* Mœnch. Meth. 290 (1794); S. F. Gray, Nat. Arr. ii. 272 (1821). *Bilderdykia Convolvulus,* Dumort. Fl. Belg. Prodr. 18 (1827). Twining or trailing, 1-2 ft. high, minutely scabrous: leaves 1-2 in. long, hastate-cordate, acuminate: fl. in axillary interrupted racemes: fruiting perianth 1½-2 lines long, equalling the somewhat opaque granulate-striate achene. A weed in cultivated lands; native of Europe, not yet prevalent in California, but already met with near Berkeley, and in the valley of the Sacramento. July-Sept.

2. **RUMEX,** *Pliny* (DOCK. SORREL). Coarse perennials (rarely annual or biennial), with leafy stems, and cylindrical obliquely truncate scarious stipules; the small green or reddish perfect or unisexual flowers fascicled or verticillate, forming panicled racemes. Perianth of 6 (rarely 4) nearly or quite distinct sepals; the outer herbaceous, spreading or reflexed; inner larger, in some becoming greatly enlarged in fruit, appressed to the 3- (or 2-) angled achene. Stamens 6; filaments very short. Styles 3 (or 2); stigmas tufted. Embryo lateral, slender, slightly curved. The docks *(Lapathum)* and sorrels *(Acetosa)* seem like good natural genera. The retention of our native sorrel *(Oxyria)* as distinct from *Acetosa* on account of its dimerous flowers and winged fruit, appears to us a relic of the empiricism of a former century.

* *Fl. perfect or polygamous; valves accrescent, often with a grain-like protuberance on the back; leaves elongated, never hastate, pinnately many-veined; herbage scarcely acidulous (except in n. 9).—* Old genus LAPATHUM; the DOCKS.

+ *Valves small (2 lines, more or less), one or more of them grain-bearing.*

++ *Valves with slender awned teeth; herbage pubescent or scabrous.*

1. R. OBTUSIFOLIUS, Linn. Sp. Pl. i. 335 (1753); Mœnch, Meth. 356 (1794), under *Lapathum,* also S. F. Gray, Nat. Arr. ii. 274 (1821). Tall

(3–5 ft.), erect, slender, somewhat scabrous: radical leaves oblong, obtuse, cordate or truncate at base, long-petioled, the blade 6–15 in. long: fl. in loose whorls, on long pedicels, these jointed below the middle: valves ovate-deltoid, 2–3 lines long, with 1–3 setaceous teeth on each side, usually only one valve grain-bearing. Naturalized, but rather sparingly, and in low lands only.

2. R. PULCHER, Linn. l. c. 336; S. F. Gray, l. c. 275, under *Lapathum*. Erect, 2–3 ft. high, with rigid branches divaricately and widely spreading: leaves scabrous beneath, the radical oblong or lanceolate (sometimes panduriform, acute, at base cordate or obtuse: fl. on short stout rigid pedicels: valves ovate, 2–3 lines long, with 4–6 rigidly awned teeth on each side. Very abundant in fields and by waysides; everywhere in our districts a troublesome weed.

3. R. maritimus, Linn. l. c. 335. Low (about 1 ft.), erect, stout, from an annual or biennial root: herbage minutely pubescent and of a pale or yellowish green: leaves linear-lanceolate, the margin somewhat crisped or undulate, short-petioled, the blade 1–4 in. long: inflorescence compact, the verticils dense: valves 1 line long, ovate lanceolate, all grain-bearing, and with 2 or 3 long-awned teeth on each side. Common near tide-water in the vicinity of Stockton.

++++ *Valves entire or only denticulate: herbage glabrous.*

4. R. CONGLOMERATUS, Murr. Prodr. Gœtt. 52 (1770). *Lapathum virgatum*, Mœnch. Meth. 355 (1794); *L. conglomeratum*, S. F. Gray, l. c. 273 (1821). Stoutish, 3–4 ft. high, leafy-paniculate above: radical leaves ovate or lanceolate, cordate, slightly undulate: pedicels short, stout and geniculate in fruit, jointed near the base: valves small, all grain-bearing, ovate-lanceolate, acute. Naturalized in many parts of the State, but perhaps nowhere troublesome or even at all common.

5. R. CRISPUS, Linn. Sp. Pl. i. 335 (1753); S. F. Gray, l. c., under *Lapathum*. Size and habit of the last, but panicles less leafy and more condensed: leaves long-petioled, truncate at base, strongly undulate or crisped: pedicels 2–4 lines long, rather slender: valves all grain-bearing, ovate or cordate, strongly reticulate. Very common in waste places.

6. R. Berlandieri, Meisn. in DC. Prodr. xiv. 45 (1856). Stout, erect, 2–4 ft. high: leaves narrowly lanceolate, very undulate, more or less acuminate, narrowed below to an abruptly cuneate or almost truncate base, 6 in. long, short-petioled: pedicels 1–2 lines long, jointed below the middle: valves ovate-lanceolate, $1\frac{1}{2}$ lines long, finely reticulate, all grain-bearing.- Closely related to the Old World *R. crispus*, this is a native American species, and is said to have been found at San Francisco.

7. R. salicifolius, Weinm. in Flora, iv. 28 (1821). Stems clustered,

ascending, 1–3 ft. high : lowest leaves oblong, upper linear-lanceolate, 3–6 in. long, acuminate, narrowed to a short petiole, not undulate, pale green : panicle open, somewhat leafy, the flowers crowded : pedicels slender, 1–3 lines long: valves ovate-rhomboid or broadly deltoid, $1\frac{1}{2}$–2 lines long, entire or denticulate, one or two of them with large whitish grains. Rather common in low grounds, both near the coast and in the interior.

++ *Valves $\frac{1}{4}$–$\frac{1}{2}$ in. long, not grain-bearing; herbage glabrous.*

8. **R. occidentalis,** Wats. Proc. Am. Acad. xii. 253 (1877). Erect, 3–6 ft. high, sparingly branched : leaves oblong-lanceolate, usually narrowing upward from the truncate or somewhat cordate base, not decurrent upon the petiole, 1 ft. long or more, scarcely undulate, usually acute : panicle narrow, elongated, nearly leafless : pedicels slender, $\frac{1}{4}$–$\frac{1}{2}$ in. long, obscurely jointed near the base : valves broadly cordate, with a very shallow sinus, becoming about $\frac{1}{4}$ in. broad, often denticulate near the base : achenes $1\frac{1}{2}$ lines long. Frequent in marshy places from Marin Co. northward.

9. **R. hymenosepalus,** Torr. Bot. Mex. Bound. 177 (1859). Stout and fleshy, 1–3 ft. high, the stems and short petioles reddish and agreeably acidulous : leaves oblong-lanceolate, scarcely rounded at base, somewhat undulate, 1 ft. long : pedicels $\frac{1}{4}$–$\frac{1}{2}$ in. long, jointed near the base : valves very thin, reddish when mature, broadly cordate, 4–6 lines wide : achene 2 lines long. From the southern part of Monterey Co. southward, and eastward to the Rio Grande, in low sandy or gravelly washes and dry beds of streams ; also a weed in cultivated lands. A link between dock and rhubarb, and sometimes used as a substitute for the latter.

10. **R. venosus,** Pursh, Fl. ii. 733 (1814) ; Hook. Fl. ii. 130. t. 174. Erect, 1 ft. high, from running rootstocks, stout and with usually a pair of leafy sterile branches equalling or surpassing the small subsessile terminal panicle : leaves on short rather slender petioles, ovate or oblong to lanceolate, 3–6 in. long, acute or acuminate, only the lowest obtuse or somewhat cordate at base ; stipules dilated and conspicuous : fruiting pedicels 4–9 lines long, jointed near the base : valves entire, cordate-orbicular with a deep sinus, $\frac{3}{4}$–1 in. broad, acutish or emarginate, rose-color, very veiny. On the eastern slope of the Sierra only ; a plant of the Interior Basin of the continent, and of the high northern plains.

* * *Glabrous perennials with reddish usually diœcious flowers; valves not grain-bearing; leaves mostly either broad and rounded, or hastate, sparingly veined; herbage tender and acid.*

Old genus ACETOSA (the Sorrels).

11. **R. paucifolius,** Nutt. in Wats. Bot. King Exp. 314 (1871). Erect, slender, 1–2 ft. high, leafy below : leaves narrowly lanceolate, or the

lowest broader, not at all hastate, 2–4 in. long, acutish, narrowed to a
slender petiole : panicle naked, its branches slender, erect : fl. in loose
fascicles ½ line long, fruiting sparingly : pedicels filiform, jointed near
the base : valves cordate-ovate, entire, nearly 2 lines long. Near Lake
Tenayo in the Sierra Nevada, *Brewer*, thence northward and eastward.

12. **R. digynus,** Linn. Sp. Pl. i. 337 (1753) ; Hill, Veget. Syst. x. 24
(1765), under *Oxyria;* Mill. Dict. 8th ed. (1768), under *Acetosa;* Wahlb.
Fl. Lapp. 101, t. 9 (1812), under *Rheum.* Stoutish and somewhat fleshy,
6–18 in. high, the stems usually several, from a perpendicular simple or
branched fleshy root : leaves mostly radical, long-petioled, round-
reniform, 1–2 in. broad : fl. perfect, greenish or reddish, in scarious
bracted fascicles forming panicled racemes, dimerous, the sepals 4,
stamens 6 and stigmas 2 : achene thin, flat, broadly winged, the wing
exserted from the two spatulate erect inner sepals, red in age. Common
in cold wet rocky places, along snow-fed streamlets etc., in the higher
Sierra ; and in like situations far northward around the whole circuit of
the northern hemisphere : a merely dimerous and wing-fruited sorrel,
formerly cultivated in northern Europe like other *Acetosa* species for
its tender and keenly acid fruit-like and wholesome herbage.

13. **R. Acetosella,** Linn. l. c. 338. *Acetosa tenuifolia,* Mœnch, Meth.
357 (1794) ; *A. repens,* S. F. Gray, Nat. Arr. ii. 276 (1821). Stems erect
from running rootstocks, slender, 6–18 in. high : leaves oblong- to
linear-lanceolate, or oblanceolate, 1–3 in. long, usually hastate, the lobes
often toothed : panicle naked, long and narrow ; fl. diœcious, small, red,
in loose fascicles ; pedicels short, jointed at top : achene small, ovate-
triquetrous, ⅔ line long. Very common, and one of the most persistent
of field and pasture weeds, multiplying excessively both by seeds and by
its rootstocks ; native of the Old World, but now naturalized in all
temperate regions of the globe.

3. **EMEX,** *Necker.* Annual herbs with alternate leaves, and axillary
solitary or clustered unisexual flowers. Staminate perianth 5–6-parted:
segments equal, spreading. Stamens 4–6 ; filaments filiform. Fertile
perianth with urceolate tube and 6 unequal lobes in 2 series, the whole
accrescent in fruit and indurated ; the outer lobes spreading and spinescent, the inner plane, erect-connivent. Fruit a triquetrous achene
enclosed in the tube of the perianth but free from it. Seed subterete :
embryo incurved.

1. **E. australis,** Steinh. Ann. Sc. Nat. ix. 195 (1838) : *E. Centropodium*
Meisn. Linnæa, xiv. 490 (1840). Glabrous ; the stout and rigid prostrate
branches 1–2 ft. long : leaves triangular-ovate, entire, 2 in. long, at base
abruptly narrowed to a long petiole : staminate fl. often clustered at the
end of a peduncle ; the pistillate sessile : fructiferous perianth $\frac{1}{3}$–$\frac{1}{2}$

in. long, thick and almost woody; outer lobes broadly subulate and thorn-like, the inner broadly ovate, mucronate. Native of South Africa and Australia; adventive on our sea-beaches; collected in fruit by the author, on a railway embankment at South Vallejo in 1874.

4. **HOLLISTERIA,** *S. Watson.* Diffuse and fragile annual with alternate cuspidate leaves and two small herbaceous free stipules at base of each. Involucre solitary, sessile in the leaf-axil, composed of 3 slightly united linear unequal obtuse bracts, 2-flowered. Flowers unequally pedicelled, with a minute scarious bract at base. Perianth turbinate, membranous, 6-cleft to the middle. Stamens 9, on the throat, included. Styles slender. Achene glabrous, ovate, triquetrous at tip. Embryo curved; the orbicular cotyledons accumbent.

1. **H. lanata,** Wats. Proc. Am. Acad. xiv. 296 (1879). Branches prostrate, 1 ft. long or more; herbage loosely woolly; leaves elliptical or oblanceolate, the lowest 1–3 in. long, the upper much smaller and more ovate, aculeate-tipped; stipules linear-subulate, 1–3 lines long; perianth woolly, 1 line long, the linear-lanceolate lobes green, with a scarious margin, the inner slightly shorter and broader. An interesting plant, perhaps somewhat local, near San Luis Obispo, *Lemmon.*

5. **NEMACAULIS,** *Nuttall.* Diffuse annuals with alternate (mostly basal and rosulate) exstipulate leaves, and small flowers without involucre clustered at the nodes of the slender branches. Flowers perfect, each with a free herbaceous bractlet. Perianth 6-cleft, colored, enclosing the achene. Stamens 3. Styles 3; stigmas capitate. Achene short-ovoid, obscurely 3-angled.

1. **N. denudata,** Nutt. Journ. Philad. Acad. 2d ser. i. 168 (1847); Greene, Bull. Torr. Club, xiv. 217: *N. Nuttallii,* Benth. in DC. Prodr. xiv. 23 (1856); Wats. Bot. Calif. ii. 16. Stems ascending, 1 ft. long, terete, glabrate, purplish; leaves narrowly spatulate, 1–3 in. long including the short petiole, densely tomentose-hairy on both sides; bractlets of the flower-clusters obovate to spatulate, 1 line long, the outer without flowers, the inner smaller, all very woolly within, glabrous without; fl. yellowish, scarcely 1½ line long, slightly exceeding their bracts, short-pedicellate, glabrous; inner segments broadest; achene 1⁄3 line long. Along sandy beaches in the southern part of the State; perhaps not as far north as Santa Barbara.

6. **ERIOGONUM,** *Michaux.* Annual, perennial or suffrutescent plants with radical or alternate or verticillate exstipulate leaves and a greatly diversified inflorescence of involucrate, mostly small and dense primary flower-clusters. Involucre campanulate, turbinate or oblong, 4–8-toothed or -lobed without awns; the pedicels few or many, more or less exserted, subtended by scarious and narrow or quite setaceous

bractlets. Perianth 6-cleft or -parted, colored, enclosing the achene. Stamens 9, upon the base of the perianth. Styles 3; stigmas capitate. Achene 3-angled, rarely 3-winged. Embryo in all Californian species more or less lateral and incurved; cotyledons foliaceous, mostly shorter than the radicle.

* *Perennials, often woody at base; involucres umbellate (rarely solitary) at summit of naked or leafy-bracted scape-like peduncles, turbinate, 4 8-toothed or lobed; perianth narrowed to a slender stipe-like base; filaments pubescent below.* ERIOGONUM proper.

+ *Involucres deeply lobed, the lobes becoming reflexed.*

++ *Peduncles erect from a branched woody base; umbel simple or compound.*

1. **E. tripodum**, Greene, Pittonia, i. 39 (1887). Leaves linear-spatulate, 1 in. long, including the short petiole, revolute, tomentose on both sides: peduncles slender, 1 ft. high, bearing a whorl of leaves above midway and these parted into about 3 elongated and nearly erect rays, of which two bear a whorl of bracts in the middle, the third being naked: each ray with a single involucre at summit: perianth yellow, densely villous, less than 2 lines long, abruptly narrowed to a very short stipitiform base. At Hough's Springs, Lake Co., *Mrs. Curran.*

2. **E. umbellatum**, Torr. Ann. Lyc. N. Y. ii. 241 (1828). More or less tomentose when young, the upper surface of the leaves, or sometimes the whole plant, glabrate in age: leaves obovate- to oblong-spatulate, 1 2 in. long, on a slender petiole: peduncle 6 15 in. high, naked, bearing a simple umbel of naked rays subtended by a whorl of leaves: lobes of the involucre shorter than the turbinate tube: fl. yellow, often reddening in age, 2 3 lines long, tapering gradually to the long stipe-like base: filaments very hairy. At considerable elevations in the Sierra, from Mt. Dana northward: one of the most common and widely dispersed species of the genus, originally discovered in the Rocky Mountains, where it is more plentiful than in California. The two following seem to have been more or less confused with it.

3. **E. speciosum**, Drew, Bull. Torr. Club, xvi. 152 (1889). Habit and foliage of the preceding, but all the parts larger, the petioles shorter: umbel bearing 2 4 elongated rays which are usually again divided, all the nodes leafy-bracted: involucre small, the lobes acute: fl. numerous, bright yellow tinged with purple, $\frac{1}{2}$ in. long, including the slender stipe-like portion: filaments slightly hairy. Gravelly banks of the South Fork of Trinity River, in the Hyampum Valley, *Chesnut & Drew.* Also doubtless along the upper Sacramento, where it may have passed for *E. umbellatum.* July.

4. **E. Tolmieanum**, Hook. Fl. ii. 134 (1840): *E. umbellatum*, var.

monocephalum, T. & G. Proc. Am. Acad. viii. 160 (1870). Woody basal branchlets short, very leafy, densely cespitose: leaves ovate, 3 lines long, narrowed to a short petiole, glabrate above; scape 2-4 in. high, bearing a whorl of bracts above midway and a single large globose flower-cluster of 1 or more small involucres: fl. yellow; the campanulate perianth abruptly narrowed to a short stipitiform base; filaments long-villous for a short space below the middle. A neat and pretty alpine species, found at Sonora Pass, *Brewer*, and again from the Scott Mountains northward; more common on the high plains of eastern Oregon. Aug.

5. **E. Torreyanum,** Gray, Proc. Am. Acad. viii. 58 (1870). Glabrous, the obovate- or oblong-spatulate leaves 1-2 in. long, coriaceous; peduncles stout, a span to a foot high, naked or with a single leaf in the middle, bearing a few-rayed umbel; lateral rays leafy-bracted in the middle, often divided: fl. 3-4 lines long, yellow or reddish, the stipitiform base very short. Subalpine in the Sierra Nevada from Silver Mountain and near Donner Lake northward.

6. **E. stellatum,** Benth. Trans. Linn. Soc. xvii. 409 (1837); Hook. Fl. ii. 131. t. 177. More or less tomentose, the stems diffuse and leafy; leaves ovate-spatulate to oblanceolate; peduncles ½-1 ft. high, bearing an umbel of 2-4 usually elongated and cymosely divided rays; the nodes all leafy-bracted: fl. yellow; stipitate base of perianth elongated. Var. **bahiæforme,** Wats. Inflorescence more compound; leaves smaller and more decidedly tomentose. Rather common in the Sierra; also apparently in the Coast Range northward.

7. **E. robustum,** Greene. Bull. Calif. Acad. i. 126 (1885). Whitish-tomentose throughout, the very thick caudex much branched, forming a broad tuft; leaves ovate, 1-1½ in. long, erect, on stout petioles 2 in. long; peduncles stout, erect, 6 in. high, bearing an ample umbel of about 5 thrice divided rays; umbels and umbellules subtended, the former by spatulate, the latter by linear-lanceolate leafy bracts 1 in. long; involucres ½ in. long: fl. cream-colored, ¼ in. long; stipe like base of perianth very short. Eastern foot-hills of the Sierra between Reno and Virginia City, Nevada, *Mrs. Curran*. Not yet found within the Californian boundary, but to be expected. July.

8. **E. compositum,** Dougl. in Benth. l. l. 410 (1837). More or less white- or yellowish-tomentose, the leaves densely so beneath; these oblong-ovate, cordate at base, acute or acutish, 1-3 in. long on rather long petioles; peduncles stout, naked, ½-1½ ft. high, nearly glabrous; umbel of 6-10 long rays, each bearing a short several-rayed umbellule, subtended by whorls of linear-oblanceolate leaflets: fl. 2-4 lines long, cream-colored or yellow, the stipe-like base relatively short. From Napa and Sonoma counties northward, at middle elevations of the Coast Range.

++ ++ *Peduncles decumbent or almost prostrate, from a simple or sparingly branched but stout caudex.*

9. **E. Lobbii**, T. & G. Proc. Am. Acad. viii. 162 (1870). Hoary when young, with a soft arachnoid tomentum : leaves oval or more rounded, 1_2 1^1_2 in. long, on stoutish petioles, thick, glabrate above in age ; peduncles 3 6 in. long, weak and reclining ; umbel simple, of few stout and short rays, subtended by 3 or 4 oblong or oblanceolate leaflets connate at base : involucres 1_2 in. long, many-flowered : fl. dull white with usually a tinge of rose, 2 3 lines long, the stipe-like base very short. Common on barren rocky or gravelly alpine summits north and south of Donner Lake. Aug. Oct.

+ + *Involucres with short nearly or quite erect teeth ; peduncles from a diffuse woody base (except in n. 9).*

10. **E. pyrolaefolium**, Hook. Journ. Bot. v. 395. t. 10 (1853). Leaves (from a short and thick, sparingly branched caudex) thick, glabrous, round-obovate to oblong, 1_4 3_4 in. broad, abruptly narrowed to a short petiole ; peduncles glabrous, 2 3 in. high, bearing a simple 2-bracteate umbel of 1 4 short-pedicellate involucres, these sinuate-toothed and villous ; fl. rose-colored, 1^1_2 2 lines long, sparingly villous on the outside. In volcanic ashes near the summit of Lassen's Peak, *Lemmon ;* also in similar places on Mt. Shasta.

11. **E. ursinum**, Wats. Proc. Am. Acad. x. 347 (1875). Densely tomentose, the peduncle and compound 6 12-rayed umbel somewhat villous ; leaves ovate, acute, 4 6 lines broad, cuneate or rounded at base, exceeding the petiole, glabrate above ; peduncles stout, 6 12 in. high ; bracts elongated, oblanceolate or linear ; involucres large and turbinate, sharply toothed ; fl. 1^1_2 2^1_2 lines long, pale yellow ; filaments very villous.- Plumas Co., *Mr. Lemmon, Mrs. Ames.* Sept.

12. **E. incanum**, T. & G. Proc. Am. Acad. viii. 161 (1870). Rather densely cespitose ; leaves oblong-spatulate or oblanceolate, 1_2 1 in. long, short-petioled, densely tomentose on both faces ; peduncles slender, 2 10 in. high ; umbel simple, of 5 8 slender rays subtended by a few small linear bracts, the central involucre sessile or the whole umbel reduced to a small head : involucres 1^1_2 lines long, strongly toothed ; fl. of a rather greenish yellow, in age often tinged with red, 1 3 lines long.- In the higher Sierra, from Mariposa Co. to near Donner Lake.

13. **E. marifolium**, T. & G. l. c. Slenderly but intricately branched at base ; leaves ovate or oblong, 1_4 1_2 in. long, usually glabrate above ; peduncle 2 12 in. high ; umbel simple, of 5 8 usually short rays, the bracts short and linear ; involucre 1 line long ; fl. brownish or yellowish. 1 2^1_2 lines long, the smaller staminate only. In the higher Sierra, from Mariposa Co. northward. A small and homely species. July Sept.

14. **E. Kelloggii**, Gray. Proc. Am. Acad. viii. 293 (1870). Stems and sterile stolons slender and cespitose, forming a broad mat ; leaves

oblanceolate, 2–4 lines long, subsessile, villous-tomentose : peduncles 2–4 in. high, with a central whorl of 3–5 foliaceous bracts and a solitary terminal naked strongly toothed involucre 2 lines long : fl. rose-colored or white, 1½–2½ lines long. In fir-woods on Red Mountain, Mendocino Co., *Kellogg & Harford.*

* * *Involucres cylindric-turbinate or prismatic, 5–6-nerved, with erect teeth, in heads or cymosely or virgately scattered on the branches; perianth abruptly contracted at base; filaments usually glabrous.*— Subgenus OREGONIUM, Wats.

← *Perennials with scape-like peduncles.*

15. **E. gracilipes,** Wats. Proc. Am. Acad. xxiv. 85 (1889). Dwarf, densely cespitose, the branches of the caudex bearing crowded oblanceolate tomentose leaves 1½ in. long or less: peduncles slender, 1–2 in. high, glandular-puberulent : involucres turbinate, tomentose, few, forming a solitary terminal head : fl. glabrous, rose-colored.— On the White Mountains, Mono Co., at the highest altitudes (13,000 ft.), *W. H. Shockley.*

16. **E. ovalifolium,** Nutt. Journ. Philad. Acad. vii. 50. t. 8 (1834) ; Pl. Gamb. 166 (1848), under *Eucycla.* Cespitose, densely white-tomentose : leaves broadly oval or oblong, acutish, 2–6 lines broad, abruptly narrowed to a long slender petiole : peduncles slender, 3–9 in. high: involucres 3–8, in a close head, 2–2½ lines long : fl. yellow, white or rose-red, 1½–2½ lines long, the outer sepals almost orbicular, the inner spatulate, obtuse or retuse. Var. **proliferum,** Wats.: *E. proliferum,* T. & G. l. c. Larger than the type, the involucres loosely cymose-umbellate. Mostly along the eastern foot-hills of the Sierra, and eastward. The variety is one of the handsomest plants of the genus.

17. **E. Kennedyi,** Porter, Proc. Am. Acad. xii. 263 (1877). Densely cespitose but scarcely woody, white-tomentose : leaves narrowly oblong, 1⅓–3 lines long, revolute : peduncles glabrous, slender and wiry, 2–4 in. high : involucres 2–10, in a dense head, somewhat tomentose, 1½ lines long, strongly nerved, the teeth short : fl. glabrous, white with red veins, 1½ lines long, the sepals all alike.—Obtained somewhere in Kern Co., *W. L. Kennedy,* 1876.

18. **E. latifolium,** Smith, in Rees Cycl. (1815) : *E. arachnoideum,* Esch. (1826). Stout, tomentose throughout, the short caudex sparingly branched and leafy : leaves oblong or oval, obtuse or acute, 1–2 in. long, rounded or cordate, or rarely cuneate at base, commonly undulate, often glabrate above, 1–2 in. long, the stoutish petiole often short and margined : peduncles stout, 6–20 in. high : bracts triangular : involucres very-many-flowered, crowded in 1–3 large terminal heads, or the peduncles more than once forked above and the heads smaller : bractlets densely villous-plumose : fl. white, the sepals broadly obovate. In

rocky or sandy places from San Simeon Bay northward to Humboldt Co., chiefly near the sea; common and variable; or possibly a complex species as here received. *E. oblongifolium*, Benth., of late referred here, is far less tomentose than the type, and has leaves narrowed at base. A plant of the State Survey (n. 6569) from Mendocino Co. is white-arachnoid-tomentose throughout, even to the involucres, and its leaves are nearly orbicular. The type is common on the San Francisco peninsula, flowering in summer and autumn.

19. **E. nudum**, Dougl. in Benth. Trans. Linn. Soc. xvii. 413 (1837); *E. auriculatum*, Benth. l. c. 412. Much taller and more slender than the last, the ovate or oblong leaves ($\frac{1}{2}$ 2 in. long) densely tomentose beneath, glabrate above; peduncle and loose panicle 1– 2 ft. high, glabrous and glaucescent, or somewhat floccose-tomentose: involucres 2 3 lines long, nearly or quite glabrous, 3 6 in each cluster; fl. glabrous or villous, 1 $1\frac{1}{2}$ lines long, white, reddish or sulphur-yellow. Var. **pauciflorum**, Wats. Panicle more diffuse, the involucres solitary or in pairs, the peduncle often inflated. Var. **oblongifolium**, Wats. *E. affine*, Benth. Often tomentose throughout; leaves attenuate to a long slender petiole; bracts sometimes foliaceous.– Common throughout the State in various forms, or possibly a complex species; the type on clayey hills and banks not far from the sea; the varieties belonging to the interior, the last one chiefly northward and commonly or always yellow-flowered. July Oct.

++ *Shrubs with fascicles of smaller leaves in the axils; involucres capitate, the clusters more or less cymose-umbellate.*

20. **E. parvifolium**, Smith, in Rees Cycl. (1815). Branched from the base and tufted, 3 ft. high, more or less white-tomentose throughout: leaves broadly ovate or oblong, $\frac{1}{2}$ $\frac{3}{4}$ in. long, on very short petioles, revolute and undulate, in age glabrate above: peduncles short; heads few and dense; fl. glabrous, white or with a tinge of rose, $1\frac{1}{2}$ lines long. On the seacoast from Santa Cruz, *Anderson*, to Santa Barbara.

21. **E. fasciculatum**, Benth. Trans. Linn. Soc. xvii. 411 (1837); *E. rosmarinifolium*, Nutt. Pl. Gamb. 164 (1848); *E. ericæfolium*, T. & G. Proc. Am. Acad. viii. 170 (1870). Stems woody and brittle, mostly 2–3 ft. high, with many fascicles of rigid, almost heath-like leaves, these $\frac{1}{2}$ $\frac{3}{4}$ in. long, oblong-linear, acute, revolute, subsessile, deep green and shining above, tomentose beneath: peduncles short, bearing a cymosely divided umbel of many sessile 5-toothed involucres; fl. white or pinkish, 1 line long. From Santa Barbara southward, mostly near the sea; flowering almost all the year round, and a favorite food-plant with bees.

22. **E. polifolium**, Benth. in DC. Prodr. xiv. 12 (1856); *E. fasciculatum*, Wats. in part. Less woody and less brittle than the last and

smaller : leaves oblong, revolute, narrowed to a short petiole, cinereous-tomentose above, the indument denser and white on the lower face : peduncles elongated and naked, bearing mostly 2 or several large heads of clustered involucres, or these not rarely cymose-umbellate : fl. as in the preceding. Sufficiently distinct from the last by its different habit, broader, thinner and cinereous foliage, etc. It is in general a plant of the interior rather than of the seaboard, and has a more northerly range than *E. fasciculatum*, being plentiful on the mountains of Kern Co., north of the Mohave Desert. July Oct.

+ + + *Leaves not fascicled; involucres scattered in open cymes.*

23. **E. microthecum,** Nutt. Pl. Gamb. 162 (1848). Shrubby at base, erect, rather slender and diffuse, 3—12 in. high, more or less white-tomentose : leaves oblanceolate to linear, $\frac{1}{2}$ 1$\frac{1}{2}$ in. long, acute, more or less revolute, white-tomentose beneath, glabrate above : peduncles short, bearing a short cyme of once to thrice subdivided branches : involucres $\frac{3}{4}$ 1$\frac{1}{2}$ lines long, attenuate at base, some of them pedunculate : fl. $\frac{1}{2}$ -1 line long, white, rose-color or yellow. A widely dispersed and variable species of the Rocky Mountain plateau, reaching our borders in the Sierras of Mono Co. and northward, in a reduced and subalpine form.

24. **E. corymbosum,** Benth. in DC. Prodr. xiv. 17 (1856). Very near the last, but stouter : leaves broader and longer, less revolute : cyme broader and shorter, with fewer involucres and rather larger flowers.- Of the same distribution as the last nearly ; reaching the eastern base of the Sierra.

25. **E. truncatum,** T. & G. Proc. Am. Acad. viii. 173 (1870). Annual, slender, floccose-tomentose throughout, 1 ft. high : leaves mostly rosulate near the base of the stem, sometimes a whorl subtending the lowest node ; blade oblanceolate, 1 in. long, attenuate to a slender petiole, the margin undulate : inflorescence very lax, in a kind of umbel of 4 6 elongated and di- or trichotomous rays : inflorescence few, oblong-turbinate, 2 lines long : fl. rose-color, 1 line. Seemingly a local species, but plentiful among the hills just at the eastern base of Mt. Diablo.

+ + + + *Involucres scattered along the branches of an open naked dichotomous panicle.*

++ *Suffrutescent perennials leafy below; herbage white-tomentose; panicle of few virgate branches.*

26. **E. elongatum,** Benth. Bot. Sulph. 45 (1844). Woody and leafy stem low (seldom 1 ft.), the simple or sparingly branched flowering stems often 2 ft. high : leaves somewhat scattered, oblong-lanceolate, or broader, acute, 1 in. long, narrowed to a short petiole, in age glabrate above : involucres distant, 2 -3 lines long, obtusely toothed : fl. 1 1$\frac{1}{2}$

lines long, white or rose-colored; achene glabrous. From Monterey southward near the coast.

27. **E. trachygonum,** Torr. in DC. Prodr. xiv. 15 (1856). Woody stems erect, rather slender, 6–10 in. high, densely clothed with the living and dead leaves; these narrowly oblanceolate, 1 in. long, narrowed to a slender petiole, the dense tomentum persistent on both faces, as on the stems and peduncles: panicle short-peduncled, 3–5 in. high, twice or thrice dichotomous; lower involucres scattered, upper more condensed, sessile, campanulate-tubular, prominently but obtusely angled, glabrous except the woolly and obtusely toothed orifice: sepals white with a green midrib, the inner longer and somewhat narrower than the outer: ovary pubescent on the angles. Abundant in dry gravel beds along Putah Creek, *Jepson, P. S. Woolsey,* and elsewhere in like situations up and down the valley of the Sacramento, and southward in the interior. Entirely distinct in habit, foliage, etc. from the Texano-New-Mexican *E. Wrightii* with which it has been confounded. Sept., Oct.

28. **E. Wrightii,** Torr. l. c. var. **subscaposum,** Wats. Woody and leafy stems very short and cespitose: leaves $\frac{1}{4}$–$\frac{1}{2}$ in. long, oblong, obtuse, short-petioled, not persistent on the old branchlets: peduncles naked and scapiform, the involucres spicate on a few wide-spread cymose branches, or fewer and capitate-congested, tomentose throughout, the teeth rigid, acute: sepals less unequal than in the last, the midvein red. High Sierra from near Donner Lake southward. Sept., Oct.

29. **E. saxatile,** Wats. Proc. Am. Acad. xii. 267 (1878). Caudex stoutish, sparingly branched, very leafy, $\frac{1}{2}$–1 ft. high: leaves obovate, obtuse, 6–8 lines broad, 1 in. long, cuneate at base, short-petioled, densely tomentose on both sides: branches of the inflorescence short, spreading: bracts subfoliaceous, triangular or oblong and acute: involucres $1\frac{1}{2}$–2 lines long: teeth acute: fl. rose-colored, 2 lines long, the sepals all spatulate-oblong and carinate, about equal, the inner appressed to the achene. In Reliz Cañon, Monterey Co., *Hickman,* and Santa Lucia Mts., *Palmer.*

++ ++ *Annuals with leaves mostly basal and rosulate.*
Involucres 2 lines long, tubular.

30. **E. virgatum,** Benth. in DC. Prodr. xiv. 16 (1856). Usually white-tomentose throughout: leaves oblong, 1 in. long on slender petioles: peduncle simple, or with only a few erect virgate branches, 1–$2\frac{1}{2}$ ft. high, the involucres remote, the 5 teeth very short: perianth 1 line long, buff or yellow: outer sepals broadly obovate, cuneately narrowed at base, the inner about as long, spatulate-oblong: achene with a minutely puberulent rather slender beak. Near Monterey, *Hartweg;* towards the

Yosemite, *Bolander*, n. 4953, a less tomentose or even somewhat glabrate state; banks of Putah Creek, Solano Co., *Jepson*, and very common farther northward in the State.

31. **E. roseum,** Dur. & Hilg. Journ. Philad. Acad. iii. 45 (1854) & Pac. R. Rep. v. 14. t. 15 (1855). Smaller, the leaves spatulate-ovate, undulate, narrowed to a long petiole: peduncle with rather divergent and stiff panicled branches: involucres remote, slightly widening upwards, rather deeply 5-toothed: perianth rose-color, ⅜ line long; outer sepals obovate-oblong, not attenuate at base. Of more southerly distribution than the last, but common as far north as Mariposa and Tulare counties.

32. **E. dasyanthemum,** T. & G. Proc. Am. Acad. viii. 177 (1870). Usually hoary-tomentose, sometimes nearly glabrate: leaves oval or rounded, 5—10 lines long, abruptly narrowed to a slender petiole: peduncle 1 ft. high, mostly rather loosely but widely branching, the branches often more or less cymose-dichotomous: involucres rather remote, not always solitary, narrowly tubular (in the type), very shortly toothed, tomentose except the prominent ribs, these glabrous: fl. scarcely exserted, erect, not numerous, 1 line long, white or rose-color; more or less densely villous on the outside. Var. **Jepsonii.** Panicle ample, as broad as high, the dichotomous branches widely spreading: involucres campanulate-tubular, very-many-flowered, the pedicels exserted and recurved; fl. rose-red. The type, originally from near Clear Lake, Lake Co., *Torrey*, is plentiful along Putah Creek in the valley of the lower Sacramento, *Jepson*, *Woolsey;* the variety, from Gate's Cañon, not far from Vacaville, is more showy and may perhaps be a distinct species. Sept., Oct.

Involucres 1—1½ lines long, usually turbinate.

33. **E. vimineum,** Dougl. in Benth. Trans. Linn. Soc. xvii. 416 (1837); DC. Prodr. xiv. 17. Seldom at all tomentose except on the lower face of the ovate or orbicular slender-stalked leaves: peduncle 1 ft. high, branched from near the base, the branches slender and virgate, or sometimes the whole inflorescence more spreading and repeatedly dichotomous: involucres very narrow and rather prismatic, the teeth very short: fl. few, rose-color or white, exserted; outer sepals obovate, the inner obovate-oblong and only half as broad. Var. **caninum.** Involucres turbinate, many-flowered, disposed in divergent-branched dichotomous cymes: fl. rose-red: outer sepals broadly obovate, the inner spatulate-oblong. Nothing quite like the type of this species of the far northeastern interior is found in middle California. The common plant of the interior of our State was named a var. *erioclados* by Bentham. Our var. *caninum*, found only at Tiburon, on dry hills, is so very unlike all else which has been called *E. vimineum* that it may take specific rank.

unless intermediate forms are found. July Sept.

34. **E. gracile,** Benth. Bot. Sulph. 46 (1844) & DC. Prodr. l. c. Slender, 1 2 ft. high, usually white-woolly throughout : leaves rosulate or scattered, ovate, oblong or oblanceolate, tomentose on both faces : panicle of few or many usually rather strict and virgate very slender branches : involucres many-flowered, turbinate, the 5 teeth stout, prominent, acutish : fl. white, rose-color or yellowish, ¾ line long ; outer sepals obovate, inner oblong. — In the interior, from the Sacramento valley southward : apt to be confounded with *E. virgatum* if one overlook the small size of the flowers and the teeth of the involucre. As here perhaps too loosely defined it embraces *E. leucoladon*, Benth., and *E. acetoselloides*, Torr., both of which may yet be found to deserve restoration.

35. **E. cithariforme,** Wats. Proc. Am. Acad. xxiii. 266 (1888). Prostrate or procumbent, branching from the base, mostly glabrous except the floccose-woolly lower face of the leaves ; these 3–4 in. long, dilated at summit, the rounded base abruptly contracted to a winged petiole, margin undulate : branches 1 ft. long, many times forked, the lower bracts foliaceous, the upper triangular : involucres glabrous, broadly turbinate, 1–1½ lines long, with broad teeth : fl. rose-color, 1 line long: sepals spatulate-obovate. — In San Luis Obispo Co., *Lemmon*.

36. **E. Plumatella,** Dur. & Hilg. Pac. R. Rep. v. 14. t. 16 (1855). Slender, 3–8 in. high : leaves rosulate very near the base of the stem, orbicular, ½ in. broad, on slender petioles : panicle diffusely and intricately branched, the branchlets (like the leaves) grayish-tomentose : involucres very short (½ line or even less): fl. few, yellow, rose-color or white, ¾ line long or more : sepals broadly ovate-cuneiform and retuse, slightly unequal. —A Nevada species, common along the eastern base of the Sierra, but first obtained in Kern Co., Calif., on Posé Creek.

* * * *Annuals; involucres pedicellate in diffuse di- or trichotomous cymose umbels or panicles; perianth not attenuate at base; filaments glabrous.* Subgenus GANYSMA, Wats.

+— *Nodes of the panicle leafy.*

37. **E. angulosum,** Benth. Trans. Linn. Soc. xvii. 406. t. 18 (1837). Grayish-tomentose, 6–18 in. high, loosely and widely branching from near the base, the branches 4–6-angled : lowest leaves ovate or rounded, cuneate or somewhat cordate at base, obtuse, often undulate, ½–1 in. long, on rather short petioles ; upper oblong or lanceolate, subsessile : pedicels of the involucres ¼–1¼ in. long, filiform : involucre hemispherical, 1–2 lines broad, many-flowered, smooth or glandular : bractlets mostly dilated and rather firm : fl. rose-color, purplish, or even greenish-white, ½ line long, not quite glabrous ; outer sepals ovate,

concave, the inner lanceolate, plane, somewhat longer. Plains of the interior, from near Sacramento southward.

38 ? **E. gossypinum**, Curran, Bull. Calif. Acad. i. 274 (1885). Stem and branches scarcely woolly, obscurely angular, 1—2 ft. high, very diffuse: lowest leaves oblong, tomentose beneath: pedicels filiform, 1—6 lines long: involucres turbinate, cleft to the middle, the lobes oblong, obtuse, villous on the inside: fl. about 5 to the involucre, the bractlets linear-spatulate, villous on the upper face; sepals nearly linear, the inner acute, slightly longer than the outer.— Plains near Bakersfield, Kern Co., *Mrs. Curran.* An ambiguous species, perhaps better referred to *Nemacaulis*, especially if the achenes be as reported, "lenticular."

+ + *Panicle leafless.*

39. **E. trichopodum**, Torr. in Emory's Rep. 151 (1848); Benth. DC. Prodr. xiv. 20. Nearly glabrous, the stem and branches vivid green: leaves round-cordate to oblong-ovate, $\frac{1}{4}$—1 in. long: peduncles branched and occasionally somewhat inflated below the nodes; branchlets and pedicels filiform, the latter almost capillary: involucres minute ($\frac{1}{3}$ line long), turbinate-campanulate, glabrous: fl. few, $\frac{1}{2}$ line long, yellowish, pubescent. San Benito Co., *Hickman;* otherwise a plant of the southeastern extremity of the State, where it is associated with *E. inflatum*, a similar species not yet heard of as within our limits.

7. **OXYTHECA**, *Nuttall.* Slender annuals, glandular-pubescent (not tomentose), with a rosulate basal tuft of leaves and a repeatedly dichotomous paniculate inflorescence. Bracts of the flowering branches foliaceous, more or less connate. Involucres small, few-flowered, more or less distinctly pedicellate, the lobes awn-tipped or unarmed. Perianth 6-parted, usually glandular-pubescent on the outside, the segments alike. Stamens 6. Achene commonly lenticular.—Apparently a natural genus, though no absolute character has been shown by which it may be distinguished from *Eriogonum.* To name involucral awns as of such value is altogether empirical; and to draw the line there excludes plants naturally inseparable from the *Oxytheca* type.

* *Bracts united into a broad rounded concave disk.*

1. **O. perfoliata**, Torr. & Gray, Proc. Am. Acad. viii. 191 (1870). Herb rather rigid, glaucous, reddish, branched from the base, 3—8 in. high: leaves 1 in. long or less, spatulate, ciliate: lowest bracts small, joined at base only, the upper large and conspicuous, perfoliate, 3-awned, netveined: involucres almost sessile, narrowly turbinate, deeply 5-cleft, 1—1½ lines long, with long awns, 4—6-flowered: fl. white, 1 line long. Eastern base of the Sierra, beyond our limits, but to be sought in Mono and Inyo counties.

* * *Bracts joined at base only, not enlarged.*

+ *Involucres with awn-tipped lobes.*

2. **O. dendroidea,** Nutt. Pl. Gamb. 169 (1848). *Bresignoa Chilensis,* Remy, in Gay, Fl. Chil. v. 292. Atl. t. 58 (1849). Very slender and diffuse, 1 ft. high: leaves linear-oblanceolate, 1½ 1½ in. long, acute, hirsute: bracts unequal, awnless, linear or linear-oblong: involucres turbinate, 1½ 1½ lines long (excluding the short awns), unequally 3–4-lobed, those in the forks on slender pedicels 1–4 lines long, the rest subsessile: fl. rose-color, 1½ line long. Also of the Nevada deserts, but doubtless occurring within the borders of our State.

     + + *Involucres without awns, bristly or naked.*

3. **O. inermis,** Wats. Proc. Am. Acad. xii. 273 (1878), and *Eriogonum vagans,* l. c. xx. 370 (1885). Slender and low, 3–6 in. high, rather diffuse: leaves broadly oblanceolate, 1 in. long, glabrous except the scabrous-ciliate margins: bracts linear-oblong, acute, awnless: involucres short-pedicelled, 4-cleft almost to the base, the oblong-lanceolate lobes 1 line long, acute but awnless: fl. rose-color, 1½ line long: achenes obtusely triangular. Supposed to have been found originally on Mt. Diablo, *Miss M. J. Bancroft,* but better known from parts of the State lying southward beyond our limits.

4. **O. hirtiflora,** Greene. Gray, Proc. Am. Acad. xii. 259 (1878), under *Eriogonum.* Glandular-puberulent and viscid, 6 in. high, erect, cymose-paniculate above: leaves 1 in. long, oblong-spatulate, with scabrous-ciliate margins and a broad red midvein: bracts hispidulous, oblong, 1¼ in. long or less, acutish: involucres awnless, 1½ line long, on slender erect or nodding pedicels 1–3 lines long: fl. 3–5, very hirsute, rose-red, 1½ line long. Foothills of the Sierra in Tuolumne and Calaveras counties, *Hooker & Gray,* also near Ione, *Parry.*

5. **O. spergulina,** Greene. Gray, Proc. Am. Acad. vii. 389 (1868), under *Eriogonum; Oxytheca Reddingiana,* Jones, Bull. Torr. Club, ix. 32 (1882). Very slender, diffusely branched from the base, 3 in. to 2 ft. high, of a dull dark green or purplish, nearly glabrous, or somewhat glandular-hispid: leaves linear-oblanceolate, 1½–2 in. long, more or less hirsute: pedicels very slender, 1¼–1½ in. long, spreading: involucres turbinate, 1¼ line long, deeply 4-cleft, glabrous, awnless, 1–2-flowered, the pedicels without bracteoles: perianth white or pinkish, 1½–1 line long, slightly glandular-puberulent at base: achene lenticular. In the middle and higher altitudes of the Sierra, from Kern Co. northward.

8. **CHORIZANTHE,** *Robert Brown.* Dichotomous annuals with few and mostly basal leaves; the branches with ternate bracts at the nodes. Involucres 1–3-flowered, sessile, more or less tubular, coriaceous or chartaceous, often corrugated or reticulate, 3–6-angled or ribbed, with as many cuspidate or rigidly awned teeth or segments. Flowers rarely

exserted, 6-parted or -cleft; bractlets minute or obsolete. Stamens 9 (rarely 6 or 3). Achenes triangular.

* *Villous or hirsute; involucres usually clustered, 6-angled and sulcate, the teeth cuspidate; bractlets obsolete: perianth 6-cleft, the stamens inserted at or near its base.* CHORIZANTHE proper.

+— *Erect or erect-spreading; involucres mostly in dense cymose clusters.*
++ *Margins of involucral lobes scarious.*

1. **C. membranacea,** Benth. Trans. Linn. Soc. xvii. 419. t. 17 (1837). Floccose-tomentose, erect, sparingly branched, with long internodes and leafy nodes, $1\frac{1}{2}$ - 2 ft. high: leaves linear, acute, 1 2 in. long: bracts similar to the leaves but cuspidate: heads sessile, solitary or few upon the branches: involucres tomentose, 2 $2\frac{1}{2}$ lines long, the limb at length dilated and with uncinate teeth: tube contracted in the middle: perianth villous, becoming $1\frac{1}{2}$ lines long, 6-parted, the segments oblong or spatulate: achene broadly triangular and rostrate-attenuate. In rocky places among the foothills and lower mountains. May.

2. **C. stellulata,** Benth. Pl. Hartw. 333 (1849). Pilose-pubescent, 3 6 in. high, umbellately branched from near the base: leaves scattered, or the upper opposite, 1 in. long, linear-oblanceolate: involucres solitary in the lower axils, capitate-congested at the ends of the branches; tube strongly 6-costate, becoming triangular, the angles ciliate or glabrous; segments equal, with not greatly dilated scarious margins, the awns recurved: perianth short-pedicellate, the segments exserted, nearly equal, obcordately lobed: achene narrowly triangular. A rare or local species, long known only from Hartweg's specimens obtained somewhere in the valley of the Sacramento, but rediscovered by the late Dr. Parry on volcanic rocks near Chico perhaps the original station. May.

3. **C. Douglasii,** Benth. Trans. Linn. Soc. xvii. 418 (1837). Dichotomously branching and widely spreading from a short and simple main stem, this bearing one or two whorls of oblong-spatulate leaves which taper to a short winged petiole: upper leaves reduced to sessile bracts: herbage hoary pubescent: involucres in small terminal clusters with setaceous bracts, oblong-campanulate, contracted above, sharply angled, transversely corrugated between the angles; teeth spreading, shorter than the tube, scarious-margined to near the uncinate tips and pinkish: perianth short-pedicellate: lobes slightly unequal, truncate, the outer cuspidate, the inner shorter and retuse: achenes narrowly winged. Apparently local in the Santa Cruz Mountains, near Felton and Ben Lomond, in sandy soil, *Parry;* first collected by Douglas, probably in the same district. A var. **albens,** Parry, l. c., more pubescent, and with white flowers, occurs in the valley of the Salinas.

4. **C. robusta,** Parry, Proc. Davenp. Acad. v. 176 (1889). Stout, erect,

6–18 in. high, dichotomously branched, the main stem below with several whorls of oblanceolate petiolate leaves; herbage hirsute, the inflorescence and growing parts almost canescently so; capitate cymes sessile and solitary in the lower forks, several and peduncled along the upper branches; bracts linear, with acerose tips; involucres oblong-campanulate, sharply angled; segments unequal, the scarious margin very narrow, purplish, the uncinate teeth not widely spreading; perianth short-pedicellate; lobes nearly equal, erose-denticulate and mucronulate.—In dry sandy soils along Monterey Bay, at Aptos, also at Alameda.

5. **C. Breweri**, Wats. Proc. Am. Acad. xii. 270 (1878). Slender, soft-pubescent, erect, with ascending branches, 2–4 in. high ; leaves ovate or rounded, 3–6 lines broad, on slender petioles; bracts linear-oblanceolate, acerose-tipped ; involucres few in the head, small ($1\frac{1}{2}$ lines long); teeth slightly unequal, scariously margined at base, stout and curved, short-awned ; perianth glabrous or villous, the segments broadly oblong, the inner ones shorter. At San Luis Obispo, on dry hillsides, *Brewer;* also in Santa Margarita Valley.

++ ++ *Lobes of involucre without scarious margins.*

6. **C. valida**, Wats. l. c. 271. Stout, 6–18 in. high, branching above, villous ; lower leaves oblanceolate, 1 in. long, on long petioles : involucres in dense heads 2–3 lines long, the lobes nearly equal, slightly spreading, the awns straight ; perianth subsessile, narrowly tubular, $2\frac{1}{2}$ lines long, villous or glabrous, cleft one-third of the length, the lobes oblong, very unequal, the shorter ones erose : filaments adnate to the middle or even higher. In Sonoma Co., near Petaluma, etc.

7. **C. Palmeri**, Wats. l. c. Stout but low (3–5 in.), villous-pubescent : leaves spatulate, 2 in. long; bracts oblanceolate, conspicuous; involucres in large dense clusters, 2 lines long ; one segment long-awned, the rest nearly equal ; perianth subsessile, glabrous, rose-color, 2 lines long, broadly lobed above, the outer lobes orbicular, inner shorter, truncate or bifid, shortly laciniate : stamens near the base. From near Monterey to San Luis Obispo, on rocky hills.

+– + *Of diffuse habit; involucres scattered, or in loose clusters.*
++ *Lobes of involucre with narrow scarious margins.*

8. **C. pungens**, Benth. Trans. Linn. Soc. xvii. 419. t. 19 (1837). Branches prostrate, 6–12 in. long, hirsute-pubescent ; leaves spatulate or oblanceolate, 1 in. long, mostly opposite ; bracts similar but narrower, acerose at apex ; involucres crowded on short lateral branchlets, $1\frac{1}{2}$–2 lines long, unequally toothed, usually margined ; teeth strongly uncinate ; perianth obconic, subsessile, shortly cleft ; segments equal, oblong, entire ; filaments more or less adnate to the lower part of the tube.—Common on sandy hills on the San Francisco peninsula, and near Monterey. May—Aug.

9. **C. diffusa**, Benth. Pl. Hartw. 333 (1849). Near the last, but slender, the branches not leafy, the whole plant much smaller ; leaves all at the very base of the stem, oblong or spatulate : involucres 1 line long, unequally toothed, the longer teeth equalling the tube, uncinate, the scarious margins broad and petaloid, pinkish ; perianth shortly cleft ; segments obtuse, nearly equal, the inner somewhat narrower : filaments inserted near the base. Sandy plains near Monterey, *Hartweg*, also near the seashore at the same place, and in the Santa Cruz Mountains, *Parry*; and Dr. Parry considered the plant a variety of *C. pungens*.

10. **C. Andersonii**, Parry, Proc. Davenp. Acad. v. 175 (1889). Branches a span long : leaves mostly radical, oblanceolate, narrowed to a winged petiole ; foliaceous bracts occasional at the lower nodes ; the loosely cymose inflorescence acerose-bracted ; involucres sharply ribbed, the intervals somewhat corrugated ; longer segments equalling the tube, the alternate ones half as long, all scarious-dilated at base and uncinate-tipped ; perianth narrowly obconic, with short spatulate equal entire apiculate lobes. Scott's Valley, near Santa Cruz, *Anderson*; also on Ben Lomond, *Parry*. July.

++ ++ *Lobes of involucre without scarious margins.*

11. **C. cuspidata**, Wats. Proc. Am. Acad. xvii. 379 (1882). Habit of *C. pungens*, leafy-bracted ; leaves narrowly oblanceolate, 1 in. long ; floral bracts acerose : involucres loosely cymose-clustered, 1 line long, 6-toothed, without scarious margins, the alternate teeth shorter, all armed with hooked awns ; perianth subsessile, pinkish ; lobes nearly equal, oblong, acutish, the strong nerve excurrent as a short cusp. This was regarded by Dr. Parry as only a common form of *C. pungens*; but by Dr. Watson's description, it should be very distinct. Sandy hills at San Francisco, *Marcus Jones, Dr. Parry*.

12. **C. Clevelandi**, Parry, Proc. Davenp. Acad. v. 62 (1884). Prostrate or assurgent, the rather few branches 2-8 in. long, villous-pubescent : leaves mostly radical, broadly oblanceolate, narrowed to a rather long and slender petiole : involucres soft-pubescent, the triquetrous tube contracted above ; segments very unequal, 3 as long as the tube, the other 3 scarcely half as long, all uncinate ; perianth shortly cleft ; outer segments broadly ovate, erose, retuse or emarginate, the inner narrow and lacerate ; stamens 3 ; anthers orbicular. A well marked species, common among the wooded hills of Sonoma and Lake counties, in clayey soil. June-Sept.

13. **C. uniaristata**, T. & G. Proc. Am. Acad. viii. 195 (1870). Prostrate or assurgent, 3-8 in. broad, cinereous with a soft pubescence : lowest leaves spathulate, obtuse, pilose beneath ; cauline narrow, recurved and pungent : involucre with short and sharply angular tube, and stout

rigid segments one of which is long and straight, the other 4 or 5 short and uncinate : perianth yellowish ; segments unequal, the outer spathulate, entire, the inner only half as large, crenate : stamens 9 ; anthers oblong. Dry hills in Monterey and San Benito counties, and southward.

+ + + *Stoutish and erect, or erect-spreading, the rather coarse branchlets very fragile at the joints; involucres scattered or loosely cymose, their segments never scarious-margined.*

14. **C. staticoides,** Benth. Trans. Linn. Soc. xvii. 418 (1837). Erect, stoutish, often 1 ft. high, with spreading branches, villous-pubescent, often purplish ; leaves (all radical) spatulate, petioled, hirsute : involucres in rather close cymes, $1\frac{1}{2}$ — 3 lines long, the alternate teeth large and almost equal ; perianth sessile, glabrous, cleft one-third the length ; outer segments oblong-lanceolate, the inner larger and obovate : stamens at base of tube ; anthers oblong. From Monterey southward.

15. **C. Xanti,** Wats. Proc. Am. Acad. xii. 272 (1878). Branching near the base, 4—12 in. high, hirsute and somewhat tomentose : leaves ovate-oblong, 2—6 lines long on slender petioles, tomentose beneath ; the lower bracts similar or linear-oblanceolate : involucres in loose cymes, tomentose ; tube 2 lines long, with prominent angles ; segments unequal, abruptly recurved and uncinate : perianth sessile, rose-color, villous, $2\frac{1}{2}$ lines long ; segments linear-oblong, entire, acutish, the alternate ones shorter : stamens 6—9, unequal ; anthers oval. From Tehachapi Pass, Kern Co., southward. June.

16. **C. Wheeleri,** Wats. l. c. Low, with spreading branches 3—4 in. long, villous and tomentose : oblanceolate leaves and bracts 1 in. long or less, tomentose beneath : involucres in small terminal cymes, glabrous, 1 line long, with short stout teeth, the alternate ones smaller : perianth sessile, $1\frac{1}{2}$ lines long, glabrous, cleft one-third the length ; segments broadly oblong, the alternate ones rather shorter and broader : stamens 6, near the base. Towards Santa Barbara, *Rothrock;* considered only a form of *C. staticoides* by Dr. Parry. It is unknown to us.

17. **C. fimbriata,** Nutt. Pl. Gamb. 168 (1848). Stem erect and simple at the leafy base, becoming much branched and widely spreading, the branches appressed-pubescent, the inflorescence villous and somewhat glandular, the whole plant purplish : leaves 1—2 in. long, obovate-spatulate, retuse or obcordate at the rather abruptly widened summit : bracts all setaceous, recurved, rigidly awned : involucres scattered in the lower forks, more or less clustered at the ends of the branchlets : tube cylindrical, pubescent, strongly ribbed; segments unequal, recurved, straight or uncinate-tipped : perianth-segments exserted, red or purple, nearly equal, with an oblong obtuse terminal lobe and deeply lacerate-

fringed margins below ; stamens inserted near the base ; anthers oval. — In Santa Barbara Co., and southward beyond our limits. June.

* * *Involucres broadly triquetrous, 3–9-toothed or lobed; teeth very unequal; tube transversely corrugated; stamens on the throat of the perianth.* — Genus ACANTHOGONUM, Torr.

18. **C. polygonoides**, T. & G. Proc. Am. Acad. viii. 197 (1870). Diffuse, prostrate, 4–10 in. broad, glabrous or sparingly pubescent, the branches very fragile at the joints ; radical leaves narrowly spatulate, obtuse, narrowed to a slender petiole which is dilated at base ; bracts scarcely foliaceous, acute ; involucres rather crowded on the short branches, broadly triangular-turbinate, 3-costate, with 3 stout and broad uncinate teeth longer than the tube, the intermediate ones very small ; perianth nearly sessile ; segments oblong, equal ; stamens 6 or 9 : filaments very short ; anthers round-oval ; achene broadly triquetrous, rostrate. Foot-hills of the Sierra near Placerville, *Rattan*, and about Chico, *Parry, Mrs. Austin.*

* * * *Glabrous or glandular, never villous-tomentose; nodes subtended by large bracts (except in n. 21); involucres coriaceo-chartaceous, the awns not uncinate; perianth bearing stamens at base.* Genus MUCRONEA, Benth.

19. **C. Californica**, Gray, Proc. Am. Acad. viii. 197 (1870) ; Benth. Trans. Linn. Soc. xvii. 416. t. 20 (1837), under *Mucronea.* Sparingly hirsute and glandular, 1 ft. high or less, whole plant red or purplish ; bracts amplexicaul or rarely perfoliate, deeply 3-lobed, the lobes cuspidate ; involucres solitary on the ultimate branchlets but at the lower nodes somewhat clustered, rather obtusely 2–3-angled and not sulcate ; segments of perianth obovate, entire. In the Coast Range from San Luis Obispo southward.

20. **C. perfoliata**, Gray, l. c. Habit of the preceding, though branched from the base, the branches decumbent ; bracts connate about the stem, forming a somewhat unilateral triangular disk, the lower 1 in. broad ; involucres scattered on the slender branchlets, $1\frac{1}{2}$–3 lines long, strongly angled and sulcate, becoming corrugated, mostly 4-toothed ; perianth-segments equal, oblong, laciniately fringed. Dry foothills of the Sierra, from Stanislaus Co. southward.

21. **C. insignis**, Curran, Bull. Calif. Acad. i. 275 (1885). Very slender, 2–4 in. high, glandular-pubernlent ; involucres turbinate, several-flowered, almost hyaline between the 5 angles, the awns straight, equal ; perianths pedicellate, exserted, villous, pale rose-color ; segments oval or oblong ; achenes lenticular. Indian Valley, near the Salinas River, *Mrs. Curran.*

9. **LASTARRIÆA,** *Remy.* A small diffuse rigid fragile annual with the aspect of *Chorizanthe* proper. Involucre 0. Perianth involucre-like, coriaceous, tubular, 5 6-cleft to the middle; the narrow teeth rigid, awned, recurved and uncinate. Stamens 3, inserted on the throat: filaments very short, with small membranous appendages intervening at their insertion.

1. **L. Chilensis,** Remy, in Gay, Fl. Chil. v. 289, t. 58 (1849). *Chorizanthe Lastarriæa,* Parry, Proc. Davenp. Acad. v. 47 & 63 (1884). Hirsute; the assurgent or ascending branches 2–6 in. long: lowest leaves linear, obtuse, hispid-ciliate; cauline in whorls of 4 or 5, unequal; perianth nearly concealed by the whorled bracts; tube triquetrous; teeth or segments 5, 3 long and 2 short; anthers small, orbicular; achene triquetrous-oblong. From the plains of the San Joaquin, near Antioch, southward.

10. **PTEROSTEGIA,** *Fischer & Meyer.* Our species diffusely dichotomous slender and flaccid (or in age somewhat wiry) annual with opposite, petiolate, exstipulate 2-lobed leaves, and small foliaceous bracts. Involucres each of a single bract shorter than the solitary sessile flower, rounded and 2-lobed, in age larger, reticulated, loosely enfolding the achene, and gibbously 2-saccate on the back. Perianth 5- or 6-parted; segments equal, oblong-lanceolate. Stamens as many or fewer, inserted at the base of the segments. Achene triquetrous.

1. **P. drymarioides,** F. & M. Ind. Sem. Petr. ii. 48 (1835). Glabrous or hirsute-pubescent; leaves obovate, obcordate or reniform-bifid, often with the lobes again 2-lobed, the lowest petiolate, the upper sessile, $\frac{1}{4}$–$\frac{3}{4}$ in. long; fl. minute; fructiferous involucre $1\frac{1}{2}$ lines long, closely enfolding the minute light brown achene. Common on rocky hills, and on sandy banks along the seashore; very variable in pubescence, form of leaves, and, according to Nuttall, embracing several species.

Order XXIII. NYCTAGINEÆ.

A. L. de Jussieu, in Annales du Museum, ii. 269 (1803).

Herbs or suffrutescent plants (ours mostly coarse and fleshy seaside herbs) with tumid joints, opposite exstipulate entire leaves and showy perfect flowers in axillary pedunculate and involucrate clusters. Involucre calyx-like, closely subtending the flower-cluster. Perianth corolla-like, campanulate, salverform or tubular, the persistent base indurated and constricted over the 1-celled 1-seeded free ovary. Stamens few, hypogynous; filaments slender; anthers small and rounded. Pistil 1, simple. Seed erect; embryo encircling a copious mealy albumen. A small family, containing a number of highly ornamental and a few medicinal plants; as closely allied to Polygoneæ as to any other family, though of very different floral structure.

NYCTAGINEÆ.

1. MIRABILIS, *Parkinson* (FOUR-O'CLOCK). Perennials with erect stems from large fusiform roots. Leaves mostly ample, not very succulent. Inflorescence axillary and terminal. Involucres calyx-like, herbaceous, 5-cleft or -parted, enlarged in fruit but not otherwise altered. Perianth tubular or narrowly funnelform, with abruptly spreading limb, vespertine. Stamens usually 5, as long as the perianth; filaments united at base. Stigma capitate, granulate. Fruit a rather large oblong nutlet, dark-colored and often obscurely ribbed.

1. **M. Fræbelii,** Greene, Bull. Calif. Acad. i. 124 (1885); Behr, Proc. Calif. Acad. i. 69 (1855), under *Oxybaphus*; *M. multiflora*, var. *pubescens*, Wats. Bot. Calif. ii. 2 (1880). Stout, decumbent, 1–2 ft. high, viscid-pubescent, in age somewhat scabrous: leaves broadly ovate, the lowest somewhat cordate, 4 in. long and almost as broad: involucre cleft to the middle; lobes acutish; fl. 5 or 6: perianth narrowly funnelform, $1\frac{1}{2}$ in. long, the limb 1 in. broad, purple, pubescent and viscid on the outside: nutlet ovate-oblong, light brown, smooth, 10-lineate. Mountains of Kern Co. and southward. July–Sept.

2. **M. lævis,** Curran, Proc. Calif. Acad. 2d ser. i. 235 (1888); Benth. Bot. Sulph. 44 (1844), under *Oxybaphus*; *M. Californica*, Gray, in Torr. Bot. Mex. Bound., 173 (1859). Stems many, rather slender, ascending from a somewhat woody base, 1–3 ft. high: herbage more or less viscid-pubescent: leaves rather fleshy, round-ovate to ovate-oblong, $\frac{1}{2}$–$1\frac{1}{4}$ in. long, obtuse or acute, short-petioled: involucres short-peduncled or almost sessile, 5-cleft, the lobes often unequal, acute; fl. 1–3: perianth rose-purple, narrowly campanulate, 5 lines long, the lobes spreading, emarginate: stamens as long as the perianth or longer: nutlet ovate, smooth, $1\frac{1}{2}$ lines long. Very common in the South, but possibly not within our limits.

2. ALLIONIA, *Linnæus*. Perennial, with opposite unequal leaves, and axillary pedunculate flowers. Involucre herbaceous, 3-parted, unchanged in fruit, 3-flowered. Perianth funnelform; limb oblique, 4–5-lobed. Stamens 3–5, nearly distinct. Stigma capitate. Nutlet ovate, compressed, smooth and convex on the inner side, the back with a double line of stipitate tubercles enclosed by a rigid inflexed and toothed margin. Embryo plicate, the cotyledons unequal.

1. **A. incarnata,** Linn. Sp. Pl. 2d ed. i. 147 (1762). Slender, prostrate; herbage of a pale green, more or less woolly-pubescent and slightly viscid: leaves ovate, $\frac{1}{2}$–$1\frac{1}{2}$ in. long, very unequal, obtuse or acute, exceeding the slender petiole: segments of the involucre concave, broadly oblong or rounded: perianth deep purple or paler, 2–4 lines long, the lobes emarginate, one much shorter than the rest: fr. $1\frac{1}{2}$ lines long, usually somewhat carinate on the convex side; teeth of the margin

variable in number and size, but usually 5 on each side, either broad or narrow, sometimes gland-tipped. Plant of very wide dissemination in Mexico and South America, common in southern California, occurring as far northward as Monterey.

3. ABRONIA, *Jussieu.* Decumbent or prostrate viscid-pubescent and rather succulent herbs with opposite and somewhat unequal leaves. Flowers in umbel-like heads on rather long axillary peduncles. Involucre of 5-15 distinct or slightly united somewhat scarious bracts enfolding the base of the heads. Perianth salverform ; tube elongated : limb of 5 or 4 emarginate or obcordate lobes. Stamens mostly 5, adnate to the tube and not exserted. Stigma linear-clavate. Fruit coriaceous, 3-5-winged, enclosing a smooth and cylindric achene. Embryo with but one cotyledon.

* *Wings coriaceous, lateral, not encircling the fruit.*
+ *Wings thin but solid; body of fruit rigid or ligneous.*

1. **A. umbellata,** Lam. Ill. i. 469. t. 105 (1791); Hook. Exot. Fl. iii. t. 194. Perennial, prostrate, rather slender, viscid-puberulent, the stems 1-3 ft. long : leaves almost glabrous, ovate to narrowly oblong, 1 -1$\frac{1}{2}$ in. long, narrowed to a slender petiole, obtuse, the margin often somewhat sinuate : peduncles 2-6 in. long : bracts of the involucre narrowly lanceolate, $\frac{1}{4}$ in. long ; head 10-15-flowered : perianth rose-purple, 6-8 lines long ; lobes emarginate : fr. 4-5 lines long, nearly glabrous, the body oblong, attenuate at each end ; wings thin, nearly as long, broadest and often truncate above, narrowing toward the base : achene 1$\frac{1}{2}$ lines long. Sandy places along the seaboard everywhere. June-Oct.

2. **A. maritima,** Nutt. in Wats. Bot. Calif. ii. 4 (1880). Perennial, prostrate or assurgent, very stout and succulent, somewhat pubescent, very viscid : leaves broadly ovate to oblong, cuneate or rounded at base, 1 in. long, on short stout petioles : peduncles little exceeding the leaves : bracts ovate-oblong : fl. deep purple, $\frac{1}{2}$ in. long : fr. viscid-pubescent, the wings somewhat coriaceous. Along beaches from near Santa Barbara southward. June-Oct.

+ + *Wings thicker, the central cavity of the fruit extending through them.*

3. **A. latifolia,** Esch. Mem. Acad. St. Petersb. x. 281 (1826) : *A. arenaria,* Menzies, in Hook. Exot. Fl. iii. t. 193 (1827). Perennial, stout and succulent, very viscid, the stems prostrate, 1-2 ft. long : leaves broadly ovate or reniform, $\frac{1}{2}$-1$\frac{1}{2}$ in. long, obtuse : peduncles usually exceeding the leaves : bracts 5, rounded to ovate or oblong, 2-4 lines long : fl. numerous, 5 or 6 lines long, bright yellow, very fragrant, the lobes emarginate : fr. 4-6 lines long, coriaceous, acute at each end ; wings usually narrow. Plentiful along the seashore from Monterey northward.

* * *Wings membranous, orbicular, encircling the fruit.*

4. **A. Crux-Maltæ,** Kellogg, Proc. Calif. Acad. ii. 71. f. 16 (1863). Annual, branched from the base, stout and succulent, 6–10 in. high, sparingly pubescent and viscid : leaves ovate-oblong, 1 in. long, narrowed to rather long petioles : peduncles shorter than the leaves : bracts lanceolate-acuminate, united at base : fl. 7–9 lines long ; tube greenish; limb rose-color, 4-lobed, the lobes deeply cleft : fr. 5–6 lines long, pubescent, coarsely reticulate-pitted, the ovate body long-stipitate : achene $2\frac{1}{2}$ lines long. Deserts of western Nevada, but found near Reno, *Sonne*, and to be expected within our State. May, June.

Order XXIV. AMARANTACEÆ.

Robert Brown, Prodromus Floræ Novæ Hollandiæ, 413 (1810). *Amaranthi*, Juss. Gen. 87 (1789).

Herbs with simple exstipulate leaves, and small inconspicuous (mostly greenish) axillary solitary or clustered perfect or unisexual flowers. Calyx of 3–5 hypogynous more or less scarious persistent sepals, occasionally with a pair of bractlets at base, generally enveloped by dry and almost chaffy bracts. Corolla 0. Stamens usually 5 or more, distinct or monadelphous. Stigmas 2 or 3, sessile on an undivided style. Fruit utricular, sometimes circumscissile, or bursting irregularly. Seed small, compressed, vertical. Embryo curved.

1. **AMARANTUS,** *Dodonæus.* Annual weeds ; leaves alternate, usually broad, veiny, and tipped with a short sharp mucro. Flowers green or purplish, in axillary spiked clusters or spikelets, the staminate usually mingled with the pistillate in the same cluster. Sepals distinct or united at base, seldom less than 3 or more than 5, more or less scarious, erect, or the tips spreading. Stamens as many as the sepals, distinct. Stigmas linear. Utricle ovate, 2–3-beaked, circumscissile or indehiscent often deciduous with the perianth.

* *Sepals distinct, oblong-lanceolate, erect; fl. monœcious.*
+ *Stout, erect; flower-clusters in naked terminal and axillary spikes; sepals 5.* AMARANTUS proper.

1. A. RETROFLEXUS, Linn. Sp. Pl. ii. 991 (1753). Stout, 1–4 ft. high, paniculately branched above ; herbage dull green, roughish and more or less pubescent : leaves ovate or rhombic-ovate, 1–4 in. long, on slender petioles not so long : fl. green, in erect or somewhat spreading nearly cylindrical spikes : bracts lanceolate-subulate, scarious except the green carinate midrib, attenuate to a rigid awn, $1\frac{1}{2}$–3 lines long : sepals narrowly oblong, mostly acute or even mucronate, exceeding the utricle: seed $\frac{1}{2}$ line broad, black and shining, with a rather obtuse margin. Gardens and waste lands ; native of tropical America.

+ + *Low, diffuse or prostrate; sepals 1–3; fl. in small axillary clusters.*
++ *Sepals 3.* Genus DIMEIANTHUS, Raf.

2. **A. albus**, Linn. Sp. Pl. 2d ed. ii. 1404 (1763). Erect, 1_2–2 ft. high, rigidly and widely branched from the base; herbage of a light green, glabrous or nearly so: leaves oblong-spatulate to obovate, 1_2–1^1_2 in. long including the slender petiole, obtuse or retuse, often crisped: spikelets 4–5-flowered: bracts subulate, rigid, pungently awned, 1–2^1_2 lines long, the lateral ones reduced or wanting: sepals oblong-lanceolate, subulate-mucronate, shorter than the somewhat rugose utricle: seed 1_3 line broad, black and shining, very sharply margined. Too well known in the prairie regions of North America, under the name of *Tumble-weed*; only occasional in California.

3. **A. blitoides**, Wats. Proc. Am. Acad. xii. 273 (1878). Somewhat succulent, weak and prostrate, the branches often 1–2 ft. long, whitish, the foliage of a rather deep shining green, glabrous or nearly so: spikelets few-flowered and contracted: bracts ovate-oblong, shortly acuminate, about equal, 1–1^1_2 lines long, little longer than the oblong obtuse and mucronulate or acute sepals: utricle smooth, little surpassing the sepals: seed 3_1 line broad, abruptly but rather obtusely margined. Very common in the Rocky Mountain region, where it is indigenous; becoming established along the railroad at Niles, Suisun, and perhaps elsewhere in the State, but an immigrant.

++ ++ *Sepals and bracts only 1 each to the fertile flower.* Genus
MENGEA, Schauer.

4. **A. Californicus**, Wats. Bot. Calif. ii. 42 (1880); Moq. in DC. Prodr. xiii^2. 270 (1849), under *Mengea*. Stems stoutish and rather fleshy, branched from the base, prostrate, the branches 1–1^1_2 ft. long, with many short lateral branchlets: leaves obovate or oblong, 1 in. long or less, including the short petiole, obtuse or acutish, with white veins and margins: fl. green or purplish, in many small dense axillary clusters: bract more or less scarious, little exceeding the utricle: sepals of staminate fl. 3_1 line long; of the fertile shorter: utricle slightly rugose, tardily circumscissile: seed 1_2 line broad, obscurely margined. In low and moist rather alkaline soils, from Monterey, *Hartweg*, to Stockton, *Sanford;* also in the interior states of Nevada, Idaho, etc.

2. **NITROPHILA**, *S. Watson*. A low perennial branching glabrous herb, with opposite amplexicaul fleshy leaves, and axillary subsessile perfect bibracteate flowers. Perianth of about 5 equal erect concave and carinate sepals. Stamens 5–7, joined at base into a narrow perigynous disk; anthers 2-celled: staminodia 0. Style short; stigmas 2, slender. Utricle subglobose, indehiscent, 1-seeded, beaked with the slender style, included within the connivent sepals.

1. **N. occidentalis**, Wats. Bot. King Exp. 297 (1871); Moq. in DC. Prodr. xiii^2. 279 (1849), under *Banalia*. Stems erect from a decumbent

base and running rootstocks, 3 -8 in. high, angular : lowest leaves broadly ovate or oblong, 2 3 lines long, the others linear, semiterete, 1½–1 in. long, acuminate, cuspidate : bracts much like the leaves but shorter, about twice the length of the flowers: fl. 1—3 in each axil, the lateral ones often pedicellate, 2 –3-bracted, the central one often bractless : sepals 1 line long, ovate, acutish, rather rigid, exceeding the stamens and style : seed ½ line broad, black and shining.- In alkaline lowlands of the interior, from near Sacramento, *Pickering*, and Lathrop, *Greene*, southward and eastward.

Order XXV. SALSOLACEÆ.

Linnæus, Classes Plantarum, 507 (1738). *Blita*, Adans. Fam. ii. 258 (1763). *Atriplices*, Juss. Gen. 83 (1789). *Chenopodeæ*, Vent. Tabl. ii. 253 (1799). *Chenopodiaceæ*, Lindl. Intr. 2d ed. 208 (1836).

Herbs or shrubs, often succulent, glabrous, pubescent, mealy or scurfy, sometimes leafless. Flowers clustered, apetalous. Perianth of a solitary bract-like sepal, or of 2 which are distinct and valvate or more or less united, or of five distinct or united at base and calyx-like, never scarious. Stamens as many as the sepals and opposite to them, or fewer ; anthers 2-celled. Ovary 1-celled, 1-ovuled, becoming an utricle or achene enclosed in the persistent perianth. Embryo annular or spiral ; albumen mealy or wanting.—A rather large order, closely connecting Amarantaceæ and Portulaceæ, hardly separable from the former except by vegetative characters ; containing many garden and field pests (goosefoot, pig weed) and as many useful plants like the beet, spinach, orach, etc. Several shrubby species of West American desert plains are valued forage plants, and the herbaceous kinds abound along the seaboard, in salt marshes, or on subsaline plains of the interior.

Hints of the Genera.

Stems nearly or quite leafless, stout, fleshy, cylindrical, articulated, - - - 8
Stems leafy ;
    Leaves fleshy, terete, - - - - - - - - - - 9
    " plane, fleshy or membranaceous ;
        Perianth of 1 bract-like sepal, - - - - - - 3
        " campanulate, 3—5-toothed, - - - - - 2
        " of fertile fl. 5-cleft or -divided, - - - 1, 4
        " of fertile fl. of 2 more or less united bracts, 5, 6, 7

1. **CHENOPODIUM,** *Tabernaemontanus* (GOOSEFOOT. PIG WEED). Herbs with alternate petiolate mostly angular foliage. Flowers small, greenish, sessile, clustered in axillary or terminal spikes or cymes, perfect, or pistillate only, bractless. Perianth herbaceous, 3 5-parted ; lobes imbricate, often carinate or crested, persistent and more or less covering the fruit, remaining green and herbaceous or becoming colored and fleshy. Stamens 5 or fewer. Styles, 2, 3 or 4, slender. Pericarp

membranous, closely investing the lenticular horizontal or vertical seed. Embryo annular, or curved around a copious albumen.

* *Annual, more or less mealy, not pubescent; seed horizontal; embryo annular.* CHENOPODIUM proper.

← *Pericarp closely persistent upon the seed.*

1. C. ALBUM, Linn. Sp. Pl. i. 219 (1753). Erect, stoutish, more or less paniculately branching, 1–4 ft. high; herbage pale green or whitish with a mealy indument: leaves petiolate, ascending, rhombic-ovate, obtuse, acute or cuneate at base, sinuate-dentate or subentire, 1–2 in. long, whiter beneath than above; flowers densely clustered in close spikes, these forming a rather strict leafless panicle: sepals of fruiting calyx carinate, completely covering the fruit: seed smooth, shining, acutely margined.—A very common weed of fields, gardens and waste places; native of Europe. June–Oct.

2. C. VIRIDE, Linn. l. c. (1753): *C. concatenatum*, Thuill. Fl. Par. 125 (1799): *C. album*, var. *viride*, Moq. DC. Prodr. xiii^2. 71 (1849). Size and general habit of the preceding, but herbage green throughout and scarcely mealy: branches and leaves more spreading: fl. and fr. scattered in loose spreading spikes. —Rather frequent in cultivated grounds among the foothills of the Sierra; from Europe like the last.

3. C. MURALE, Linn. l. c. Stoutish and rather low, often with many decumbent or ascending branches from the base: herbage dark green, rather succulent, the growing parts very mealy: leaves petiolate, ascending, ovate-rhomboid, unequally and sharply toothed: fl. in rather dense axillary nearly leafless cymes: fruiting calyx nearly closed, the sepals slightly carinate: seed opaque, punctate-rugose, sharply margined.—A more common weed than either of the preceding, preferring rich soil, but thriving everywhere; very hardy, often flowering and fruiting throughout even our whole winter season.

4. C. VULVARIA, Linn. l. c. 220. Rather slender and diffuse, 1 ft. high or more; herbage somewhat pale and mealy, very ill-scented: leaves petiolate, ascending, obtuse or acutish, entire, 1 in. long: fl. in dense leafless spicate clusters: fruiting calyx closed, not carinate: seed shining but delicately puncticulate, rather sharply margined. Abundantly naturalized in the vicinity of Sacramento, though not otherwise known within the State.

← ← *Pericarp separating readily from the seed.*

5. C. Fremonti, Wats. Bot. King Exp. 287 (1871). Erect, slender, 1–2 ft. high, whitish-mealy: leaves broadly triangular-hastate, ¼–1 in. long, obtuse or abruptly acute, truncate or cuneate at base, the upper narrower and from oblong to linear-lanceolate: fl. in small clusters upon slender open-panicled branchlets: sepals strongly carinate: seed smooth

and shining. — Plant of the interior Basin, reaching our borders along the eastern base of the Sierra.

6. **C. leptophyllum**, Nutt. in DC. Prodr. xiii². 71 (1849), under *C. album*; Wats. Proc. Am. Acad. ix. 94 (1874) : *C. album*, var. *leptophyllum*, Moq. in DC. l. c. Erect and strict, simple or branching, $1_{\frac{1}{2}}$–3 ft. high ; herbage white-mealy or glabrate : leaves lanceolate or linear, entire, $1_{\frac{1}{2}}$–1 in. long, acute, usually mucronate, short-petioled : fl. in short dense clusters formed into close or interrupted spikelets ; sepals acute, strongly carinate ; seed black and shining. — Same range as the preceding.

* * *Herbage not mealy, glandular-pubescent and aromatic; seed horizontal (except in n. 10); embryo curved.* Genera AMBRINA and BOTRYDIUM, Spach.

7. **C. BOTRYS**, Linn. Sp. Pl. i. 219 (1753). *Botrydium aromaticum*, Spach. Phaner. v. 299 (1836). Annual, erect, often widely branching, 1–2 ft. high, glandular-pubescent and highly aromatic : leaves ovate or oblong, 1–2 in. long, sinuate-pinnatifid, the lobes often toothed : fl. scattered in very numerous slender axillary cymose panicles : sepals acute, loosely investing the fruit : pericarp persistent : seed $1_{\frac{1}{3}}$ line broad, thick-lenticular, black and shining. — Frequent in the interior of the State ; native of S. Europe, and commonly called *Jerusalem Oak*.

8. **C. ANTHELMINTICUM**, Linn. l. c. 220; Spach, l. c. 298, under *Ambrina*. Perennial, stems stoutish, decumbent, 1—2 ft. long ; herbage light green, glandular-puberulent, pleasantly aromatic : leaves thin, oblong, narrowed at base, obtuse, sinuate-serrate or sometimes remotely dentate, 1 in. long or less : inflorescence a terminal leafless panicle of dense but slender spikes : sepals not carinate, completely enclosing the fruit : seed smooth and shining, obtusely margined. Not rare among the foothills of the Sierra from Shasta Co. southward, by waysides.

9. **C. AMBROSIOIDES**, Linn. l. c. 219 ; Spach, l. c. 297, under *Ambrina*. Annual, erect or ascending, 2–3 ft. high, deep green, glabrous or slightly scabrous, the foliage occasionally puberulent : leaves oblong, attenuate at each end, acutish, remotely sinuate-toothed or entire, the uppermost and floral linear-lanceolate : inflorescence loosely spicate and leafy : fruiting perianth completely closed : seed smooth and shining, obtusely margined.— Very common by waysides and in waste lands at the outskirts of cities and villages along the seaboard ; said to be native of tropical America, but in our district too hardy, flowering and fruiting all the year round and becoming suffrutescent. It is less aromatic than *C. anthelminticum*, and manifestly distinct from it, though there seem to be natural hybrids between them where they meet.

10. **C. CARINATUM**, R. Br. Prodr. 407 (1810) ; Moq. DC. Prodr. xiii². 81 (1849), under *Blitum*. Annual, slender, diffusely branched from the

base, the branches 6-12 in. long; herbage pubescent, glandular and aromatic: leaves ovate, oblong or lanceolate, 1 in. long or less, sinuate-pinnatifid, deep green above, paler beneath and somewhat glaucous: fl. glomerate in the axils: stamen 1: fruiting perianth only partly enclosing the vertical utricle; segments obtusely carinate, or at least thickened on the back: seed black and shining, the margin acute. Introduced from Australia, and frequent in the foothills of the Sierra, at Ione, etc.

* * * *Glabrous or slightly mealy; seed vertical, more or less exserted from the more gamophyllous perianth; embryo annular.* Old genus BLITUM.

11. **C. Californicum,** Wats. Bot. Calif. ii. 48 (1880); also Rev. Chenop. 101 (1874), under *Blitum*. Stems several from a long fusiform perennial root, stout, decumbent, mostly simple, 1-3 ft. high; herbage light but rather dull green, the young parts a little mealy: leaves broadly triangular-hastate, 2-3 in. long, truncate or with sinuses at base, acuminate, sharply, unequally, and often deeply sinuate-dentate: fl. in dense clusters in a long simple terminal spike: perianth campanulate, rather deeply 5-toothed, enfolding the utricle only loosely: pericarp persistent: seed somewhat compressed, $\frac{3}{4}-1$ line broad. Common both on the seaboard and in the interior; the native American counterpart of the Old World *C. Bonus Henricus*, to which it was in early days referred.

12. **C. glaucum,** Linn. Sp. Pl. i. 220 (1753); Koch, Fl. Germ. 608 (1837), under *Blitum*. Annual, stout and rather fleshy, erect with ascending branches, $\frac{1}{2}$-1 ft. high: leaves ovate to oblong-lanceolate, 1 in. long, obtuse, petiolate, remotely and rather coarsely dentate, glabrous and green above, paler and mealy beneath: fl. in axillary spiked clusters: perianth small, with rounded lobes, not quite concealing the vertical, or as often horizontal utricle. In the Suisun marshes on elevated and dry ground; apparently indigenous, though possibly introduced from Europe, where it is a common barnyard weed.

2. **ROUBIEVA,** *Moquin*. Herb perennial, glandular, heavy-scented, with alternate pinnatifid leaves, and flowers few or solitary in the axils. Perianth bractless, deeply campanulate, 3-5-toothed, at length saccate and contracted over the fruit, 3-5-nerved, net-veined. Stamens 5, included. Styles 3, somewhat lateral, exserted. Pericarp membranous, glandular-dotted, deciduous. Seed vertical, lenticular; embryo annular around a copious albumen.

1. R. MULTIFIDA, Moq. Ann. Sc. Nat. 2d ser. i. 293 (1834); Linn. Sp. Pl. i. 220 (1753), under *Chenopodium*. Stems several from an oblong or fusiform root, prostrate, 1 ft. long or more, branching and leafy: herbage pale-green, glandular-puberulent, aromatic: leaves 1-1½ in. long, lanceolate to linear, short-petiolate, deeply pinnatifid with narrow lobes, the nerves beneath very prominent: fl. in dense glomerules: fruiting

perianth reticulate-nerved; segments ovate, obtusish; pericarp whitish and with scattered glandular dots: seed subrostellate, obtusely margined, dark brown, shining and minutely punctate-rugose.—Native of South America; credited to Plumas Co., *Mrs. Ames*, as adventive or naturalized.

3. MONOLEPIS, *Schrader.* Annuals, with the habit and foliage of *Chenopodium*, but the perianth consisting of a single scale-like or bract-like sepal (or this to be regarded as a mere bract subtending an achlamydeous flower). Stamen 1. Styles 2, filiform. Pericarp membranous, persistent upon the vertical compressed seed. Embryo annular; albumen copious.—In aspect wholly like *Chenopodium*, to which the genus may as well be united as *Munyea* to *Amarantus*.

1. **M. Nuttalliana,** Greene. Rom. & Schult. Mant. i. 65 (1822), under *Blitum: M. chenopodioides*, Moq. in DC. Prodr. xiii. 85 (1849). *Blitum chenopodioides (?)* Nutt. (1818), not Linn. Branches many, decumbent or almost prostrate, ¼—1 ft. long; herbage deep green, the growing parts mealy: leaves lanceolate-hastate, ½—1 in. long, entire or remotely sinuate-dentate, acute or obtuse, cuneate at base, the upper floral subsessile: flower-clusters axillary, dense, sometimes reddish: sepal foliaceous and fleshy, oblanceolate or spatulate, often exceeding the fruit: pericarp somewhat fleshy, becoming dry and favose-pitted, adherent: seed lenticular or reniform, ½ line broad.—Alkaline soils along the eastern base of the Sierra.

2. **M. spathulata,** Gray, Proc. Am. Acad. vii. 389 (1868). Smaller and more slender than the last, more diffuse and leafy, but leaves smaller, narrowly oblanceolate or spatulate, entire, ½ in. long: sepal rarely exceeding the fruit: pericarp minutely papillose, separating from the minute shining seed.—A rare species of the more volcanic districts of the Sierra Nevada; Mono Lake, and in Sierra Co.

4. BETA, *Columna* (BEET). Rather coarse glabrous biennials, with alternate leaves, the radical large and long-petioled, the floral reduced and sessile. Flowers fascicled in the axils and spicate-congested along the paniculate branches, connate at base, perfect. Sepals 5, inserted on the margin of a concave receptacle, imbricate. Stamens 5, opposite the sepals, the filaments subulate. Ovary partly inferior and encircled by a disk-like margin of the receptacle; style short, the 2 or 3 branches stigmatose on the inside. Fruit partly adnate to the receptacle, and enclosed by the thicked and somewhat fleshy sepals.

1. **B. vulgaris,** Linn. Sp. Pl. i. 222 (1753). Stout, 2—4 ft. high: radical leaves often 1 ft. long including the stout petiole, commonly with prominent nerves and a more or less undulate margin, the outline oblong

or oval: inflorescence 1-3 ft. long. Escaped from gardens to moist
lands bordering the salt marshes, where it is becoming a common weed
in some places.

5. **ATRIPLEX,** *Pliny* (ORACHE.) Herbs or shrubs, mealy or scurfy,
monœcious or diœcious: inflorescence axillary and glomerate, or terminal
and spicate or panicled. Staminate perianth bractless, 3-5-parted,
enclosing as many stamens. Pistillate fl. bibracteate, without perianth
or rarely with 2-4 distinct hyaline sepals; the bracts erect, appressed,
distinct or more or less united, their margins often becoming dilated, the
surface sometimes in age thickened indurated and muricate. Fruit
compressed, utricular. Seed vertical. Embryo annular, around copious
albumen. A large and perplexing genus of plants in eastern N. America:
the first group not naturally separable from *Chenopodium*; but some of
the species so different from others in fruit as to almost demand the
reinstatement of several genera which have latterly been rejected.

* *Monœcious annuals, somewhat succulent and mealy; bracts distinct or
nearly so, ovate-oblong to broadly triangular or hastate.*

1. **A. hastata,** Linn. var. **oppositifolia,** Moq. DC. Prodr. xiii^2. 95
(1849): *A. oppositifolia*, DC. Rapp. i. 12 (1808). Rather slender, with
divaricate and somewhat decumbent branches 2-3 ft. long, or stouter
and erect with ascending branches; herbage mealy, not very succulent:
leaves triangular-hastate or deltoid, mostly entire, all the lower opposite:
flower-clusters small, spicate: bracts small, triangular, entire or denticulate $\frac{1}{4}$ in. long: seed 1 line long, dark colored.— Common along the
borders of brackish marshes at Petaluma, and elsewhere to the westward
of San Francisco Bay.

2. **A. patula,** Linn. Sp. Pl. ii. 1053 (1753). Stout and succulent,
mostly erect, 1 ft. high, with few ascending branches; herbage deep
green, only the growing parts somewhat mealy: lowest leaves often
opposite, broadly lanceolate, sometimes with hastate base: inflorescence
more or less leafy at base: bracts rhombic-ovate, thick and subcoriaceous, often $\frac{1}{2}$ in. long. Very common in salt marshes and near beaches.

3. **A. phyllostegia,** Wats. Proc. Am. Acad. ix. 108 (1879); Torr. Bot.
King. Exp. 291 (1871), under *Obione*. Erect, with short ascending
branches, 6-18 in. high, pale and mealy-scurfy: leaves alternate,
rhombic-ovate or -lanceolate, acuminate, with salient narrow hastate
lobes toward the base, otherwise entire, 1 in. long: almost diœcious: fl.
mostly axillary: staminate calyx 5-parted: bracts linear-lanceolate or
broader, acute or acuminate, $\frac{1}{4}$-$\frac{1}{2}$ in. long, foliaceous, the sides
indurated in fruit, 3-nerved, the lateral nerves often bituberculate: seed
brownish, $\frac{1}{2}$ line broad; radicle nearly superior. In subsaline soil near
Lathrop, and southward.

4. **A. spicata,** Wats. l. c. Stout, erect, 1–2 ft. high, sparingly branching, mealy : leaves alternate, rhombic-ovate, acute, coarsely and irregularly sinuate-toothed, 2 in. long, attenuate to a short petiole : fl. densely spicate, the 4-sepalous calyx usually staminate, but not rarely pistillate and with a horizontal seed : bracts of pistillate fl. ovate, acute, little enlarged in fruit, partly coherent at base 1½ lines long : seed black, ½ line broad : radicle inferior.— Alkaline soil among the foothills of the Mt. Diablo Range, on Marsh's Creek, also near Livermore, and on low plains of the Sacramento near Chico : a common weed in fields, and one of several plants in which the supposed distinctions between *Atriplex* and *Chenopodium* fail.

* * *Herbs or shrubs, seldom succulent or mealy, but silvery-scurfy; bracts mostly rounded and more or less completely united, naked or variously appendaged or winged, frequently hard and nut-like in fruit.—*
Genus OBIONE, Gaertn.
← *Monoecious annuals.*

5. **A. argentea,** Nutt. Gen. i. 198 (1818) : Moq. Chenop. Enum. 76 (1840), under *Obione*. Stout, erect, ½–1½ ft. high, diffusely branching, the lower branches often decumbent ; herbage densely mealy-scurfy : leaves alternate (except the lowest), triangular-hastate to rhombic-ovate, acute or obtuse, ½–2 in. long, mostly sessile : staminate fl. in terminal spicate clusters, or in the upper axils ; calyx deeply 5-cleft : fertile fl. short-pedicellate in axillary clusters ; bracts when mature 2–4 lines long, rhombic-ovate, united, indurated and spongy, margined except at base, bifid at apex, sharply and deeply toothed : sides usually roughened with irregular herbaceous projections or with a double toothed crest : seed 1 line broad. Eastern slope of the Sierra, from Sierra Co. southward, in alkaline soils.

6. **A. nodosa,** Greene, Pittonia, i. 40 (1887). Stout, branched from the base, 1 ft. high, mealy and apparently scabrous ; leaves broadly rhomboid : fruit-clusters borne at the enlarged nodes of the widely and irregularly branching stem : pedicels stout, thickened under the bracts ; these united and forming an almost globose fruit 2 lines in diameter, 3-lobed at summit, the sides covered with lichenoid spongy projections.— Near Antioch, *Mrs. Curran;* unknown except in the fruiting state, but remarkable for the nodose stem and branches, and the subglobose rough fruits.

7. **A. expansa,** Wats. Proc. Am. Acad. ix. 116 (1874). Erect, but with many decumbent and widespread branches, these 1–2 ft. long ; herbage silvery-scurfy and slightly mealy : leaves triangular and somewhat hastate, acute, 1 in. long or more, sessile : flowering branches leafy and virgate ; staminate spikes slender, interrupted, naked above : fruiting

bracts compressed, the sides usually unappendaged and strongly reticulate. Same range as *A. argentea*, though reported on our side of the Sierra, in Santa Barbara Co., and therefore to be expected farther north.

8. **A. coronata**, Wats. l. c. 114. Stout, erect, 1–2 ft. high, branching and leafy, mealy : leaves lanceolate, entire, $\frac{1}{2}$–1 in. long, acute or acuminate, attenuate to a short petiole or sessile : flower clusters axillary, androgynous ; fruiting bracts strongly compressed, orbicular, 2 lines broad, united, surrounded by an herbaceous gash-toothed margin as broad as the body, the sides now and then somewhat muricate : seed $\frac{3}{4}$ line broad. Plains of the lower San Joaquin and southward.

9. **A. bracteosa**, Wats. l. c. 115 ; Dur. & Hilg. Pac. R. Rep. v. 13. t. 14 (1855), under *Obione*. Stout, branched from the base, 2–3 ft. high, mealy : leaves thin, sessile, lanceolate, acute or acuminate, $1\frac{1}{2}$–1 in. long, sinuate-dentate or the uppermost entire : staminate fl. in dense globose clusters in a naked terminal simple or branching spike ; calyx deeply 5-cleft : fruiting bracts in small axillary clusters, cuneate-orbicular, 1–$1\frac{1}{2}$ lines broad, the upper rounded margin irregularly toothed ; sides smooth or muricate ; seed less than $\frac{1}{2}$ line broad.— From the plains of Tulare Co. southward.

10. **A. Coulteri**, Dietr. Syn. v. 537 (1852) ; Moq. DC. Prodr. xiii². 113 (1849), under *Obione*. Erect, 1–2 ft. high, slender, virgate, rigid, branched and mealy : leaves lanceolate, attenuate at each end, mucronulate, entire, rather thick, $\frac{1}{2}$ in. long, 1 line broad or less : fruiting bracts in axillary clusters, cuneate-orbicular, 1 line broad, the rounded margin reaching nearly to the base, and with short blunt teeth : seed $\frac{1}{2}$ line broad. A rare and long lost species, collected only by Coulter, now sixty years since, perhaps near Monterey ; the species possibly belonging to the shrubby group.

+ + *Diœcious shrubby species (except n. 11).*

11. **A. Californica**, Moq. DC. l. c. 98 (1849). Branches many, slender and wiry, prostrate, from a short and thick oblong or fusiform perennial root ; herbage densely mealy : leaves ovate- to linear-lanceolate, 3–8 lines long, entire, acute, the lowest opposite : flower-clusters all axillary, the upper ones more staminate, the calyx of these deeply 4-cleft : fruiting bracts rhombic-ovate, membranous, distinct, $1\frac{1}{2}$ lines long, somewhat convex : seed $1\frac{1}{2}$ line broad.— On the seacoast, and along the edges of salt marshes, from near San Francisco and Alameda, southward. The rather succulent roots are yellow, and have the flavor of beets.

12. **A. lentiformis**, Wats. Proc. Am. Acad. ix. 118 (1874) ; Torr. Sitgr. Rep. 169. t. 14 (1853), under *Obione*. Diffusely branched, 2–12 ft. high, the branches terete, the branchlets divaricate, rigid and somewhat

spinescent, closely scurfy : leaves ovate- to oblong-rhombic or somewhat hastate, cuneate at base, $\frac{1}{2}$–$1\frac{1}{2}$ in. long : staminate calyx 5-parted : fruiting bracts orbicular, 1–2 lines broad, strongly compressed, united to above the middle, the free margins obscurely crenate: seed dark, $\frac{2}{3}$ line broad. From Tulare Co., southward and eastward.

13. **A. Breweri**, Wats. l. c. 119. Diffusely branched but erect, 6 ft. high, grayish-puberulent ; the branches terete, often long and flexuous : leaves ovate-oblong or somewhat rhomboid, cuneate at base, obtuse or acutish, 1–2 in. long : staminate calyx deeply 4-cleft : fruiting bracts ovate or orbicular, united at the margin to the middle, entire, convex, 1–$1\frac{1}{2}$ lines broad. Near the sea at Santa Barbara and southward ; but also in Monterey Co., on the Salinas River, *Abbott*.

14. **A. leucophylla**, Dietr. Syn. v. 536 (1852) ; Moq. DC. Prodr. xiii². 109 (1849), under *Obione*. Stout, shrubby, but the stem and branches flexible and mostly reclining, 1–2 feet long ; plant hoary-scurfy throughout : leaves thick, broadly obovate, obtuse or acutish, cuneate at base, sessile, 3-nerved, $\frac{1}{2}$–$1\frac{1}{2}$ in. long : staminate fl. in dense clusters in short terminal spikes ; calyx large, 5-cleft : fruiting bracts in axillary clusters 2–4 lines long, rhombic-ovate, united, spongy, the sides 2-crested, the narrow margin entire or obscurely toothed.– On sand beaches of San Francisco Bay, and along the seacoast southward.

6. **EUROTIA**, *Adanson*. (WHITE SAGE). Low shrubs stellate-tomentose, with alternate entire leaves, and diœcious flowers in spicate terminal clusters. Staminate flowers bractless ; calyx 4-parted : stamens 4, with slender exserted filaments. Pistillate flowers bibracteate, without perianth. Fruiting bracts obcompressed, united, becoming enlarged and membranaceous, densely hairy, not winged or appendaged. Styles 2, somewhat hairy, exserted. Fruit oblong-ovate, sessile, the pericarp membranous and rather firm, pubescent. Seed vertical, obovate, with a simple membranous testa. Cotyledons broad and green ; radicle inferior.

1. **E. lanata**, Moq. Chenop. Enum. 81 (1840) ; Pursh, Fl. ii. 602 (1814), under *Diotis*. Branched from the base and erect, 1 ft. high or more, white-tomentose, becoming reddish with age, the branches strict, leafy : leaves linear to narrowly lanceolate, obtuse, $\frac{1}{2}$–$1\frac{1}{2}$ in. long, obtuse, the margins revolute : calyx-lobes ovate, acute, hairy : fruiting bracts lanceolate, 2–3 lines long, adorned with 4 dense tufts of long white hairs, and beaked above with 2 short erect horns : utricle filling the cavity and loosely enveloping the seed, which is $1\frac{1}{2}$ lines long. –A common forage shrub of the Great Basin and Rocky Mountain regions ; reaching our borders along the eastern base of the Sierra.

7. **GRAYIA**, *Hooker & Arnott*. Slightly scurfy or mealy shrubs, with alternate entire leaves, and small diœcious or monœcious flowers in

axillary clusters or terminal spikes. Staminate flowers bractless ; calyx 4-parted ; stamens 4 or 5, with short subulate filaments. Pistillate flowers bibracteate, without perianth. Bracts membranous, strongly obcompressed, joined into an orbicular sac with a small naked orifice at apex, enlarged in fruit, net-veined and wing-margined. Styles 2. Pericarp thin and membranous. Seed vertical, orbicular, with thin membranous testa. Radicle inferior.

1. **G. spinosa,** Moq. DC. Prodr. xiii^2. 119 (1849); Hook. Fl. ii. 127 (1840), under *Chenopodium: G. polygaloides,* H. & A. in Hook. Ic. iii. 271 (1840, but later in the year). Erect, branching and spinescent, 1-3 ft. high : leaves glabrous, or when young mealy, somewhat fleshy, oblanceolate or spatulate to obovate, $\frac{1}{2}$-1$\frac{1}{2}$ in. long, obtuse or acute, narrowed at base: staminate fl. in axillary clusters ; pistillate terminal and in simple or branching spikes : fruiting bracts 3-6 lines broad, sessile, glabrous, thin, white or reddish, coherent below the pedicel of the ovary: styles slender, at first exserted : seed nearly central, $\frac{2}{3}$ line broad.— In alkaline soils east of the Sierra both northward and southward.

8. **SALICORNIA,** *Tournefort* (SAMPHIRE). Herbs or shrubs with cylindrical fleshy jointed and apparently leafless branches. Flowers very simple, in threes at the joints of the spike-like ends of the branches : the lateral ones of each trio often only staminate. Perianth of 4-5 distinct or variously united sepals, at length spongy-thickened about the fruit. Stamens 1 or 2. Styles 2 or 3, short. Pericarp membranaceous, adherent to, or free from the vertical seed.

* *Branches and flowers opposite.* SALICORNIA proper.

1. **S. ambigua,** Michx. Fl. i. 2 (1803); Moq. DC. Prodr. xiii^2. 151 (1849), under *Arthrocnemum.* Perennial, decumbent, often rooting at the base, usually freely branching, $\frac{1}{2}$-1$\frac{1}{2}$ ft. high : spikes not thicker than the sterile parts of the branches, $\frac{1}{2}$-2 in. long : perianth sac-like, with an anterior opening (formed of 2 sepals united above and below), enclosing the fruit : pericarp membranous, adherent to the obovate-oblong seed, this $\frac{1}{6}$ line long, pubescent. Plentiful in salt marshes everywhere along the seaboard. Some forms appear to be annual, and the species as here received may be complex — embracing several. The subject calls for special study and investigation.

* * *Branches alternate and flowers spirally arranged in the spikes.*—
Genus SPIROSTACHYS, Wats.

2. **S. occidentalis,** Greene. Wats. Bot. King Exp. 293 (1871), under *Halostachys,* and Proc. Am. Acad. ix. 125 (1874), under *Spirostachys.* Shrubby, diffusely branched, the main stem erect, often 5 ft. high, with a close and smooth gray bark : scale-like crowded and fleshy leaves broadly triangular and acute, amplexicaul, often nearly obsolete : fl.

densely spiked ; perianth of 4 or 5 concave carinate sepals more or less united ; pericarp free from the oblong seed, this ½ line long or less. Plentiful in alkaline soil at Byron Springs; also in Tulare Co. In characters of the flower and fruit this group differs greatly from *Salicornia* proper; but there are wider differences, of the same kind, within the limits of *Atriplex* as now received, not to speak of the great diversities of habit between true *Atriplex* and the *Obione* section; consistency therefore forbids the dismemberment of the old *Salicornia*, at least until *Obione* and other genera shall have been restored.

9. **SUÆDA,** *Forskaal* (SEA BLITE). Saline herbs or shrubs, with alternate fleshy linear entire leaves, and axillary sessile usually perfect flowers. Perianth minutely bracteolate, 5-cleft or -parted, fleshy; lobes unappendaged, more or less carinate, crested or winged, enclosing the fruit. Stamens 5. Styles 2, 3 or 4, short and thick. Pericarp membranous, free or slightly adherent to the vertical or horizontal lenticular seed. Testa shining, black and crustaceous. Embryo spiral; albumen scant.

* *Annuals.*

1. **S. diffusa,** Wats. Proc. Am. Acad. ix. 88 (1874). Erect, ½–1½ ft. high, with elongated usually flexuous branches, glabrous or more or less pubescent; leaves semiterete, narrow at base, acute or acuminate, $\frac{1}{2}$–1 in. long, the floral ones shorter, rather distant on the branches; clusters 2–4-flowered; perianth cleft below the middle, the segments not carinate or appendaged; seeds usually vertical, ½ line broad, very smooth.—In alkaline soils east of the Sierra Nevada; but also on the western side near Fort Tejon, *Blake.*

2. **S. depressa,** Wats. l. c.; Pursh. Fl. i. 197 (1814), under *Salsola.* Low and usually decumbent, with short ascending branches; leaves semiterete, broadest at base, ¼–1 in. long, the floral ones shorter and oblong, or ovate, or ovate-lanceolate, rather crowded; perianth cleft to the middle; lobes somewhat unequal, acute, one or more of them strongly carinate or crested; seed vertical or horizontal, ½ line broad, very lightly reticulate.— East of the Sierra only.

* * *Perennials, often woody at base.*

3. **S. Californica,** Wats. l. c. 99 (1874). Stout, 2–3 ft. high, very leafy, glabrous or somewhat pubescent; leaves broadly linear, subterete, not wider at base, ½–1 in. long, acute, crowded on the branchlets; fl. large, 1–4 in each axil; perianth cleft nearly to the base; lobes not appendaged; seeds vertical or horizontal, nearly 1 line broad, faintly reticulate.—Apparently confined to the vicinity of sand beaches about San Francisco Bay, and seldom seen.

ORDER XXVI. **PORTULACEÆ.**

Jussieu, Genera Plantarum, 313 (1789).

More or less succulent herbs, with entire leaves and regular complete flowers which open in sunshine only. Sepals 2 (in *Lewisia* 4–8), sometimes cohering at base. Petals 5, (in *Lewisia* 8–16) often united at base. Stamens commonly 5 (3–∞), opposite the petals, hypogynous, perigynous or epipetalous; filaments distinct; anthers versatile. Ovary 1-celled, with few or many ovules on a central placenta. Seeds commonly strophiolate; embryo slender, curved or coiled around a mealy albumen. A small family, intimately related to both Caryophylleæ and Salsolaceæ; containing some valued ornamental plants, and several weedy species. The tender herbage of many, like purslane and Claytonia, recommends them as potherbs.

1. **PORTULACA,** *Lobelius* (PURSLANE). Fleshy annuals, with axillary and terminal yellow or rose-colored flowers. Sepals 2, united below and coherent with the base of the ovary; the limb free and deciduous. Petals 4–6. Stamens 7–20, perigynous with the petals. Style 3–8-cleft. Capsule circumscissile, opening by a lid. Seeds small.

1. P. OLERACEA, Linn. Sp. Pl. i. 445 (1753). Prostrate, glabrous, the herbage usually reddish or purplish; leaves flat, obovate, obtuse; sepals acute, carinate; petals $1\frac{1}{2}$–2 lines long, yellow; stigmas 5; capsule 3–5 lines long; seeds black, dull, minutely tuberculate. Native of S. Europe, and a very common weed of eastern N. America; already frequent in California.

2. **P. pilosa,** Linn. l. c. Stems ascending; linear and subterete leaves with long white hairs in their axils; fl. several, terminal; sepals membranaceous, not keeled, acute; petals 2–3 lines long, bright red; stamens 15–25; stigmas 5 or 6; seeds black, tuberculate.— Obtained on the State Survey in "dry sandy soils near the Soda Springs on the upper Sacramento," but not since reported.

2. **LEWISIA,** *Pursh*. Low acaulescent fleshy perennials, with thick fusiform roots, and short 1-flowered scapes. Sepals 4–8, broadly ovate, unequal, persistent, imbricate. Petals 8–16, large and showy. Stamens ∞. Style 3–8-parted nearly to the base. Capsule circumscissile at base, the upper and deciduous part more or less valvate-cleft. Seeds ∞, black and shining.—This genus should perhaps include *Calandrinia*, which is distinguishable only by the disepalous calyx.

1. **L. rediviva,** Pursh. Fl. ii. 368 (1814); H. & A. Bot. Beech. 344. t. 86; Hook. f. Bot. Mag. t. 5395; *L. alba*, Kell. Proc. Calif. Acad. ii. 115. fig. 36. Leaves densely clustered on the short thick caudex, linear oblong, glabrous, glaucous; scapes little exceeding the leaves, jointed at

the middle, and with 5–7 subulate scarious bracts whorled at the joint: sepals 6–8, broadly ovate, scarious-margined, 1½–3¼ in. long: petals 12–15, oblong, ½–1 in. long, pinkish or white: stamens 40 or more: capsule broadly ovate, ¼ in. long.—On Mt. Diablo, at the summit, *Brewer*, and in hills east of Napa, *Greene* (1874); common far to the northward and eastward of California.

2. **L. brachycalyx**, Engelm. Proc. Am. Acad. vii. 400 (1868). Leaves spatulate or almost linear: scapes jointless, bibracteate at base, shorter than the leaves: sepals 4, herbaceous, ¼ in. long: petals 7–9, oblong, 1½ in. long: stamens 10–15: capsule shorter than the calyx.—In beds of disintegrated granite on the eastern slope of the Sierra, in Fresno Co., at 8,000 ft. altitude, *Muir*.

3. **CALANDRINIA**, *Humboldt, Bonpland & Kunth*. Sepals 2 only, subequal, persistent. Petals 3–10. Stamens 3–25, apparently always hypogynous. Capsule 3-valved from the summit, or circumscissile at base.

* *Caulescent annuals; capsule 3-valved.*—CALANDRINIA proper.

1. **C. Menziesii**, T. & G. Fl. i. 197 (1838); Hook. Fl. i. 223. t. 70 (1833), under *Talinum*: *C. caulescens*, var. *Menziesii*, Gray, Proc. Am. Acad. xxii. 277 (1887) in part. Rather slender and diffuse, the branches 3–6 in. long: leaves linear-spatulate, mostly radical and long-peduncled; the upper and floral reduced and glandular-ciliate: sepals ovate, acuminate, the margins and sharp keel glandular-ciliate: corolla little exceeding the sepals, white or bright purple: stamens 3–10; seeds broadly ovoid, shining. From Santa Barbara, northward, through the Mt. Diablo Range, to Oregon. A small depressed glandular-ciliate and small flowered species, apparently quite distinct from the next. Apr., May.

2. **C. elegans**, Spach, Phaner. v. 232 (1836): *C. pulchella*, Lilja, Linnæa, xvii. 109 (1843): *C. speciosa*, Lindl. Bot. Reg. t. 1598 (1833), not of Lehm. Larger and stouter than the last, glabrous, the decumbent and ascending branches often 1 ft. long, flowering throughout: sepals ovate, acute or acuminate, less sharply carinate, the keel and margins entire or with a sparse short and flattened but in no wise glandular ciliation: stamens 10–15: corolla twice the length of the calyx, ³⁄₁ in. broad when expanded, bright rose-red: seeds larger, nearly orbicular. Very common throughout the Bay region and elsewhere in the State: passing currently for *C. Menziesii*. Apr. June.

3. **C. Breweri**, Wats. Proc. Am. Acad. xi. 124 (1876). Habit of the preceding but still larger, the ascending branches often more than 1 ft. high, glabrous: pedicels rather remote, in fruit deflexed: sepals broadly ovate, truncate at base, surpassed by the long-conical (1½ in. long) capsule: seeds dull, tuberculate. Collected only by Brewer, on

the Santa Inez Mountains; but to be expected in the southern extension of the Mt. Diablo Range.

* * *Acaulescent perennials; capsules circumscissile at base.*

Subgenus PACHYRHIZEA, Gray.

4. **C. Grayi,** Britton, Bull. Torr. Club, xvii. 312 (1890). *Talinum pygmæum,* Gray, Am. Journ. Sci. xxxiii. 407 (1862), also *Calandrinia pygmæa* in Proc. Am. Acad. viii. 623 (1873), not of Müller (1858). Glabrous; leaves linear, 1–2 in. long, with short and broadly winged subterranean petioles; scapes mostly simple and 1-flowered, 1–2 in. high, with a pair of small scarious bracts; sepals suborbicular, 2–3 lines long, glandular-toothed; petals red; capsule obtuse, nearly equalling the calyx. — In the Sierra Nevada at 8,000 ft., from Mt. Lyell northward.

5. **C. Nevadensis,** Gray, Proc. Am. Acad. viii. 623 (1873). Near the last, but larger; scapes 1–3 in. high, with a pair of larger and foliaceous bracts, 1–3-flowered; sepals entire; petals white. In the high Sierra with the last, though of more northerly range, from Summit, Cisco, etc.

4. **CLAYTONIA,** *Gronovius.* Glabrous herbs, often glaucous. Leaves radical except an involucral pair (sometimes united) under the racemose or subumbellate inflorescence of the usually scapiform peduncles. Sepals 2, persistent. Petals 5, equal, commonly united by their short claws. Stamens 5, hypogynous (when the petals are distinct), or each joined to the claw of its petal. Capsule membranaceous, ovoid or globose, 3-valved, elastically dehiscent, each valve elastically involute, ejecting the rather few black and shining seeds. Plants inseparable from the next genus except by the scapiform stems and involucrate inflorescence.

* *Perennials, with deep-seated tubers.*

1. **C. lanceolata,** Pursh, Fl i. 175. t. 3 (1814); *C. Caroliniana,* var. *sessilifolia,* Torr. Pac. R. Rep. iv. 70 (1857). Radical leaf lanceolate; cauline pair sessile, oblong or lanceolate to linear, 1–2 in. long; raceme nearly sessile, few-flowered, often with a scarious bract at base; sepals ovate, acutish; petals 2–4 lines long, emarginate or obcordate, rose-color or nearly white. From near Cisco in the Sierra, *Kellogg, Rev. Dr. Boute,* northward.

2. **C. triphylla,** Wats. Proc. Am. Acad. x. 345 (1875). Tuberous root small; stem slender, 2–3 in. high, bearing a pair, or a whorl of 3 narrowly linear leaves 2 in. long; fl. small, in a sessile or pedunculate compound raceme; bracts minute; petals oblong, 2 lines long, exceeding the rounded sepals. Sierra Nevada, at rather high altitudes, from Yosemite northward.

* * *Fibrous-rooted species.*

← *Perennials; pedicels axillary to manifest bracts (except in n. 5).*

3. **C. Sibirica**, Linn. Sp. Pl. i. 204 ? (1753); Sims, Bot. Mag. t. 2243; Haw. Syn. 11 (1812), under *Limnia*. Stems 1—1½ ft. high: herbage almost dark green, disposed to blacken in drying: radical leaves lanceolate to rhombic-ovate, acute or acuminate, 1–2 in. long, long-petioled; cauline sessile, distinct, ovate or ovate-lanceolate, acute, not indistinctly parallel-veined: raceme very lax, the fl. on long pedicels: petals 4 lines long, rose-purple, retuse or emarginate at summit, at base narrowed to a distinct claw. In open swamps (never in "cool woods") from near Point Bonita Light House in Marin Co., to Mendocino: perhaps farther northward in California, and also in Alaska. The derivation of original *C. Sibirica* is somewhat obscure, but the present plant answers, better than any other Californian species, to the accounts given of it.

4. **C. alsinoides**, Sims, Bot. Mag. t. 1309 (1810). Root sometimes (or always) annual: stems 4–10 in. high; herbage light green, unchanged in drying: radical leaves broadly ovate, abruptly acute, veinless or somewhat feather-veined; cauline pair sessile, distinct, very broadly ovate: petals 2 lines long, white, united at base but not unguiculate. Common in moist woods, from Humboldt Co., and perhaps Mendocino, northward.

5. **C. asarifolia**, Bong. Veg. Sitch. 157 (1833): *C. Neradensis*, Wats. Bot. Calif. i. 77 (1876): *C. cordifolia*, Wats. Proc. Am. Acad. xvii. 365 (1882). Stems clustered at summit of a slender creeping rootstock, 4–10 in. high; herbage rather succulent, dark-colored when dry: radical leaves cordate-reniform or ovate, orbicular, or obovate, 1 in. broad, abruptly narrowed to a long petiole; cauline pair sessile, oblong-ovate: raceme short, lax, bractless: petals 4 lines long, white, broadly spatulate, with narrow claws. A common species at the distant north, but found in the Trinity Mountains, Humboldt Co., *Marshall*.

++ ++ *Annuals; racemes bractless except at base, recurred.*

++ *Herbage light green, not glaucous.*

6. **C. perfoliata**, Donn, in Willd. Sp. ii. 1186 (1798). Stems 2–16 in. high: radical leaves from deltoid-cordate, deltoid or rhomboidal to rhombic-lanceolate, 1–2 in. long, on long petioles; cauline pair joined into a more or less orbicular perfoliate nearly plane or strongly concave disk $\frac{1}{3}$–4 in. broad: raceme short-peduncled or sessile, with an ovate acute or acuminate small foliaceous bract at base: petals 1–2 lines long, white; blade linear-oblong, retuse or emarginate; claws united at base and stamens epipetalous: fruiting calyx 2 lines long, twice the length of the subglobose 3-seeded capsule: seed $\frac{3}{4}$ line long, round-oval, black and shining but depressed-granular (under a strong lens) and with a small white strophiole. Var. (1) **carnosa**. Stout and low: the whole herbage very succulent: fruiting calyx $\frac{1}{4}$ in. long: seed nearly orbicular,

1½ lines broad. Var. (2) **angustifolia.** Quite like the type save that the lowest radical leaves are linear, almost without distinction of blade and petiole, the later ones somewhat broader and lanceolate : involucre truncate and with acute angles on the upper side (opposite the deflection of the pedicels) rounded on the other. Var (3) **amplectens.** Like smaller states of the type ; but involucral pair of leaves united on one side only, forming a 2-lobed bract which is narrowed to a short petiole sheathing the base of the short raceme. The most prevalent of Californian winter annuals, attaining its best development in the shade of oaks and laurels among the hills ; in open grounds much smaller ; in sandy soil near the sea usually reduced and depressed. The first variety is peculiar to the Mt. Diablo region, growing in open grounds, in fields and waste places. The second grows along with the type everywhere, and is remarkably different from it in that its earliest leaves are linear, only the later ones widening to the lanceolate, thus reversing the common order ; for in the type the earliest leaves are broader than long, only the later ones being somewhat narrower. This may perhaps be a species. It can hardly be referred to *C. parviflora*, a plant which belongs to Washington and British Columbia, and has a different foliage, always linear, an equilateral involucre, etc. Our third variety belongs to middle elevations of the Sierra. The flowering season of the species is from March to July, according to locality.

++ ++ *Herbage glaucous, in age flesh-colored.*

7. **C. gypsophiloides,** F. & M. Ind. Sem. Petr. ii. 33 (1835) : *C. spathulata*, Gray, Proc. Am. Acad. xxii. 282, excl. var., not of Dougl. Very pale and glaucous, 4–10 in. high : radical leaves linear, one-half or one-third as long as the slender scapes ; cauline pair short and united on one side to form a quadrate disk-like involucre, or longer, lanceolate-acuminate and less perfectly united: raceme peduncled, many-flowered ; pedicels scattered, often 1 in. long : petals rose-purple, thrice the length of the calyx, cuneate-oblong, deeply emarginate, unguiculate at base and united around the ovary : seed dull to the unaided eye, under a lens roughened with a low and rounded but smooth and shining tuberculation. On northward slopes and at the summits of the higher mountains of the Coast Range, from Tamalpais and Mt. Diablo northward ; plentiful in its localities, and the most beautiful of all *Claytonias*. Mar., Apr.

8. **C. spathulata,** Dougl. in Hook. Fl. i 225. t. 74 (1833) : *C. exigua* & *tenuifolia*, T. & G. Fl. i. 200 & 201 (1838) ; *C. spathulata*, var. *tenuifolia*, Gray, Proc. Am. Acad. xxii. 282. Low, densely tufted and fleshy, 1–3 in. high : scapes little exceeding the linear leaves ; involucral leaves lanceolate or linear, more or less dilated at base and there connate on one side, equalling or exceeding the short raceme : petals white or purplish, little longer than the sepals, truncate or rounded at apex : seed oval,

$1\frac{1}{2}$ line long, black and shining, the polished low tuberculation appearing under a lens as a kind of reticulation. Common on ledges of rock and gravelly summits of low hills along the seaboard. Douglas' type, figured in Hooker's Flora, is not with us, but is a northern plant with short involucral leaves connate on both sides. If ours should be specifically distinct, its name will be *C. exigua*. Most unlike *C. gypsophiloides*, this is a very inconspicuous and homely plant, usually appearing in the shape of a compact hemispherical tuft of glaucous succulent foliage, the small flowers seldom rising above the leafy mass, the involucral leaves commonly quite surpassing the raceme.

5. **MONTIA**, *Micheli*. Annuals, or by stolons or bulblets perennial. Leaves opposite or alternate. Flowers few or many in axillary racemose clusters, or in a single terminal raceme. Calyx, corolla, capsule and seeds as in *Claytonia*; but segments of corolla often unequal and stamens reduced to 3. Seeds sometimes 1 or 2 only.

* *Leaves opposite.*
+ *Annuals.*

1. **M. fontana**, Linn. Sp. Pl. i. 87 (1753). Stems slender, erect, ascending or procumbent, 1-4 in. long : leaves narrowly oblanceolate or spatulate, dilated and somewhat connate at base, $\frac{1}{4}$ -$\frac{3}{4}$ in. long : corolla white, minute, little exceeding the calyx and seldom expanding, the petals unequal, united at base : seed minute, roundish, dull black, but under a lens shining and covered with an almost echinate murication.— Common and variable ; the coarser form inhabiting the margins of streamlets and shores of muddy pools ; the smaller and nearly prostrate state found on dry ground under growing grain in rather low fields.

2. **M. Hallii**, Greene. Gray, Proc. Am. Acad. xxii. 283 (1887), under *Claytonia*. Larger than the last, 3'-6 in. high, simple or sparingly branched from the base: leaves in 2 or 3 pairs, oblanceolate or spatulate, $1\frac{1}{2}$ -1 in. long : corolla twice the length of the calyx ; petals equal : seeds 1 or 2, muriculate.— In Plumas Co., *Lemmon*, and in southern Oregon ; a species just intermediate between *M. fontana* and *Chamissonis*.

+ + *Perennial, stoloniferous and bulbilliferous.*

3. **M. Chamissonis**, Greene. Esch. in Spreng. Syst. i. 790 (1825), under *Claytonia*. *C. stolonifera*, C. A. Mey., *C. aquatica*, Nutt. and *C. flagellaris*, Bong. Stems weak, decumbent, 4-12 in. high, stoloniferous at base, and bearing bulblets at the ends of short branchlets or in the lower axils : leaves in few pairs, oblanceolate or spatulate, $\frac{1}{2}$ -$1\frac{1}{2}$ in. long : racemes few-flowered ; pedicels elongated : calyx minute, obtuse: petals white, obovate-oblong, rounded and entire, three or four times the length of the calyx : seed rather opaque, under a lens tuberculate-

roughened. Common in very wet places in subalpine swamps throughout California. The seeds are exactly those attributed to his *M. lamprosperma* by Chamisso, and the species may possibly be identical.

* * *Leaves alternate.*
+— *Annuals.*

4. **M. linearis,** Greene. Dougl. in Hook. Fl. i. 222. t. 71 (1833), under *Claytonia*. Stems erect, branching, 3–6 in. high: leaves narrowly linear, 1–2 in. long, clasping at base: racemes rather numerous: calyx large (2 lines or more), broad and obtuse: petals little exceeding the calyx, white, unequal: stamens 3, epipetalous: seed large ($\frac{3}{4}$ line broad), orbicular, compressed, obtusely margined, smooth and shining. From near Napa, *Bigelow*, and Sierra Co., *Lemmon*, far northward.

5. **M. diffusa,** Greene. Nutt. in T. & G. Fl. i. 202 (1838), under *Claytonia*. Diffusely and dichotomously branched, 6 in. high, leafy and floriferous throughout: leaves ovate or deltoid, acute, $\frac{1}{2}$–1 in. long: racemes numerous, few-flowered; pedicels very slender: calyx minute: corolla a little longer, pale rose-color: seeds compressed, minutely cancellate, obtusely margined. An Oregonian species apparently rare; said to have been collected in California by *Kellogg & Harford*.

+— +— *Perennial (?), viviparous by deciduous axillary leafy buds.*

6. **M. parvifolia,** Greene. Moç. in DC. Prodr. iii. 361 (1828), under *Claytonia*. Slender, succulent, 4–10 in. high: leaves borne on a short stem or caudex $\frac{1}{2}$–1 in. high, ovate or lanceolate, 1 in. long or less, including the slender petiole: racemose peduncles elongated, leafy below, the nodes in age bearing bud-like plantlets: calyx minute: petals rose-color, 2–4 lines long: seeds mostly solitary in the capsules, oval, shining. From Calaveras Co., and Donner Lake, northward in the Sierra, on moist rocks. Plant with the fleshiness and something of the aspect of a *Sedum*.

6. **CALYPTRIDIUM,** *Nuttall*. Glabrous and rather succulent herbs, with alternate leaves, and small ephemeral flowers in scorpioid solitary or clustered scorpioid spikes. Sepals 2, broadly ovate or cordate-orbicular, scarious, usually persistent. Sepals 2–4. Stamens 1–3. Style bifid. Capsule membranaceous, 2-valved, 6–12-seeded. A small genus, almostly exclusively Californian, and uncommonly well-defined for a genus of this family.

* *Annuals; branches leafy; stamen 1.*

1. **C. monandrum,** Nutt. T. & G. Fl. i. 198 (1838). Branches depressed, 2–6 in. long: leaves spatulate or linear, 1 in. long or more: sepals and petals each 2, about a line long, the latter in age coherent and borne calyptra-like upon the summit of the long-exserted linear capsule. From the Santa Clara Valley and Fort Tejon southward.

2. **C. roseum,** Wats. Bot. King. Exp. 44. t. 6. (1871). Diffuse and nearly prostrate branches 3–6 in. long: leaves oblong-spatulate, obtuse: sepals nearly orbicular, unequal: petals 2, minute, round-obovate, narrower at base: capsule oblong-ovate, not exceeding the calyx.– Sierra Co., *Lemmon*, and southward along the eastern base of the mountains.

3. **C. tetrapetalum,** Wats. Proc. Am. Acad. xx. 356 (1885). Branches erect or ascending from a more or less decumbent base, leafy up to the short dense spikes: leaves broadly spatulate, 1–3 in. long: sepals round-reniform, conspicuously nerved and scariously margined, 2–4 lines broad, exceeding the 4 oblong or round-ovate petals: stigmas broad; nearly sessile: capsule oblong, 3 lines long, 12–20-seeded. Lake Co. and Sonoma, *Torrey, Greene, Rattan, Simonds.* A rather local species of the geyser district of the Coast Range; in the "Botany of California" referred to *C. roseum.*

* * *Biennials or annuals; leaves radical; stamens 3.*

4. **C. paniculatum,** Greene, Bull. Torr. Club, xiii. 144 (1886); Kell. Proc. Calif. Acad. ii. 187 (1863), under *Spraguea.* Plant of the same habit and aspect as the preceding, but larger (4–8 in. high), the spicate racemes densely panicled: sepals 4 lines broad, 3 lines long: seeds reniform.– Not yet found in California, but the original station, "West of Virginia City" is not far from the State line.

5. **C. umbellatum,** Greene, l. c.; Torr. Pl. Frem. 4. t. 1 (1850) under *Spraguea.* Stems several, from a fleshy slender-fusiform biennial root, ascending or erect, about 1 ft. high: leaves mostly radical, in a rosulate tuft, spatulate or oblanceolate, 1–4 in. long including the petiole; the cauline reduced to bracts; an involucre of small scarious bracts subtending the terminal whorl of dense nearly sessile spikes: sepals scarious and pinkish or flesh-color, 2–4 lines broad, about equalling the oblong-obovate petals: capsule round-ovate, compressed, surmounted by a long style, few-seeded: seed obliquely oval.– In the Sierra Nevada from Yosemite northward; also in the Coast Range near Santa Cruz, *Anderson,* and mountains of Humboldt Co., *Chesnut & Drew.* A plant of rocky or gravelly places, at high altitudes chiefly. July–Oct.

6. **C. nudum,** Greene, Pittonia, i. 64 (1887). Root annual, fleshy-fibrous: leaves all radical: scapes 3–6 in. high, naked, terminated by a compact orbicular capitate-congested cluster of short spikes: petals narrowly spatulate: stamens long-exserted; anthers linear, yellow: fr. unknown. A common annual of the Donner Lake district, apparently quite distinct from *C. umbellatum.*

Order XXVII. **CRASSULACEÆ.**

De Candolle, Bull. Philom. n. 49. p. 1 (1801). Part of *Succulentæ*, Vent. Tabl. iii. 271 (1799).

Succulent herbs with exstipulate leaves. Flowers perfectly symmetrical, cymosely arranged. Sepals 3–20, more or less united at base. Petals as many, inserted in the bottom of the calyx, distinct or cohering below to form a gamopetalous corolla. Stamens as many or twice as many as the petals, when of the same number alternate with them; filaments distinct, subulate. Ovaries as many as the petals, opposite to them, forming a whorl, each with or without a hypogynous scale at base. Fruit follicular. Seeds attached to the margins of the suture, small, albuminous.

1. **TILLÆA,** *Micheli.* Small and slender fleshy glabrous annuals. Leaves opposite, entire. Flowers minute, axillary, white or pinkish. Sepals and petals 3–5, distinct or united at base. Stamens as many Carpels distinct; styles short-subulate. Follicles 1–several-seeded. Seeds striate lengthwise.

* *Fl. clustered; petals acuminate; carpels 1–2-seeded.*

1. **T. minima,** Meiers, Chil. ii. 530 (1826) ; *T. leptopetala*, Benth. Pl. Hartw. 310 (1849). Simple or with few or many ascending branches, 1–3 in. high : herbage very light green when young, in age reddish : internodes short : leaves ovate or oblong, obtuse, 1 line long, connate : fl. in short axillary panicles, mostly subsessile, occasionally some with long pedicels : sepals 4, $1\over 2$ line long, acute, nearly or quite equalled by the linear-lanceolate acuminate petals ; carpels acute, not longer than the petals. Very common in clayey or sandy soils in the hilly districts everywhere. Mar.–May.

* * *Fl. solitary; petals oval or oblong; carpels several-seeded.*

2. **T. Drummondii,** T. & G. Fl. i. 558 (1840) : *T. angustifolia*, Nutt. l. c. ? Stems very slender, dichotomous, diffuse, rooting at some of the lower nodes, 1 in. long or more : leaves oblong-linear, slightly connate : pedicels at length equalling or exceeding the leaves · petals red, fully equalling the obtuse carpels, and twice or thrice the length of the calyx-lobes.– Common in moist low places in wheat fields near Suisun. May.

3. **T. Bolanderi,** Greene. *T. angustifolia*, var. (?) *Bolanderi*, Wats. Bot. Calif. i. 208 (1876). Stems stoutish, simple, 2–5 in. long, the lower portion with long internodes and rooting at the nodes : leaves linear or linear-oblong, acutish, subterete, slightly connate: fl. short-pedicellate, the pedicel in fruit elongated and surpassing the leaves : petals oblong, acutish, equalling the carpels, more than twice the length of the ovate calyx-segments. In characters of flower and fruit much like the

preceding, but widely different in all other particulars. Frequent on muddy shores about San Francisco. May.

2. **SEDUM,** *Columna* (STONE-CROP). Glabrous perennials or annuals. Flowers in cymes, mostly secund. Sepals 4 or 5, united at base. Petals as many, distinct. Stamens twice as many. Carpels distinct, or rarely connate at base, few- or many-seeded.

* *Perennial with flat serrate leaves, and flowers in a compact compound cyme.*

1. **S. roseum,** Scop. Fl. Carn. i. 326 (1772); Linn. Sp. Pl. ii. 1035 (1753), under *Rhodiola*. *Rhodia officinarum*, Crantz, Inst. i. 191 (1766). *Sedum Rhodiola*, DC. Fl. Fr. iv. 386 (1805). Stems simple, erect, 2–10 in. high, from a thick rose-scented root : leaves alternate, oblong-lanceolate, acute, rarely entire, 1_2–$1\frac{1}{2}$ in. long : cyme sessile, 1–2 in. broad : fl. on short naked pedicels, usually 4-merous, diœcious, dark-purple in age : sepals oblong : petals linear-oblong, $1\frac{1}{2}$ lines long : carpels becoming 3 lines long, short-beaked. In wet soil in the higher Sierra ; also in subarctic America and in Europe.

* * *Perennials with entire leaves, and flowers secund upon the branches of a forked cyme.*

2. **S. spathulifolium,** Hook. Fl. i. 227 (1833). Glaucous and often pulverulent : stems 4–6 in. high, ascending from a branched and rooting caudex : leaves flat, obovate or spatulate, obtuse, 6–10 lines long : fl. 3 lines long : petals yellow, lanceolate, acute, twice longer than the ovate acute sepals. Rocky places on the northward slopes of hills and mountains from San Francisco and Berkeley northward.

3. **S. Oreganum,** Nutt. T. & G. Fl. i. 559 (1840). Resembling the last, but not glaucous : fl. larger (4–5 lines long); petals pale rose-color, narrowly lanceolate, acuminate ; sepals acute. From Mendocino Co. northward.

4. **S. obtusatum,** Gray, Proc. Am. Acad. vii. 342 (1868). Habit of the above, scarcely glaucous : leaves spatulate or cuneate, the uppermost oblong : fl. loosely cymose, pedicellate : petals yellowish, oblong-lanceolate or ovate, twice longer than the broad obtusish sepals. In the Sierra Nevada from the Yosemite northward.

5. **S. radiatum,** Wats. Proc. Am. Acad. xviii. 193 (1883) : *S. Douglasii*, Wats. Bot. Calif. not Hook. Stems 3–6 in. high, decumbent at base from a branching rooting caudex : leaves oblong or oblong-ovate, obtuse or acutish, somewhat clasping by the narrower base, $\frac{1}{4}$–$\frac{1}{2}$ in. long : fl. sessile ; sepals short, triangular : petals yellow, narrowly lanceolate, acuminate, 3 lines long : carpels broad, the beaks abruptly divergent. Coast Range, from Monterey Co. to Mendocino and Trinity.

* * * *Annual, with flowers cymose.*

6. **S. pumilum**, Benth. Pl. Hartw. 310 (1849). Slender, simple or branching, 1 - 3 in. high: leaves 1—2 lines long, ovate-oblong; fl. sessile in sparingly branched cymes; calyx-lobes minute, triangular; petals yellow, linear, acute, 1½ lines long: follicles short, 1-seeded, the seed erect, filling the cavity. -Valley of the Sacramento, *Hartweg*; also at Placerville, *Rattan*, and the Marysville Buttes, *Jepson*.

3. **COTYLEDON**, *Nicander*. Succulent herbs coarser than *Sedum* and larger, but quite like them in all other respects save that the petals are more or less united into a tube, and the follicles erect or suberect rather than spreading.

1. **C. Nevadensis**, Wats. Bot. Calif. i. 212 (1876). Acaulescent, glaucous: leaves obovate to oblanceolate, somewhat rhomboidal, acute or acuminate, the larger 2 - 4 in. long: flowering branches 6 - 10 in. high, with scattered lanceolate to broadly triangular acute bracts; inflorescence a spreading compound cyme; pedicels ¼ — ¾ in. long: sepals ovate, acute, 2 lines long or less: petals lanceolate, acute, 5 lines long, yellow or reddish: carpels ovate-oblong, 3 lines long.- Yosemite and northward.

2. **C. cæspitosa**, Haw. Misc. Nat. 180 (1803). Nearly or quite acaulescent, scarcely glaucous but dull green: leaves ovate-oblong to oblong-lanceolate, acute, the larger 1½ - 3 in. long: flowering branches ½ - 1 ft. high, with broadly triangular-ovate clasping bracts: inflorescence a short and rather close compound cyme; pedicels short and stout, subtended by broad bracts: sepals ovate, 2 lines long or less: petals yellow, broadly lanceolate, acute, 4 - 5 lines long: carpels ovate-oblong, about 3 lines long. From near San Francisco northward.

3. **C. laxa**, B. & W. Bot. Calif. i. 212 (1876); Lindl. Journ. Hort. Soc. iv. 292 (1849), under *Echeveria*. Nearly acaulescent, very glaucous: leaves lanceolate, sharply acuminate, the larger 3 -5 in. long: flowering branches 1 -2 ft. high, slender, with scattered leafy bracts, of which the lower are narrowly lanceolate, the upper shorter and broader: inflorescence of 2 - 4 simple secund racemes 3 - 5 in. long: floral bracts small; pedicels 2 - 3 lines long: sepals ovate, acute, 2 lines long: petals yellow, oblong-lanceolate, acute or acuminate, 5 - 7 lines long: carpels ovate-oblong, 4 lines long.—From near Monterey southward in the Coast Range.

4. **C. farinosa**, Baker, Refug. Bot. i. t. 71 (1869); Lindl. l. c. (1849), under *Echeveria*. Short-caulescent, more or less white-farinose: leaves rather flaccid, ascending, lanceolate, acuminate, the larger ones 2 - 4 in. long, acute: flowering branches 6—10 in. high, with scattered broadly ovate to lanceolate clasping bracts: inflorescence a short and close

compound cyme ; bracts ovate-lanceolate ; pedicels 1 3 lines long ; sepals broadly lanceolate, ¼ in. long ; petals yellow, oblong-lanceolate, mostly acuminate, 4 6 lines long ; carpels ovate-oblong, ¼ in. long.— Near Monterey and Sonoma ; also in the foothills of the Sierra.

5. **C. Palmeri,** Wats. Proc. Am. Acad. xiv. 292 (1879). Caulescent, not mealy or glaucous ; leaves reddish, lanceolate, acuminate, 2 in. long, 8 or 9 lines wide at base and gradually tapering to the very acute apex, the margin obtuse ; flowering stem with broadly triangular-ovate leafy bracts ; inflorescence of a few simple spreading secund racemes, somewhat glaucous ; pedicels ¼ – 1½ in. long ; calyx broad ; sepals triangular-ovate, 2 lines long ; petals pale lemon-yellow, 5 6 lines long ; carpels 4 lines, at length somewhat spreading, with divergent styles. At San Simeon Bay, *Palmer*.

6. **C. Lingula,** Wats. l. c. 293. Habit of the preceding ; leaves oblong, acute, 2 3 in. long, 1 in. broad ; stems 1½–2 ft. long, the branches of the cyme less spreading ; pedicels very short ; sepals narrower and longer ; carpels ¼ in. long, somewhat spreading, with straight styles. Habitat of the last.

Order XXVIII. SAXIFRAGEÆ.

Ventenat, Tabl. du Reg. Veget. iii. 277 (1799). *Saxifragæ*, Juss. (1789).

Herbs (*Ribes* shrubby) with simple alternate usually exstipulate leaves, the petiole often stipulaceously dilated at base. Stems mostly simple below, commonly leafless and scape-like. Inflorescence mostly either cymose, racemose or paniculate. Calyx of about 5 sepals, often more or less coherent below and united to the base of the ovary. Petals as many or 0. Stamens 5 or 10, perigynous or hypogynous. Ovary of about 2 carpels more or less cohering by their inner faces, commonly distinct and diverging at apex ; style often wanting and stigmas sessile on the tips of the lobes of the ovary. Fruit capsular or follicular (in *Ribes* baccate). Seeds many, small, albuminous.—A considerable family, analogous to Rosaceæ, but with few and definite stamens, and albuminous seeds ; more closely related to Crassulaceæ, from which they are most readily known by the dicarpellary pistil and different (seldom succulent) herbage.

Hints of the Genera.

Shrubs; fruit berry-like, - - - - - - - - - - 11
Herbs, or low alpine undershrubs ;
    Petals 0; capsule obcordate, compressed, - - - - - 10
    " 5, undivided ;
        Stamens 10, - - - - - - - - - 1, 5
        " 5, - - - - - - - - 2, 3, 6, 9
        " 3, - - - - - - - - - - 4
    Petals 5, toothed or cleft ;
        Stamens 10, - - - - - - - - - 5, 7
        " 5, - - - - - - - - - - 8

1. **SAXIFRAGA,** *Pliny* (SAXIFRAGE). Short-stemmed or stemless plants with simple leaves, their petioles commonly sheathing at base. Flowers in cymose thyrsoid or panicled clusters or solitary. Sepals distinct, or at base conjoined to each other and the base of the ovary. Petals entire, imbricate in bud. Stamens 10, inserted with or below the petals, on the base of the calyx, or between it and a fleshy disk. Carpels 2, usually partly united, dehiscent by the inside of the divergent beaks. Seeds with thin coat and no wing or appendage. Herbaceous or more enduring plants of rocky woods or alpine heights, margins of cold streams, etc.; very diverse in habit; perhaps belonging to several natural genera.

* *Herbaceous; scapes slender, paniculate; calyx free from the ovary, reflexed; petals unguiculate; filaments filiform or clavate.*

+ *Annual; leaves cuneate at base; filaments filiform.*

1. **S. bryophora,** Gray, Proc. Am. Acad. vi. 533 (1865). Very slender. 2–8 in. high: leaves not rosulate, but scattered up and down the lower part of the stem, 1 in. long, spatulate-oblong, acutish, entire, veinless: panicle lax, the branches bearing lateral pedicellate deflexed leafy plantlets and a single terminal flower: petals oblong-ovate, 2 lines long including the claw; blade white with a pair of round yellow spots at base: stamens nearly equalling the petals; filaments filiform, somewhat flattened toward the base; anthers red-purple: carpels nearly distinct: styles 0. -Common in the high Sierra from Mt. Whitney to Donner Lake; the root apparently of but one year's duration, but the propagation is by leafy bulblets more than by seeds. Aug.- Oct.

+ + *Perennials; leaves broad not cuneate at base; filaments more or less clavate.*

2. **S. Marshalli,** Greene, Pittonia, i. 159 (1888). Scape and leaves from a short crown, sparingly glandular-pubescent: leaves 1 in. long or more, on somewhat flattened petioles of 1–2 in., oblong, obtuse, the base abruptly acute, or nearly truncate, or subcordate, the margin closely beset with sharp triangular teeth: scape 8–16 in. high, rather loosely paniculate at summit: petals 1½ lines long, oval or oblong, white with a pair of oval green spots near the base, very shortly unguiculate: stamens equalling the petals; filaments strongly clavate.- Hoopa Valley, Humboldt Co., *C. C. Marshall,* April, 1888: also obtained a year later, on Rogue River, Oregon, in large form, by Mr. Howell.

3. **S. Mertensiana,** Bong. Veg. Sitch. 141 (1833): *S. heterantha,* Hook. Fl. i. 252. t. 78 (1833). Scape and leaves from a scaly-bulbous base, glandular-pubescent, ½–1 ft. high: leaves thin and pale, round-cordate, crenately or incisely many-lobed, ¾–1½ in. broad, on long petioles which are scarious-dilated at base: cymose panicle loose, the branches often flowering at apex and bearing granular bulblets down the

sides: fl. as in the last, but petals longer, scarcely unguiculate, and stamens with petaloid-dilated filaments: capsule inflated-ovate.—From Sonoma Co., northward in the Coast Range.

4. **S. æstivalis,** Fisch. Ind. Sem. Petr. ii. 37 (1835): *S. arguta,* Don. Saxifr. 356? (1822): *S. punctata* of Bot. Calif., not Linn. Scape and leaves from a short thick creeping rootstock, glabrous or pubescent, deep green or reddish, slightly fleshy: leaves on long petioles, reniform or round-cordate, equally and deeply dentate: petioles scarcely dilated except at the insertion: panicle narrow: petals oval, obtuse, unguiculate, white, with a pair of yellow spots at base of blade: some of the dilated filaments antherless and petaloid: capsule oblong.—Margins of alpine brooks in the high Sierra, and common far northward; variable in size, mostly larger than other species of the group.

* * *Stems short, cespitose, very leafy; leaves evergreen.*

5. **S. ledifolia,** Greene, Pittonia, ii. 101 (1890): *S. Tolmiei,* Gray, Bot. Calif. i. 195, not T. & G. Branches stoutish, ascending, very leafy: leaves spatulate-oblong, obtuse, entire, 6–8 lines long: peduncles terminal at the ends of leafy shoots, stout, 3–4 in. high, cymosely 5–15-flowered: calyx nearly free from the ovary, the almost distinct sepals erect: petals lanceolate, white, twice the length of the sepals: filaments dilated at summit: capsule large, purplish.—Moist open ground on alpine summits of the Sierra Nevada. Aug., Sept.

* * * *Stout stemless riparian herb; leaves broad, peltate.*

6. **S. peltata,** Torr. in Benth. Pl. Hartw. 311 (1849), and Bot. Wilkes Exp. 309. t. 5 (1862); Hook. f. Bot. Mag. t. 6074. *Leptarrhena inundata,* Behr, Proc. Calif. Acad. i. 45 & 57 (1855). Scape (vernal) and leaves (appearing later) from stout horizontal rootstocks 1—3 in. thick; peduncles and petioles 1—3 ft. high: blade of leaf orbicular, centrally peltate, 6–14 in. broad, 9–14-lobed, membranaceous, with a short-funnelform concavity over the insertion upon the petiole: fl. cymose-panicled, large, rose-color or nearly white: calyx 5-parted, nearly free from the ovary, the segments spreading in fruit: petals round-oval, 2-3 lines long, persistent and at length larger: filaments subulate: follicles distinct, divergent, turgid-ovate.—Along rocky margins of rivers and streams in the foothills and at middle elevations in the Sierra Nevada: the rootstocks mostly submersed; scapes and flowers appearing in spring, the leaves some weeks later. One of the largest and most remarkable of saxifrageous herbs; the fleshy rhizomes said to be palatable and nutritious insomuch as to have been in requisition for food with the aborigines of the mountain districts. Apr. June.

* * * * *Stemless; root corm-like; calyx campanulate, free from the ovary.*

7. **S. Parryi,** Torr. Bot. Mex. Bound. 69. t. 25 (1859). Scapes (autumnal) and leaves (vernal) 2–6 in. high, slender, pubescent; leaf-

blade rounded-subcordate, slightly crenate-lobed and -toothed, 1 in.
broad or less ; petioles short : cyme lax, 3 7-flowered : calyx only
slightly 5-lobed, the campanulate brown-nerved tube nearly enclosing
the 2-lobed capsule : petals ovate or spatulate, inserted by short claws
nearly in the sinuses of the calyx, white, marked with purple or
brownish veins : filaments slender-subulate, inserted below the petals :
styles slender, in fruit exserted from the calyx.- A very remarkable
species, long supposed to be peculiar to the dry hills of San Diego Co.,
there sending up its scapes and unfolding its flowers in November or
December, the leaves appearing later in winter. It was found in our
part of the State near Ione, by Dr. Parry, in the autumn of 1887.

* * * * * *Perennial, fibrous-rooted, the scape and leaves from a short
crown; calyx partly adherent to the ovary; petals not
unguiculate; filaments not clavate.*

8. **S. Californica,** Greene, Pittonia, i. 286 (1889): *S. Virginiensis,* Gray,
Bot. Calif. i. 195, not Michx. Leaves few, rather thick, reddish-veined,
sparsely glandular-villous, oval, oblong or elliptical, 1 2 in. long, on
broad petioles of ½—1 in.; margin coarsely crenate to repand-denticulate,
rarely either sharply dentate or nearly entire, tomentose-ciliolate : scape
6 -18 in. high, loosely cymose-paniculate : calyx nearly free from the
ovary, the sepals reflexed : petals oblong, thrice the length of the sepals,
white or rose-tinted : filaments subulate, inserted under the edge of an
elevated perigynous disk which equals the summit of the ovary.— Plentiful throughout the State, on cool northward slopes of both ranges of
mountains. In the Sierra Nevada the leaves are conspicuously toothed.
In the Coast Range they vary from crenate to almost entire. Mar.--May.

9. **S. nivalis,** Linn. Sp. Pl. i. 405 (1753). Plant deep green or purplish,
more or less viscid-pubescent : leaves oblong-ovate or spatulate-obovate,
1 in. long or less, coarsely dentate, or crenate. or almost entire : scape
1— 6 in. high : fl. crowded or distant in a terminal cluster : calyx-lobes
ovate, longer than the tube, erect : petals white, oblong or spatulate,
exceeding the calyx ; stamens filiform : styles very short : ovary and
fruit usually purple.---Common at alpine heights in the Sierra. June-Oct.

10. **S. integrifolia,** Hook. Fl. i. 249. t. 86 (1833). Plant light or dark
green, more or less roughened with a short glandular pubescence : leaves
varying from ovate to oblong-lanceolate and -spatulate, acute or obtuse,
2 –5 in. long, entire, remotely denticulate or more distinctly dentate or
crenate : scape 1—3 ft. high : fl. in small clusters in an interrupted
thyrse or panicle : petals obovate-spatulate, dull white, exceeding the
reflexed calyx-lobes : filaments short, broadly subulate : stigmas sessile
on the conical, at length widely divergent beaks of the ovary.— In swamps
of the middle Sierra ; extremely variable, and possibly a composite

species, some states of which would pass for *S. Pennsylvanica* but for their broad white petals. Other forms are very different. June -Sept.

2. **BOYKINIA,** *Nuttall.* Perennial herbs with erect leafy stems, and corymbose or paniculate cymes of white flowers ; leaves round-reniform, palmately lobed or toothed, the teeth with callous-glandular tips ; the petioles stipularly dilated at base. Calyx 5-lobed ; lobes valvate, but early open in the bud ; the tube more or less adherent to the ovary. Petals 5, entire, imbricate or convolute in bud. Stamens 5, short, alternate with the petals. Capsule 2-celled, dehiscent down the beaks. Seeds minutely granulate or papillose.

* *Leaves exstipulate; flowers secund, the corolla slightly irregular, the petals narrow.*—Genus THEROFON, Raf.

1. **B. elata,** Greene. Nutt. T. & G. Fl. i. 575 (1840), under *Saxifraga*. *B. occidentalis*, T. & G. l. c. 577. Slender, 1—2 ft. high, glabrous or glandular-pubescent, the bases of the petioles bearing brown bristly hairs : leaves thin-membranaceous, 5—7-lobed, 1—3 in. broad : calyx-lobes lanceolate-triangular ; tube oval and urceolate in fruit : petals cuneate-oblong, obtuse, persistent and in age recurved : seeds elongated-oblong, acute at one end, dark brown, rather densely tuberculate.— Shady banks and rocky margins of streams in the Coast Range from Santa Barbara to British Columbia ; one of our finest saxifrageous plants. The corolla has just the irregularity seen in most species of *Pelargonium*. May Aug.

2. **B. rotundifolia,** Parry, Proc. Am. Acad. xiii. 371 (1878). Near the preceding but larger, 2 3 ft. high ; leaves 2—4 in. broad, crenately incised or toothed ; stem and petioles villous as well as glandular : flowers short-pedicelled on the few and elongated branches of the inflorescence : petals small, little exceeding the calyx-lobes : calyx-tube in fruit broadly urceolate.—In the San Bernardino Mountains, but to be expected in Kern Co.

* * *Stipules manifest; flowers corymbose-cymose at summit of stem; corolla regular, the petals broad.*—Genus HEMIEVA, Raf.

3. **B. major,** Gray, Bot. Calif. i. 196 (1876) : *B. occidentalis*, var. *elata*, Gray, Proc. Am. Acad. viii. 383 (1868). Coarse and stout, 2 -3 ft. high, somewhat glandular-scabrous : leaves 4—8 lines broad, almost pedately 5—9-cleft and coarsely toothed ; petioles abruptly dilated at base, the cauline short, with broad foliaceous stipules : calyx-lobes triangular : petals white, obovate.—In the Sierra Nevada, from Mariposa Co. northward. Aug., Sept.

4. **B. ranunculifolia,** Greene. Hook. Fl. i. 246. t. 83 (1833), under *Saxifraga*. Rather slender, 1 ft. high, glandular-pubescent above, nearly glabrous below : lower leaves $1\frac{1}{2}$—1 in. broad, 3-parted, the segments

cuneiform, obtusely cleft; cauline few, simply 3-lobed, or more reduced and simple: axils of the lower bulbiferous: fl. in a simple corymb; petals obovate, white, twice the length of the acute calyx-lobes: calyx campanulate in fruit. On Spanish Peak, Plumas Co., *Mrs. Austin*, and northward.

3. **BOLANDRA**, *A. Gray.* Herbs with the foliage and habit of *Boykinia* proper, but flowers fewer, petals narrower and purplish, and ovary wholly free from the broad campanulate calyx. Seeds minute, with a thin loose coat.

1. **B. Californica**, Gray, Proc. Am. Acad. vii. 341 (1868). Erect, rather slender, 1 ft. high or less, glabrous, bearing granular bulblets at base of stem: leaves thin; the lower round-reniform, 5-lobed, on long slender petioles; upper sessile or clasping, incised or toothed, the floral reduced to ovate or lanceolate entire bracts: fl. solitary at the ends of paniculate branches: calyx broad, somewhat inflated, truncate at base: petals slender, subulate, recurved, persistent: capsule membranaceous, enclosed in the calyx. — On wet rocks in the Yosemite and northward.

4. **TOLMIEA**, *Torrey & Gray.* Perennial, with radical leaves (these in age gemmiparous at base of the blade) and scapiform racemose stems. Calyx funnelform, free from the ovary, membranaceous, gibbous at base: the 5 lobes somewhat unequal; tube in age splitting down one side. Petals 5, filiform, inserted in the sinuses of the calyx, recurved, persistent. Stamens 3, inserted in the throat of the calyx opposite the upper lobe and the two lateral ones; filaments short; anther-cells confluent. Ovary elongated-oblong or clavate, attenuate at base, cleft above, 1-celled, with 2 parietal placentae; styles slender; stigmas capitellate. Capsule at base attenuate to a stipe, dehiscent between the divergent equal beaks. Seeds globose, the close firm coat muriculate.

1. **T. Menziesii**, T. & G. Fl. i. 582 (1840); Pursh, Fl. i. 313 (1814), under *Tiarella;* Hook. Fl. i. 237. t. 80 (1833), under *Heuchera.* Hispidly pubescent, 1—2 ft. high: radical leaves round-cordate, more or less lobed and toothed: stem with 2 or 3 small leaves: raceme loose, $\frac{1}{2}$ - 1 ft. long; fl. and capsule $\frac{1}{2}$ in. long, greenish or purplish.—Forests of Mendocino Co., *Bolander,* and northward.

5. **TELLIMA**, *Robert Brown.* Perennial herbs, with leaves chiefly radical, round-cordate, toothed or palmately divided, their petioles stipulaceously dilated at base. Flowers in a simple terminal raceme. Calyx campanulate or turbinate, 5-lobed, free from the ovary, or coherent with it at base or even to above the middle; the short triangular lobes valvate in bud. Petals 5, laciniate-pinnatifid, or 3—7-lobed, or entire, distant and sometimes involute in bud. Stamens 10, short, included. Ovary short, 1-celled, with 2 or 3 parietal placentae; styles 2 or 3, very short; stigmas capitate. Capsule conical, opening between the short

beaks. Seeds very numerous, with a close coat. An unsatisfactory genus, of two very diverse types, perhaps better received as distinct genera; but the *Lithophragma* species are very closely analogous to typical *Saxifraga;* the typical *Tellima* is as near to the older genus *Mitella*, to which, at the first, it was referred.

* *Corolla regular, the petals green, sessile by a broad base, laciniately pinnatifid; styles and placentæ 2.*—TELLIMA proper.

1. **T. grandiflora,** Dougl. Bot. Reg. t. 1178 (1828); Pursh, Fl. i. 314 (1814), under *Mitella*. Stoutish, 1–2 ft. high, from rather coarse tufted rootstocks; herbage rough-hirsute: leaves round-cordate, more or less lobed, 2–4 in. broad: calyx $1/4$–$1/2$ in. long, inflated-campanulate: seeds light brown, oval, strongly rugose-pitted.—Wooded hills, or sometimes in open ground, from Santa Cruz northward. May, June.

* * *Corolla slightly irregular, the petals white or pinkish, entire or lobed or toothed, short-unguiculate; styles and placentæ 3.—*
Genus LITHOPHRAGMA, Nutt.

← *Calyx turbinate, the tube more or less coherent with the ovary.*

2. **T. affinis,** Boland. Catal. 11 (1870); Gray, Proc. Am. Acad. vi. 534 (1865), under *Lithophragma*. Stems one or several from a slender horizontal or ascending tuberiferous rootstock, commonly $1/2$ ft. high, scabrous-hirsute: radical leaves very few, round-reniform, slightly lobed, 1 in. broad; cauline relatively broader, 3-lobed to the middle, the lobes coarsely toothed: calyx $2\frac{1}{2}$ lines long; pedicels rather longer: lower (3) petals 4 or 5 lines long, 3-toothed, the upper (2) narrower and a trifle shorter, entire: styles short, not exserted from the calyx: seeds oblong, dark brown, faintly striate-pitted or almost smooth.—Frequent on shady hillsides almost throughout the State.—Apr.—June.

3. **T. Cymbalaria,** Walp. Rep. ii. 372 (1843); T. & G. Fl. i. 585 (1840), under *Lithophragma*. Stems very slender, almost filiform, 1 ft. high or less: radical leaves $1/2$ in. broad or less, 3-lobed, the lobes rounded, entire; cauline usually only an opposite pair, 3-parted: fl. very few, on filiform pedicels thrice the length of the calyx: calyx-lobes broad and short: petals spatulate-obovate, entire, the 2 upper smaller, with shorter and broader blade and more pronounced claw.— An apparently rare species of the South (Santa Barbara, etc.); the fruit not known.

4. **T. tenella,** Walp. Rep. ii. 371 (1843); Nutt. T. & G. Fl. i. 584 (1840), under *Lithophragma*. Slender, 2–10 in. high, purple, roughish with a minute glandular pubescence: lower leaves parted into 3–5 cuneiform toothed lobes, $1/2$ in. broad, the cauline few, smaller, all the axils bearing minute granular bulblets: pedicels few, ascending, as long as the obconical calyx: petals large, pinkish, 3–5-parted into linear divisions.— In the Sierra Nevada from Donner Lake northward. June.

← ← *Calyx campanulate, with truncate or rounded base nearly or quite free from the ovary.*

5. **T. scabrella,** Greene, Pittonia, ii. 162 (1891). Slender, closely glandular-scabrous, 1 ft. high : leaves small ; lowest round-reniform and 3 -5-lobed, or 3-cleft or -parted, in age bearing each a rather large purple bulblet in the axil ; cauline 3 or 4, alternate, deeply 3-cleft or -parted : pedicels nearly or quite equalling or even exceeding the calyx ; this with a rounded and obtuse base : petals entire ; the 2 upper oblong, obtuse, shorter and broader than the others, all with exserted slender claws : capsule very short, included ; styles manifest, glabrous : seeds muriculate. — In dry ravines among the pine forests of the higher mountains of Kern Co. ; also at the Marysville Buttes, *Jepson*. Equally related to *T. cymbalaria* and *heterophylla*. June.

6. **T. heterophylla,** H. & A. Bot. Beech. 346 (1840); T. & G. Fl. i. 584, under *Lithophragma*. Slender, 1 ft. high, scabrous-hirsute : lowest leaves ¾-1 in. broad, with 5 shallow rounded lobes : cauline more deeply 3-lobed or -parted : pedicels very short, the broad truncate-based calyx appearing almost sessile : petals (at least the lower 3) obtusely 3-lobed : styles glabrous : seeds muriculate. — Common in the Coast Range.

7. **T. Bolanderi,** Boland. Catal. 11 (1870); Gray, Proc. Am. Acad. vi. 535 (1865), under *Lithophragma*. Near the last but larger, often 2 ft. high, more hirsute : radical leaves $1\frac{1}{2}-2\frac{1}{2}$ in. broad; cauline more divided : petals 3 -4 lines long, obovate or oval, the upper entire, the lower often with a lateral tooth on each side : seeds muricate-scabrous.— Southern slope of Mt. Diablo, *Brewer*, and Mendocino Co., *Bolander;* a somewhat obscure species, easily confounded with the preceding.

6. **TIARELLA,** *Linnæus*. Perennial herbs with simple or 3-foliolate alternate more or less distinctly stipulate leaves, and a terminal panicle or raceme of small white flowers. Calyx 5-parted, the lobes valvate. Petals 5, entire, unguiculate. Stamens 10, inserted with the petals into the base of the calyx ; anthers with 2 parallel cells. Ovary 1-celled, compressed, the two valves early separating and becoming unequal, one becoming lanceolate-elongated, the other remaining short. Seeds few at the base of each placenta.

1. **T. unifoliata,** Hook. Fl. i. 238. t. 81 (1840). Pubescent, the flowering stems 6—15 in. high : leaves thin, ovate-cordate, rounded or triangular, 3—5-lobed, the lobes crenate-toothed ; the radical ones long-petioled : the cauline few, small, short-petioled : panicle narrow and raceme-like : petals small and almost filiform. From San Mateo Co., *Kellogg*, northward, in shady mountain woods.

7. **MITELLA,** *Tournefort* (MITRE-WORT). Small perennials with slender rootstocks, radical leaves, and scapose stem with a simple raceme of small usually green flowers. Calyx short ; the broad tube coherent

with the base of the ovary and dilated above it; lobes valvate in bud, spreading in flower. Petals 5, inserted on the dilated throat of the calyx, pinnately parted or palmately 3-cleft, their divisions, in the first group, almost capillary. Stamens 10 or 5, very short; anthers cordate or reniform, 2-celled. Ovary short and broad; becoming a globular or depressed capsule opening across the summit.

* *Petals green, pinnately parted.*

1. **M. Breweri,** Gray, Proc. Am. Acad. vi. 533 (1865). Leaves shortpetioled, 2–3 in. broad, round-reniform, doubly and incisely crenate: scape naked, a span high: calyx spreading, the limb with very shallow lobes, or merely 5-undulate: stamens 5, alternate with the pectinatepinnatifid petals.—Forests of the middle and higher Sierra, from Mariposa Co. northward.

2. **M. ovalis,** Greene, Pittonia, i. 32 (1887). Leaves thickish, 2 in. long, oval or oblong, obtuse, cordate at base, with closed sinus, the margin with shallow crenate lobes and mucronulate teeth; upper face sparsely hirsute with curved hairs; petioles ferruginous-hirsute with deflexed hairs: scape 1 ft. high, glabrous or nearly so; pedicels very short: calyx-lobes short and broad: petals pinnately parted into 3–5 linear lobes: stamens 5.—Mendocino Co., *Bolander.* This, the *M. trifida* of the "Botany of California," is very distinct from the plant of Graham; for that has deep calyx-lobes and trifid white petals.

* * *Petals white, palmately trifid.*

3. **M. diversifolia,** Greene, l. c. Leaves thin, ovate or orbicular in general outline, but with 3 or 5 shallow angular lobes, these entire, and the whole margin ciliolate: scape 1 ft. high: calyx-lobes small, shallow, whitish: petals white, cuneate-oblanceolate, palmately trifid at the abruptly widened apex: stamens 5.—Summit of Trinity Mountains, near lingering snow-drifts, July, 1886, *C. C. Marshall.*

8. **HEUCHERA,** *Linnæus* (ALUM-ROOT). Perennial herbs, with leaves and flowering stems from a short branching caudex, the former longpetioled, palmately veined, roundish cordate, slightly lobed. Stems somewhat scapiform, bearing few alternate reduced leaves and a panicle or thyrse of cymose-dichotomous clusters of white or greenish or rosecolored flowers. Calyx campanulate, 5-lobed, the tube coherent with the ovary below; lobes obtuse, imbricate in bud. Petals 5, small, entire. Stamens 5, alternate with the petals; anthers 2-celled. Capsule 1-celled, with 2 parietal placentæ; 2-beaked, dehiscent between the beaks. Seeds horizontal, oval, muriculate or hispidulous.

1. **H. rubescens,** Torr. Stansb. Rep. 388. t. 5 (1853). Leaves thickish, 1 in. broad or less, crenately lobed and toothed: scape 6–10 in. high,

nearly leafless and almost glabrous : fl. loosely panicled : calyx 2 lines long, oblong-campanulate, tinged with rose-color ; linear petals and filiform filaments white or pinkish.—At higher altitudes in the Sierra, thence eastward. July—Oct.

2. **H. micrantha,** Dougl.; Lindl. Bot. Reg. xv. t. 1302 (1830). Leaves thin, 1—3 in. broad, ovate-cordate, 5—9-lobed, hairy on the veins beneath : stem villous, bearing a few small leaves and a loose panicle often $1\frac{1}{2}$ ft. long ; calyx campanulate, 1—2 lines long, acute at base, shorter than the slender pedicels, puberulent ; narrowly spatulate petals and slender filaments white, well exserted. Common in shady ravines both of the Coast Range and the Sierra. May—July.

3. **H. pilosissima,** F. & M. Ind. Sem. Petr. v. 36 (1838) : *H. hirtiflora*, T. & G. Fl. i. 582 (1840). Hirsute with rusty and viscid spreading hairs : leaves 1—3 in. broad, round-cordate, obtusely lobed and crenate : stem 1—$2\frac{1}{2}$ ft. high, naked or few-leaved, rather densely and thyrsoidly paniculate : calyx densely hairy, subglobose, the tube rounded, the lobes incurved ; filaments and narrowly spatulate petals little exserted. Var. **Hartwegi,** Wats. Stems 2—3 ft. high : panicle more open : the whole plant, and especially the calyx, less hairy.—In the Coast Range, and apparently not common ; at all events seldom seen.

9. **PARNASSIA,** *Tournefort* (Grass-of-Parnassus). Glabrous stemless perennials, with entire petioled exstipulate leaves and simple 1-flowered scapes. Calyx 5-parted : the base free from or adnate to the base of the ovary, somewhat imbricate in bud. Petals 5, oval or oblong, imbricate in bud, white, with conspicuous green veins, widely expanding, tardily deciduous. Stamens 5, alternating with the petals, and with as many clusters of short gland-tipped sterile filaments. Ovary ovate, 1-celled, with 3 or 4 parietal placentæ ; stigmas as many, closely sessile each directly over its corresponding placenta. Capsule 3—4-valved from the apex, the valves placentiferous in the middle. Seeds with a thickish and somewhat winged loose testa, and little or no albumen.

1. **P. Californica,** Greene, Pitt. ii. 102 (1890) : *P. palustris*, var. *Californica*, Gray, in Bot. Calif. i. 202 (1876). Radical leaves ovate or ovate-oblong, 1—2 in. long, tapering from the broad and sometimes slightly rounded base to a long or short petiole : scapes 1—2 ft. high, the very small sessile but not clasping leaf borne much above the middle : petals oval or obovate, sessile, entire, $\frac{3}{4}$ in. long : sterile filaments about 20 in each set, united to the middle, each tipped with a conspicuous antheroid protuberance. In wet places at considerable elevations in the Sierra. *P. palustris* is not likely to occur within our limits. It is very distinct from the present species by its cordate radical leaves, and by that of the scape being also large, cordate-clasping and inserted low, near the radical ones.

2. **P. fimbriata**, Banks; Sims & Kœn. Ann. Bot. i. 391 (1805). Leaves and scapes from thick branching rootstocks: leaves from reniform to cordate-ovate, 1 in. broad: scapes 1 ft. high, slender, bearing a sessile leaf above the middle: petals obovate or oblong, narrowed to a broad claw which is coarsely fringed on both margins: sterile filaments completely united and forming a merely lobed carinate scale, or sometimes free above.— Has been found near Mt. Shasta, and also in the San Bernardino Mountains: hence to be expected in the high Sierra within our limits.

10. **CHRYSOSPLENIUM**, *Tournefort*. Low fleshy glabrous herbs, with petiolate crenate exstipulate leaves, and small axillary short-pedicelled greenish flowers. Calyx-tube adnate to the ovary; lobes 4 or 5, obtuse. Petals 0. Stamens 8—10, very short, on the margin of a disk. Ovary 1-celled, 2-lobed above; styles 2, short, recurved. Capsule compressed, obcordate, 2-valved at top, with 2 parietal placentæ, ∞-seeded.

1. **C. glechomæfolium**, Nutt.; T. & G. Fl. i. 589 (1840). Stems slender, depressed or ascending, rooting at the lower joints: leaves opposite, or the upper alternate, roundish or ovate, abruptly cuneate at base, crenately dentate, 2—6 lines long, about equalling the petioles: fl. 1—1½ lines long, exceeding the pedicels: seeds large for the plant, ovate, brown, shining.--In wet soil, Humboldt Co., *Marshall*, and northward.

11. **RIBES**, *Fuchs*. Shrubs, with alternate palmately lobed often resinous-glandular or viscid leaves; the stipules when present adnate to the petiole. Flowers racemose (rarely solitary) on short leafy shoots from lateral buds; pedicels subtended by a bract and usually bibracteolate about midway. Calyx-tube adnate to the globose ovary and more or less produced above it, 5-lobed (4-lobed in n. 21), the lobes commonly spreading or reflexed and usually colored. Petals 5, mostly smaller than the calyx-lobes, inserted in or near the sinuses. Stamens as many as the petals, alternate with them; anthers short. Ovary 1-celled; placentæ 2, parietal; styles 2, more or less united; stigmas terminal. Fruit a berry, crowned with the withered remains of the flower. Seeds with minute embryo in a firm albumen.

* *Unarmed; leaves convolute in bud; calyx-tube narrow and elongated.--*
Genus CHRYSOBOTRYA, Spach.

1. **R. tenuiflorum**, Lindl. Bot. Reg. t. 1274 (1829); Greene, Gard. and Forest, iii. 198. *Chrysobotrya Lindleyana*, Spach, Phaner. vi. 151 (1838). Shrub 5—10 ft. high, nearly glabrous, glandless: leaves light green, 3—5-lobed at apex, not at all cordate: racemes ∞-flowered; bracts green and conspicuous: fl. bright yellow, scentless; calyx salverform, the tube 1½ in. long or more, thrice longer than the oval lobes: berry glabrous, amber-colored and translucent, acidulous when ripe.—In both the Coast

Range and the Sierra, but chiefly northward ; apparently uncommon within our limits ; on the Salinas, *Klee*. Called *R. aureum* in the "Botany of California ;" but that has spicy-fragrant flowers, a purple opaque sweetish fruit, and is not found on the Pacific Coast.

* * *Unarmed; leaves plaited in the bud; calyx-tube broader.*—
RIBES proper (Currant).

2. **R. bracteosum**, Dougl.; Hook. Fl. i. 233 (1833). Shrub 4–10 ft. high ; branches glabrous : leaves 3–9 in. broad, 5–7-cleft, glabrous, at least in age, but resinous-dotted ; lobes ovate or narrower, acute or acuminate, coarsely and doubly serrate ; petioles long : raceme ∞-flowered, 3–6 in. or at length 1 ft. long ; bracts persistent, from filiform to spatulate, or the lower larger and passing into leaves : fl. greenish-white ; calyx-tube saucer-shaped, the lobes roundish : berry black, resinous-dotted, $\frac{1}{3}$ in. in diameter.—Woods of Mendocino Co., *Bolander*, thence northward to Alaska.

3. **R. cereum**, Dougl.; Bot. Reg. t. 1263 (1829). Shrub 1–3 ft. high, with many short stout branches, minutely pubescent, resinous-dotted and glutinous : leaves 1 in. broad, rounded or reniform, rather obscurely 3-lobed, crenately toothed or incised : racemes compact, short-peduncled, 3–5-flowered : calyx white, often with a greenish or pinkish tinge ; tube cylindrical, $\frac{1}{2}$ in. long ; lobes short, ovate, recurved : petals orbicular : berry scarlet, translucent, the pulp very firm, sweet, but with a disagreeable resinous flavor.—The commonest species of the Rocky Mountain region, reaching the eastern, and even the western slope of the Sierra Nevada within our borders ; rather ornamental, whether in flower or fruit, but the fruit of no value.

4. **R. viscosissimum**, Pursh, Fl. i. 163 (1814) ; Hook. Fl. t. 76. Shrub 2–6 ft. high, with straggling branches and smooth dark brown bark : leaves thin, 1–4 in. broad, round-cordate, moderately lobed, both faces and the petioles clothed with glandular-viscid hairs ; stipules foliaceous : racemes ascending, somewhat corymbose : fl. large, greenish-white : calyx-tube campanulate, the oblong-ovate lobes scarcely spreading : petals white, smaller than the calyx-lobes : berry oblong, $\frac{1}{3}$ in. long, black, glandular-hirsute.—Forests of the higher Sierra, in dry rocky places, as far south as Mariposa Co., but more common northward.

5. **R. sanguineum**, Pursh, l. c. 164 (1814) ; Smith, Rees Cycl. xxx. (1815) ; Lindl. Bot. Reg. t. 1349. Shrub 4—8 ft. high ; young branches dull-red and densely soft-puberulent : leaves cordate, 3–5-lobed, thickish, 2—3 in. broad, very soft-tomentulose on both faces ; the ascending petioles 2 in. long, gradually dilated and very coarsely ciliate at base : racemes short-peduncled, ascending, ∞-flowered, dense : calyx not bracteolate at base, campanulate-funnelform, 6 lines long, deep rose-red,

cleft below the middle: petals white, spatulate-oblong, shorter than the calyx-lobes; berry small, oval, blue with a heavy bloom; pulp firm, black, insipid.—From Del Norte Co. and perhaps Humboldt, northward to British Columbia; perhaps not within our limits; replaced in middle California by the three following closely related species, or subspecies.

6. **R. glutinosum**, Benth. Trans. Hort. Soc. i. 476 (1835); Walp. Rep. ii. 360: *R. sanguineum*, var. *glutinosum*, Gray, Bot. Calif. i. 207 (1876). Near the last but larger, 6–15 ft. high; the bark of young branches pale and shining, sparsely scabro-puberulent: leaves thin, 3–5 in. broad, glutinous when young, glabrous or more or less pubescent in age; petioles divaricate, very abruptly dilated at base and obscurely ciliolate: racemes long-peduncled, pendulous, very many-flowered: calyx with two conspicuous but caducous bracteoles at base, cleft scarcely to the middle, from pale pink to rose-color: berry large, globose, blue with a dense bloom, and glandular-hispid; pulp black, dry, insipid.—Common on moist banks of streams, and around springy places, at low altitudes in the Coast and Mt. Diablo Ranges, chiefly or exclusively in the middle section of the State, not in the Sierra. Too unlike *R. sanguineum* to be a mere variety of it; yet hardly more than a geographical subspecies, and remarkable, as a currant, for the long interval between the flowering of the shrub and the ripening of its fruit. In flower from January (or even Dec.) to March. Fruit not ripe until August or September. They who have described the berries of these shrubs as "bitter" must have made their test before it had become mature. No fruits can be more absolutely tasteless when ripe.

7. **R. malvaceum**, Smith, Rees Cycl. xxx. (1815): *R. tubulosum*, Esch. Mem. Acad. Petersb. x. 283 (1826). *R. sanguineum*, var. *malvaceum*, Gray, l. c. Shrub low and compact, 3–6 ft. high; growing branches canescently tomentulose, glabrous and red when mature: leaves thick, 1–2 in. broad, strongly rugulose and somewhat scabrous above, more or less densely white-tomentose beneath; the slight stipular dilatation of the petiole only obscurely ciliolate: racemes short-peduncled, ascending, dense; pedicels and ovaries whitish-tomentose: calyx-tube subcylindrical, abruptly dilated and broadest just above the ovary; segments short, spreading, the whole rose-color: petals white, roundish or subreniform: berry oval, ⅓ in. long, purple, glaucous; pulp soft and sweet.—On dry open hills of the Coast Range, from Bolinas Ridge, *Drew*, southward; very common in San Mateo Co., and Monterey; occasional in Contra Costa and Alameda. Species exceedingly well marked in habit, foliage, flower and fruit. Fl. Mar. Apr. Fr. Apr. May.

8. **R. Nevadense**, Kellogg, Proc. Calif. Acad. i. 63 (1855): *R. malvaceum*, Kell. l. c. 46, not of Smith: *R. sanguineum*, var. *variegatum*, Wats. partly. Rather slender, loosely branching, 3–6 ft. high: bark of young

branches shining, glabrous, of the older flaky and deciduous: leaves
2–4 in. broad, very thin, not rugulose, bright green and glabrous above,
paler beneath with a sparse tomentulose pubescence; stipular base of
petiole clothed with very long coarse hairs each of which is sparsely
pilose: racemes short and dense (1 in. long), pendulous on slender
peticles of 2 in. or more: fl. small, rose-red: calyx-tube urceolate,
broadest just above the ovary, only 1½ lines long, the spreading segments
as long or longer: berry small, globose, black, densely glaucous; pulp
soft, very sweet. At middle elevations in the Sierra Nevada, in open
groves of Sequoia, from Kern Co., *Palmer & Wright* (n. 101), and the
Calaveras Big Trees, *Greene*, to Placer Co. A most distinct species, of
peculiar habitat; the foliage thinner and more nearly glabrous than
that of *E. glutinosum*, which is its analogue of the Coast Range. In all
three of the last preceding the bark of the stem and older branches is
close, smooth, dark brown and white-dotted, like that of young birches.
In the present species it is flaky and deciduous. Doubtless the *R. Wolfii*
of regions east of the Sierra, in Nevada and Colorado, is another member
of this group of American *Black Currants*.

* * * *Thorny; leaves plaited in the bud; fl. few in the cluster, or
solitary.*—Gooseberry.

+ *Fl. 5-merous; calyx-lobes reflexed.*—Genus GROSSULARIA, Philip Miller.

9. **R. lacustre,** Poir. Suppl. ii. 856 (1811), var. **molle,** Gray, Bot.
Calif. i. 206. Shrub small and depressed, the spreading branches ½–2
ft. long, bristly or naked, armed with short triple or multiple thorns
under the leaf-fascicles: leaves ½–1 in. broad, deeply 5-parted, the
lobes incisely toothed and cleft, pubescent: racemes 3–9-flowered: fl.
greenish-white; calyx saucer-shaped, ¼ in. broad, its short lobes
rounded; petals small; stamens very short: berry globose, red, ⅓ in.
thick, more or less glandular-hispid: pulp juicy, acidulous.—At rather
high altitudes of the Sierra, in rocky shades; quite intermediate between
Currant and Gooseberry.

10. **R. oxyacanthoides,** Linn. Sp. Pl. i. 201 (1753). Mostly glabrous,
2–4 ft. high; thorns small, single or triple: leaves roundish, deeply
5-lobed; lobes incised and coarsely toothed: peduncles mostly shorter
than the pedicels of the 2 or 3 flowers: fl. greenish-white, 3–4 lines long;
calyx-lobes oblong, equalling the campanulate tube, little exceeding the
cuneate-obovate petals, about equalling the stamens; style cleft, villous
below, longer than the stamens: berry of middle size, glabrous, black,
pleasant.—This widely distributed species occurs within our limits only
toward the higher parts of the Sierra, and chiefly on the eastward slope.

11. **R. divaricatum,** Dougl. Trans. Hort. Soc. vii. 515 (1830); Bot.
Reg. t. 1359: *R. villosum,* Nutt.; T. & G. Fl. i. 547 (1840). Nearly gla-

brous : stems clustered, the widely spreading branches 5—12 ft. long ; thorns single or triple : leaves roundish, 3—5-lobed ; the lobes incisely toothed : peduncles elongated, slender, drooping, 3—9-flowered ; pedicels with a small broad bract at base : fl. $1\over 3$ in. long ; calyx green without, dark livid purple within, the oblong-linear lobes exceeding the campanulate tube: petals white, fan-shaped, plane, the margins convolutely overlapping : filiform villous filaments and deeply cleft style long-exserted : berry small, glabrous, black, agreeable. - Along streams and on northward slopes throughout the Coast Range, from Santa Barbara northward. Our description is drawn from the Californian shrub, which differs not a little from the Oregonian type of Douglas' species, and is the *R. villosum* of Nuttall—possibly a subspecies. Fl. March. Fr. June.

12. **R. leptanthum**, Gray, Pl. Fendl. 53 (1859). Shrub 3—4 ft. high, glabrous, not bristly ; subaxillary spines usually solitary : leaves about $1\over 3$ in. broad, 5-cleft, the lobes incised : peduncles short, deflexed, 1—3-flowered : fl. white, $1\over 2$ in. long ; calyx with slender cylindrical tube and spathulate lobes of about equal length ; petals short ; stamens exserted ; style glabrous, undivided : berry small, black, unarmed, glabrous.— Common in New Mexico ; thence westward to the eastern Sierra Nevada.

13. **R. velutinum**, Greene, Bull. Calif. Acad. i. 83 (1885) : *R. leptanthum*, var. *brachyanthum*, Gray, Bot. Calif. i. 205. Stout and rigid, 4—6 ft. high, with strongly recurved branches, these not prickly, but with solitary stout axillary thorns : leaves very small, deeply 5-cleft, the lobes 3-cleft, these and the growing branchlets densely velvety-pubescent or almost glabrous : peduncles short, deflexed, 2-flowered : fl. white or pinkish, velvety on the outside ; calyx-tube campanulate, as broad as long (2 lines), the segments rather longer : ovary white-villous or almost glabrous : berry small, black, velvety-pubescent or glabrate. Common in the more arid mountain districts, chiefly eastward and northward, beyond our limits ; but also at Tehachapi, Kern Co. Variable in respect to pubescence ; but well marked in floral character and general habit.

14. **R. quercetorum**, Greene, l. c. (1885). The many rigid recurved glabrous branches forming a very compact bush 3—5 ft. high and of equal diameter ; very leafy and wholly glabrous : the subaxillary spine solitary : leaves small, 5-cleft, the lobes narrow, cuneiform, 3-lobed at summit ; petioles slender, 1 in. long ; peduncles slender, short, deflexed, 2—4-flowered : fl. light yellow, very fragrant : calyx-tube cylindraceous, 2 lines long, about equalled by the linear-oblong reflexed segments, these a little longer than the petals ; stamens very short and included ; style glabrous, undivided : ovary glabrous : berry large ($1\over 2$ in. thick), globose, glabrous, dark red, pulpy and agreeable.—Very common about El Paso de Robles, and in the interior valleys and low hills of Monterey and San Luis Obispo counties. Fl. March. Fr. May.

15. **R. ambiguum**, Wats. Proc. Am. Acad. xviii. 193 (1883). Glandular-pubescent and villous; subaxillary spines short: leaves 1 2 in. broad, 5-lobed and incised : fl. mostly solitary, 1/2 in. long or less, greenish, more or less villous : stamens equalling or barely exceeding the white petals ; anthers very small, light-colored, glabrous, obtuse at both ends : fr. large, densely spinose. On the Scott Mountains and northward ; but doubtless to be found within our limits.

16. **R. Marshallii**, Greene, Pittonia, i. 31 (1887). Near the last, but glabrous : fl. 1 in. long ; calyx-segments oblong-linear, spreading or reflexed, equalling or exceeding the tube, dark-purple ; petals 2 3 lines long, salmon-color, rather thin, manifestly involute ; filaments slender. exserted ; anthers oblong, obtuse at both ends, glabrous ; ovary bristly.— Summit of Trinity Mountains, near lingering snow-drifts, July, 1886 ; in flower only. Possibly a glabrous and large-flowered state of the preceding.

17. **R. Victoris**, Greene, Pittonia, i. 224 (1888). Shrub 5 ft. high : branches commonly very prickly ; subaxillary spines triple, rather slender : leaves (not very deeply 5-lobed) and growing branchlets pubescent and viscid : pedicels short, deflexed, with 1 or 2 persistent bracts and as many short-pedicellate greenish flowers 1/2 in. long : calyx-tube short-campanulate, much exceeded by the green (occasionally livid-purplish) lobes ; petals 1 1/2 lines long, white, thinnish, involute, acute and more or less toothed at apex ; filaments stoutish, little surpassing the petals ; anthers large, subsagittate, mucronate ; ovary glandular-hispid.—By streams in the Coast Range, but not common ; at Lagunitas, in Marin Co., *Chesnut & Drew;* Rutherford Cañon, and near Calistoga, *Parry.* Possibly identical with *R. occidentale*, H. & A., a shrub which can not be identified by the very inadequate diagnosis given in the Botany of Beechey's Voyage.

18. **R. Californicum**, H. & A. Bot. Beech. 346 (1840). Shrub 2 4 ft. high, with very rigid and flexuous glabrous branches : subaxillary spines ternate, short, stoutish : leaves small, 3—5-lobed and incised, sparsely glandular-puberulent when young, not at all viscid or heavy-scented, in maturity glabrous : peduncles very short, 1 3-flowered ; the very short pedicels each with a small round-ovate bract beneath : calyx-tube very short, the reflexed lurid-purple ligulate segments thrice as long ; petals white, thick, strongly involute, truncate and erose-toothed at summit ; filaments stout, thrice the length of the petals, the anthers ovate-oblong. mucronate, reddish ; style simple ; ovary glandular-hispid : berry large. prickly. One of a goodly number of very clear species which herbarium writers have confused with *R. Menziesii*. This is very common, on open slopes and along streams in the Oakland Hills ; remarkable for its very early flowering, its short-jointed zigzag branches, small glabrous scentless foliage, etc. One can hardly be positive that it is the plant of Hooker &

Arnot; neither can one have much doubt. The description, as far as it goes, applies well. Fl. Feb., Mar. Fr. June.

19. **R. subvestitum,** H. & A. l. c. (1840): Tall leafy open and rather handsome shrub 5—10 ft. high; branches usually more or less setose-hispid: subaxillary spines 3 or 4, rather slender: leaves more or less glandular-pubescent, very viscid and heavy-scented: peduncles 1—3-flowered; pedicels elongated, the small bract persistent: calyx-tube broadly campanulate, 1½ lines long, the red-purple reflexed segments nearly twice as long: petals white-waxy, truncate, entire, strongly involute; filaments well-exserted: ovary densely glandular-hairy: berry large, as densely clothed with short stiff gland-tipped hairs; pulp soft, sweet. Very common in the Coast Ranges from at least Sonoma Co. to Monterey: perhaps a variety of the next, but the differences constant, the flower very handsome. Mar. Apr.

20. **R. Menziesii,** Pursh, Fl. ii. 732 (1814); Lindl. Bot. Reg. xxxiii. t. 56: *R. ferox*, Smith, Rees Cycl. xxix. (1815). Size and habit of the last; branches strongly hispid, or varying to glabrous: leaves more than 1 in. broad and of greater length than breadth, deeply 3-cleft, the lobes coarsely incised, usually soft-pubescent beneath, seldom or never viscid: peduncles slender, pendulous, 1—2-pedicellate above the middle, the bracts small, persistent: fl. ¾ in. long; calyx of a rich red-purple, pubescent exteriorly, the tubular-funnelform tube about half as long as the ligulate reflexed segments: petals large, thickish, truncate, involute, cream-color or whitish: filaments subulate, not exserted, only the large linear-oblong mucronate white anthers borne beyond the petals: ovary densely echinate: fruit very prickly; pulp not ill-flavored.—From Humboldt Co. to Santa Barbara, but in the Bay region seldom met with outside of Marin Co.; perhaps confluent with the preceding, though the typical forms of the two appear abundantly distinct.

21. **R. amictum,** Greene, Pittonia, i. 69 (1887): *R. Menziesii*, V. & R. Contr. U. S. Herb. i. 2, not Pursh. Shrub 3—4 ft. high, with rigid flexuous widely spreading glabrous or merely pubescent branches; subaxillary spines triple, short and stoutish: leaves small (less than 1 in.), glabrous or pruinose-pubescent, 3—5-lobed, often broader than long: peduncles 1—2-flowered; pedicels not at first apparent, the base of the flower enclosed by the large cucullate and more or less caducous bract: calyx dark crimson or red-purple, ½—¾ in. long; tube nearly cylindrical, about equalled by the segments: filaments subulate, little exceeding the truncate erose-dentate involute white petals; anthers ovate-oblong, mucronate, rose- or deep-red; ovary echinate; berry large (½ in. or more), usually strongly armed with stout prickles; pulp agreeable.—Very common from Oregon to Kern Co. Calif., but apparently only in the mountains at some distance inland; abundant at middle elevations

of the Sierra everywhere; formerly confused with *R. Menziesii*, from which it is very distinct. Owing to its inland habitat it was not obtained by the early explorers of the coast. Its large elongated dark red flowers, conspicuous cucullate bracts, and very large fruits, mark it as an excellent species, notwithstanding its great variability in respect to pubescence; for while some specimens are almost hoary, even to the calyx, others are wholly glabrous. But the branches appear to be always destitute of the bristly hairs or soft prickles which, in other allied species are almost invariably present. The large well-flavored reddish fruits are rarely almost free from the prickles; while in the more tomentulose forms the prickles themselves, as well as the surface of the fruit, are sparsely villous-hairy. In the shrub of our southern foothill regions, the bracts appear to be persistent, the pedicels elongating after flowering.

← ← *Fl. 4-merous; calyx-lobes erect.*—Genus ROBSONIA, Berlandier.

22. **R. speciosum**, Pursh, Fl. ii. 732 (1814); Bot. Reg. t. 1557: *R. stamineum*, Smith, Rees Cycl. (1815): *R. fuchsioides*, Berl. Mem. Soc. Genev. (1828). *Robsonia speciosa*, Spach, Phaner. vi. 181 (1838). Shrub 6–10 ft. high, with long leafy red-bristly branches: subaxillary spines 3, united at base: leaves subcoriaceous, dark green, very smooth and shining above, rounded and 3-lobed; lobes short, crenately-toothed: peduncles pendulous, 2–5-flowered: fl. bright red, often 2 in. long from the base of the ovary to the tips of the long-exserted stamens; calyx cylindraceous, the 4 (rarely 5) lobes erect; anthers oval, small: ovary bristly: berry small, rather dry, densely prickly.—The most beautiful species of the genus, and the only one in our flora which is evergreen: frequent along the seaboard from Monterey southward. Mar.—May.

ORDER XXIX. **PHILADELPHEÆ.**
Don; Edinb. Phil. Journ. 133 (1826). Part of *Myrtaceæ*, Juss. (1789).

Shrubs or undershrubs, with opposite exstipulate toothed leaves, and cymose-paniculate or axillary and solitary white flowers. Calyx with turbinate tube adherent to the ovary; the 4–6-parted limb valvate in the bud. Petals as many as the calyx-lobes, alternate with them, convolute or imbricate in bud. Stamens 8–∞ in 1 or 2 series. Styles and stigmas several, the former, and sometimes the latter more or less coalescent. Fruit capsular, 4–10-celled, loculicidally or septicidally dehiscent. Seeds numerous, on axial placentæ, mostly pendulous, elongated; testa thin: albumen fleshy.

1. **PHILADELPHUS**, *Ruppius* (MOCK-ORANGE. SYRINGO). Leaves ovate or oblong, short-petioled. Flowers 4-merous (sometimes terminal one 5-merous), more or less clustered terminally and in the upper axils.

showy. Petals obovate or roundish, convolute in bud. Stamens 20—40; filaments subulate or filiform. Styles 3—5, united at base, or almost to the top: stigmas oblong, introrse, sometimes connate. Capsule 3—5-celled, loculicidally 3—5-valved from the apex; the valves in age 2-parted. Seeds very many, pendulous, oblong, the thin loose testa reticulate, usually prolonged at both ends.

1. **P. Lewisii,** Pursh, var. **Californicus,** Gray, Bot. Calif. i. 202 : *P. Californicus*, Benth. Pl. Hartw. 309 (1849). Shrub 3—6 ft. high, nearly or quite glabrous : leaves ovate or ovate-lanceolate, 1—2 in. long, nearly or quite entire : fl. in a pedunculate naked cluster : calyx-lobes about twice the length of the tube : styles distinct at apex only; stigmas narrow.—On banks of streams among the foothills of the Sierra from Mariposa Co. northward. May, June.

2. **P. Gordonianus,** Lindl. Bot. Reg. xxiv. Misc. 21 (1838), & xxv. t. 32 (1839). Shrub 8—10 ft. high, with spreading or recurved branches, sparsely pubescent or almost glabrous : leaves ovate or oblong-ovate, 2—4 in. long, coarsely and remotely serrate-toothed : fl. in loose axillary clusters somewhat leafy at base; petals ¾—1 in. long; styles distinct to the middle; stigmas more or less dilated; calyx-lobes twice as long as the tube.—In the Coast Range, from Mendocino Co. northward.

3. **P. serpyllifolius,** Gray, Pl. Wright. i. 77 (1852). Shrub stout, rigid, branching, 3—5 ft. high, older branches whitish, striate : leaves ovate-oblong, ½—1 in. long, entire, 3-nerved, pale on both faces with a minute appressed pubescence : fl. subsessile, solitary or in threes at the ends of the short, rigid branchlets, cream-color or white : calyx silky-pubescent, the lobes ovate; style very short; stigma thick, 4-lobed : capsule 4-celled.—White Mountains, Mono Co., *Shockley.*

2. **CARPENTERIA,** *Torrey.* An evergreen shrub, with elongated subcoriaceous denticulate leaves, and terminal pedunceled cymes of few large mostly 6-merous flowers. Calyx 5—6-parted, only the very short tube adnate to the ovary. Petals rounded, concave, convolute-imbricate. Stamens about 200; filaments filiform, somewhat distinctly gathered into 5 or 6 fascicles. Styles united throughout; the 5—7 stigmas distinct. Capsule 5—7-celled, dehiscent septicidally. Seeds nearly as in *Philadelphus.*

1. **C. Californica,** Torr. Pl Frem. 12. t. 7 (1850); Greene, Pitt. i. 66. Shrub 4—6 ft. high, the many stems forming a dense clump; bark pale, becoming shreddy, nearly glabrous on young shoots : thickish persistent leaves broadly lanceolate, remotely denticulate, in age more or less revolute, 2—4 in. long, short-petioled, glabrate above, pale and tomentulose beneath : peduncle leafless : corolla nearly rotate, 2—3 in. broad, pure white : stamens with small bright yellow anthers : seeds oblong, with

short obtuse appendage at each end.—A rare shrub, scarcely known in the wild state, inhabiting the Sierra Nevada below midway of the State. In cultivation at Berkeley flowering in May, but perfecting no fruit.

3. WHIPPLEA, *Torrey*. Low diffuse hairy undershrub, with slender stem, slightly petioled leaves, and terminal naked-peduncled clusters of small white flowers. Calyx white like the petals, 5-cleft; tube adnate to lower part of the ovary. Petals 5, ovate or oblong, narrowed at base. Stamens usually 10; filaments subulate. Ovary 3–5-celled, with a single ovule in each cell; styles distinct; stigmas introrse. Capsule septicidally parting into distinct cartilaginous 1-seeded portions which open ventrally only. Seeds oblong, with a short obtuse appendage at each end.

1. **W. modesta**, Torr. Pac. R. Rep. iv. 90. t. 7 (1857). Stems slender, trailing, 1 ft. long or more: leaves thin, ovate or oval, somewhat toothed or entire, 1 in. long or less: peduncle slender, bearing a close few-flowered cluster: fl. 2 lines broad or less: calyx-tube nearly hemispherical: capsule globular; styles deciduous from it. Borders of thickets, or in deep woods, through the Coast Range from Monterey northward. The only species; though a small *Fendlera*,— *F. Utahensis* (Wats.), Greene has erroneously been referred to this genus.

Order XXX. EPILOBIACEÆ.

Ventenat, Tabl. du Reg. Veget. iii. 483 (1799). ONAGRARIÆ, Juss. Ann. du Mus. iii. 315 (1804). *Onagræ*, Adans. (1763).

Herbs, often with hard shrubby-looking stems shedding a thin papery outer bark. Leaves simple, usually alternate, entire, toothed or pinnatifid. Flowers axillary to the leaves, or in merely bracted or naked racemes or spikes, rarely panicled; usually 4-merous. Calyx-tube partly or wholly adherent to the ovary; lobes valvate in bud. Petals borne on the throat of the calyx-tube, or at the sinuses of the lobes, convolute in bud. Stamens 2–8. Styles single; stigma capitate or 4-lobed; ovary 2 or 4-celled. Seeds naked or appendaged; albumen none.

Hints of the Genera.

Free portion of the calyx deciduous from the ovary;
   Fruit 4-celled, capsular;
      Seeds comose at apex, - - - - - - - 1, 2
      " not comose,
         Fl. yellow or white, - - - - - 3, 4
         " purple or rose-color, - - - - - 5—7
   Fruit indehiscent, - - - - - - - - - - - 8, 9
Calyx-lobes persistent on the ovary, - - - - - - - - - 10

1. **EPILOBIUM**, *C. Gesner* (WILLOW-HERB). Tube of calyx little prolonged beyond the ovary; limb deeply 4-cleft, campanulate or funnel-

EPILOBIACEÆ.

form, or 4-parted to the base with the lobes spreading, deciduous. `Petals 4, spreading or erect, often emarginate or bifid, purple or white. Stamens 8, the 4 alternate ones shorter; anthers elliptical or roundish, fixed near the middle. Stigma oblong, clavate, or with 4 spreading or revolute lobes. Capsule mostly linear, 4-sided, 4-celled, 4-valved. Seeds numerous, ascending; the summit bearing a tuft of long white hairs.

* *Perennials (often stoloniferous) or annuals; fl. small, regular; petals ascending; stamens and style erect.*—EPILOBIUM *proper*.

+— *Annuals, with terete stems; leaves alternate (except the lowest).*

1. **E. minutum**, Lindl.; Hook. Fl. i. 207 (1833). Diffusely branched from the base, the mostly decumbent or ascending branches ½–1 ft. long, puberulent: leaves ovate-lanceolate or lanceolate, entire or repand-denticulate, ½–¾ in. long: fl. solitary in all the axils, very small: petals obcordate, white or with a tinge of rose: 4 long stamens equalling the style: stigma clavate, the lobes at length expanded and fimbriate: capsule pedicellate, about 1 in. long, more or less crenate: seeds rather few, smooth; the coma very deciduous.—In the Coast Range almost throughout the State, on dry hills in the wooded sections. Apr.–June.

2. **E. paniculatum**, Nutt.; T. & G. Fl. i. 490 (1840). Erect at base, slenderly paniculate-branched above, 1—10 ft. high, from wholly glabrous to minutely and densely glandular-pubescent: leaves narrowly lanceolate or linear, obscurely serrulate, 1–2 in. long, with smaller ones fascicled in the axils, the floral reduced to subulate bracts: corolla cruciform; the rose-colored petals quadrate-oblong, abruptly and often deeply notched, rose-purple and veiny, 1—2 lines long, rotate-spreading: capsule pedicellate, 1 in. long, attenuate at each end, often arcuate: seeds minutely papillose. Var. **jucundum**, Trel.: *E. jucundum*, Gray, Proc. Am. Acad. xii. 57 (1876). About 2 ft. high, stouter, panicle condensed and thyrsoid: petals ⅓–½ in. long, broadly obcordate, only ascending (not rotate-spreading).—The type extremely common seaward throughout California and far northward as well as eastward; in the Bay region commonly 5—6 ft. high, not rarely 10 ft. The variety– or subspecies confined to the interior valley, from near Sacramento, *Drew*, northward to Oregon. July—Nov.

+— +— *Perennials, often stoloniferous; leaves mostly opposite (except the upper and floral).*

3. **E. Franciscanum**, Barbey; Bot. Calif. i. 220 (1876); Trel. N. Am. Epil. 90. t. 15. Very stout, simple, or rather closely paniculate-branched, 2–4 ft. high, pubescent with soft short glandular hairs: stem reddish, subterete, but with delicate sharp angles running down from the leaf-bases: only the alternate upper and floral leaves strictly sessile, the lowest with a very short but distinct stout petiole, these 2–4 in. long,

oblong-lanceolate, rounded at base, serrulate : racemes dense, notably leafy-bracted, the large red-purple or pale flowers appearing somewhat corymbose : petals ½ in. long or more, deeply emarginate : capsule 2 in. long or more : seed obovoid-oblong, acutely pointed at base, the hyaline papillæ forming close longitudinal lines. —Plentiful in springy places, along streamlets and shores of ponds about San Francisco, at Point Lobos, Mountain Lake, etc., thence northward to Oregon and Alaska. Flowers large and mostly bright-colored ; the herbage not well described as "hoary," in the "Botany of California;" for the delicate pubescence seldom if ever imparts a shade of color to the plant. June—Dec.

4. **E. Watsoni,** Barbey, l. c. 219 ; Trel. l. c. 16. Size of the preceding, but not stout, the terete stems with less marked lines, somewhat hoary with a soft pubescence : leaves oblong-lanceolate, rather obtuse, denticulate, rounded to short-winged petioles : fl. not crowded, suberect in the axils of the more reduced and acute upper leaves, rose-red ; petals elongated-obcordate: seeds more coarsely granulate-striate. On Russian River, Sonoma Co., and perhaps common along the seaboard northward ; also in Solano Co., on Alamo Creek, *Jepson*.

5. **E. holosericeum,** Trel. N. Am. Epil. 91. t. 17 (1891). Loosely branched, at least the upper leaves and branches canescent with subappressed hairs ; leaves oblong-lanceolate, obtuse or sometimes acute, remotely serrulate, attenuate, or abruptly contracted and then cuneately narrowed, to short petioles : fl. small, scattered on the elongated branches, pale : mature capsules on peduncles equalling the floral leaves : seeds short-beaked, very minutely papillose-striate.—Of the southern part of the State, coming within our limits in Kern Co., according to the author of the species.

6. **E. adenocaulon,** Hausskn. Bot. Zeitsch. 119 (1879), var. **occidentale,** Trel. l. c. 95: *E. coloratum*, Bot. Calif. partly, not Muhl. Tall, with paniculate ascending branches and long internodes ; branches, inflorescence and capsules glandular-pubescent: leaves ovate- or triangular-lanceolate, ascending, abruptly rounded to short-winged petioles, prominently denticulate, the floral small, acute at both ends : fl. small : capsule slender, short-pedicellate : seed elongated, obovoid, very minutely striate.—Common in both the Coast Range and the Sierra, by streams and about springy places.

7. **E. Californicum,** Hausskn. Monogr. 260 (1884) ; Trel. l. c. 96. t. 26. Tall, slender, more sparingly branched, glabrous below ; pubescence of the buds, pods, etc. of coarse ascending, not glandular hairs : leaves lanceolate, acutish, rather remotely serrulate, short-petiolate : fl. scattered : fruiting peduncles slender, almost equalling the floral leaves : capsules nearly glabrous : seeds almost beakless.—Apparently along the seaboard only, and less common than the last.

8. **E. brevistylum,** Barbey; Bot. Calif. i. 220 (1876); Trel. l. c. 100. t. 30. Stem stoutish, terete, 10—18 in. high, simple or at summit sparingly branching, marked with 2—4 decurrent lines : leaves mostly opposite, sessile, broadly lanceolate, denticulate with small rigid teeth : fl. small : inner row of stamens very short ; the outer exceeding the style : seeds tapering above.--In Sierra Co., *Lemmon*, and northward.

9. **E. exaltatum,** Drew, Bull. Torr. Club, xvi. 151 (1889); Trel. l. c. 95. t. 24. Stems simple below, only loosely and corymbosely paniculate at the very summit, terete, glabrous : leaves all opposite except the floral, thin, rather pale, ovate- or oblong-lanceolate, acute, denticulate, 2 in. long, abruptly short-petiolate : inflorescence glandular-puberulent : fl. large, rose-purple ; the obcordate petals 4—5 lines long : capsules on slender pedicels equalling the floral leaves : seeds small, linear-oblong, very minutely papillose-striate. –On Grouse Creek, Humboldt Co., *Chesnut & Drew*, and northward. An excellent species of peculiar habit, and large flowers for this group ; the stigma 4-lobed. Aug. Sept.

10. **E. ursinum,** Parish ; Trel. l. c. 100. t. 31 (1891). Slender, 1½—1 ft. high, pilose with white hairs, the inflorescence minutely glandular-pubescent ; stem terete, with long internodes : leaves small, only the lowest opposite, ovate-lanceolate, denticulate, serrulate or nearly entire, abruptly rounded to the sessile base : fl. few ; petals white or very pale : capsules on very slender peduncles of more than half their length, glabrate : seeds rather rough-papillose. Var. **subfalcatum,** Trel. l. c. 101. t. 32. Densely tomentose or pilose up to the glandular inflorescence : leaves narrower sometimes falcate, more cuneate at base, more remotely and inconspicuously denticulate.--Presumably common in middle California somewhere ; but no locality given, except "San Bernardino Co.," for either type or variety ; communicated to the author from the Calaveras Big Trees, 1888, *Wm. Rieger*.

11. **E. glaberrimum,** Barbey ; Bot. Calif. 220 (1876) ; Trel. l. c. 104. t. 38. Nearly simple up to the short inflorescence, commonly 2 ft. high, glabrous, glaucous, the stems terete and slender : leaves oblong-lanceolate, obtusish, repand-denticulate or nearly entire, the lowest short-petiolate : petals 2 lines long, rose-purple or paler : capsules very slender, straight or somewhat arcuate, often long-pedicelled : seeds rather roughly papillose-striate.—Common along streams and ditches at middle and lower elevations of the Sierra. June- Sept.

12. **E. Hornemanni,** Reichenb. Ic. Crit. ii. 73 (1824) : Trel. l. c. 105. t. 41. Stems stoloniferous at base, ascending, simple, 3—10 in. high, the inflorescence pubescent, and also the stem along the decurrent lines : leaves small, ascending, ovate-oblong, obtuse, from almost entire to remotely serrulate, the lower cuneately narrowed, the upper usually

rounded to the short petioles : fl. few, small, erect ; petals very pale, or deeply colored ; capsules slender, erect, on slender peduncles about equalling the floral leaves ;- seeds abruptly short-appendaged, from smoothish to notably rough. Not known as Californian except in n. 1417, *Brewer*, which came from Mt. Shasta or near there.

13. **E. alpinum,** Linn. Sp. Pl. i. 348 (1753), in part ; Trel. l. c. 108. t. 44. Like the last, but with less obvious pubescence : leaves thin and delicate, all gradually narrowed to the slender petioles ; fl. white or rose-tinted ; capsules very slender, on peduncles often stoutish and as long as the capsules : seeds smooth, with a very manifest beak. Scarcely known as Californian, yet likely to be found anywhere in the high Sierra.

14. **E. Oregonense,** Haussku. Monogr. 276 (1884) ; Trel. l. c. 109. t. 25. Low, simple and stoloniferous, the stem erect to the summit ; inflorescence sparingly glandular-pubescent, the plant otherwise glabrous ; leaves crowded below, remote and very small above, suberect, ovate-lanceolate, or the uppermost linear, very obtuse, remotely denticulate, somewhat tapering to the sessile base : fl. few, erect, small, deep purple : capsules stoutish, surpassing the summit of the stems, their peduncles far exceeding the floral leaves : seeds smooth, obtuse, beakless. A northern species, to which Prof. Trelease refers somewhat doubtfully certain alpine plants collected by *Bolander* (n. 4965) and *Lemmon*, along our northeastern borders.

15. **E. anagallidifolium,** Lam. Encycl. ii. 376 (1786) ; Trel. l. c. 110. t. 47. Very low, cespitose by the numerous stolons, the very slender stems ascending, nodding at summit ; pubescent in lines, the inflorescence somewhat glandular : leaves small, uniformly distributed, narrowly ovate or oblong, rather obtuse, entire or sparingly denticulate, cuneately narrowed. the lowest wing-petioled : capsule narrowly linear, long and slender, the peduncles long or short : seeds somewhat obovoid-fusiform, short-beaked. Common by alpine brooklets in Washington and Oregon ; reaching Mt. Shasta, and doubtless coming within our limits northward.

16. **E. obcordatum,** Gray, Proc. Am. Acad. vi. 532 (1865) ; Trel. l. c. 84. t. 6. Somewhat cespitose, the stems stoutish, ascending or decumbent, 3–10 in. long : herbage glabrous, very pale and glaucous, the inflorescence glandular-puberulent : leaves all opposite, broadly ovate, 1_2 –3_4 in. long, repand-denticulate, rounded to very short winged petioles : fl. few, large, slender-peduncled, in the axils of the scarcely reduced upper leaves : petals obcordately 2-lobed, 1_2 in. long or more, rose-purple : capsule 3_4–1 in. long, subclavate, pedicellate, few-seeded : seeds finely papillate.--Common on moist slopes below perpetual snow, etc., near the highest parts of the Sierra Nevada, from Tulare Co., northward ; our most beautiful species. July- Oct.

* * *Perennial; fl. large, somewhat irregular; style and stamens declined.*--
Genus CHAMÆNERION, Tourn., S. F. Gray.

17. **E. spicatum,** Lam. Fl. Fr. iii. 482 (1778); Trel. l. c. 80. t. 1. Stout and simple, 2–6 ft. high, glabrate below, the inflorescence canescently puberulent: leaves alternate, lanceolate, acute, entire, short-stalked, 3–6 in. long, deep green above, pale beneath: inflorescence racemose, the buds deflexed: calyx cleft nearly to the base: petals bright purple, unguiculate, $\frac{1}{2}$ in. long or more: style surpassing the stamens, hairy at base: capsule short- or long-stalked, 1 in. long or more.—In the Sierra Nevada, apparently throughout the State, but far less common than in more northerly regions. July—Sept.

2. **ZAUSCHNERIA,** *Presl.* Perennial herbs (not at all suffrutescent), spreading by subterranean shoots. Leaves opposite, except the upper and floral. Flowers racemose along the leafy branches, large, scarlet. Calyx-tube globose-inflated just above the ovary, thence becoming narrow-funnelform, 4-lobed, within bearing 8 small scales, 4 erect and 4 deflexed. Petals 4, little exceeding the calyx-lobes, obcordate or deeply cleft. Stamens 8, the 4 alternate ones shorter; anthers linear-oblong, attached by the middle. Stigma peltate or capitate, 4-lobed. Capsule slender-fusiform, obtusely 4-angled, 4-valved, ∞-seeded. Seeds comose.

1. **Z. Californica,** Presl. Rel. Hænk. ii. 28. t. 52 (1835). Erect or decumbent, 1–3 ft. high, canescent with a minute but dense tomentose pubescence: leaves linear-lanceolate, $\frac{3}{4}$–$1\frac{1}{2}$ in. long, entire or denticulate, thickish, seldom at all feather-veined: fl. $1\frac{3}{4}$ in. long; calyx-tube narrow-funnelform, twice the length of the linear-lanceolate segments, these surpassed by the deeply cleft petals: capsule nearly glabrous, distinctly pedicelled: seeds oblong-obovate. In the Coast and Mt. Diablo Ranges, from Lake Co. southward, on dry open ground.

2. **Z. latifolia,** Greene, Pittonia, i. 25 (1887): *Z. Californica,* var. *latifolia,* Hook. Bot. Mag. t. 4493 (1850). Decumbent, seldom 1 ft. high, occasionally canescent with a villous and more or less glandular pubescence, more commonly nearly glabrous: leaves from broadly ovate to ovate-lanceolate, $\frac{1}{2}$–1 in. long, very acute, more or less serrate-toothed, thin, conspicuously feather-veined: fl. 1 in. long; calyx-tube narrowly cylindrical for about 2 lines above the globose base, thence widening abruptly to a funnelform throat, the whole not longer than the petals: capsule subsessile, glabrous. In moist ground in the Sierra Nevada, at considerable altitudes; also at the eastern base of the Mt. Diablo Range; scarcely more than a subspecies of the first, but of different and far more extensive geographical range. June—Nov.

3. **Z. tomentella,** Greene, l. c. 26. Size of the last; canescent with a short coarse somewhat tomentose pubescence extending even to the calyx and capsules: leaves ovate-lanceolate, acute, entire or toothed, thickish, feather-veined: petals half as long as the narrow calyx-tube

which widens gradually from the globose base: capsules sessile; seeds somewhat pyriform. — An obscure plant of the Sierra Nevada, only once or twice collected: inflorescence peculiarly strict and virgate.

3. **ŒNOTHERA**, *Linnæus* (EVENING PRIMROSE). Herbs exceedingly diverse in habit. Leaves alternate. Flowers yellow, white or purplish, axillary, spicate or racemose. Calyx-tube prolonged above the ovary, mostly deciduous; segments commonly coherent after as before the expansion of the corolla save as parting by a single one of the four sutures to liberate the expanding petals. Petals 4, mostly vespertine as to time of opening, and evanescent, usually obcordate or flabelliform. Stamens 8, equal, or those opposite to the petals shorter; anthers various. Ovary 4-celled, ∞-ovuled; style filiform; stigma 4-lobed or capitate. Capsule from membranaceous to woody, more or less perfectly 4-valved and dehiscent, or indehiscent. Seeds in 1 or 2 rows in each cell, horizontal or ascending, naked, often slightly margined.—A complicated genus, by Spach and others divided into several, perhaps with good reason.

* *Calyx greatly prolonged beyond the ovary, deciduous from it; stamens nearly equal, anthers versatile; stigma-lobes linear; capsule coriaceous.*

+– *Flowers yellow, in a leafy spike, erect in bud, vespertine; tips of the calyx-lobes free; capsule narrowly oblong, sessile, straight; seeds in 2 rows in each cell.*—Genus ONAGRA, Tourn., Spach.

++ *Coarse annuals or biennials.*

1. **Œ. Hookeri**, T. & G. Fl. i. 493 (1840). Biennial; stem red, stout, angular, 3—6 ft. high: herbage canescently pubescent and somewhat villous: leaves lanceolate, sessile, acute, obscurely denticulate: calyx-tube 1¼ in. long, the segments nearly as long: petals nearly 1½ in., obcordate, very pale yellow, turning to rose-color: filaments slender, elongated; stigma-lobes yellow, spreading: capsule ¾ in. long, sessile, quadrangular, with plane sides, canescent throughout with a fine close pubescence: seeds chestnut-brown, only ½ line long, not wing-angled, delicately striate.—Common in river bottoms and often in dry places in the southern counties.

2. **Œ. Jepsonii**. Erect, 3—5 feet high, canescently pubescent when young, the older parts, and especially the capsules, hirsute: leaves rather thin, lanceolate, denticulate: calyx-tube 1¼ in. long; segments only ½ in., their tips very short, not contiguous: petals less than ½ in. long, light yellow: filaments subulate, short, the long anthers exserted: style short; stigma-lobes green, not widely spreading: capsule slender, 1¼ in. long, tapering from below the middle to apex, scarcely angular, the valves with a broad prominent midvein, separating at apex only: seeds dark-colored, sharply angled.—Along the Sacramento River in Solano Co., *Jepson*, and near Sacramento, *Dr. Pyburn*.

3. Œ. GRANDIFLORA, Ait. Hort. Kew. ii. 2 (1789): *Œ. biennis*, var. *grandiflora*, Lindl. Bot. Reg. t. 1604 (1833). Erect, 3-5 ft. high; stem and inflorescence scabrous and sparsely hirsute; the ovate-lanceolate denticulate leaves minutely and sparsely pubescent: calyx-tube 1 2 in. long, the segments almost as long, their slender tips elongated: petals obcordate, 1½- 2 in. long, yellow, turning to deeper yellow: filaments filiform, declined: style shorter than the petals, the linear stigma-lobes ¼ in. long, yellow: capsule obtusely quadrangular, slightly tapering from near the base, the valves with a strong midrib: seeds sharply angled. Common in cultivation, and sparingly naturalized about Oakland, Alameda etc.; differing essentially from *Œ. biennis* in its annual root, large almost scentless flowers, declined stamens etc.

++ ++ *Perennial.*

4. **Œ. arguta.** Stems rather slender, decumbent, about 1 ft. high, from a perennial root: herbage puberulent: leaves linear-lanceolate, saliently dentate, 2-4 in. long, 3-4 lines broad, the cauline broadest at the sessile somewhat clasping base: calyx-tube 1½ in. long: petals as long, deeply obcordate, bright yellow turning to orange; anthers filiform, about equalled by the style; stigma-lobes linear, yellow. - In moist places near Monterey, *Michener*, and southward apparently at considerable elevations in the mountains. Very distinct from all our annual species of the group.

+- +- *Flowers diurnal, white, fading pinkish, nodding in bud; capsules sessile, long and narrow; seeds in 1 row in the cell.* -

Genus BAUMANNIA, Spach.

5. **Œ. albicaulis,** Nutt. Gen. i. 245 (1818); Pursh, Fl. ii. 733 (1814), excl. descr.; Nutt. Fras. Cat. (1813), name only. *Baumannia Nuttalliana*, Spach. Phaner. iv. 352 (1835). Stem erect or decumbent, from perennial running rootstocks, simple or branched, 1-3 ft. high: herbage glabrous or pubescent, the stem and branches with a smooth shining white bark: leaves linear to oblong-lanceolate, entire, repand-denticulate, or toward the base pinnatifid, 1-3 in. long; calyx-tube 1 in. long; tips of the lobes free in the bud: petals white, becoming pinkish, 1 in. long, entire or emarginate: anthers ½ in. long, on filiform filaments: capsule 1 1½ in. long, nearly linear, though slightly tapering from base to summit: seeds terete, 1 line long. Common along the eastern base of the Sierra Nevada north and south; a handsome species, but the flowers ill-scented.

6. **Œ. Californica,** Wats. Bot. Calif. i. 223 (1876): *Œ. albicaulis*, var. *Californica*, Wats. Proc. Am. Acad. viii. 582 (1873). Perennial and white-stemmed like the last, but low and stout, hoary-pubescent and somewhat villous; leaves oblanceolate, acuminate, mostly petiolate, sinuately toothed or irregularly pinnatifid: fl. larger; calyx-tube longer, the segments somewhat villous: petals obcordate: capsule 2 in. long, slightly

tapering: seeds oblong, turgid, obtusely angled.—Interior of the State, from Sacramento, *Dr. Pyburn*, southward; flowers said to be fragrant.

7. **Œ. trichocalyx,** Nutt.; T. & G. Fl. i. 494 (1840). Erect, often simple, white-stemmed, 6–18 in. high, from a biennial root: leaves all pinnatifid: calyx almost woolly, the tips of the lobes not free in the bud: petals obcordate: capsule thickened at base: seeds ovate-oblong, somewhat compressed. Eastern base of the Sierra.

+ + + *Nearly or quite acaulescent; fl. vespertine, the buds erect; capsule ovate, ribbed or wing-angled; seeds in 2 rows in the cells.—*
Genera PACHYLOPHUS and LAVAUXIA, Spach.

8. **Œ. cæspitosa,** Nutt.; Fras. Cat. (1813); Sims, Bot. Mag. t. 1593 (1814); Nutt. Gen. i. 246 (1818): *Œ. scapigera*, Pursh, Fl. i. 263 (1814). *Pachylophus Nuttallii*, Spach, Phaner. iv. 365 (1835). Stem very short and stout, 2–6 in. high, from a biennial root: leaves oblanceolate, acute, petiolate, irregularly sinuate-toothed or nearly entire, glabrous or somewhat villous: calyx-tube four times the length of the segments, the tips of these not free in the bud: petals broadly obcordate, 1–1¼ in. long, white, turning to rose-color: capsules sessile, ovate-oblong, attenuate above, 1–1½ in. long, the margin of the valves tuberculate-crested: seeds oval-oblong, with a narrow groove along the ventral side, minutely tuberculate on the back.—Very common in the Rocky Mountain region, doubtless reaching our eastern borders.

9. **Œ. triloba,** Nutt.; Bart. Fl. N. Am. ii. 37. t. 49 (1823); Sims, Bot. Mag. t. 2566 (1825); Journ. Philad. Acad. ii. 118 (1826?). Stem very short or obsolete: root biennial: leaves runcinate-pinnatifid, petiolate, nearly glabrous: calyx-tube very long and slender; segments with tips free in the bud: petals yellow, broadly obovate, 1 in. long, 3–5-nerved: capsule sessile, oblong or obovate, ¾ in. long, with broadly winged angles, the sides at length ribbed and reticulate: seeds angled and minutely tuberculate.—Truckee Valley and northward, east of the Sierra.

* * *Acaulescent; fl. diurnal, yellow, erect in bud; calyx-tube filiform above the ovary; stamens alternately long and short; stigmas capitate; seeds in 2 rows in the cells of the capsules.*

+ *Perennials; calyx-tube persistent; capsules not winged.—*
Genus TARAXIA, Nutt.

10. **Œ. Nuttallii,** T. & G. Fl. i. 507 (1840). Canescently pubescent: leaves petiolate, oblanceolate, acuminate, 2–6 in. long, pinnatifid: the segments unequal, usually rounded or obtuse: calyx-tube 1–2½ in. long; segments somewhat shorter: petals ½ in. long: capsules narrow, attenuate upwards, 6–10 lines long, obtusely quadrangular: seeds oblong, terete, 1 line long, obscurely lineolate.—Eastern base of the Sierra from near Carson City northward.

11. **Œ. heterantha,** Nutt. Journ. Philad. Acad. vii. 22 (1834). *Jussiæa acaulis,* Pursh, Fl. i. 304 (1814). Glabrous: leaves oblong-lanceolate, acute or acuminate, entire or slightly repand-denticulate, 6 in. long : calyx-tube 1 –3 in. long : petals ¼—½ in. long : capsules ovoid-oblong, narrowed at each end, nearly 1 in. long, rather acutely angled: seeds minutely pitted. Var. **taraxacifolia,** Wats. Proc. Am. Acad. xiii. 589 (1873). Leaves larger, more or less lyrately pinnatifid.—In moist mountain meadows of the Sierra Nevada, chiefly on the eastward slope.

12. **Œ. ovata,** Nutt. T. & G. Fl. i. 507 (1840). Sparingly pubescent: leaves mostly oblong-lanceolate, entire or denticulate, often somewhat undulate, 3—8 in. long : calyx-tube 1 –4 in. long : petals ½ –¾ in. long: capsules subterranean, chartaceous, 1 in. long, tapering above, scarcely dehiscent : seeds ovoid-oblong, smooth.--Very common in open grounds, from Marin and Solano Counties to San Luis Obispo ; only the very earliest leaves ovate. Feb. –Apr.

+ + *Annual; calyx-tube deciduous; capsules winged.*

13. **Œ. graciliflora,** H. & A. Bot. Beech. 341 (1840) ; Hook. Ic. t. 338. Herbage green and pilose : leaves linear, entire or obscurely denticulate: calyx-tube not longer than the leaves ; segments short : petals 3—5 lines long, obcordate, turning greenish : capsule hard-coriaceous, 1½ in. long or less, angled at base, 4-winged above, the wings obliquely truncate and hairy ; seeds smooth. -Hillsides and plains of the interior from Butte Co. to Monterey.

* * * *Caulescent annuals (rarely more enduring); calyx-tube obconic or short-funnelform; stigma capitate; capsules sessile, mostly contorted; seeds in 1 row.-* Genus SPHÆROSTIGMA, F. & M.

+— *Maritime plants, with short primary axis (this flowerless after the early months) bearing crowded elongated narrow leaves, and radiating decumbent or prostrate shrubby-looking flowering branches with broader shorter foliage; capsules angular, sessile, contorted.*

++ *Flowers yellow, turning greenish.*

14. **Œ. virescens,** Hook. Fl. i. 214 (1833) : *Œ. cheiranthifolia,* var. *suffruticosa,* Wats. Proc. Am. Acad. viii. 592 (1873). Silvery-canescent with a short and dense appressed pubescence, also apparently slightly viscid : assurgent branches 1--2 ft. long, stoutish, purplish, appressed-pubescent : leaves from spatulate- and linear-oblong to ovate-cordate, usually entire, 1 in. long or more : petals 6 –8 lines long : anthers linear-oblong, fixed below the middle, recurved or somewhat contorted in age : capsule pubescent, rather slender, attenuate at apex. From Monterey to San Diego.

15. **Œ. nitida,** Greene. Pittonia, i. 70 (1887). The decumbent branches ½ –1 foot long, very rigid : leaves spatulate or oblanceolate, glabrous

and rather fleshy, dark green and lustrous : petals 1/2 in. long : anthers linear-oblong, fixed almost in the middle : capsule 10 lines long, stout, coriaceous, glabrous, acutely angled : seeds black, obovate, acute at base, compressed, smooth but dull. —First noticed on San Miguel Island, but found along the shores of Monterey Bay, *Abbott*.

++ ++ *Flowers turning red or tawny.*

16. **Œ. spiralis,** Hook. Fl. i. 219 (1833) : *Œ. cheiranthifolia*, Bot. Calif. Radiating branches stout, procumbent, 1 3 ft. long : leaves from spatulate to ovate-cordate, 1 3 in. long, entire or dentate, more or less hirsute : petals 4 6 lines long : anthers linear-oblong, fixed in the middle : capsule acutely quadrangular, hirsute : seeds ovate, acute at base, compressed, dark brown. Plentiful on the sand hills of San Francisco and southward, flowering almost throughout the year. Our plant must be very distinct from the Chilian *Œ. cheiranthifolia*, which has a different mode of growth, pellucid-punctate and glaucescent herbage, short roundish anthers, etc.

17. **Œ. micrantha,** Hornem. "Hort. Hafn. Suppl." (1819); Walp. Rep. ii. 77 (1843), under *Sphærostigma*. Size and habit of the last, but more slender and hirsute, the small calyx densely hairy : leaves from narrowly oblanceolate to linear-oblong, 2 - 4 in. long, acutish, more or less undulate : petals only 1—2 lines long, entire or emarginate : capsule 4-angled, contorted, rather slender, gradually attenuate upwards, usually more than 1 in. long, sparsely but stiffly hirsutulous.—At San Francisco, near the Presidio, etc., and southward along the coast.

+— +— *Plants not maritime, erect at base and with ascending branches; capsules narrow, less contorted (except in n. 18); flowers yellow, turning red.*

++ *Radical leaves narrow and petiolate, the cauline broad, sessile; capsules sharply angled, much contorted.*

18. **Œ. hirtella.** Stoutish, erect, simple, or with a few ascending branches from the base, 6—10 in. high, the herbage purplish, short-hirsute : radical leaves oblanceolate, denticulate, 1½ in. long ; cauline ovate, sessile, ½ in. long, coarsely toothed and more or less undulate or crisped : petals 1 line long or more : capsules hirsute, narrow, attenuate upwards, once or twice coiled : seeds pale, smooth, more or less regularly rhombic-ovate. —Common in the hill country away from the sea, from Lake Co. and Solano southward. Heretofore made a part of *Œ. micrantha*, a maritime species, with the trailing habit of the preceding group, which may be identical with the South Californian and maritime *Œ. bistorta*. The present plant is strictly erect, flowering in early summer and soon disappearing.

++ ++ *Annuals, without radical leaves; branches many, slender, leafy; capsules narrowly linear, slightly or not at all contorted.*

19. **Œ. strigulosa,** T. & G. Fl. i. 512 (1840); F. & M. Ind. Sem. Petr. ii. 50 (1835), under *Sphærostigma*. Slender, erect-spreading, 1_2–1 ft. high, all but the older parts clothed with short appressed or incurved white hairs; leaves ½ in. long, linear-lanceolate, acutish, denticulate, subsessile: petals broadly obovate, 1½ lines long, yellow, turning deep red: anthers roundish, basifixed: capsule about ¾ in. long, sessile, straight or arcuate, not contorted, scarcely attenuate at apex. Var. (1) **epilobioides.** Strictly erect, with ascending somewhat virgate branches: pubescence neither white nor appressed, but spreading and hirsute: pods longer and more slender. Var. (2) **pubens,** Wats. l. c. 594. With stouter and decumbent branches all from the base: almost canescently hirsute: pods more than 1 in. long, linear-clavate.—The type of this variable species appears to be restricted to the seaboard in the Bay district, and is common at San Francisco, Alameda, etc. The first variety is of the interior, from Oregon to San Diego, at which last place it comes out to the seaboard. The second belongs to the desert regions east of the Sierra. This looks like a distinct species; and possibly even the first should be admitted in that rank. They would be almost or quite as good species as the commonly received species of *Gayophytum*, to which genus the present group is very closely related.

20. **Œ. campestris.** *(Œ. dentata,* Bot. Calif. i. 226, not Cav. Branched from the base and bushy, 6–10 in. high and as broad, more or less hirsute-pubescent throughout: leaves linear-lanceolate, 1 in. long, dentate: petals very broadly cuneate-obovate, 4–5 lines long, turning brick-red: anthers linear-oblong, ¾ line long, fixed toward the middle and versatile: pods more than 1 in. long, narrowly linear, slightly incurved. Var. **cruciata** (Wats. l. c. under *Œ. dentata).* Petals half as large, narrowly obovate or oblong, often emarginate.—Common on the plains from Antioch southward. The variety, not seen by me, may possibly represent the S. American *Œ. dentata;* but it is certain that the type can not so be referred; for that, from the description, must have callous-tipped calyx-lobes, and the short, rounded and basifixed anthers of the other members of this group.

+ + + *Flowers white or rose-colored, in a nodding spike, (spike erect in n. 24); capsules terete or obtusely angled, much contorted.*

21. **Œ. alyssoides,** H. & A. Bot. Beech. 340 (1840); Hook. Ic. t. 339. Simple, or with ascending branches from the base, 3–12 in. high, canescently puberulent: leaves oblanceolate or oblong-lanceolate, narrowed into a slender petiole, entire or repand-denticulate, 1–2½ in. long; the floral similar though smaller: spike elongated, many-flowered: petals rounded, rose-purple, 2–4 lines long: capsule 1 in. long, slender, attenuate above, contorted: seeds ash-colored, minutely pitted.—Eastern base of the Sierra.

22. **Œ. Boothii,** Dougl.; Hook. Fl. i. 213 (1833). Stouter than the last, villous and viscid : leaves ovate to lanceolate : capsules broader : seeds brownish, angled, very minutely tuberculate. From near Mono Lake, *Chesnut & Drew*, northward, on the eastward slope of the Sierra only.

23. **Œ. decorticans,** Greene. H. & A. Bot. Beech. 343 (1840), under *Gaura*. (*Œ. gaurœflora*, T. & G. Fl. i. 510 (1840). Erect, stout, 1_2 2 ft high, glandular-puberulent above : the bark white, loose, exfoliating : leaves lanceolate to narrowly oblanceolate, attenuate into a petiole, denticulate : spike many-flowered, elongated, and often also crowded : calyx-tube narrowly funnelform, equalling the petals : petals 1_4 in. long, white or pink : capsules 8-15 lines long, stoutish below, narrowed above to a slender beak : seeds dark, 1 line long, angled.- From the valley of the Sacramento to Monterey Co. and far southward.

24. **Œ. Nevadensis,** Kellogg, Proc. Calif. Acad. ii. 224. f. 70 (1863) ; Curran, Bull. Calif. Acad. i. 137. Lower and relatively stouter than the last, the spike short, crowded and erect : leaves oblong-lanceolate, irregularly repand-dentate, the blade decurrent upon the petiole and even to the stem, which is wing-angled : bracts of the inflorescence petiolate : calyx-tube very narrowly funnelform, nearly equalling the ovary : petals ovate or obovate, short-unguiculate : anthers oblong, versatile : capsule stout, tapering from base to apex, quadrangular, tortuous: seeds oblong.- Western Nevada, and doubtless within our borders.

* * * * *Caulescent annuals; leaves mostly radical; stigmas capitate; capsules pedicellate, linear or clavate, obtuse, not contorted.—*
Genus CHYLISMIA, Nutt.

25. **Œ. scapoidea,** T. & G. Fl. i. 506 (1840). Erect, simple, or branching from the base, 1_2 2 ft. high, glabrous or puberulent : leaves lyrate-pinnate or undivided, petiolate : bracts of the nodding raceme small or 0 : calyx-tube funnelform, 1--2 lines long : petals longer, obovate, entire, yellow (sometimes purplish) : capsules glabrous, $\frac{1}{2}$ 1 in. long, on spreading pedicels of $\frac{1}{2}$ in.- From Mono Lake northward, along our eastern borders.

* * * * *Paniculate branching annuals, without radical leaves; calyx-tube nearly obsolete; stigmas capitate; capsules sessile, elongated, linear, refracted.-* Genus EULOBUS, Nutt.

26. **Œ. leptocarpa,** Greene, Pittonia. i. 302 (1889). *Eulobus Californicus*, Nutt.; T. & G. Fl. i. 515 (1840). Erect, with somewhat virgately panicled branches, 2 - 5 ft. high ; glabrous, glaucescent : leaves linear, 1- 2 in. long, sinuately pinnatifid or coarsely and divaricately toothed : calyx-tube less than $\frac{1}{2}$ line long : petals 4 5 lines long, yellow, turning red : stamens unequal, the shorter with globose, the longer with oblong anthers : capsules 3--4 in. long, deflexed : seeds smooth, ovate-oblong.

3-angled, $\frac{2}{3}$ line long. Not reported north of Santa Barbara, but to be expected in San Luis Obispo Co. if not in Monterey. The only very near allies of this species, *Œ. crassiuscula* and *Œ. angelorum*, are Lower Californian, and have sessile contorted capsules.

4. **GAYOPHYTUM,** *A. Jussieu*. Erect very slender diffusely branching annuals, with alternate linear entire leaves and axillary small white or purplish flowers. Calyx-tube not prolonged above the ovary, the 4-parted deciduous limb reflexed. Petals 4. Stamens 8, the alternate ones usually minute and sterile; filaments filiform; anthers subglobose, fixed near the middle. Ovary oblong or linear, compressed, 2-celled; stigma capitate or clavate. Capsule membranaceous, clavate, 2-celled, 4-valved. Seeds few or many, in one row in each cell, smooth, naked, mostly oblong.

1. **G. ramosissimum,** T. & G. Fl. i. 513 (1840). Glabrous, or the inflorescence puberulent, erect, with spreading branches, $\frac{1}{2}$ -1$\frac{1}{2}$ ft. high: leaves 1 in. long or less: fl. $\frac{1}{2}$ line long, mostly near the ends of the branches: capsule 2–3 lines long, on pedicels of about the same length, often deflexed, 3–5-seeded.—Common species of the Rocky Mts. and Great Basin, reaching our borders in Mono and other eastern counties.

2. **G. diffusum,** T. & G. l. c. Nearly glabrous, branching dichotomously and widely, 1–3 ft. high: fl. 2–4 lines broad, white, fading purplish: capsules $\frac{1}{4}$–$\frac{1}{2}$ in. long, exceeding the pedicels, the cells 4–8-seeded. The most common Californian species; abundant at middle elevations, in open pine and sequoia woods from Calaveras Co. southward.

3. **G. racemosum,** T. & G. l. c. 514. Glabrous or canescent with a short appressed pubescence, $\frac{1}{2}$–1$\frac{1}{2}$ ft. high, the branches mostly simple and elongated: fl. $\frac{1}{2}$ line long, axillary the whole length of the branches: capsules linear, sessile or short-pedicelled, 8–10 lines long, ∞-seeded: seeds erect in the cells.—In Humboldt Co., and northward.

4. **G. lasiospermum,** Greene, Pittonia, ii. 164 (1891). Erect, very slender, 1–2 ft. high, with numerous dichotomous branches; glabrous except a scant pubescence on the flower-buds: corolla 1$\frac{1}{2}$ lines long, rose-color, turning purple: capsules on capillary erect pedicels, torulose, few-seeded: seeds canescent with an appressed silky pubescence.—Known thus far only from the Coast Range in the northern part of San Diego Co., but probably occurring within our limits.

5. **G. pumilum,** Wats. Proc. Am. Acad. xviii. 193 (1883). Near *G. racemosum*, but smaller and less branching, seldom 6 in. high, glabrous or nearly so: fl. less than a line broad: capsule erect, short-pedicelled, $\frac{1}{2}$ in. long, the numerous seeds oblique in the cells. Said to occur from San Bernardino northward throughout the State.

5. **GODETIA**, *Spach*. Erect simple or branching annuals. Leaves alternate, entire or denticulate. Flowers mostly purple, showy, in leafy spikes or racemes. Calyx-tube above the ovary obconic or short-funnelform, deciduous. Petals 4, broad, sessile, entire, emarginate or cleft, diurnal and lasting for two days or more. Stamens 8, unequal, the filaments opposite the petals shortest ; anthers perfect, elongated, attached by the base, erect or arcuate-recurved. Ovary 4-celled, ∞-ovuled ; style short ; stigma-lobes short, linear or roundish. Capsule ovate to linear, 4-sided, coriaceous, loculicidally dehiscent. Seeds ascending or horizontal, in 1 or 2 rows, obliquely angled, the upper part tuberculate-margined.

* *Flowers in a strict dense spike; capsule ovate or oblong.*

+— *Tips of the calyx-lobes not free in the bud; sides of capsule not 2-costate; seeds in 2 rows in each cell.*

1. **G. grandiflora**, Lindl. Bot. Reg. xxvii. Misc. 61 (1841) & xxviii. t. 61. *Œnothera Whitneyi*, Gray, Proc. Am. Acad. vii. 340 (1868). Puberulent : stem stout, simple or with a few short branches near the summit, 1–2 ft. high ; leaves lanceolate, 2–3 in. long, acute at each end, short-petioled, obscurely repand-denticulate or entire : calyx-tube turbinate, 4–6 lines long : petals 1–2 in. long, emarginate, pale purple with often a crimson spot in the middle or toward the base : stigma-lobes linear, 1/4 in. long : capsule oblong to linear, 8–15 lines long, puberulent, 4-toothed at apex.—Humboldt and Mendocino counties.

2. **G. purpurea**, Wats. Bot. Calif. i. 229 (1876); Curtis, Bot. Mag. t. 352 (1796), under *Œnothera*. *G. Willdenoviana*, Spach, Phaner. iv. 387 (1835). Stem erect, 6–15 in. high, puberulent : leaves oblong or lanceolate-oblong, obtusish, entire, glaucescent : calyx-tube funnelform, as long as the segments : petals broadly obovate, 1/2 in. long or more, crenulate, deep purple : stamens much shorter than the petals : stigma lobes broad and short, dark purple : capsule ovate-oblong, 1/2–3/4 in. long, hairy, the sides nearly flat, with a strong midvein. Mr. Watson attributes to this species two rows of seeds in each cell of the capsule. No such plant has been recognized by the present writer ; neither has he seen the plant figured as the type of *G. purpurea;* nevertheless, it should be common in some part of California.

+— +— *Tips of the calyx-lobes slightly free in the bud; capsule 2-costate on at least two of the sides (except in n. 3); seeds in 1 row in each cell.*

3. **G. decumbens**, Spach, Phaner. iv. 388 (1835) ; Lindl. Bot. Reg. t. 1221 (1829), under *Œnothera; G. lepida*, var. *parviflora*, Wats. Stems ascending, 1 1/2 ft. high, branching, sparsely hirsute : leaves ovate-lanceolate, entire, pubescent, glaucescent : fl. sessile in the axils, and shorter than the leaves, red-purple : calyx-tube very short, half as long as the

segments: filaments of the alternate stamens very short: stigma whitish, appearing capitate by the close recurving of the broad lobes: capsule tomentose, somewhat quadrangular and tapering, the sides with a midrib but no furrows.—Another obscure species, perhaps not occurring in middle California: doubtless wrongly referred to the next, for the capsules, if true to the descriptions, are essentially different.

4. **G. lepida,** Lindl. Bot. Reg. t. 1849 (1836); H. & A. Bot. Beech. 342 (1836), under *Œnothera*. Stem erect, branching above, pubescent with short appressed hairs: leaves ovate-lanceolate, entire, slightly pubescent: calyx-tube obconical, very short, greatly surpassed by the segments: petals rounded and emarginate at apex, pale purple, with a dark red cuneate spot at summit: stigma purple, cruciform: capsule ovate-oblong, sessile, closely ribbed and sulcate, densely white-villous. Very common, apparently throughout the State.

5. **G. micropetala,** Greene, Pittonia, i. 32 (1887). Erect, slender, simple, 1–3 ft. high, puberulent: leaves 1 in. long, narrowly lanceolate, entire, sessile: spike rather short: calyx-tube scarcely 2 lines long: segments 4 lines, the slender elongated tips twisted in the bud: petals linear-lanceolate, only 3 lines long, entire or erose: stigma purple, the lobes broad and short: capsule sessile, $^3_4$ in. long, linear-oblong, abruptly pointed, hirsute, the alternate sides bicostate.—Hills of Contra Costa Co., in the Mt. Diablo Range, near Walnut Creek, *Greene*, and about Martinez, *Frank Swett*.

6. **G. Arnottii,** Walp. Rep. ii. 88 (1843); T. & G. Fl. i. 503 (1840), under *Œnothera*; *G. lepida*, var. *Arnottii*, Wats. Nearly glabrous, slightly glaucous, 1–2 ft. high, densely flowered at the leafy summit: leaves mostly opposite, except the floral, oblong-lanceolate, obtuse or acutish, obscurely denticulate: calyx-tube very short: corolla deep purple: stigma purple, the lobes oval: capsule cylindrical-conic, bicostate on the sides, glabrous.—Common in the Sacramento Valley and southward.

7. **G. albescens,** Lindl. Bot. Reg. xxviii. t. 9 (1842). Stem erect, rigid, pubescent, the branches very short, crowded at the summit: leaves glabrous, glaucous, lanceolate, entire: fl. sessile, densely crowded among the upper leaves on the short branchlets: calyx-tube funnelform, as long as the segments: petals obcordate, $\frac{1}{2}$ in. long, pale purple, with a small darker spot in the centre: stigma-lobes narrow, greenish: capsule oblong, 8-sulcate, acuminate, villous: seeds roundish, scabrous.—Lake and Solano Counties, to Monterey.

** *Flowers in loose spikes or racemes; capsules mostly linear; seeds in 1 row.*

← *Racemes erect in bud; calyx-lobes distinct and reflexed in flower; capsules sessile.*

8. **G. Williamsonii,** Wats. Bot. Calif. i. 230 (1876); Dur. & Hilg. Pac. R. Rep. v. 7. t. 5 (1855), under *Œnothera*. Erect, sparingly branching, rather slender, 1 ft. high, canescently puberulent ; leaves linear, entire, sessile ; calyx villous ; its tube funnelform ; tips of the lobes free in the bud ; petals nearly 1 in. long, lilac with yellow base and a deep purple spot in the centre ; stigma-lobes oblong, yellow ; capsules 6–8 lines long, attenuate from the base upward, bicostate on the sides, puberulent. Sierra Nevada, from Placer Co. to Kern.

9. **G. quadrivulnera,** Spach, Phaner. iv. 389 (1835); Dougl. in Lindl. Bot. Reg. t. 1119 (1827), under *Œnothera*. Very slender, 1–2 ft. high, puberulent ; leaves linear or linear-lanceolate, entire or slightly denticulate ; calyx-tube obconic, 2–3 lines long ; petals purplish with a dark spot at summit, 3–6 lines long ; stigma-lobes short, purple ; capsules 5–10 lines long, attenuate at apex, bicostate at the alternate angles. Common toward the coast everywhere.

10. **G. viminea,** Spach, Nouv. Ann. Mus. iv. 389 (1835); Dougl. in Hook. Bot. Mag. t. 2873 (1829), under *Œnothera;* Bot. Reg. t. 1220. Stem erect, 1–3 ft. high, glabrous ; leaves linear-lanceolate, entire, narrowed at base, 1–2 in. long, puberulent ; calyx-tube 2–4 lines long ; petals purplish with a dark spot at summit, $\frac{3}{4}$–$1\frac{1}{4}$ in. long ; stamens short, subequal ; stigma-lobes linear-oblong, purple ; capsules 1–$1\frac{1}{2}$ in. long, pubescent, slightly bicostate on the sides. From the middle parts of the State northward ; the original, from Oregon, said to be "glaucous," but probably rather pale with a minute indument.

11. **G. tenella,** Wats. l. c. (1876) ; *G. Cavanillesii*, Spach, Phaner. iv. 390 (1835). *Œ. tenella*, Cav. Ic. iv. t. 396. f. 2 (1797). Puberulent, slender, erect, $\frac{1}{2}$–$1\frac{1}{2}$ ft. high ; leaves linear, acute or obtuse, mostly entire, more or less narrowed at base, $\frac{1}{2}$–2 in. long ; calyx-tube obconic, 1–3 lines long ; petals 3–5 lines, deep purple ; stigma-lobes purple ; capsule linear, attenuate at apex, 8–14 lines long, quadrangular, the sides not costate but the midvein usually prominent.—Common toward the coast throughout the State.

+ + *Racemes nodding in the bud; calyx-lobes united and turned to one side under the open corolla; stigma yellow or white (except in n. 12); capsule pedicellate.*

12. **G. pulcherrima,** Greene, Pittonia, ii. 217 (1891). Slender, erect, 1–3 ft. high, puberulent ; leaves linear, acute at each end, nearly entire ; calyx-tube very broad and short, almost cyathiform ; petals $1\frac{1}{4}$ in. long, cuneate-obovate, truncate or retuse at the cross summit, lilac and streaked with white veins above, whitish at base, dotted throughout with elongated spots of dark crimson ; stamens equal ; filaments lilac ; anthers white ; style long ; stigma-lobes purple, obovoid ; capsule linear, 1 in. long, pedicellate, scarcely costate.—From Fort Tejon southward.

13. **G. amœna**, Lilja, Linnæa, xv. 265 (1841); Lehm. Ind. Sem. Hamb. 8 (1821), and Nov. Act. Leop. xiv. 811. t. 45 (1828), under *Œnothera: G. rinosa*, Lindl. Bot. Reg. t. 1880 (1836). Erect, slender, 1–2 ft. high, puberulent: leaves lanceolate or oblanceolate, entire or denticulate: calyx-tube obconic, 2–4 lines long: petals 8–15 lines long, white, pink or purple, with a dark purple spot near the base: filaments rather stout; anthers deep crimson, the vacant upper end white or yellowish: stigma-lobes linear: capsule 1–1½ in. long, narrowed at each end; pedicel 2–6 lines long. From Monterey northward; the white-flowered and typical form seldom seen.

14. **G. rubicunda**, Lindl. Bot. Reg. t. 1856 (1836). Near the preceding, but often 4 ft. high: calyx-tube longer, funnelform: petals purple, with an orange spot at base: anthers orange-red, the empty end bright yellow: capsules sessile, scarcely attenuate at apex. Of more northerly general range than the preceding; perhaps not distinct from it; it is nevertheless quite as possible that three or four species formerly distinguished are confusedly embraced by these two names.

15. **G. Bottæ**, Spach, Monogr. Onagr. 73 (1835); T. & G. Fl. i. 505 (1840), under *Œnothera*. Erect, sparingly branching, 1–2 ft. high, puberulent or glabrous: leaves linear-lanceolate, entire or sparingly denticulate, petiolate, 1–2 in. long: calyx-tube short-obconic: petals light purple, ½–1 in. long: filaments slender, and style elongated: stigma-lobes short, yellow: capsule attenuate at each end, 10–15 lines long; pedicel ½–¾ in. long.—In the southern Coast Range from Monterey southward.

16. **G. epilobioides**, Wats. Bot. Calif. i. 231 (1876); Nutt.; T. & G. Fl. i. 511 (1840), under *Œnothera*. Habit of the preceding, but more slender, usually 1 ft. high; glabrous or tomentose-puberulent: leaves linear or linear-lanceolate: calyx-tube 1–2 lines long: petals 3–6 lines, dull white, with or without a tinge of rose: stigma-lobes short: capsule acuminate, attenuate to a short pedicel or sessile, ½–1 in. long.—Said to occur throughout the State; but we doubt its existence except from Santa Barbara Co. southward, where it is common.

17. **G. hispidula**, Wats. l. c., and Proc. Am. Acad. l. c., under *Œnothera*. Erect, simple, a few inches high, often 1-flowered: pubescence hispidulous: leaves narrowly linear, 1–2 in. long: calyx-tube 2–3 lines: petals purple, ½–1 in. long: filaments slender: stigma-lobes linear: capsules ⅓–¼ in. long, attenuate at apex, below abruptly contracted to a short pedicel. A little known species of the valleys of the Sacramento and San Joaquin.

18. **G. biloba**, Wats. Bot. Calif. i. 231 (1876); Durand, Pl. Pratt. 87 (1855), under *Œnothera*. Slender, erect, 1–2 ft. high, sparingly branching,

nearly glabrous: leaves linear to narrowly lanceolate, 1 2 in. long, obscurely denticulate, the lower on long and slender petioles: petals light-purple, 1_2 3_4 in. long, cuneate-obovate, deeply 2-lobed: capsule puberulent, 1_2 3_4 in. long, attenuate at apex, narrowed at base into a short pedicel. -From Nevada Co. to Tuolumne in the Sierra, at low altitudes; also plentiful in the Briones Hills of the Mt. Diablo Range in Contra Costa Co. not far back of Martinez. A beautiful plant, and a near relative of *Clarkia Xantiana*, from which genus *Godetia* is perhaps not naturally separable.

6. **CLARKIA**, *Pursh.* Erect sparingly branched annuals, with alternate petiolate leaves, racemose or spicate purple flowers nodding in the bud. Calyx-tube more or less prolonged above the ovary, deciduous. Petals 4, unguiculate, often lobed or cleft. Stamens normally 8, but those opposite the petals often sterile or rudimentary, or sometimes wanting; anthers oblong or linear, fixed by the base. Ovary 4-celled; style elongated; stigma with 4 broad spreading lobes. Capsule linear, attenuate above, coriaceous, straight or somewhat curved, 4-angled, 4-celled, 4-valved to the middle. Seeds angled or margined.

* *Petals 3-lobed.*

← *Calyx-tube obconical; 4 stamens rudimentary.* - Typical CLARKIA.

1. **C. pulchella**, Pursh, Fl. i. 260. t. 11 (1814); Lindl. Bot. Reg. t. 1100. Stem 1 1½ ft. high, puberulent: leaves linear-lanceolate or linear, 1 - 3 in. long, entire, glabrous: petals 1½—3½ in. long, with 3 broad divergent lobes, the claw with a pair of recurved teeth: perfect stamens with a linear scale on each side at the base; the rudimentary ones filiform: stigma-lobes equal, dilated: capsule 1 in. long or less, 8-angled, on a spreading pedicel 2 -3 lines long: seed obliquely cubical, minutely tuberculate.—Plumas Co., *Mrs. Austin*, and northward.

← ← *Calyx-tube elongated and almost filiform; stamens 4 only.*—
Genus EUCHARIDIUM, F. & M.

2. **C. concinna**, Greene, Pittonia, i. 140 (1887). *Eucharidium concinnum*, F. & M. Ind. Sem. Petr. ii. 11 (1835); Meyer, Sert. Petr. t. 12; Lindl. Bot. Reg. t. 1962. Simple, or with a few subcorymbose branches, 1 -2 ft. high, glabrous or puberulent: leaves ovate, entire: calyx-tube almost filiform, 1 in. long: corolla regular; petals ½ -3½ in. long, cuneate-obovate, 3-lobed, the middle lobe broadest, little longer than the others: filaments subulate; anthers recurved after dehiscence, somewhat villous: stigma subpeltate, the lobes short, rounded: capsule subcylindrical, in maturity obscurely quadrangular, acutish.- Coast Range, from Mendocino Co. to Santa Barbara. May, June.

3. **C. grandiflora**, Greene. *Eucharidium grandiflorum*, F. & M. Ind. Sem. Petr. vii. 48 (1840); Sert. Petr. t. 13. Near the preceding, but

diffusely branching from the base: corolla larger, irregular, the 3 upper petals approximate, ascending, the lower one remote from these and declined, the middle lobe of each attenuate to a claw and far surpassing the others. -Very common in the Mt. Diablo Range.

4. **C. Breweri,** Greene, Pittonia, i. 141 (1887); Gray, Proc. Am. Acad. vi. 532 (1865), under *Eucharidium: C. Sareana,* Greene, l. c. 140. Glabrous, 1–2 ft. high: leaves lanceolate, entire, short-petioled: calyx-tube slender, more than 1 in. long, abruptly dilated at base: corolla irregular; petals round-obcordate, with a linear-spatulate middle lobe proceeding from the deep, or rather shallow sinus and far exceeding the others: filaments clavate: anthers densely white-villous along the sutures, erect after dehiscence as before: capsule sessile, 1 in. long, curving away from the stem: seeds large, tuberculate, conspicuously winged. On Mt. Oso, Stanislaus Co., *Brewer,* and near the Geysers in Sonoma Co., *Dr. Saxe.* Very erroneously described in the "Botany of California" as having "a narrow subulate lobe in the deep sinus."

* * *Calyx-tube obconic; petals never 3-lobed; stamens 8, all perfect.—*
Genus PHÆOSTOMA, Spach.

5. **C. Xantiana,** Gray, Proc. Bost. Soc. vii. 145 (1859). Glabrous, glaucescent, 1–3 ft. high, stoutish, sparingly branching above: leaves narrowly lanceolate or linear, entire or denticulate, ashy-puberulent: petals 2-lobed with a subulate tooth in the sinus; the claw short and broad: stigma-lobes broadly oval. Fort Tejon and southward.

6. **C. elegans,** Dougl.; Lindl. Bot. Reg. t. 1575 (1833): *C. unguiculata,* Lindl. l. c. xxiii., under t. 1981 (1837). Glabrous or puberulent, reddish and glaucous, erect, 1–6 ft. high, simple or somewhat branching, stout and rigid: leaves broadly ovate to linear, repand-dentate: petals entire, the rhomboidal limb about equalling the linear claw: filaments with a densely hairy scale on each side at base: capsule $\frac{1}{2}$–$\frac{3}{4}$ in. long, stout, sessile, 4-angled, somewhat curved, often hairy. -On open or half-shaded hillsides of both the Coast Range and the lower Sierra.

7. **C. rhomboidea,** Dougl.; Hook. Fl. i. 214 (1833); Lindl. Bot. Reg. t. 1981. *Opsianthes gauroides,* Lilja, Linnæa, xv. 261 (1841). Puberulent or glabrous, 1—3 ft. high, rather slender: leaves thin, entire, oblong-lanceolate to ovate, 1–2 in. long: petals with rhomboidal limb and short broad claw which is often broadly toothed: filaments with hairy scales at base: capsules pedicellate, 8–12 lines long, 4-angled, glabrous, curved near the base. Of wider range than the last; equally common.

7. **BOISDUVALIA,** *Spach.* Annuals, rigid and leafy, rather low (except the first species); the leaves alternate, sessile. Flowers small, purple, in leafy-bracted spikes. Calyx-tube funnelform above the ovary,

deciduous; the lobes not reflexed in flower. Petals 4, obovate-cuneiform, sessile, 2-lobed. Stamens 8, all perfect, unequal; filaments slender, naked at base; anthers oblong, fixed near the base. Ovary 4-celled, several-ovuled; stigma-lobes short, somewhat cuneate. Capsule membranaceous, ovate-oblong to linear, nearly terete, acute, dehiscent to the base. Seeds in 1 row in the cell. A very distinct genus in habit and aspect, but separable from *Œnothera* only by the erect calyx-lobes, and from *Epilobium*, to which it is even more nearly allied, only by the naked seeds. Fl. late summer and early autumn.

1. **B. densiflora**, Wats. Bot. Calif. i. 233 (1876); Lindl. Bot. Reg. t. 1593 (1833), under *Œnothera*. *B. Douglasii*, Spach, Monogr. Onagr. 80. t. 31. f. 2 (1835). Stoutish, sparingly branching, 1–3 ft. high, soft-pubescent throughout: lower leaves lanceolate, acuminate, serrate-toothed; the floral broader, entire: flowers in rather loose terminal spikes: calyx 1½–3 lines long, half as long as the purple petals: capsules ovate-oblong, glabrous or villous, 2–4 lines long; cells 3 6-seeded, the partitions separating from the valves and adhering to the placenta: seeds nearly a line long. Var. **imbricata**. Less canescent than the type, the whole plant larger and coarser: spikes thick and dense, the capsules concealed under the very broad, acute closely imbricated bracts. The type, figured by Lindley, is not in middle California, but must be Oregonian. The species is represented in the Bay region only by the variety, which is a plant of very different aspect.

2. **B. stricta**, Greene. Gray, Proc. Am. Acad. vii. 340 (1868), under *Gayophytum*. *B. Torreyi*, Wats. Bot. Calif. i. 233 (1876). *Œnothera Torreyi*, Wats. Proc. Am. Acad. viii. 600 (1873). Canescent with a short stiff spreading pubescence; plant slender, seldom 1 ft. high: leaves lanceolate or linear, narrow at base, entire or denticulate, the floral not differing from the others except as being smaller: flowers in a loose simple spike, minute; capsules linear-acuminate, 4–6 lines long; cells 6—8-seeded; seeds ½ line long or less, ovate. Frequent in the Coast Range from Santa Clara Co. northward.

3. **B. cleistogama**, Curran, Bull. Calif. Acad. i. 12 (1884). Very pale and glaucescent, glabrous or hispidulous; 4–10 in. high, rather slender: leaves ovate-lanceolate, ½–1½ in. long, remotely serrate: fl. small, rose-red, the earliest ones cleistogamous: capsule rather coriaceous: seeds numerous. Species well marked by its very pallid herbage, coriaceous capsules, etc.; common on the Sacramento plains from Chico southward, also in Lake Co., *Simonds*, and near Petaluma, *Greene*. B. GLABELLA, the only other species, is smaller than this, glabrous and not pallid. It may be found within our borders northeastward, but is mainly Oregonian.

8. **GAURA**, Linnæus. Herbs with alternate leaves, and terminal spikes or racemes of scarcely regular white or rose-colored flowers.

Calyx-tube prolonged beyond the ovary, deciduous; the lobes spreading or reflexed. Petals 4, with claws. Stamens 8; filaments with or without a scale-like appendage at base; anthers oval. Ovary 4-celled, with 1 or 2 pendulous ovules in each cell; stigma 4-lobed or discoid. Fruit hard and nut-like, dehiscent at apex only, or not at all. Seeds 1 or several.

* *Filaments with a scale-like appendage on the inside below; stigma 4-lobed, surrounded by a ring or indusium.*—GAURA proper.

1. **G. parviflora**, Dougl.; Lehm. Pugill. ii. 15 (1830). Annual, erect, 1—5 ft. high; pubescence dense, spreading, very soft: leaves ovate to lanceolate, repand-denticulate: fl. small, in dense strict terminal spikes; petals spatulate-oblong, scarcely unguiculate, shorter than the calyx-lobes, rose-red: anthers oval, versatile: fr. 3—4 lines long, obscurely 4-angled at summit, 4-nerved, about 2-seeded, indehiscent.—Upper valley of the San Joaquin and southward.

* * *Filaments naked; stigma discoid.*—Genus HETEROGAURA, Rothr.

2. **G. heterandra**, Torr. Pac. R. Rep. iv. 87 (1857). *Heterogaura Californica*, Rothr. Proc. Am. Acad. vi. 354 (1864). Annual, slender, erect, glabrous, except the younger parts, which are slightly puberulent: leaves very thin, ovate-lanceolate, entire or obsoletely sinuate-dentate, petiolate: spikes panicled, loose: the lowest flowers leafy-bracted: petals narrowly spatulate, 2 lines long, purple: anthers round-cordate, basifixed; those opposite the petals borne on shorter filaments, lanceolate, acute, sterile: fr. 2 lines long, obovoid, 4-angled.—On shady banks in the woods of the Sierra Nevada from Yuba Co. southward; also at Fort Tejon. June—Aug.

9. **CIRCÆA**, *Lobelius* (ENCHANTER'S NIGHTSHADE). Erect slender perennials, with opposite petiolate thin leaves, and small white flowers in terminal and lateral racemes; fruit on slender spreading or deflexed pedicels. Calyx-tube little produced above the obovoid ovary, the base nearly filled by a cup-shaped disk; limb 2-parted, deciduous. Petals 2, obcordate. Stamens 2, alternate with the petals; anthers small, rounded. Ovary 1- or 2-celled; ovules 1 in each cell, ascending. Fruit pear-shaped, indehiscent, covered with hooked prickles.

1. **C. Pacifica**, Ascherson & Magnus, Bot. Zeit. xxix. 392 (1871); *C. Lutetiana*, Boland. Cat. 12, not of Lobel. Glabrous, simple, $1/2$—1 ft. high, from a perennial slender running rootstock: leaves ovate, rounded or cordate at base, acuminate, repand-denticulate, 1—$2\frac{1}{2}$ in. long, on slender petioles about as long: racemes bractless: fl. $\frac{1}{2}$ line long; calyx white, with very short tube: fr. 1 line long, rather loosely clothed with soft hairs which are merely curved above, scarcely hooked, 1-celled, 1-seeded.—Yosemite and northward in the mountains; inhabiting cool moist shades, by springs and streamlets.

10. LUDWIGIA, *Linnæus.* Herbs of various habit (ours creeping and aquatic or riparian), with entire leaves, and axillary or spicate colorless or yellow 4 5-merous flowers. Calyx-tube prismatic or cylindrical, not produced beyond the ovary; lobes 4 or 5, persistent. Petals as many, or 0. Stamens as many or twice as many. Ovary broad at apex and usually flattened, or crowned with a conical style-base; stigma capitate, 4 5-grooved. Capsule 4 5-celled, dehiscent by lateral slits or terminal pores. Seeds very many, minute.

* *Leaves opposite; fl. 4-merous, apetalous.—*Genus ISNARDIA, Linn.

1. **L. palustris,** Ell. Sk. i. 211 (1821); Linn. Sp. Pl. i. 120 (1753), under *Isnardia.* Glabrous; stems creeping or floating, 4 10 in. long: leaves all opposite, oval or ovate, acute, ½ 1 in. long, tapering to a short petiole: fl. sessile, 1 in each axil: petals rarely present, minute, reddish; capsule oblong, 2 lines long or less, somewhat 4-angled.—Common on muddy banks and shores of streams and ponds in the Sacramento and San Joaquin valleys; also in the mountain districts adjacent on both sides.

* * *Leaves alternate; fl. 5 merous, with large yellow caducous petals.—*
Genus JUSSIÆA, Linn.

2. **L. diffusa** (Forsk, Ægypt. Arab. 210, under *Jussiæa*), var. **Californica,** Greene. *Jussiæa repens,* var. *Californica,* Wats. Bot. Calif. i. 217 (1876). Perennial, the stout floating stems 1 ft. to 2 yards long: herbage altogether glabrous: leaves obovate to obovate-oblong and even lanceolate, obtuse or acute, 1—2¼ in. long, on petioles of 1_2—1 in.: stipules gland-like or slightly scale-like: fl. 6 8 lines broad, deep yellow; the petals obtuse, but not obcordate: fr. 1 in. long, spongy and indehiscent; the pedicel 1_2 in. or more.—Plentiful, forming extensive floating masses, covering the surface of stagnant waters in the interior, from Lathrop and Stockton to above Sacramento; flowering in early summer. The plant is far from agreeing with the description of *L. (Jussiæa) repens;* and, following a suggestion from Baron von Mueller, I find it more nearly at agreement with *L. (Jussiæa) diffusa,* which occurs in Australia.

ORDER XXXI. **HALORAGEÆ.**
Robert Brown, in Flinder's Voyage, 17 (1814).

A small order, not very distinct from the last: the plants herbaceous and mostly aquatic, with small inconspicuous usually apetalous flowers sessile in the axils of leaves or bracts. Calyx, in fertile flowers, adnate to the ovary, its limb there short or obsolete. Fruit indehiscent and nut-like, 1 4-celled, with a single seed suspended in each cell. Cotyledons small and short. Albumen copious.

1. **HIPPURIS,** *Linnæus.* Erect stoutish but low perennial aquatics. Stem simple, short-jointed, with linear entire leaves in whorls of 8 or 12. Calyx-tube globular; the limb entire. Petals 0. Stamen 1; filament subulate. Ovary 1-celled; style becoming filiform and elongated, stigmatic throughout. Fruit oblong-ovoid, nut-like, 1-seeded.

1. **H. vulgaris,** Linn. Sp. Pl. i. 4 (1753). *Limnopeuce vulgaris,* Vaill. Mem. de l'Acad. 15 (1719); Ray, Syn. Meth. 136 (1724). Stem ½—1 ft. high; herbage glabrous: leaves ½—1 in. long, acute: calyx ½ line long: style and stamen rather conspicuous: nutlet nearly 1 line long. — In shallow ponds and pools, and about springy places, from the seaboard to the high Sierra; but not often met with in California.

2. **MYRIOPHYLLUM,** *Matthiolus* (WATER-MILFOIL). Aquatic perennials. Leaves usually verticillate, sometimes opposite or alternate; the submersed ones pinnately divided into capillary or filiform segments; the emersed ones pectinate or toothed or entire and bract-like. Flowers axillary, commonly unisexual; the staminate with a very short calyx-tube, and 2—4-lobed limb or none. Petals 2—4. Stamens 4—8. Calyx of pistillate fl. with a more or less deeply 4-grooved tube and 4 minute lobes or none. Styles 4, short, often plumose and recurved. Fruit somewhat drupaceous, quadrangular, when ripe splitting into 4 one-seeded carpels.

1. **M. spicatum,** Linn. Sp. Pl. ii. 992 (1763). Stems often many feet long, growing in deep waters, branching above: flowers in emersed short-peduncled verticillate spikes 2—3 in. long; bracts reduced and inconspicuous; submersed leaves in whorls of 4 or 5; petals 4, deciduous: stamens 8: carpels rounded on the back, with a deep wide groove between them.—Perhaps not rare in California, but known only in Mountain Lake, San Francisco, where it is abundant and of rank growth in two or three feet of water. July.

2. **M. hippuroides,** Nutt.; T. & G. Fl. i. 530 (1840); Morong, Bull. Torr. Club, xviii. 245. Stems 4—8 in. long, growing in mud or shallow water, the emersed branches erect, simple, leafy, flowering throughout their length: submersed leaves in whorls of 4 and 6, with 6—8 pairs of capillary pinnæ; emersed ones often alternate, linear-lanceolate, serrate or dentate, or the uppermost entire; the lowest often pinnatifid: petals often pinkish and somewhat persistent: stamens 4: carpels carinate and somewhat roughened; deep grooves between them. —Rather common in Lake and Marin counties; also at Stockton. June.

3. **CALLITRICHE,** *Columna.* Small and slender, growing in water or on moist shaded ground. Leaves opposite, linear, spatulate or obovate. Flowers solitary in the axils, subtended by a pair of falcate or lunate membranous bracts, mostly consisting of a single stamen and pistil.

Filaments elongated; anthers reniform, the cells ultimately confluent. Styles 2, filiform, papillose. Fruit sessile or pedunculed, 4-celled, more or less carinate or winged on the margins, 4-lobed, the lobes united in pairs, forming 2 discs with a groove between them, at maturity parting into 4 compressed carpels, each 1-seeded.

* *Fruit pedicellate.*

1. **C. marginata,** Torr. Pac. R. Rep. iv. 135 (1857); Mor. Bull. Torr. Club, xviii. 235. Usually terrestrial and very small; when aquatic the submersed leaves linear, 1-nerved, passing gradually into the emersed, which are oblanceolate or spatulate, 3-nerved : styles elongated, reflexed, deciduous : mature fruit on slender pedicels, often buried in the mud, deeply emarginate at both ends, broader than high, the margins of the thick carpels widely divergent and narrowly winged. - Common in low grounds, among growing grain, etc., from San Mateo and Alameda counties northward; commonly terrestrial, but burying its fruit in the wet earth. June.

* * *Fruit sessile.*

2. **C. palustris,** Linn. Sp. Pl. ii. 969 (1753), essentially : *C. verna*, Linn. Fl. Suec. 2d ed. 2 (1755) : *C. aquatica*, Huds. Fl. Angl. 439 (1762) : *C. pallens*, S. F. Gray, Nat. Arr. 555 (1821). Usually aquatic, with linear retuse or bifid submersed leaves, and spatulate or obovate emersed ones, these rounded or truncate or retuse at apex, narrowed into a margined petiole, and profusely dotted with stellate scales : fr. oblong, with a small apical notch, and narrow-winged above, deeply grooved between the lobes. Credited to middle California generally; but most of the specimens perhaps belong to the next.

3. **C. Bolanderi,** Hegelm. Bot. Ver. Brandenb. x. 116 (1868 ?); Mor. l. c. 238. Coarser than *C. palustris*, with floating leaves obovate or rhombic-obovate : styles twice as long as the fruit, subpersistent : fr. roundish or obcordate with acute or obtuse closely approximated margins. In vernal ponds and pools (terrestrial states not seen), from Contra Costa Co. and Placer, northward. Apr. -June.

4. **C. stenocarpa,** Hegelm. l. c. 114. Mor. l. c. 237. Floating leaves obovate, rounded and entire at apex, 3-nerved, tapering to a short margined petiole, with stellate scales; the submersed linear, all of a pale or dull green : styles erect, twice as long as the fruit, deciduous : fr. much compressed, sharply winged, round-obovate, abruptly and deeply emarginate. - In the Sierra Nevada, near Summit Station, *Greene*, and near Mt. Stanford, *Sonne;* in cold deep pools in swampy meadow-lands.

5. **C. autumnalis,** Linn. Fl. Suec. 2d ed. 2 (1755) : *C. virens*, Gold. Act. Mosq. v. 119 (1817); S. F. Gray, Nat. Arr. ii. 556. Submersed,

bright green, internodes short: leaves linear-lanceolate, broader and clasping at base, retuse or bifid at apex, destitute of stellate scales: styles about equalling the fruit, deciduous: fr. occasionally short-peduncled, orbicular, or somewhat longer than broad, the lobes wing-margined.—A northern species, sometimes growing in running water; Sierra Co., *Lemmon.*

Order XXXII. CERATOPHYLLÆ.

S. F. Gray, Natural Arrangement of British Plants, ii. 554 (1821).

Represented by a single species of the genus

CERATOPHYLLUM, *Linnæus* (HORNWORT). Aquatic herbs, with rigid verticillate leaves, these usually pinnatifid and the segments toothed. Flowers clustered in the leaf-axils, involucrate, unisexual. Involucre multifid. Calyx and corolla wanting. Stamens 14—20. Ovary ovate, 1-celled; style filiform, incurved. Fruit a small nutlet; the seed pendulous. Albumen 0; cotyledons 4, verticillate, 2 larger than the others; plumule conspicuous, compound.

1. **C. demersum,** Linn. Sp. Pl. i. 992 (1753): *C. apiculatum,* Cham. Linnæa, iv. 503. t. 5 (1829). Stem 1–2 ft. long, nearly glabrous; internodes short: leaves in whorls of 6 or 8; the linear segments acute, aculeate-toothed: achene 2 lines long or more, elliptical, somewhat compressed, short-stipitate, with a short spine or tubercule on each side near the base, not margined: style equalling the achene.—At San Francisco, *Chamisso,* and at Clear Lake, *Simonds.*

Order XXXIII. SALICARIÆ.

Adanson, Familles des Plantes, ii. 232 (1763); Juss. Gen. 330 (1789). LYTHRARIÆ, Juss. Dict. Sc. Nat. xxvii. 453 (1823).

Herbs (as to our few species), with entire leaves, and axillary or spicate mostly 5-merous purplish (rarely apetalous) flowers. Calyx tubular, enclosing the ovary but free from it; the petals and definite stamens borne on the throat of it. Style 1. Capsule mostly 1-celled by the vanishing of the thin partitions. Seeds numerous, small, on a central placenta, exalbuminous.—Family closely allied to Epilobiaceæ, but necessarily separated on account of the superior ovary.

1. **LYTHRUM,** *Linnæus.* Calyx cylindrical, 10–12-angled or -striate, 10–12-toothed; the teeth alternately long and erect and shorter and incurved. Petals 5 or 6, inserted on the throat of the calyx-tube alter-

nately with the erect teeth. Stamens from the middle or the base of
the calyx-tube, as many or twice as many as the petals. Style filiform :
stigma capitate.

* *Petals minute, pale.*

1. **L. Hyssopifolia,** Linn. Sp. Pl. i. 447 (1753). Annual, simple or
branching, erect, 4—10 in. high ; herbage pale, glabrous : lowest leaves
opposite : fl. subsessile in the axils of the alternate leaves, very small,
whitish or pale-purple. Not rare in the Coast Range, from Humboldt
Co. southward throughout the State; also, in a large form, in the
interior, near Stockton, etc. June—Aug.

2. **L. adsurgens,** Greene, Pittonia, ii. 12 (1889). Stoloniferous perennial, the 5-angled branches 1—3 ft. long, decumbent or assurgent ;
herbage pallid, glabrous, slightly succulent: calyx $2\frac{1}{2}$ lines long, 12-striate,
the striæ at length widening below : petals pale purple. Plant very near
the preceding in all points except its great size and perennial stoloniferous habit. Common in wet places near the Bay, at West Berkeley, etc.

* * *Petals larger, bright red-purple.*

3. **L. Californicum,** T. & G. Fl. i. 482 (1840). Stoloniferous perennial,
the roots spreading near the surface of the ground : stem erect, 2—3 ft.
high, simple below, paniculately branching above : lower leaves lanceolate ; upper and floral linear : striæ of the calyx not wing-margined :
teeth very short. Common in marshy land ; also along streams, and in
springy places, both in the mountains and around San Francisco Bay.

4. **L. Sanfordi,** Greene, Pittonia, ii. 12 (1889). Perennial, not stoloniferous, the stout contorted roots deep-seated and more or less horizontally spreading like rootstocks : stem erect, 1—2 ft. high, branched
from the base, acutely 5- or 6-angled : herbage deep green, glabrous :
leaves all alternate, linear-oblong, sessile : petals 6, bright purple,
showy : calyx minutely and very acutely 12-carinate ; teeth short, triangular, the main ones nearly equalled by the intervening processes.—In
dry fields along the lower San Joaquin, near Stockton, *Sanford*, and at
Fresno, *Bioletti;* a troublesome plant in vineyards and orchards, owing
to the tenacious vitality of the roots, which send up shoots and form new
plants when cut in pieces by the plow.

2. **AMMANNIA,** *Houston.* Glabrous opposite-leaved annuals ; the
flowers 2 or more in each axil. Calyx subglobose, more or less distinctly
4-angled, 4-toothed, usually with horn-shaped appendages alternating
with the teeth. Petals 4, purplish, small and deciduous, sometimes
wanting. Stamens 4—8. Capsule globular.

1. **A. coccinea,** Rottb. Progr. n. 4 (1773) : *A. latifolia,* Bot. Calif. not
Linn. Erect, stoutish, $\frac{1}{2}$—2 ft. high, with few spreading branches :

stem 4-angled : leaves linear-lanceolate, 1 3 in. long, with a broad auricled base : fl. 1 5 in each axil, mostly sessile : calyx 1½ lines long, in fruit becoming 2 lines broad : petals small, bright purple : capsule bursting irregularly.—Common along the rivers and smaller streams of the interior ; on the Sacramento, *Jepson;* at Stockton, *Sanford.*

2. **A. humilis,** Michx. Fl. i. 99 (1803). *Rotala ramosior,* Koehne. Smaller than the preceding : leaves linear-oblanceolate, not auricled at base but tapering, sometimes short-petiolate : fl. 1 3 in each axil : calyx globular, the accessory teeth as long as the lobes or shorter : petals small, purplish : capsule globular, dehiscent septicidally. -Habitat of the preceding, but less frequent. In deference to a technicality of the fruit structure, the species has lately been transferred to *Rotala,* but unadvisedly, as we think.

Order XXXIV. LOASEÆ.

Jussieu, Annales du Museum, v. 18 (1804).

Plants analogous in some respects, though scarcely allied, to Epilobiaceæ. Herbage clothed with stinging or jointed and barbed hairs. Bark of the brittle stems often white and deciduous. Leaves without stipules. Calyx-tube adnate to the 1-celled ovary. Stamens often very numerous, and some of the outer petaloid. We have but the following rather polymorphous genus.

MENTZELIA, *Plumier.* Erect rigid and rough often fragile annuals and biennials. Leaves alternate, mostly coarsely toothed or pinnatifid. Flowers solitary or cymose, large or very small, yellow or white. Calyx-tube cylindrical, ovoid or turbinate ; the 5-lobed limb persistent. Petals 5 or 10. Stamens ∞, inserted on the throat of the calyx ; filaments free, or in clusters opposite the petals, filiform, or the outer more or less dilated and without anthers. Ovary truncate at summit, 1-celled ; ovules horizontal or pendulous, in 1 or 2 rows on the 3 linear parietal placentæ. Capsule mostly cylindrical, and opening irregularly at the summit. Seeds angled or compressed.

* *Annuals, small-flowered; petals 5 only; stamens rather few; capsule linear; seeds few, not flattened, irregularly angled, opaque, usually minutely tuberculate.*--Genus TRACHYPHYTUM, Nutt.

+— *Seeds prismatical, with grooved angles.*

1. **M. dispersa,** Wats. Proc. Am. Acad. xi. 137 (1876). Slender, 1 ft. high : leaves narrowly lanceolate, sinuate-toothed or entire, the uppermost often ovate : fl. clustered near the ends of the many branchlets : calyx-lobes 1 line long, shorter than the petals : filaments not dilated :

capsules narrowly linear-clavate, 1_2-3 in. long: seeds very often in a single row, short-prismatic, the 3 angles grooved, the sides only faintly tuberculate. -Yosemite Valley and northward, chiefly to the eastward of the mountains. It seems probable that *M. Veatchiana*, Kell. may be identical with this; in which case that name must be adopted, as long antedating *M. dispersa*.

2. **M. affinis**, Greene, Pittonia, ii. 103 (1890). Near the last, but stouter, often 2 ft. high, simple and leafy below, widely branching above: leaves lanceolate, deeply sinuate-pinnatifid: fl. scattered, 1_2 in. broad: calyx-lobes attenuate-subulate, 2-3 lines long: capsule 1 in. long, almost linear, hispid with short stiff hairs which have a pustulate base: seeds prismatical with grooved angles but relatively shorter than in the last, and more tuberculate. Plains of the San Joaquin, and far southward.

3. **M. micrantha**, T. & G. Fl. i. 535 (1840); H. & A. Bot. Beech. 343. t. 85 (1840), under *Bartonia*. Rather slender, 1-2 ft. high, simple below, corymbosely and rather compactly dichotomous above: leaves ovate, acute or acuminate, entire or serrate- or sinuate-toothed, 1-2 in. long: fl. almost minute, shorter than the floral leaves: petals 5, oval, exceeding the calyx-lobes: 5 of the filaments petaloid, with emarginate apex: capsules cylindrical, or nearly so, 3-6 lines long, few-seeded: seeds prismatical, the length about twice the breadth, the base often oblique, the angles with a very shallow groove, the sides only faintly tuberculate. Rather rare, but found at Clear Lake, also on Mt. Hamilton, and far southward.

+ + *Seeds irregularly angled.*

4. **M. albicaulis**, Dougl.; Hook. Fl. i. 222 (1833); Hook. l. c., under *Bartonia*; *M. Veatchiana*, Kell. *fide*, Wats. Stem low, branching from the base or only corymbose, rather slender, 1_2-1 ft. high, white, shining and almost glabrous: leaves remote, sessile, lanceolate, from deeply sinuate-pinnatifid to nearly entire: fl. small, solitary or in loose clusters, not bracted: petals obovate, little exceeding the subulate-lanceolate calyx-segments: filaments all subulate-filiform: capsules narrow-cylindrical: seeds rather numerous (20-40), 1_2 line long, tuberculate, irregularly and obtusely angled. Eastern slope of the Sierra; also at Ft. Tejon.

5. **M. congesta**, T. & G. Fl. i. 534 (1840). Habit and foliage of the last: flowers clustered at the ends of the branches, half hidden by broad toothed bracts which are scarious at base: petals 1_4-1_2 in. long: filaments all filiform: capsule clavate, 1_2 in. long: seeds nearly a line long, tuberculate. East of the Sierra only.

6. **M. gracilenta**, T. & G. l. c. Less branching, 1-$1 1_2$ ft. high, stouter: leaves narrowly lanceolate, pinnatifid with many narrow lobes.

or only sinuate-toothed : fl. usually clustered, not conspicuously bracteate : calyx-lobes 2 5 lines long : petals obovate to oblanceolate, rounded or acutish at apex, 4—8 lines long : filaments subulate-filiform : capsule slightly clavate-dilated, 1_2 - 1 in. long : seeds in 3 rows, angular, minutely tuberculate. 2_3 line long.—Plains and foothills of the interior, from the Sacramento southward ; also east of the Sierra.

7. **M. nitens.** Loosely diffuse, the lower branches decumbent, 1 -2 ft. long, all clothed with a glabrous very white shining bark ; leaves few, the internodes elongated : fl. solitary in the upper forks, and somewhat clustered at the ends of the branches, 1 in. broad : petals oblong-obovate, obtuse or emarginate : stamens short ; filaments subulate ; anthers oblong : seeds tuberculate, sharply and very sinuously angled.— Near Benton, Mono Co., *W. H. Shockley*; possibly perennial.

8. **M. pectinata,** Kellogg, Proc. Calif. Acad. iii. 40. f. 9 (1868). Stem low and mostly simple, 4—8 in. high, clothed like the leaves with a rather dense barbed pubescence : leaves rather deeply and closely pinnatifid : fl. deep yellow, 1 in. broad ; petals mostly obcordate, with a minute cusp in the sinus : stamens very many, half as long as the petals ; filaments filiform, or slightly subulate-dilated at base ; anthers small, orbicular : seeds unknown.—Marysville Buttes, *Jepson*, and southward along the foothills of the Sierra. Possibly all the so-called *M. gracilenta* of the interior of California may be of this very genuine species. It is a very beautiful plant, resembling the next.

9. **M. Lindleyi,** T. & G. Fl. i. 533 (1840) : *M. Bartonia*, Steud. Nom. i. 189 (1840). *Bartonia aurea*, Lindl. Bot. Reg. t. 1831. Slender, simple or bushy-branched, 1 -3 ft. high : leaves ovate to narrowly lanceolate, 2—3 in. long, from pectinate-pinnatifid to coarsely toothed : fl. axillary and terminal : calyx-lobes rather broadly lanceolate, $\frac{1}{2}$—$\frac{3}{4}$ in. long : fl. vespertine : petals obovate, abruptly acuminate or cuspidate, 1 in. long or more, golden yellow : filaments many, very slender, unequal, the longest almost equalling the petals ; anthers minute, oval : capsule 1 in. long or more : seeds angular, tuberculate. -Common in the Mt. Diablo Range, near Livermore, on Mt. Hamilton, etc., on dry open hillsides.

* * *Flowers large; petals 5 or 10; stamens very numerous, the outer petaloid-dilated; seeds many, in double rows on the 3 placentæ, horizontally flattened and winged.*— Genus NUTTALLIA, Raf.

10. **M. lævicaulis,** T. & G. Fl. i. 535 (1840) ; Dougl.; Hook. Fl. i. 221. t. 69 (1833), under *Bartonia*. Biennial, stout, branched above, 2 -3 ft. high ; stem white, scarcely roughened : leaves lanceolate, sinuate-toothed, 2 8 in. long : fl. sessile on short branches, 3—4 in. broad, light yellow, diurnal : calyx-tube naked ; segments 1 in. long or more : petals oblanceolate, acute, almost equalled by the numerous stamens : capsule

1¼ in. long, 3—4 lines in diameter ; seeds minutely tuberculate, 1½ lines broad. Very common in the mountain districts at low altitudes, and on the plains near the foothills.

ORDER XXXV. **CUCURBITACEÆ.**

Haller, Enumeratio Methodica Stirpium Helvetiæ, 34 & 505 (1742); Juss. Gen. 393 (1789).

Herbs, tendril-bearing, trailing or climbing, the herbage commonly scabrous and more or less succulent. Flowers axillary to the alternate leaves, solitary or clustered, unisexual. Calyx-tube coherent with the ovary ; limb of 5 lobes or teeth. Corolla with petals more or less united into a cup or tube. Stamens 5, more or less united ; anthers 2-celled, or one of them 1-celled. Ovary 2—3-celled ; style often wanting ; stigma 3—5-lobed. Fruit large, fleshy. Seeds large, usually compressed, exalbuminous ; cotyledons fleshy, foliaceous or hypogeous.

1. **CUCURBITA,** *Pliny* (SQUASH. PUMPKIN). Our species perennial, prostrate. Leaves cordate, lobed. Flowers solitary. Calyx-tube campanulate, 5-lobed. Corolla campanulate, 5-cleft to the middle or lower, the lobes recurved. Sterile fl. with stamens at the base of the corolla : filaments distinct ; anthers more or less united, flexuous. Fertile fl. with 3 rudimentary stamens ; ovary oblong, with 3 placentæ and many horizontal ovules ; style short ; stigmas 3, 2-lobed. Fruit fleshy, indehiscent, in our species with a hard shell-like rind.

1. **C. fœtidissima,** HBK. Gen. et Sp. ii. 123 (1817) : *C. perennis* (James), Gray, Journ. Bost. Soc. vi. 193 (1850). Root perennial, large, fusiform : stems many feet long, trailing : leaves fleshy, scabrous, whitish beneath, triangular-cordate, acute, the slight lobes rounded or angled, mucronate-denticulate : petiole shorter than the blade : tendrils 3—5-cleft : fl. 3—4 in. long, yellow : corolla-lobes obtuse, mucronate : calyx-tube ⅓ in. long, the linear lobes as long : fr. globose, 2—3 in. thick, smooth, yellow, on a slender peduncle 1—2 in. long ; shell filled with a fibrous bitter pulp : seed thin, obovate, 4—5 lines long, obtusely margined. -From San Joaquin Co. southward, on low plains ; herbage very heavy-scented.

2. **C. palmata,** Wats. Proc. Am. Acad. xi. 137 (1876). Smaller, canescent with a short scabrous indument and a more appressed pubescence on the foliage : leaves thick, 2—3 in. broad, of round-cordate outline, palmately 5-cleft to the middle ; lobes lanceolate, acuminate, often obtusely toothed near the base : fl. 3 in. long, on stout peduncles : corolla-lobes acutish : calyx-tube 1 in. long, the teeth broader than in

the last: fr. globose: seeds 5 lines long. Near Fresno, *Bioletti*, and southward.

3. **C. Californica,** Torr.; Wats. l. c. Canescent with a short white stiff pubescence: leaves thick, 5-lobed, 2 in. broad; lobes triangular, acute or acuminate, mucronate: tendrils slender, parted to the base: fl. small, 1 in. long or more: calyx 4–5 lines, the linear teeth 2 lines.—Fresno Co., *Bioletti*, and northward to the Sacramento Valley.

2. **MICRAMPELIS,** *Rafinesque* (BIG-ROOT). Rather slender membranous-leaved trailing or climbing herbs, with simple tendrils, and small white flowers, the fertile solitary and the sterile racemose or panicled from the same axil. Calyx-tube broadly campanulate, the teeth very small or obsolete. Corolla rotate or campanulate, deeply 5–7-lobed, with elongated papillose segments. Sterile fl. with stamens at base; filaments short, united: anthers distinct or more or less coherent. Fertile fl. pedicellate, with or without abortive stamens. Ovary globose or oblong, bristly, 2–4-celled: cells 1–4-ovuled: style short; stigma 2–3-parted or -lobed. Fruit prickly on the outside, fibrous and watery-pulpy within, dehiscing somewhat irregularly near the apex. Seeds large, ovoid or more rounded, more or less compressed, encircled by a mere marginal line; hilum linear, acute; cotyledons thick, remaining within the integuments after germination. All our species perennials with very large fleshy fusiform roots.

* *Leaves rather longer than broad; corollas rotate.*

1. **M. fabacea,** Greene, Pittonia, ii. 129 (1890); Naud. Ann. Sc. Nat. xii. 154 (1859), under *Echinocystis*. *Megarrhiza Californica*, Wats. Proc. Am. Acad. xi. 138 (1876). (Glabrous or nearly so, the younger parts with scattered short curved hairs: stem climbing, 10–30 ft. long: leaves 2–6 in. broad, of round-ovate general outline, more or less deeply and angularly 5–7-lobed; lobes abruptly acute, mucronate, the sinuses obtuse: sterile fl. 15–30 in slender racemes, the pedicels 1–2 lines long; corolla 3–4 lines broad, of a dull or greenish white: fertile fl. 5–6 lines broad, without abortive stamens: ovary globose, densely echinate, 2-celled, 4-ovuled: fr. globose, 2 in. long, densely covered with stout pungent spines $\frac{1}{2}$–1 in. long: seeds 4, obovoid, 10 lines long, 6 lines broad. Var. **agrestis.** Stems 2–4 ft. long, prostrate or merely trailing; leaves and fruits much smaller, the latter armed only sparsely with very short almost innocuous spines. The type is common all along the seaboard, growing in thickets and climbing high over shrubs and small trees. The variety occupies the open plains of the interior, and is a weed in the grain fields of the valley of the San Joaquin. Fl. Jan.–Apr.

2. **M. macrocarpa,** Greene, l. c., and Bull. Calif. Acad. i. 188 (1885), under *Echinocystis*. Size habit and foliage of the above, but fruits more

than twice as large, equally spinescent, 4-celled, 14-seeded; 12 of the seeds arranged imbricately in the cells, the other two lying horizontally across the base of the fruit, both attached to the same side. Southern, probably occurring north of Santa Barbara.

* * *Leaves broader than long; corollas campanulate.*

3. **M. Marah,** Greene, l. c.; Wats. Proc. Am. Acad. l. c., under *Echinocystis*. *Marah muricata*, Kell. Proc. Calif. Acad. i. 38 (1854). Herbage more or less scabrous, rather succulent: stems 3–30 ft. long: leaves reniform or round-cordate, 3–6 in. broad, more pedately lobed than in the last: racemes of sterile flowers 1_2–1 ft. long: corolla 1_2 in. long or more, campanulate, clear white: fertile fl. with abortive stamens; pedicel slender, 2–6 lines long: ovary oblong-ovate, acuminate, more or less clothed with soft spines, 2–3-celled: ovules 1–4 in each cell, ascending or horizontal, attached to the outer side of the cell: fr. ovate-oblong, 4 in. long, attenuate at each end, more or less muricate with short weak spines: seeds horizontally placed, somewhat elliptical or nearly orbicular, compressed, 1 in. long and about half as thick.—Shady banks, or open northward slopes, trailing or high-climbing; common about Mt. Tamalpais, also in Alameda and Contra Costa counties among the hills.

4. **M. Watsonii,** Greene, l. c.; Cogn. in DC. Monogr. Phaner. iii. 819 (1881), under *Echinocystis*. *E. muricata*, Kell. Proc. Calif. Acad. i. 57 (1854). *Megarrhiza muricata*, Wats. Slender, not succulent, glaucous, somewhat scabrous: stems 6—8 ft. long: leaves broad-reniform, 5--7-lobed, 2–4 in. broad, the lobes broader above and sharply sinuate-toothed or -lobed: sterile racemes few-flowered: fl. small, white: fertile fl. without abortive stamens, on slender pedicels 1–2 in. long: ovary glabrous or muricate: fr. nearly globose, 1 in. thick or somewhat larger, naked or with a few weak spines near the base, 2-celled, 2–4-seeded: seed nearly globose, $\frac{1}{2}$ in. thick, attached to the outer side of the cell, marginless or nearly so.—Foothills on both sides of the interior valley, from Calaveras Co. to the Marysville Buttes. Mar. May.

Order XXXVI. ARISTOLOCHIACEÆ.

Lindley, Introd. to Nat. Syst. 2d ed. 205 (1836). Aristolochiæ, Adans.

Shrubs or perennial herbs, with alternate entire mostly cordate or reniform exstipulate leaves, and solitary apetalous perfect flowers. Perianth lurid-purple or greenish, with a valvate regularly or irregularly 3-lobed limb; the tube more or less adnate to a 6-celled ovary, which becomes a 6-valved capsule, or a berry. Stamens 6--12, on the ovary, and more or less adnate to the style; anthers extrorse. Styles usually 6, united at the base. Seeds in 1 or 2 rows in each cell.

1. ARISTOLOCHIA, *Dioscorides* (PIPE-VINE). Perianth very irregular; tube inflated above the ovary, deciduous from it. Anthers 6, sessile and adnate to the short simple style. Stigma 3–6-lobed or -angled. Fruit capsular, 6-angled, 6-valved, septicidally dehiscent.

1. **A. Californica,** Torr. Pac. R. Rep. iv. 128 (1857). A deciduous climbing shrub, 6–10 ft. high, more or less densely pubescent with short silky hairs: leaves ovate-cordate, obtuse or acutish, 2–4 in. long, short-petioled: peduncles slender, 1–2 in. long, with a small cordate or obovate bract in the middle: calyx-tube broadly saccate and doubled upon itself, 1–1½ in. long from the base to the top of the curvature, ½ in. broad, little contracted at the throat; limb bilabiate, the upper lip of 2 broad obtuse lobes, with a disk-like thickening on the inner side: anthers contiguous in pairs under each of the 3 broad obtuse stigma-lobes: ovary linear-clavate, pubescent: capsule spongy-coriaceous, obovate, attenuate to a slender base, 1½ in. long, 6-winged: seeds cuneate-obovate, deeply concave on the upper side; raphe prominent, spongy.—Not very common, but met with here and there from Monterey to Napa in the Coast Mountains, and in the interior along the upper Sacramento; flowers dull purple, appearing in April.

2. ASARUM, *Dioscorides* (WILD GINGER). Perianth regular, campanulate; limb 3-cleft, persistent, the tips of the segments infolded in the bud. Stamens 12, nearly free from the styles, the alternate ones shorter; connective continued beyond the anthers, pointed. Styles 6, more or less united. Capsule globose, fleshy, irregularly dehiscent. Seeds large, thick, in 2 rows in each cell.

1. **A. Hartwegi,** Wats. Proc. Am. Acad. x. 346 (1875). Acaulescent, with creeping aromatic rootstocks, the branches of these bearing 2 or 3 scarious sessile caducous bracts, and 2 long petioled leaves, and a pedunculate flower in the axil of the lower leaf: leaves 2–6 in. long, reniform-cordate, with large rounded auricles, acutish, glabrous and mottled above, the margin ciliate: peduncle stout, ½–1 in. long: ovary ½ in. broad: calyx-lobes ovate, narrowed to a linear apex, 1–1½ in. long: filaments stoutish, nearly free from the style; anthers 1 line long, the produced connective setaceous, 1–2 lines long: styles short, nearly distinct, scarcely equalling the anthers: seeds 2 lines long, ovate.—Forests of the Sierra, at 4,000 to 7,000 ft., forming tufts.

2. **A. caudatum,** Lindl. Bot. Reg. under t. 1399 (1831). More slender, with longer rootstocks, sparingly pubescent with long floccose hairs: leaves cordate-reniform, somewhat cucullate, acutish or obtuse, 2–4 in. long, sparingly pubescent above: peduncles slender, 6–15 lines long: ovary 4 lines broad: calyx-lobes oblong, with long-attenuate apex, 1–2½ in. long: filaments stout, the free apex of the connective shorter

than the anther: styles united, equalling the stamens: seeds 1½ lines long, ovate. Woods of the Coast Range from Santa Cruz northward.

3. **A. Lemmoni,** Wats. Proc. Am. Acad. xiv. 294 (1879). Much like the last, but the leaves not cucullate, rounded at summit, thin, not mottled, nearly glabrous above ; fl. smaller ; calyx-lobes obtuse, or only acute, not caudate ; seeds narrowly ovate. In the Sierra Nevada, from Plumas Co. southward to Fresno ; the whole plant delightfully aromatic.

ORDER XXXVII. **FICOIDEÆ.**

A. L. de Jussieu, Genera Plantarum, 315 (1789).

Very succulent herbs or shrubs. Leaves plane, triquetrous or terete, without stipules. Calyx-tube coherent with the ovary ; the lobes usually 5, unequal, foliaceous. Petals very many and linear or 0. Stamens 5 -∞, with slender filaments, inserted on the calyx-tube. Styles 4—20. Fruit 4 -20-celled, dehiscent stellately across the summit, or circumscissile, or indehiscent. Seeds usually numerous and minute.

1. **MESEMBRYANTHEMUM,** *Breyne.* Flowers large, terminal. Calyx-tube adnate to the ovary. Petals and stamens very numerous. Fruit structurally capsular, but in ours juicy and baccate.

1. **M. æquilaterale,** Haw. Misc. Nat. 77 (1803). Perennial, glabrous, glaucescent, the stout prostrate stems several feet long, the short flowering branches ascending ; leaves opposite, very fleshy, triquetrous with linear sides, 1 3 in. long: fl. solitary, subsessile, 1½ in. broad, bright rose-purple : calyx-tube turbinate, ½ in. long or more : the larger lobes as long : stigmas 6 -10: fr. large, fragrant, edible.—Very common on banks and cliffs near the sea ; also Australian and Chilian.

2. **SESUVIUM,** *Linnæus* (SEA PURSLANE). Flowers small, axillary and terminal. Calyx-tube free from the ovary ; lobes 5, apiculate on the back near the top, scarious-margined, often purplish within. Petals 0. Stamens 5 -∞, inserted at the top of the calyx-tube. Styles 3 5. Fruit ovate-oblong, 3--5-celled, circumscissile about at the middle. ∞-seeded.

1. **S. Portulacastrum,** Linn. Syst. Nat. 10 ed. 1058 (1759). Stems prostrate, 1 ft. long or more : leaves linear- to oblong-lanceolate, ½—1½ in. long, acute or obtuse : fl. sessile or pedicellate : calyx 3 5 lines long: the lobes purple : stamens ∞.—Valley of the San Joaquin, near Lathrop, *Bioletti,* and southeastward, in moist alkaline soil.

3. **TETRAGONIA,** *Linnæus.* Perennial, with alternate plane fleshy leaves and axillary greenish apetalous flowers. Calyx 4-cleft, adherent

to the ovary, 4–8-horned ; the lobes yellowish within. Stamens several. Styles 3–8; ovary 3–8-celled. Fruit osseous, nut-like, indehiscent, 3–8-celled, the cells 1-seeded.

1. **T. expansa**, Murr. Comm. Gœtting. vi. 13. t. 5 (1783). Leaves petiolate, rhombic-ovate, acute or acuminate, entire, more or less crystalline-papillose. 1–2 in. long : fl. sessile, 1–3 in each axil : fr. 4-horned, about $\frac{1}{3}$ in. long, scarcely as broad.—Common along the beaches of San Francisco Bay, both in Marin and Alameda counties ; perhaps even more widely dispersed on this coast, where it is apparently native.

Order XXXVIII. DATISCEÆ.

Robert Brown, in Denham's Travels 25 (1826).

With us represented by a species of

DATISCA, *Linnæus*. Stout glabrous dioecious perennials. Leaves laciniate-pinnatifid ; the segments coarsely toothed. Flowers axillary, subsessile, fascicled. Calyx of sterile fl. very short, with 4–9 unequal lobes. Stamens 10–25 ; filaments short. Calyx of pistillate fl. with ovoid tube somewhat 3-angled, 3-toothed, the stamens when present 3, alternate with the teeth. Styles 3, bifid, opposite the teeth, the linear lobes stigmatic on the inner side. Capsule oblong, coriaceous, 1-celled, opening at apex between the styles. Seeds ∞, small, in several rows on the 3 parietal placentæ ; embryo cylindrical, in the axis of small albumen.

D. glomerata, B. & W. Bot. Calif. i. 242 (1876); Presl. Rel. Hænk. ii. 88. t. 64 (1835), under *Tricerastes*. Erect, 3–6 ft. high, simply or sparingly branching : leaves of ovate or lanceolate outline, acuminate, 6 in. long ; the floral shorter : fl. 4–7 in each axil of the long leafy raceme, the fertile mostly perfect : anthers subsessile, 2 lines long, yellow : styles exceeding the ovary : capsule oblong-ovate, 3–4 lines long, slightly narrowed toward the truncate triangular 3-toothed summit.—Very common along mountain streams from Lake and Amador counties southward.

Order XXXIX. CISTOIDEÆ.

Ventenat, Tableau du Reg. Veget. iii. 219 (1799). Cistineæ, DC. Prodr. i. 263 (1824). Cisti, Juss. (1789).

In Asia an extensive family, of which we have one species.

HELIANTHEMUM, *Valerius Cordus*. Low, branching, suffrutescent. Leaves alternate, simple, entire. Flowers perfect, regular. Sepals mostly 5, unequal, persistent. Petals 5, yellow, fugacious. Stamens ∞, hypogynous; filaments filiform; anthers short. Style 1, short, deciduous. Capsule ovoid, 1-celled, few- or many-seeded; the seeds borne on the middle of the valve.

H. scoparium, Nutt.; T. & G. Fl. i. 152 (1838). Plant a bushy tuft of slender almost leafless green branches, 1 ft. high; glabrate, or stellate-pubescent: the few leaves narrowly linear, $\frac{1}{3}$ 1 in. long: fl. on slender pedicels, solitary or cymose at the ends of the branches: sepals 3 lines long, acuminate, the 2 outer linear and much shorter: petals 4 lines: stamens about 20: capsule equalling the calyx. Common on dry bushy hills, from Lake Co. southward.

Order XL. VIOLARIEÆ.

De Candolle, Flore Française, iv. 801 (1805). VIOLACEÆ, S. F. Gray, Nat. Arr. ii. 667 (1821); Lindl. Synops. 35 (1829). VIOLÆ, Juss. (1789).

Represented by a fair number of species of the principal genus of the order.

VIOLA, *Pliny* (VIOLET). Low perennial herbs, with alternate leaves of involute vernation, foliaceous persistent stipules, and 1-flowered axillary peduncles. Flowers 5-merous, often of two kinds; the earlier complete and conspicuous but often sterile; the later with rudimentary petals, cleistogamous, producing numerous seeds. Sepals unequal, more or less auricled at base, persistent. Petals unequal, the lower one often spurred at base. Stamens hypogynous, the adnate anthers connivent over the pistil, broad, often coherent, the connectives of the two lower often bearing spurs which project into the spur of the petal. Ovary 1-celled, with 3 parietal placentæ; style clavate; stigma 1-sided. Capsule 3-valved; the valves bearing the seeds along the middle. Seeds rather large, with a smooth hard testa, and a large straight embryo in fleshy albumen.

* *Leaves and peduncles all from a subterranean rootstock.*

1. **V. blanda,** Willd. Hort. Berol. t. 24 (1806). Rootstock short and erect, or longer and ascending, at length producing runners: leaves round-cordate or reniform, $\frac{1}{2}$- 2 in. broad, glabrous or minutely pubescent, obscurely crenate-toothed: peduncles 1- 4 in. high: fl. white, the lower petals purple-veined, nearly beardless, 3- 4 lines long; spur short.— Wet meadows in the Sierra, at middle elevations.

2. **V. obliqua,** Hill, Hort. Kew. 316. t. 12 (1769): *V. cucullata,* Ait. Hort. Kew. iii. 288 (1789). Rootstock stoutish, branching, not stoloniferous: leaves glabrous or pubescent, cordate with a broad sinus, the lowest often reniform, the later acute or acuminate, crenate-toothed, cucullate

when young: peduncles 3—10 in. high: fl. deep or pale purple or violet; petals 5—8 lines long, the lateral and often the lowest bearded; spur short and thick.—Sierra Co., *Anderson*, *Lemmon*.

3. **V. ODORATA**, Rencalm, Specim. 141. t. 140 (1611); Linn. Sp. Pl. ii. 934 (1753). Rootstock stout, branching, stoloniferous: leaves round-cordate, obtuse, crenate, more or less villous or glabrate, on petioles 3—10 in. long: peduncles shorter than the leaves: fl. large, violet, fragrant.—Occasionally spontaneous, as an escape from the gardens.

* * *Stems short or elongated, leafy.*
+— *Leaves undivided; flowers not yellow.*

4. **V. canina**, Linn. var. **adunca**, Gray, Proc. Am. Acad. viii. 377: *V. adunca*, Smith, Rees Cycl. xxxvii (1817). Scarcely stoloniferous, mostly tufted and low, 2—6 in. high, glabrous or puberulent: leaves ovate or ovate-oblong, with subcordate or almost truncate base, obtuse, or rarely acutish, obscurely crenate, ½—1½ in. long; stipules narrowly lanceolate, lacerate-toothed: fl. rather large, violet turning to red-purple; lateral petals bearded; spur variable, much shorter than the petals, or quite as long, usually straight and obtuse, sometimes curved and even acute.—In a low compact almost acaulescent state, with small leaves and large long-stalked rich violet flowers, common on grassy hilltops along the seaboard from San Francisco northward, flowering in February and March. Less frequent in the Sierra Nevada from Mariposa Co. northward; the plant either subacaulescent and with short blunt spur, or several inches high, with spur varying from acute to obtuse and from long to short in the same localities. In the mountain forms the flowers appear to be paler, and the leaves are always larger. The leaves are less cordate and the whole plant more slender than in any Old World varieties of the species; and perhaps the *V. adunca* of Smith may be proven valid.

5. **V. ocellata**, T. & G. Fl. i. 142 (1838). Erect or ascending, ½—1 ft. high, nearly glabrous, or pubescent: leaves cordate or cordate-ovate, acutish, crenate, 1—2 in. long; stipules scarious, entire or slightly lacerate: petals 5—7 lines long, the upper ones white within, deep purple without, the others pale yellow-veined with purple, the lateral ones with a purple spot near the base, and slightly bearded.—From Monterey northward, in woods of the Coast Range; a very beautiful plant and the Pacific Coast analogue of *V. Canadensis*.

6. **V. cuneata**, Wats. Proc. Am. Acad. xiv. 290 (1879). Slender, 3—12 in. high, glabrous: leaves rhombic-ovate, acute, attenuate into a slender petiole, somewhat crenately toothed above: petals deep purple, with more or less white, 4—6 lines long, beardless, the broad short spur yellowish: capsule glabrous.—Humboldt Co. and northward.

+— +— *Leaves undivided; fl. yellow within, often brown-purple without.*

7. **V. glabella,** Nutt.; T. & G. Fl. i. 142 (1838). Stems slender from a creeping rootstock, nearly or quite leafless below, 5-12 in. high; minutely pubescent or glabrous: radical leaves on long, the cauline on short petioles, reniform-cordate to cordate, acute, crenately toothed or crenulate, 1-4 in. broad; stipules usually small and scarious, entire or serrulate: fl. bright-yellow, ½ in. long: petals more or less purple-nerved, the lateral ones bearded: capsule obovate-oblong, 4-5 lines long, abruptly beaked. In wet shades at middle elevations in the mountains, from Santa Cruz and Fresno counties northward to Alaska.

8. **V. pedunculata,** T. & G. l. c. 141; Bot. Mag. t. 5004. Stems 2-6 in. long, mostly prostrate or assurgent; almost glabrous or puberulent: leaves rhombic-cordate, usually almost truncate at the broad base, obtuse, coarsely crenate, ½-1½ in. long; stipules foliaceous, narrowly lanceolate, entire or incised: peduncles erect, greatly exceeding the leaves, 4-8 in. high, conspicuously bibracteolate: fl. 1 in. broad or more, golden-yellow, the upper dark-brown on the outside, the others purple-veined within; lateral petals bearded: capsule oblong-ovate, 4-6 lines long, glabrous. On low hills, in open ground, from the interior near Vacaville, *Jepson*, to San Francisco, and southward along the seaboard to San Diego. The most showy of our violets.

9. **V. sarmentosa,** Dougl.; Hook. Fl. i. 80 (1833). Stems prostrate, more or less creeping, slender, sparsely leafy; slightly pubescent: leaves rather thick and persistent, reniform, round-cordate or ovate, ½-1½ in. broad, finely crenate, deep green above, often rusty beneath, usually punctate with dark dots: peduncles slender, elongated: fl. light-yellow, not large. In woods of the Coast Range from Monterey northward.

10. **V. purpurea,** Kell. Proc. Calif. Acad. i. 56 (1855): *V. aurea,* Kell. l. c. ii. 185, f. 54 (1863). Stems clustered, from a branching perpendicular root, 2-6 in. high: pubescence very scant but under a lens hispidulous, somewhat retrorse or at least spreading: herbage rather succulent, in early stages purple, except the upper surface of the leaves: leaves from broadly ovate to lanceolate, tapering to the petiole, entire or more or less coarsely and often somewhat crenately toothed: peduncles little exceeding the leaves: petals 3-5 lines long, light yellow within, dark purple externally: capsule almost globular, 3 lines long, pubescent. Var. **pinetorum.** *V. pinetorum,* Greene, Pitt. ii. 14 (1889). Root or rootstock more or less horizontal or ascending, 1 ft. long or less: stems 3—10 in. high; the short hispidulous retrorse pubescence not scant: leaves mostly lanceolate or linear-lanceolate, the broader coarsely sinuate-toothed, the others merely sinuate or almost entire: peduncles slender, elongated, bearing smaller flowers which are purple without and of a very pale yellow within, the whole fading with a bluish tinge: capsules

round-obovate, cinereously hispidulous-puberulent.—Californian allies or subspecies of *V. Nuttallii*, the widely distributed yellow violet of the Great Basin and Rocky Mountain regions. The type, running into many forms, is of the Mt. Diablo Range and the Sierra Nevada northward: the variety is of the southern Sierra, from Kern Co. to San Bernardino. It is found in dry pine woods of the higher elevations.

+ + + *Leaves divided or lobed; fl. yellow.*

11. **V. lobata**, Benth. Pl. Hartw. 298 (1849). Stoutish, erect, $\frac{1}{2}$–1 ft. high, from an erect rootstock, leafy to the summit; puberulent or nearly glabrous; leaves of reniform or cordate outline, 2–4 in. broad, the cauline short-petioled, all palmately cleft into 5–9 narrowly oblong lobes, the central lobe largest or longest; some of the radical leaves less lobed or only coarsely toothed: petals 6 lines long, yellow, the upper brownish externally, the lateral slightly bearded: capsule 5–6 lines long, acute. Var. **integrifolia**, Wats. Leaves deltoid, acuminate, evenly crenate-serrate, not at all lobed.—Sierra Nevada and inner Coast Range; widely dispersed. The plant with undivided foliage is probably distinct. Its leaf-outline is wholly different.

12. **V. Hallii**, Gray, Proc. Am. Acad. viii. 377 (1872). Stems low, numerous, from a somewhat deep-seated erect rootstock: glabrous, pale, leafy throughout: leaves 3-parted, decurrent upon the petiole; segments lanceolate, entire, acute; stipules foliaceous, lanceolate or oblong, sub-laciniate: 2 upper petals dark purple, the others yellow; spur very short. In Humboldt Co., *Rattan, Chesnut & Drew*, and northward.

13. **V. Douglasii**, Steud. Nom. ii. 14 (1841); Greene, Pitt. ii. 14: *V. chrysantha*, Hook. Ic. t. 49 (1837), not of Schrader (1834). Root and clustered stems of the last, only the leaves and flowers appearing above ground; more or less pubescent: leaves large, bipinnately dissected into long linear or oblong segments; stipules lanceolate, entire or toothed: peduncles 2–5 in. long, equalling or exceeding the leaves: petals 5–7 lines long, golden yellow, the 2 upper brownish-purple externally: capsule 5 lines long, acute.—Throughout the State, chiefly in the interior, on open plains and hillsides.

14. **V. Sheltonii**, Torr. Pac. R. Rep. iv. 67. t. 2 (1857). Nearly or quite glabrous: leaves of broad-cordate outline, ternately compound, or 3-parted; the divisions lobed or cleft into oblong or linear segments: peduncles about equalling the leaves: fl. not half as large as in the last, yellow, veined with purple.—A slender species of the woodlands of the northern Sierra; Nevada and Plumas counties, etc.

+ + + + *Leaves divided; fl. not yellow.*

15. **V. Beckwithii**, T. & G. Pac. R. Rep. ii. 119. t. 1 (1855). Foliage of the preceding, but plant more condensed and low, appearing as if

acaulescent, the stems mostly very short: peduncles about equalling the leaves: petals 4 7 lines long, very broad and rounded above; the upper deep purple, the others lilac with a yellowish base, the lateral ones bearded, the lowest emarginate: stigma bearded at the sides: capsules obtuse. In the northern Sierra; a beautiful species, wrongly described in the "Botany of California" as with yellow lower petals: and *V. montana*, Kell. is erroneously referred here. It is a small *V. lobata*.

Order XLI. RESEDACEÆ.

De Candolle, Théorie Élémentaire, 214 (1813).

Herbs with alternate exstipulate leaves, and terminal racemes or spikes of small nearly colorless but often fragrant flowers. Sepals 4 6, often somewhat united at base, unequal, herbaceous, persistent, open in the bud. Torus bearing a rounded and glandular hypogynous disk which is produced posteriorly between the petals and the stamens. Petals 4 6, open in the bud, the lamina often lacerate or palmately parted. Stamens 3 20, inserted on the disk; anthers oval, fixed by the middle, introrse. Ovary 1-celled, 3 4-lobed, of 3 4 carpels which at apex are distinct and divergent; stigmas sessile, minute, alternate with the parietal placentæ. Fruit membranous, 1-celled, open long before maturity. Seeds reniform, exalbuminous.

1. RESEDA, *Pliny* (MIGNONETTE. DYER'S WEED). Characters of the genus almost those of the order. Two Old World species, fugitives from the flower gardens, are here and there spontaneous with us.

1. R. ALBA, Linn. Sp. Pl. i. 449 (1753). A tall stout sparingly branching perennial, with long spikes of whitish flowers: leaves deeply pinnate: sepals 5 or 6: petals as many, all equal, 3-cleft.

2. R. ODORATA, Linn. Sp. Pl. 2 ed. i. 646 (1762). Annual; leaves oblanceolate or spatulate, often undulate: spike or raceme short in fl., elongated in fr.: fl. greenish, the large anthers dull red: petals parted into about 6 spatulate-linear segments. This, the fragrant Mignonette, far more common in cultivation than the other, is less frequently met with in waste places.

2. OLIGOMERIS, *Cambessedes*. Annual, with narrow linear entire leaves, and very small nearly colorless flowers in terminal slender spikes. Sepals 4, lateral. Petals 2, next the axis, free or united at base, entire or 2 3-lobed, persistent. Disk obsolete. Stamens 3 8; filaments united at base. Capsule 4-angled, 4-beaked, ∞-seeded.

1. **O. subulata,** Boiss. Fl. Orient. i. 435 (1867); Del. Fl. Ægypt. 15 (1813), under *Reseda: O. glaucescens,* Camb. Jacquem. Voy. iv. 24. t. 25

(1834). *Ellimia ruderalis*, Nutt.; T. & G. Fl. i. 125 (1838). Glabrous, 5–10 in. high, branching from the base : leaves somewhat succulent, often fascicled, $1\frac{1}{2}$–1 in. long : fl. minute, subtended by small bracts : capsules in long loose spikes, depressed-globose, about $1\frac{1}{2}$ lines thick, angled and sulcate, shortly 4-beaked. –A seaside and alkaline-desert herb; not known north of Santa Barbara.

Order XLII. CAPPARIDEÆ.

Ventenat, Tabl. du Règ. Vég. iii. 118 (1799). CAPPARIDES, Juss. (1789).

Herbs or shrubs, with more or less heavy-scented and pungent-flavored herbage, alternate simple or compound usually exstipulate leaves, and complete flowers in bracted racemes. Sepals or calyx-lobes 4. Petals 4. Stamens 6–8, mostly unequal, usually inserted on the very base of the calyx, sometimes hypogynous. Ovary and fruit commonly stipitate, composed of 2 closely united carpels. Fruit various ; a 1-celled silique with many seeds on 2 placentæ, or the 2 valves 1-seeded and separating from the axis as nutlets. Seeds globose or reniform, exalbuminous.

1. **ISOMERIS,** *Nuttall.* A low stoutish and rigid glaucous and puberulent shrub, with 3-foliolate leaves, and rather large yellow flowers in short bracteate terminal racemes. Calyx persistent, 4-cleft ; lobes ovate, acuminate. Petals sessile, oblong, equal. Torus fleshy, dilated. Stamens 6, on the torus. Pod large, inflated, coriaceous, obovate-oblong, indehiscent. Seeds few, large, smooth.

1. **I. arborea,** Nutt.; T. & G. Fl. i. 124 (1838) ; Torr. Bot. Mex. Bound. t. 4. *Cleome Isomeris*, Greene, Pitt. i. 200 (1888). Common southern seaboard shrub, scarcely more than a species of *Cleome;* probably not found many miles north of Santa Barbara.

2. **CLEOME,** *Linnæus.* Ours erect branching annuals, with palmately 3–7-foliolate leaves (the leaflets entire), and yellow flowers in bracteate racemes. Sepals 4, sometimes united at base. Petals sessile or unguiculate. Stamens 6, on the small torus. Pod oblong or linear, 2-valved ; valves deciduous from the slender placentæ. Seeds ∞, round-reniform.

1. **C. lutea,** Hook. Fl. i. 70. t. 25 (1830) ; Bot. Reg. xxvii. t. 67. Glabrous or slightly pubescent, 1–2 ft. high : leaflets 5, linear- to oblong-lanceolate, 1–2 in. long, acute, short-petiolulate, equalling the petioles ; stipules setaceous, caducous ; bracts simple, bristle-tipped : fl. bright yellow, corymbose, the raceme elongated in fr.; petals 3–4 lines, exceeding the ovate-lanceolate sepals: stamens long-exserted: pod 6–15 lines long, 2 lines broad, acute at each end ; stipe and pedicel each $1\frac{1}{2}$ in. long. –Valleys of northwestern Nevada and doubtless within California.

2. **C. platycarpa**, Torr. Bot. Wilkes Exp. 235. t. 2 (1873). Size and habit of the last, but pubescent and glandular: leaflets 3, broadly oblong to lanceolate, 6–8 lines long, obtuse or acutish: sepals linear-setaceous: pod $\tfrac{3}{4}$ in. long, $\tfrac{1}{3}$ in. broad: seeds 10–12: style 2 lines long. Range of the preceding.

3. **CLEOMELLA**, *De Candolle*. Habit of *Cleome*; flowers smaller; leaves 3-foliolate. Pod short and few-seeded, ovoid or rhomboidal, the valves often produced laterally into horn-like appendages.

1. **C. obtusifolia**, Torr. Frem. Rep. 311 (1845). Pubescent, very much branched from the base, $\tfrac{1}{2}$–2 ft. high: leaflets oval or oblong, $1\tfrac{1}{4}$–$1\tfrac{1}{2}$ in. long, equalling the petioles, glabrous above; stipules long and fimbriate: fl. small, in leafy racemes: sepals ovate, lacerate-ciliate: petals 1–2 lines long: pods 2–5 lines broad, the valves acutely and often slenderly horned: style very slender, 2 lines long: stipe 3 lines, reflexed upon the pedicel. Common along the northern and western borders of of the Mohave Desert, but said to have been found in the first place near Sacramento, *Fremont*, but there may be an error as to that locality.

4. **WISLIZENIA**, *Engelmann*. Habit, foliage and flowers of *Cleomella*, but the pod didymous; valves contracted each upon its one seed and forming a nut-shell-like covering to it, nerved and reticulated; style elongated.

1. **W. refracta**, Engelm. Wisliz. Rep. 14 (1848); Gray, Pl. Wright, i. 11. t. 2 (1853). Erect, branching, 1–2 ft. high: leaflets 3, obovate to oblanceolate, 5–9 lines long, usually surpassing the petioles: racemes dense, in age elongated: petals 1 line long: stamens and ovary exserted: cells of the ovary 2-ovuled: fr. $1\tfrac{1}{2}$ lines broad; the divergent obovate reticulate valves separated by a perforated partition: style persistent, 1–2 lines long: stipe 2–3 lines, strongly refracted upon the pedicel.— Common at Lathrop, and near Sacramento; also at the south.

Order XLIII. CRUCIFERÆ.

Adanson, Fam. ii. 409 (1763); Jussieu, Gen. 237 (1789).

Herbaceous or rarely suffrutescent plants with watery pungent juice, alternate exstipulate leaves, and usually racemose white or yellow or sometimes purple flowers. Sepals 4, imbricate, deciduous. Petals 4, often unguiculate, the laminæ spreading in the form of a cross (unequal, and differently arranged in many of our *Streptanthi*), hypogynous, deciduous. Stamens almost always tetradynamous, *i. e.*, 4 long, 2 short (except in some *Streptanthi*, where they are in 3 unequal pairs, and in *Athysanus* and *Heterodraba*, which have them all of equal length); in some species reduced to 4 and even 2, always hypogynous (except in *Subularia*). Fruit usually a silique or silicle of 2 valves which separate from a central partition formed by the united placentæ. Seeds attached to the outer edge of the placental partition all around, usually forming a single row under the valves; albumen 0; embryo curved, the relations of radicle and cotyledons various.

Hints of the Genera.

Inflorescence bractless; pods stipitate, - - - - - - - 12, 13
" " pods not stipitate;
Pods 2-celled, 2-valved, regularly dehiscent from below;
" short, flattened parallel with the broad partition, - - 1, 2, 4
" " flattened contrary to the narrow partition, - - 20—23
" ovoid or globose, or the valves distended, - - - - 3, 7
" longer, flattened or subterete, or 4-angled, beaked, - 8, 9, 14, 18
" long, " " " beakless, 10, 11, 13, 15—17, 19
" of 2 short opposite indehiscent 1-seeded cells, - - - 25
" longer, transversely jointed or indehiscent, - - - 26, 27
" orbicular, 1-seeded, indehiscent, - - - - - 6, 24
" ovoid, several-seeded, indehiscent, - - - - - 5
Inflorescence leafy or bracted; pods without partition, dehiscent from the
apex or indehiscent, - - - - - - - - 28

1. **PLATYSPERMUM**, *Hooker*. Low annual, with pinnatifid radical leaves, and slender 1-flowered scapes. Flowers very small, white. Silicle oval, about 8-seeded. Seeds flat, with a broad scarious wing; cotyledons accumbent. The genus *Leavenworthia* of the Atlantic slope of the Continent is very nearly related to this.

1. **P. scapigerum**, Hook. Fl. i. 68. t. 18 (1829). Glabrous, 2–3 in. high: leaves somewhat runcinately pinnatifid: sepals equal, spreading: petals very small, scarcely unguiculate, obovate, white: pods about $1\frac{1}{3}$ in. long. Sierra Co., *Lemmon*, and northward, usually in rather moist open ground. Apr.

2. **ALYSSUM**, *Dioscorides*. Low perennials, with simple leaves, and more or less of a stellate pubescence. Flowers racemose. Sepals equal. Petals white or yellowish. Longer filaments often toothed. Pod orbicular; valves convex, nerveless. Seeds 1 or 2 in each cell, slightly margined; cotyledons accumbent.

1. A. ALYSSOIDES, Gouan, Hort. Monsp. 321 (1762); Linn. Sp. Pl. ii. 652 (1753), under *Clypeola; A. calycinum*, Linn. Sp. Pl. 2 ed. ii. 908 (1763). Annual, branching from the base, decumbent, $\frac{1}{2}$–1 ft. high; herbage canescent: leaves linear-spatulate, $1\frac{1}{2}$–1 in. long: raceme rather slender, the white or pale yellow petals little exceeding the sepals: pod orbicular, with a thin margin, slightly emarginate, $1\frac{1}{2}$ lines broad, little exceeding the persistent sepals, 4-seeded, on spreading pedicels a line long: style $\frac{1}{2}$ line.– Said to be naturalized about the Bay of San Francisco, but we have not seen it.

2. A. MARITIMUM, Lam. Encycl. i. 98 (1783); Linn. Sp. Pl. ii. 652 (1753), under *Clypeola*. Perennial, decumbent, the numerous branches 1 ft. long or less; herbage softer than in the last, ostensibly glabrous, under a lens showing a few appressed hairs: fl. 2 lines long, the broad white petals twice the length of the deciduous sepals: pod orbicular, 1 line broad, nearly glabrous, 2-seeded: pedicels slender, spreading. The Sweet Alyssum is spontaneous in many places, especially on the San Francisco peninsula.

3. **PHYSARIA**, *A. Gray*. Low herbs (our species perennial) silvery-canescent with a dense close stellate pubescence, entire or pinnatifid leaves, and racemose yellow flowers of middle size; calyx more or less persistent as in *Alyssum*. Pods globose or ovoid, or the valves more inflated and distended, nerveless. Style long, persistent. Seeds few, in 2 rows, flattened, rarely somewhat margined: cotyledons accumbent.

* *Pods didymous.*— Typical PHYSARIA.

1. **P. didymocarpa**, Gray, Gen. Ill. i. 162 (1849); Hook. Fl. i. 49. t. 16 (1829), under *Vesicaria*. Leaves rosulately crowded on a short crown or caudex, broadly spatulate, occasionally somewhat lyrate, those of the decumbent flowering branches oblanceolate, entire: racemes short: pods 2–6 lines broad, *i. e.*, more or less widely didymously-inflated, the partition narrow or nearly obsolete.—East of the Sierra, thence common to Colorado.

* * *Pod globose or ovate.*—Genus LESQUERELLA, Wats.

2. **P. montana**, Greene. Gray, Proc. Philad. Acad. 58 (1863), under *Vesicaria*. Habit of the preceding: rosulate radical leaves orbicular or obovate, long-petioled, those of the branches oblanceolate or spatulate, entire, or with few teeth: fl. 3 lines long: pods oblong-ovoid, $2\frac{1}{2}$ lines long, on slender recurved pedicels; style a third shorter. On Lassen's Peak, *Lemmon*, thence eastward and northward.

4. **DRABA**, *Dioscorides*. Small plants of various habit, annual or perennial. Leaves entire or toothed. Flowers small, white or yellow. Sepals equal. Filaments mostly flattened, without teeth. Pods oval, oblong or linear, flat; valves nearly flat, nerveless or faintly 1-nerved. Seeds in 2 rows in each cell, wingless; cotyledons accumbent.

* *Slender annuals or biennials; leaves mostly radical.*

1. **D. stenoloba**, Ledeb. Fl. Ross. i. 154 (1842): *D. hirta*, var. *siliquosa*, Ch. & Schl. Linnæa, i. 23 (1826). Radical leaves oblanceolate, $\frac{1}{2}$–1 in. long, rather thin, acute, scarcely at all toothed, sparsely clothed with stiff more or less branching or stellate hairs: stem 4–12 in. high, loosely racemose almost throughout, glabrate above, canescently pubescent below: petals bright or pale yellow, only 1–$1\frac{1}{2}$ lines long, obtuse: pods

3 5 lines long, linear, acute at each end, glabrous, on spreading pedicels 2 4 lines long; style 0. In the Sierra Nevada at 7,000 to 10,000 ft., from Mono Pass and Yosemite northward.

2. **D. crassifolia,** Grah. Edinb. Phil. Journ. 182 (1829). Habit of the last, but smaller, glabrous, the stem racemose only above the middle, thus seeming more scape-like, the leaves thickish and a trifle fleshy, narrowly oblanceolate or linear, $\frac{1}{4}$ 1 in. long, rarely with 1 or 2 teeth, occasionally ciliate: petals yellow, scarcely a line long, little exceeding the calyx: pods lanceolate, acute at each end, 3 -4 lines long, on pedicels nearly as long; style 0.- Habitat of the last.

* * *Stouter plants, often alpine and dwarf, mostly perennial, very leafy below.*

3. **D. aureola,** Wats. Bot. Calif. ii. 430 (1880): *D. aurea*, Wats. l. c. i. 28, not Vahl. Occasionally only biennial, but stout, erect, stellate-pubescent, the simple or sparingly branching stem 2 4 in. high: leaves crowded at base, spatulate, obtuse, entire, $\frac{1}{4}$ $\frac{1}{2}$ in. long: raceme short and dense: fl. bright yellow: pods broadly oblong, rather obtuse at each end, pubescent, 2 -4 lines long, the ascending pedicels rather shorter; style short and stout. In the typical form this is known only from Lassen's Peak, *Lemmon, Horace Davis, Chesnut & Drew*. The Mt. Dana plant referred here by Mr. Watson is considered distinct by Mr. Lemmon who has seen both in the living state.

4. **D. Lemmoni,** Wats. l. c.: *D. alpina*, var. *algida*, Bot. Calif. not Regel. Densely cespitose and dwarfed; the crowded leaves spatulate or oblong-obovate, ciliate and pilose, the hairs simple or forked: scape-like stem $\frac{1}{2}$–6 in. high: petals yellow, $1\frac{1}{2}$- $2\frac{1}{2}$ lines long, much exceeding the broad obtuse sepals: pods ovate to broadly lanceolate, somewhat twisted, on slender spreading pedicels; style short, stout.— High peaks of the Sierra Nevada.

5. **D. eurycarpa,** Gray, Proc. Am. Acad. vi. 520 (1865). Dwarf, densely cespitose, stellate-tomentose; the crowded leaves $\frac{1}{2}$ in. long, spatulate, entire: scape few-flowered, 1 2 in. high: pod ovate, 5–10 lines long, acute, beaked with a long slender style.- Found only by *Brewer*, on a dry summit, at 11,500 ft. altitude, above Sonora Pass in the Sierra.

6. **D. Douglasii,** Gray, l. c. vii. 328 (1868). Scapes many, from a branching leafy caudex, $\frac{1}{2}$ $1\frac{1}{2}$ in. high, pubescent, corymbose: lowest leaves ovate; uppermost obovate or spatulate, 1 2 lines long, entire, pale, glabrous or with simple hairs, the margin hispid-ciliolate: petals white or yellowish, 2 lines long, exceeding the broad obtuse sepals: pod ovate-oblong, acutish at each end, puberulent, 2 lines long, beaked with

a slender style; cells 2-seeded. Sierra Valley, *Lemmon*, and southward along the borders of Nevada, in high mountains. Of this species we find no mention made in Mr. Watson's recent account of the genus (1888).

7. **D. subsessilis**, Wats. Proc. Am. Acad. xxiii. 255 (1888). Very dwarf, the short branches of the caudex forming a broad mat, finely stellate-pubescent; leaves oblong, obtuse, not ciliate; scapes very short; fruiting raceme about 1 in. long; petals small, white, scarcely exceeding the yellowish ovate sepals; pods broadly ovate-elliptical, 2 lines long, short-pedicellate, pubescent; style very short and thick. — White Mountains, Mono Co., at 13,000 ft. alt., *Shockley*.

8. **D. Breweri**, Wats. l. c. 260. Biennial or perennial, densely stellate-pubescent, the shortly branching caudex bearing leafy stems 1–3 in. high; lowest leaves crowded, oblong or linear-oblong, obtuse, entire or sparingly toothed, 2–4 lines long, sometimes ciliate at base; cauline 2–4, oblong-ovate; fl. small, white; sepals oblong, herbaceous; pods linear-oblong, 2–3 lines long, obtusish, pubescent, short-pedicellate; style very short or 0. — Mt. Dana, at 12,000 ft., *Brewer*, and on the White Mountains, *Shockley*.

9. **D. glacialis**, Adams, Mem. Soc. Mosc. v. 106; DC. Syst. ii. 338 (1821). Dwarf and cespitose; leaves all in a rosulate tuft, linear-lanceolate, entire, rigid and carinate, more or less densely stellate-pubescent, sometimes ciliate at base; scape with few not small yellow flowers; pods ovate to ovate-oblong, acute, usually pubescent; style very short. — Not rare in the higher Sierra; readily identified by the linear rigid carinate leaves.

5. **HETERODRABA**, *Greene*. Slender diffuse annual, leafy only near the base, the elongated branches unilaterally racemose throughout. Leaves simple, toothed. Sepals equal. Petals without claw. Stamens 6 but equal, 3 on either side of the orbicular compressed ovary. Pod several-seeded, 2-celled by a very thin and filmy partition, indehiscent. Cotyledons accumbent.

1. **H. unilateralis**, Greene, Bull. Calif. Acad. i. 72 (1885); M. E. Jones, Bull. Torr. Club, ix. 124 (1882), under *Draba*. Pubescent with rigid short branching hairs; leaves cuneate-obovate, coarsely few-toothed above the middle, $1/2$–1 in. long; branches horizontal and trailing or prostrate, $1/2$–2 ft. long, in age rigid and wiry; pods on short rigid deflexed pedicels, 2 lines long, $1 1/2$ lines wide, stellate-pubescent and hispidulous, twisted when mature. Abundant in fields among growing grain, from Colusa Co. southward along the bases of the mountains to the San Joaquin and Livermore valleys; also in Lower California, where it was first detected by Mr. Jones. An interesting plant of early spring.

whose remarkable peculiarities of floral structure exclude it from *Draba* and link it very closely with *Athysanus*.

6. ATHYSANUS, *Greene*. Habit and character of the preceding, save that the very small orbicular and straight pods are 1-celled and 1-seeded.

1. **A. pusillus,** Greene, Bull. Calif. Acad. i. 72 (1885); Hook. Ic. t. 42 (1837), under *Thysanocarpus;* probably including *T. oblongifolius*, Nutt. Stems filiform, branching from the base, the branches mostly ascending, unilaterally racemose throughout: leaves few, ovate, sparingly toothed, ½ in. long: fl. minute, often apetalous: pods lenticular, more or less uncinate-hispid, scarcely a line long, rather acute at base; style very short.—Common on hillsides throughout western California. Mar., Apr.

7. CAMELINA, *Ruellius*. Erect herbs, sparingly branching, with clasping or sagittate leaves, and terminal loose racemes of small yellowish flowers. Sepals equal. Petals entire. Filaments without teeth. Silicle obovate or globose, beaked with a persistent style. Seeds several in each cell, oblong, marginless; cotyledons incumbent.

1. **C. sativa,** Crantz, Austr. 10 (1762); Linn. Sp. Pl. ii. 641 (1753), under *Myagrum*. Pubescent, 1½–2 ft. high: leaves lanceolate, sagittate at base, nearly entire: pods pyriform with acute base. Formerly extensively cultivated in Europe for the oil of its seeds; now become a weed in fields of grain in many countries; not yet well established in California, at least in our part, but found in a field at Berkeley by *Mr. Chesnut* in 1887.

8. SMELOWSKIA, *C. A. Meyer*. Dwarf alpine perennials, with pinnatifid leaves, and white or pinkish flowers. Sepals equal, somewhat spreading. Petals entire. Silique short, attenuate at each end, commonly somewhat 4-angled, the valves being strongly 1-nerved and carinate. Seeds in 1 row, oblong, not margined; cotyledons incumbent.

1. **S. calycina,** C. A. Mey.; Ledeb. Fl. Alt. iii. 170 (1831); Desv. Journ. Bot. iii. 168 (1813), under *Hutchinsia;* Stephan; Willd. Sp. iii. 433 (1800), under *Lepidum*. Cespitose, stoutish, 2–6 in. high, white-tomentose to almost glabrous: leaves mainly radical, with long slender petioles, pinnate or pinnatifid; segments linear to oblong, entire or sparingly lobed; calyx villous: petals 2 lines long: pod ¼–½ in. long, a line wide, 4-angled, varying to almost flat, attenuate at each end, beaked with the short style and broad stigma: seeds 2—8, a line long. —Lassen's Peak, *Lemmon, Chesnut & Drew*, and northward; also in Colorado, and in Europe.

9. PARRYA, *Robert Brown*. Perennials, with leaves mostly radical, rather firm in texture if not even somewhat fleshy, entire or toothed.

Flowers rose-color or purple. Calyx nearly equal at base. Petals entire, unguiculate. Pod broadly linear or ensiform, acuminate, nearly flat; valves with a prominent nerve. Seeds rather few, margined or not so; cotyledons various.

1. **P. Menziesii,** Greene. Hook. Fl. i. 60 (1830), and Bot. Beech. 322. t. 75, under *Hesperis*. *Phœnicaulis cheiranthoides*, Nutt.; T. & G. Fl. i. 89 (1838); *P. Menziesii*, Greene, Bull. Torr. Club, xiii. 143 (1886). *Cheiranthus Menziesii*, Wats. Bot. King, 14 (1871). Caudex branching, partly subterranean and densely clothed with the persistent petioles of former seasons: radical leaves broadly oblanceolate or narrowly obovate, obtuse or abruptly acute, entire, 2–4 in. long, canescent with a stellate pubescence: cauline reduced to bracts: scape-like stems several from each branch of the caudex, decumbent, 6–10 in. high: calyx 2 lines long: corolla 4–5 lines, bright red-purple: anthers oblong: pods 1–2 in. long, broad, not carinate, narrowed to the slender style: seeds oval, marginless.—In the Sierra, northward chiefly. Apr., May.

10. **ARABIS,** *Linnæus*. Sepals erect, equal, or two of them slightly saccate at base. Petals white or purple, with narrow claw and flat blade. Stamens tetradynamous; anthers short, straight, ovate or oblong, scarcely emarginate at base. Stigma entire or 2-lobed. Pod linear, compressed: valves nerveless or lightly 1-nerved. Seeds in 1 or 2 rows, flattened, often winged; cotyledons accumbent.—As received, this is an altogether conventional genus embracing plants very diverse in aspect, the groups probably without real congeneric affinity.

* *Biennials; pods linear, ascending or erect.*

1. **A. glabra,** Weinm. Cat. Dorp. 18 (1810); Linn. Sp. Pl. ii. 666 (1753), under *Turritis*: *A. perfoliata*, Lam. Encycl. i. 219 (1783). Stout, usually simple, 2–5 ft. high; lowest leaves and base of stem hirsute or hispidulous, the plant otherwise glabrous, glaucous: lower leaves spatulate, 2–4 in. long, sinuate-pinnatifid or toothed, ciliate at least on the petioles; cauline ovate or ovate-lanceolate, entire, clasping by a sagittate base: petals dull-white or greenish-white, 2–3 lines long, little exceeding the sepals: pods erect, usually even appressed to the stem, 2–4 in. long, less than a line wide, straight on pedicels 3–4 lines long: style short: seeds in 2 rows, narrowly winged or wingless. Common, and of rank growth in western California; of wide distribution both in Europe and N. America.

2. **A. repanda,** Wats. Proc. Am. Acad. xi. 122 (1876). Stem stout, 2 ft. high, with few-spreading somewhat flexuous branches; pubescent with coarse and loose branched hairs: leaves obovate to oblanceolate, obtuse, 3–4 in. long, coarsely sinuate- or repand-toothed, narrowed to a broad petiole: calyx $1\frac{1}{4}$ lines long, little surpassed by the pinkish petals: pods ascending, 3 in. long, 1–$1\frac{1}{2}$ lines wide; style very short:

seeds in 1 row, orbicular or ovoid, very thin and flat, broadly winged. - Yosemite, and mountains near Tehachapi, thence southward.

* * *Perennials; pods erect or ascending.*

3. **A. platysperma**, Gray, Proc. Am. Acad. vi. 519 (1865). Stems several, from a suffrutescent base, slender, 4–12 in. high; canescent with a minute stellate pubescence: lower leaves oblanceolate or spatulate, entire, 1 in. long; cauline small, sessile, not auricled at base: petals rose-color, 2–3 lines long: pods few, straight, erect, 1–2 in. long, 2 lines wide, acuminate; style 0: seeds in 1 row, with a broad thin wing. Yosemite and northward at alpine or subalpine elevations.

4. **A. Lyallii**, Wats. l. c. Habit of the preceding, but herbage either bright green or glaucous, only the lower part pubescent: radical leaves on slender petioles, oblanceolate, acute, entire; cauline oblanceolate, clasping by a sagittate base: petals 3 lines long, pinkish, twice the length of the sepals: pods straight, narrowly linear, 1–3 in. long; seeds in 2 rows, narrowly winged.— In the Sierra from Mono Pass northward.

5. **A. blepharophylla**, H. & A. Bot. Beech. 321 (1840); Bot. Mag. t. 6087. Stems stout, tufted, 4–12 in. high; herbage deep green, glabrous or sparsely pilose-pubescent: lower leaves obovate to broadly spatulate, 1–2 in. long, entire or sparingly sinuate-toothed, strongly ciliate; cauline oblong, sessile, obtuse or acutish: sepals usually purplish: petals of a rich red-purple, $\frac{1}{2}$ in. long or more: pods 1–1$\frac{1}{2}$ in. long, 1–1$\frac{1}{2}$ lines wide, suberect, beaked with the short stout style: seeds in 1 row, a line wide, wingless or narrowly margined. Common on rocky hills, from San Francisco to Monterey, and a most beautiful species. Feb.—Apr.

* * * *Perennials (except n. 10); pods spreading, recurved or deflexed.*

6. **A. Breweri**, Wats. Proc. Am. Acad. xi. 123 (1876). Low, tufted, rather rigid, 2–10 in. high, canescent with a dense stellate pubescence, and hirsute above with nearly simple hairs: radical leaves spatulate, 1 in. long, short-petioled, entire; cauline ovate-oblong, sessile, not sagittate, acute, $\frac{1}{2}$–$\frac{3}{4}$ in. long: petals rose-purple, 1–4 lines long, well exceeding the purplish sepals: pods spreading or recurved and rather crowded at the summit of the stem, 1$\frac{1}{2}$–2$\frac{1}{2}$ in. long, scarcely a line wide: seeds narrowly winged.—Inner Coast and Mt. Diablo Ranges, from Mendocino and Lake counties to Mt. Hamilton; usually growing on rocks. Apr.—June.

7. **A. Lemmoni**, Wats. l. c. xxii. 467 (1887): *A. canescens*, var. *stylosa*, Wats. Bot. Calif. ii. 431 (1880). Size and habit of the last, or smaller, stellate-pubescent below, glabrous and glaucous above: lower leaves spatulate-oblanceolate, 1$\frac{1}{2}$–$\frac{3}{4}$ in. long, rarely with a few teeth, the petiole sometimes ciliate; cauline sessile, auriculate: fl. small, rose-

color, the sepals pubescent; pods short-pedicellate, spreading or somewhat ascending, glabrous, curved, 1–2 in. long, $^2_3$ line wide, more or less attenuate to a short style or sessile stigma: seeds in 1 row, orbicular, narrowly winged.—A northern montane species, found on Lassen's Peak, *Lemmon*, and in Plumas Co., *Mrs. Austin*.

8. **A. arcuata,** Gray, Proc. Am. Acad. vi. 187 (1863); Nutt.; T. & G. Fl. i. 77 (1838), under *Streptanthus*. Stems more or less tufted, simple, 1–2 ft. high, hirsute or more or less tomentose with branching hairs; lower leaves numerous, oblanceolate, acute, entire or coarsely serrate-toothed, the petioles slender; cauline linear-lanceolate, 1–2 in. long, auricled or sagittate at the base; petals bright purple, 1_2 in. long or less, the sepals half as long, often colored; pods rather scattered on the upper part of the stem, the short pedicels divaricate, the silique arcuate-recurved, 3–4 in. long, barely a line wide; style 0; seeds narrowly or not at all winged. In dry soil on rocky hills near Fort Tejon and northward, apparently in the Sierra chiefly. Mar.–May.

9. **A. Holboellii,** Hornem. Fl. Dan. xi. t. 1879 (1825); *A. retrofracta,* Grah. Edinb. Phil. Journ. 344 (1829). *Turritis patula,* Grah. l. c. Stems usually solitary, seldom 2 or 3, rather slender, 1_2–2 ft. high, simple or with a few erect branches, more or less stellate-pubescent, with or without hirsute hairs below, sometimes glabrous, the upper part of the plant usually so, and glaucous; lower leaves spatulate, entire or toothed; cauline broadly oblanceolate, sagittate and clasping at base, 1_2–1 in. long; petals twice the length of the sepals, white, rose-color or purple, 3–4 lines long; pods 1–4 in. long, 1_2–1 line wide, glabrous, strongly reflexed, acutish; seeds as in the last. A Rocky Mountain species, reaching our borders only on the eastward slope of the Sierra.

10. **A. Bolanderi,** Wats. Proc. Am. Acad. xxii. 467 (1887). Biennial, with erect or ascending branches from the base, 1–2 ft. high, more or less stellate-pubescent throughout: lower leaves not known; cauline lanceolate, entire, 1–2 in. long, auriculate at base: fl. rose-color: fruiting pedicels slender, spreading, 1–2 lines long: pods glabrous, straight, obtuse, with a broad sessile stigma, 1–1½ in. long, the valves 1-nerved to the middle; seeds narrowly winged. In the Sierra, at Yosemite or Mono Pass, collected only by *Bolander*, and in the State Survey confused with n. 9, from which it is distinguished by its different habit and biennial duration.

* * * * *Low annual, with aspect of* CARDAMINE.

11. **A. Ludoviciana,** C. A. Mey. Ind. Sem. Petr. ix. 60 (1842). *Cardamine Ludoviciana* (Nutt.), Hook. Journ. Bot. i. 191 (1834); Nutt. MSS. under *Sisymbrium*. Nearly glabrous, branched from the base and the branches ascending, 6–10 in. long: leaves all pinnately parted into

oblong or linear few-toothed or entire segments: fl. small, white: pods spreading on short pedicels, flat, rather broadly linear, 1 in. long: seeds orbicular, wing-margined. Banks of the lower San Joaquin, near Bethany, etc.; manifestly indigenous both here, and in San Diego Co., as well as on the Atlantic slope of the continent. Perhaps better placed under *Cardamine*, notwithstanding the want of elasticity in the dehiscence of the pod; for it is very incongruous with *Arabis* as now generally received.

11. STREPTANTHUS, *Nuttall*. Mostly annuals, often stout and tall, and the few branches loosely racemose throughout. Leaves pinnatifid or toothed, rarely entire, except the cauline, and these mostly sagittate and clasping. Calyx whitish or colored, open or closed, often irregular, 2 or all of the sepals saccate at base. Petals with broad channelled claw and (in our species) a narrow usually more or less undulate limb. Stamens either tetradynamous or in 3 unequal pairs, the uppermost pair often with filaments united; anthers elongated, sagittate at base, curved in age. Pod from flat and thin to subterete; valves 1-nerved or rarely carinate. Seeds more or less flattened, margined or marginless. — Genus not at all definitely distinguished from *Arabis*, and as here given, with *Caulanthus* in part, and *Stanfordia* included, far more naturally constituted than *Arabis* as received. The great diversities, and the strange irregularities of floral structure in certain members of the Euclisia section, would have the weight of generic characters if the whole group were less uniform in respect to habit and sensible qualities. All the species have a mild sweet herbage, much like cabbage, for which the larger sorts were used as a substitute in early times. The typical *Streptanthi*, with broad blade to the petals, are of the Southern Atlantic states and have no representatives with us.

* *Flowers regular; sepals spreading or loosely erect; stamens tetradynamous, no pair of filaments united; pods subterete; seeds not margined.*
— Genus STANFORDIA and part of CAULANTHUS, Wats.

1. S. Californicus, Greene. Wats. Bot. Calif. ii. 479 (1880), under *Stanfordia*. Branched from the base, the ascending branches 1 ft. long: herbage glabrous and glaucous, with a scant pubescence on the basal part of the plant: lower leaves oblanceolate with dilated petiole, obtuse, crenately toothed, 2 in. long; upper lanceolate, sessile and clasping, sharply serrate: fl. in a loose raceme on hairy pedicels 2–3 lines long: calyx deep purple, 3–4 lines, equalling the whitish petals, these consisting of a broad fleshy claw which is cordate at base, and a very short and narrow almost obsolete limb: stamens not exserted; anthers linear-oblong: pod thickish, slightly compressed, 1 in. long: seeds wingless: cotyledons 3-lobed.—Abundant on the plains of the San Joaquin from Tulare southward; a showy plant, thoroughly congeneric with the next. The petals so far from being "without claws," are all claw, or with a merely rudimentary blade; in this respect differing only in degree from the following in all of which the claw is the broader and principal part of the petal.

2. **S. inflatus,** Greene. Wats. Proc. Am. Acad. xvii. 364 (1882), under *Caulanthus*. Annual, erect, stout, the mostly simple stem fistulous and inflated above the middle, 1-2 ft. high, sparingly hispid or glabrous: leaves ovate to oblong, all sessile and auriculate, acutish, entire, 1-3 in. long: fl. large, purple; the glabrous sepals somewhat saccate at base, 3-4 lines long; the petals with broad claw and ligulate blade: pods subterete, 3-4 in. long, on ascending pedicels of 2-4 lines: stigma sessile, deeply bifid. On the Mohave Desert; but also along the railway in Kern Co., near Bakersfield, etc., where it may have been accidentally introduced.

3. **S. Coulteri,** Gray; Wats. Bot. King. 27 (1871); Wats. l. c., under *Caulanthus*. Erect, mostly 3-5 ft. high, sparingly branched above, more or less hispid: radical leaves broadly spatulate, sinuate-toothed; cauline oblong-lanceolate, clasping with cordate base, the uppermost entire: sepals 3-5 lines long, acute, hispid: pod straight, subterete, 3-4 in. long, 1½ lines broad, pendent upon the hispid pedicel, beaked by the stout style; stigma 2-lobed.- Very common on the plains below Fresno, near Tulare, Bakersfield, etc., associated with n. 1, but ranging further southward. Apr.

* * *Sepals erect or connivent; stamens often in 3 unequal pairs; pods compressed (subterete in n. 5).*- Subgenus EUCLISIA, Nutt.

+- *Calyx regular; corolla cruciform or nearly so; stamens all distinct (except in n. 5).*

4. **S. heterophyllus,** Nutt.; T. & G. Fl. i. 77 (1838). Size and habit of n. 3; hispid below, glabrous above, not glaucous: radical leaves irregularly pinnatifid, lower cauline similar, but the uppermost narrowly lanceolate and from remotely toothed to entire, sagittate-clasping: fl. pendulous: sepals dark purple: petals purple with whitish undulate margins: pods 3-5 in. long, pendulous, compressed, narrow: seeds ½ line long, narrowly winged. Mountains of Kern Co., and southward.

5. **S. Parryi.** Stoutish, simple or with few branches 1-3 ft. high, glabrous except a sparse hispidulous pubescence on the veins of the radical leaves, and at the base of the stem, pale and glaucous throughout: lowest leaves linear-oblong, 2 in. long, saliently toothed; cauline lanceolate, sagittate, entire: raceme long and loose: fl. ½ in. long, the somewhat shorter deflexed pedicels rather densely hirsutulous: calyx dark purple in the bud, fading to flesh-purple in fl.; the well exserted strongly crisped petals white, with dark purple veins: stamens in 3 unequal pairs, the filaments united for three-fourths of their length in the longer

pairs; anthers all fertile, linear-sagittate: pods 3-4 in. long, little compressed, apparently somewhat quadrangular. Apparently common in Monterey and San Luis Obispo counties, *Parry, Lemmon;* a handsome species, related to *S. Californicus* and *S. inflatus,* notwithstanding the erect sepals and the cohesion of the filaments in each of the two long pairs of stamens.

6. **S. cordatus,** Nutt.; T. & G. Fl. i. 77 (1838). Perennial, stoutish, glabrous, glaucous, the clustered stems 1-2 ft. high: leaves thick, usually coarsely toothed at least at the obtuse apex; the radical ones spatulate-oblong, the upper cordate or ovate-lanceolate, obtuse or acute, with broad rounded and auricled base: fl. 3—4 lines long: sepals broad, obtusish, colored purplish: petals exserted, greenish or yellowish or sometimes purplish: pods broadly linear, 2—4 in. long, compressed, straight, spreading or ascending on short pedicels: seeds flattened and winged.—A common species of the Rocky Mountains, reaching California in the Sierra Nevada and coming westward even to Kern Co., near Tehachapi; closely connecting with *Arabis* through *A. repanda*.

7. **S. barbatus,** Wats. Proc. Am. Acad. xxv. 125 (1889). Perennial, but stems apparently solitary as if from a horizontal rootstock, simple or branched, rather slender, 1 ft. high or more; herbage glabrous, glaucous: leaves all cauline, subequal, very numerous and rather crowded, broadly cordate and clasping, of oval outline, obtusish, entire or at apex toothed, $\frac{1}{2}$-1 in. long: raceme rather strict and dense: fl. small, dark purple; sepals obtusish, setosely bearded at tip: petals little exserted: pods spreading or recurved, 1-2 in. long, $1\frac{1}{2}$ lines wide, compressed: seeds narrowly margined.—River banks, in sandy or gravelly ground, along the upper Sacramento and its tributaries; perhaps common there, but rarely collected.

8. **S. tortuosus,** Kellogg, Proc. Calif. Acad. ii. 152, fig. 46 (1863). Annual, erect, sparingly paniculate-branching above, 1-3 ft. high, glabrous, glaucous: lower leaves oblong, narrowed to a winged petiole, 2-3 in. long, repandly toothed; middle cauline spatulate-oblong or -obovate, auriculate-clasping, rather remotely serrate-toothed; uppermost round-cordate or cordate-oblong, mostly entire, abruptly or even cuspidately acute, 1-$1\frac{1}{2}$ in. long, clasping by a deep closed sinus: sepals purplish, 3-5 lines long, slenderly acuminate, the attenuate tips reflexed in flower, exceeded by the purple-veined yellowish petals: 4 longer stamens about equal (one pair sometimes connate, *Kell.*); anthers all equal, linear-sagittate: torus greatly enlarged under the fruit: pod very narrow, 3-6 in. long, falcate and recurved, the pedicel spreading or reflexed: seeds winged. In the Sierra about Yosemite and southward.

9. **S. orbiculatus.** *S. Breweri,* Gray, partly, and a part of *S. tortuosus,*

Wats., not Kell. Near the preceding, but low ($\frac{1}{2}$–$1\frac{1}{2}$ ft. high), diffusely branched from the base: lowest leaves round-obovate, very obtuse or even truncate, crenately or more remotely and repandly toothed, abruptly narrowed to a petiole as long as the blade; middle cauline obovate-spatulate, auricled and clasping; uppermost orbicular, mostly entire, obtuse: sepals purple, 2–3 lines long, acute but not acuminate, at length petaloid-dilated, undulate and whitish at the recurved tips: stamens in 3 unequal pairs, but the upper pair much the longest: torus dilated: pods 2 in. long, falcate on ascending pedicels, strongly torulose: seeds wingless, though sometimes distinctly but very narrowly margined.—Common in the Sierra from Donner Lake northward; also on Mt. Diablo, *McLean*. Usually mixed with *S. tortuosus* in the herbaria, from which its dissimilar habit, and very different floral structure well distinguish it. But it is the "n. 1" of Gray's *S. Breweri*, and by modern rules would retain that name; but Dr. Gray's description was drawn wholly from his ns. 2 and 3; hence I continue that use of *S. Breweri* which the specific character warrants.

10. **S. diversifolius**, Wats. Proc. Am. Acad. xvii. 363 (1882). Erect, slender, branching above, $1\frac{1}{2}$ ft. high, glabrous: cauline leaves pinnately divided with 1 or 2 pairs of narrowly linear lobes, the upper nearly or quite entire, those of the branches broadly cordate, clasping and entire, $\frac{1}{2}$ in. long or less: racemes few-flowered: sepals pale, 2–3 lines long: blade of petals rather broad, purple-veined: pods very narrow, $1\frac{1}{2}$–$2\frac{1}{2}$ in. long, strongly reflexed.—Apparently a local species of the Cosumne Creek in Sacramento Co., *Rattan*.

++ *One or two pairs of stamens with filaments connate; petals unequal, the upper pair larger and parallel, the lower smaller and divergent; plants all annual.*

+ *Calyx not manifestly irregular.*

11. **S. Breweri**, Gray, Proc. Calif. Acad. iii. 101 and Am. Acad. vi. 184 (1864), as to descr., and numbers "2" and "3" of specim. (see n. 9 *supra*). Branched from the base, 1–2 ft. high, leafy at base chiefly: leaves broadly ovate or obovate, acute or acuminate, sessile, dentate, 1–3 in. long, thickish: the cauline reduced and bract-like, lanceolate, acuminate, entire or denticulate, somewhat auricled and clasping: racemes elongated and the fl. scattered, short-pedicelled: sepals purple, acuminate, glabrous or pubescent: petals exserted: 2 pairs of filaments connate, those of the uppermost and longest joined to the summit, and the anthers reduced to less than half the size of the others: pod $1\frac{1}{2}$–$2\frac{1}{2}$ in. long, less than 1 line wide, erect or ascending, mostly somewhat incurved. In the Mt. Diablo Range southward, on Mt. San Carlos, etc. Apparently not collected since Brewer's time, and the district is a neglected one.

12. **S. barbiger**, Greene, Pittonia, i. 217 (1888). Slender, rather

widely branching, 1–1½ ft. high, pubescent or glabrous: cauline leaves linear, entire, scarcely auriculate: fl. subsessile, 3 lines long: sepals greenish, the rather acuminate tips becoming whitish, petaloid and recurved, the whole calyx commonly bristly-hairy, but often glabrous: petals white: filaments dark purple, the three pairs very unequal, the uppermost connate almost to the summit, their anthers much reduced and seemingly sterile: pods 1–2 in. long, narrowly linear, recurved.—In Lake Co., *Simonds,* and common in Napa Co., at Miravalle near St. Helena, *Greene.*

++ ++ *Calyx irregular, three sepals more or less connivent behind the upper petals, the fourth separated from these and somewhat deflexed; 1 pair of filaments connate.*

13. **S. niger,** Greene, Bull. Torr. Club, xiii. 141 (1886). Branching loosely from near the base and above, 1–3 ft. high, glabrous, glaucous: leaves linear, 2–3 in. long, the lowest with narrow pinnate gland-tipped lobes or teeth, the upper entire, auriculate-clasping: racemes loose, flexuous: pedicels ascending, 1 in. long: calyx 3 lines long, of a very dark metallic shining purple; sepals ovate-cymbiform, the 3 upper slightly separate from the lowest, and connivent at apex: blade of petals very slender, white: upper pair of filaments connate almost throughout, their anthers small and sterile: pod 2 in. long, erect or ascending, nearly straight: seed narrowly winged.—Hills above the ferry landing at Tiburon; found only by the author, in April and June, 1886, now seemingly extinct there, and not elsewhere detected.

14. **S. albidus,** Greene, Pittonia, i. 62 (1887). Stouter than the last, equally glabrous and glaucous, even the cauline leaves with callous-tipped prominent teeth, the base sagittate-clasping: racemes not flexuous: pedicels short: sepals 3–4 lines long, white, with purple base: petals ½ in. long, the lamina ample, crisped, white, with purple veins: upper pair of filaments united to the tip, their anthers small but polliniferous: fr. unknown. On hillsides not far from San Jose, *Sister Anna Raphael, Mr. Rattan.* A handsome species, in floral structure quite distinct from *S. niger,* to which it is most related.

15. **S. Mildredæ.** Slender, much branched from the base, 1–1½ ft. high, more or less pilose-hispid: lower leaves coarsely and sinuately toothed; cauline linear-lanceolate, entire, clasping: racemes somewhat flexuous, not secured: fl. small, very dark metallic-purple: petals with small, slenderly attenuate white-margined purple blade: upper pair of filaments almost wholly united, their anthers reduced to mere rudiments and closely approximate, the other 4 stamens much shorter and little unequal: pods 3 in. long, slender, arcuate-spreading on the short pedicels: seeds oval, the upper half narrowly margined.—Common on Mt. Hamilton; dedicated to Miss Mildred Holden, in whose herbarium the

plant was first seen; but it is the *S. glandulosus*, at least in part, of the State Survey collectors, probably not of Hooker, which seems to be the next.

16. **S. glandulosus,** Hook. Ic. 40 (1836): *S. peramœnus,* Greene, Bull. Torr. Club. xiii. 142 (1886). Pubescence and sinuately toothed foliage of the last, but larger, 1—2½ ft. high: racemes more or less inclined to be secund: fl. very large, bright red-purple: sepals ½ in. long, ovate-cymbiform, carinate, 3 strongly connivent at tip, the fourth hanging loosely apart from the others: petals well-exserted, white-margined: upper pair of filaments connate above the middle, thence rather widely divergent, their anthers smaller than the others, but not greatly reduced, apparently sterile: pod 3 in. long, a line wide, arcuate-recurved: seed narrowly winged. Very common on clayey hillsides and banks, from Mt. Diablo and near Berkeley to Monterey: the most beautiful of all our cruciferous plants when in flower: the irregularity of the calyx not at all indicated in Hooker's plate, and his description imperfect. Named in reference to gland-tipped teeth of foliage: but these pervade the group.

17. **S. hispidus,** Gray, Proc. Calif. Acad. iii. 101 and Am. Acad. vi. 184 (1864). Stiff-hirsute or hispid throughout, only 3—6 in. high, branching: lowest obovate- or cuneate-oblong, coarsely and somewhat incisely toothed, the teeth obtuse; cauline narrower, scarcely clasping: raceme short, loose, the fl. at length recurved: sepals red-purple with white-petaloid tips, half as long as the similarly colored petals: pods hispid, 1½—2 in. long, 1 line wide, straight, erect: seeds winged.— Still known only from the summit of Mt. Diablo, but perhaps not rare in the southern extension of that range, which still remains too little explored.

18. **S. secundus.** Slender, sparingly branched above, 1—2 ft. high; the long pinnately toothed or lobed lower leaves hispid-strigose, the cauline leaves lanceolate, sagittate, entire or toothed, and, with the branches, pedicels and pods, sparsely hispidulous with spreading short hairs: racemes rather dense, wholly secund: fl. flesh-color, 4 lines long: sepals sharply carinate, the keel hispid-ciliolate, the short tips greenish, the remote lower one distinctly, the opposite uppermost one obscurely unguiculate: petals, especially the upper pair, with ample purple-veined crisped limb: upper pair of filaments connate to near their tips, the free parts scarcely divergent, the anthers reduced in size, but polliniferous: pods 2 in. long, very slender, falcate-recurved on the divaricate pedicels, the valves carinate-veined: seeds small, wingless.— Northern base of Mt. Tamalpais. A very beautiful pale-flowered species, of well defined habit and floral character, especially as regards the calyx, with its very narrow and downwardly attenuate upper and lower sepals, the laterals being broad, and yet at tips converging toward the middle upper one in the

manner of this whole subgroup. So far as known, the plant is quite local; and we have not seen it growing there since 1886.

++ ++ ++ *Calyx very irregular, the uppermost sepal greatly enlarged, in æstivation conduplicate over the others.*

19. **S. polygaloides,** Gray, Proc. Am. Acad. vi. 519 (1865); Greene, Pitt. ii. 46. Very slender, simple or virgately branched, 1–2 ft. high, glabrous: lower leaves unknown; cauline 1–2 in. long, linear, involute to appear filiform; sagittate at base: fl. subsessile, only the yellowish calyx conspicuous: upper sepal round-obovate, ¼ in. long and nearly as broad, the several times smaller lateral and lower oblong-lanceolate, acuminate: petals minute, scarcely exceeding the shortest sepals: upper pair of filaments twice the length of the others, connate throughout, their anthers rudimentary, sterile: pod 1–1½ in. long, ½ line wide, reflexed on the very short pedicels, nearly or quite straight, attenuate upward to the short style: seeds narrowly winged.—Rocky hills in the Sierra Nevada from Grass Valley southward to Amador Co., where it is very common.

12. **THELYPODIUM,** *Endlicher.* Coarser than *Streptanthus*, often biennial; the racemes often shorter and condensed. Calyx green, whitish or purplish; sepals equal at base. Petals with narrow claw and flat linear to obovate limb, exserted, white, yellowish or rose-color. Stamens tetradynamous; filaments never connate; anthers sagittate at base, curved. Pod usually long, linear, terete or slightly compressed, sessile or short-stipitate. Seeds in 1 row, oblong, somewhat compressed, not winged.—An excellent genus as to the typical species, but passing toward *Streptanthus* by easy gradations from terete to slightly compressed pods, and from narrow to slightly dilated claws of the petals. The differences in the relation of radicle and cotyledons sometimes not even of specific importance.

1. **T. integrifolium,** Endl.; Walp. Rep. i. 172 (1842); Nutt.; T. & G. Fl. i. 96 (1838), under *Pachypodium.* Biennial, stout, 3–6 ft. high, corymbosely paniculate-branching above, glabrous, glaucescent: radical leaves often 1 ft. long, oblong-elliptical, long-petioled, entire; cauline sessile, lanceolate-oblong or narrowly lanceolate, 1–2 in. long: fl. crowded and almost corymbose at the ends of the branches: sepals 1½–2½ lines long: petals spatulate-obovate, pale rose-color; claw exceeding the sepals: stamens exserted: fruiting racemes short, dense: pod ½–1½ in. long, slender, somewhat torulose; stipe 1 line long: radicle of seed placed midway between the edge and the middle of one of the cotyledons.—In subsaline moist places near Tehachapi, *Greene*, thence eastward and northward, chiefly beyond our borders.

2. **T. brachycarpum,** Torr. Bot. Wilkes Exp. 231. t. 1 (1862); Gray,

Proc. Am. Acad. vi. 520 (1865). Biennial, erect, sparingly and virgately branching, 1-5 ft. high, pubescent or glabrous: radical leaves oblanceolate or spatulate, pinnatifid or toothed; cauline narrow, sagittate and clasping, entire or toothed: fl. in long dense racemes: petals narrowly linear, white, 3-4 lines long; pod 3/4-1 in. long, acuminate with the slender style, ascending on short pedicels.—In the Mono district and northward along our eastern borders; perhaps also on the upper Sacramento.

3. **T. Lemmoni,** Greene, W. Am. Sc. iii. 156 (1887). Annual, stout, 3-6 ft. high, glabrous and glaucous: lower leaves 3-6 in. long, including a short petiole, 3 in. broad, coarsely and angularly lobed below, repand-toothed above; upper lanceolate, nearly entire, sessile by a narrow base: racemes 1 ft. long, rather loose: petals whitish, 3 lines long, exceeding the lilac-purple spreading sepals: stamens exserted: pods 2 in. long, acuminate, slender, somewhat torulose, not stipitate, ascending, on pedicels of 1, 2 lines.- Fields of San Luis Obispo Co., *Lemmon;* also abundant among growing grain near Tracy on the lower San Joaquin.

4. **T. procerum,** Greene. Brewer in Gray, Proc. Am. Acad. vi. 519 (1865), under *Streptanthus;* Wats. Bot. King, 27, under *Caulanthus*, Annual, stoutish, branched from near the base, 3-7 ft. high, glabrous except at base: lower leaves petiolate, coarsely pinnatifid; upper lanceolate, sessile, acuminate: racemes long and lax: fl. greenish or yellowish white, 4-5 lines long, on ascending pedicels half as long: pod very slender, terete, 3-5 in. long, less than a line wide, acuminate, erect or somewhat spreading.—In fields along the northern and eastern base of Mt. Diablo and southward in the same range.

5. **T. flavescens,** Wats. Bot. King, 25 (1871). *Streptanthus flavescens*, Torr. Pac. R. Rep. iv. 65 (1857), not Hook. Stoutish, or rather slender, sparsely pilose-hispid below: lower leaves elongated, petiolate, sinuately toothed; upper entire, sessile, not auricled: raceme long and lax: fl. yellowish or rarely purplish, 4-5 lines long: sepals narrow and with the pedicels hispidulous: petals long-exserted, with linear and narrow claw; the blade dilated: pod 1½ in. long, nearly terete, sparsely hispidulous, erect or somewhat spreading. Very common in fields of the lower Sacramento, east and north of Vacaville, *Jepson;* originally from near Benicia, *Bigelow.*

6. **T. Hookeri.** *Streptanthus flavescens*, Hook. Ic. t. 44 (1836). Size, habit and whole aspect of the preceding, but lower leaves often pinnatifid, though as often sinuate-toothed; inflorescence the same, also size and color of fl., but sepals broader, less spreading, glabrous: petals with a rather broad claw and relatively narrow blade: pods 2 in. long, slender, terete, erect.- Common in the Mt. Diablo Range, especially in

the region east of the Livermore Valley, thence to the original station, i. e., Monterey. Although this and n. 5 have been placed in different genera, they are with difficulty held distinct as species. The only difference is in the petals; and by these the present plant would stand well in *Streptanthus* if its habit and narrow terete pods were not those of the annual Thelypods precisely.

7. **T. rigidum**, Greene, Pittonia, i. 62 (1887). Stoutish and very rigid, 1–3 ft. high, with few wide-spread branches: hispidulous below, glabrous above, deep green, not glaucous: lower leaves somewhat lyrately pinnatifid; upper oblong-lanceolate and laciniate-toothed: fl. yellowish, small, rather crowded and subsessile, the fruiting raceme long and loose: pods 1¼ in. long, nearly sessile, ascending or somewhat spreading or curved, rigid, sharply tipped with a short style. Eastern base of the Mt. Diablo Range, from near Antioch southward, chiefly on clayey hillsides.

8. **T. lasiophyllum**, Greene, Bull. Torr. Club, xiii. 143 (1886); H. & A. Bot. Beech. 321 (1840), under *Turritis*. *Sisymbrium reflexum*, Nutt. Pl. Gamb. 183. *Thelypodium neglectum*, M. E. Jones, ? Am. Nat. xvii. 875. Glabrous or more or less hirsute below, ½–6 ft. high, usually stoutish, rather rigidly erect, simple, or sparingly branching above the middle: leaves 2–4 in. long, pinnatifid with divaricate toothed segments, or the upper only sinuate-toothed: petals white or yellowish, 1½–2½ lines long: pod slender, nearly terete, 1–2 in. long, short-pedicellate, straight or somewhat curved, ascending or strongly deflexed. Common, and, as to size and pubescence, and the attitude of the pods, very variable. The common form at San Francisco is small, early flowering, and has suberect pods. In the Coast Range the plant is often a yard high or more, late flowering, with pods straight and strongly deflexed. On the plains east of the Mt. Diablo Range grows in great abundance a plant here referred which differs in being glabrous, with pods more or less curved, often spreading only, sometimes deflexed. All these need further examination; and *T. neglectum* may prove to be one of them.

13. **STANLEYA**, *Nuttall*. Stout perennials, with coarse and rather thick glaucous foliage. Flowers yellow, large, in long racemes. Sepals equal at base, spreading. Petals narrow, unguiculate. Stamens elongated; anthers linear, at length closely coiled. Pistil stipitate; becoming a long linear pod, with 1-nerved valves. Seeds in 1 row, oblong, not winged; cotyledons linear, incumbent. A remarkable genus, chiefly belonging to the Great Basin, with the aspect of *Cleome* in the preceding family, but genuinely cruciferous in technical character.

1. **S. pinnata**, Britt. Trans. N. Y. Acad. viii. 62 (1889); Pursh, Fl. ii. 739 (1814), under *Cleome: S. pinnatifida*, Nutt. Gen. ii. 71 (1818); Gray, Gen. Ill. i. 154. t. 65. Stems several from a somewhat woody base, 2–8

ft. high, simple: lower leaves coarsely and irregularly more or less lyrately pinnatifid, the segments few, oblong: upper lanceolate, entire, with slender petiole: calyx 3–4 lines long: petals half longer, their claws and the stipe of the ovary somewhat pubescent: pod 2 in. long, 1 line wide, curved attenuate to a slender stipe $\frac{1}{2}-\frac{3}{4}$ in. long, exceeding the slender pedicels. —From the valley of the Arkansas westward, reaching our borders in Kern Co., *Heermann*, and extending even into Santa Barbara Co., *Torrey*.

14. **CARDAMINE**, *Dioscorides*. Annuals or perennials of woods or moist places; rootstock often tuberous. Stems mostly simple, often very sparingly leafy. Flowers white or purplish, in short racemes. Sepals equal. Petals unguiculate. Silique elongated, linear, compressed, beaked or pointed, the valves plane, almost nerveless, more or less elastically dehiscent. Seeds compressed, not margined.

* *Without fleshy or tuberous rootstocks; leaves all pinnate.*

1. **C. oligosperma**, Nutt.; T. & G. Fl. i. 85 (1838). Annual, erect, slender, $\frac{1}{2}$–1 ft. high, nearly or quite glabrous: leaflets small, in 3–5 pairs, roundish, 1–6 lines long, often obtusely 3–5-lobed, petiolulate: petals white, 1–$1\frac{1}{2}$ lines long, twice the length of the calyx: pods few, $\frac{1}{2}$–$\frac{3}{4}$ in. long, $\frac{1}{3}$ line wide, short-beaked, not becoming dry, the mature valves, while yet green-herbaceous, separating elastically and falling in a close coil; cells about 8-seeded.—Common on shady banks along streams and in open groves. Mar., Apr.

2. **C. Gambelii**, Wats. Proc. Am. Acad. xi. 147 (1876). Perennial, glabrous or sparsely hirsute, erect and stoutish, 1–2 ft. high: leaflets 4–6 pairs, ovate-oblong to linear, sessile, entire or sparingly toothed, acute, $\frac{1}{2}$–1 in. long: fl. white, on slender pedicels: petals 4 lines long, twice the length of the sepals: pods narrowly linear, ascending, 1 in. long, equalling the reflexed pedicel; beak 1 line long, slender.—Common in marshy ground near the sea at Santa Barbara, etc.

3. **C. Breweri**, Wats. Proc. Am. Acad. x. 339 (1875). Stems from a running rootstock, erect from a decumbent base, $\frac{1}{2}$–$1\frac{1}{2}$ ft. high, glabrous, or slightly pubescent at base: leaflets 1–2 pairs, rounded or oblong, the terminal much the largest, $\frac{1}{2}$–1 in. long, entire or coarsely sinuate-toothed or lobed, obtuse, often cordate at base, the very lowest often simple and cordate-reniform: petals white, 2 lines long: pods 8–15 lines, obtuse or scarcely beaked with a short style, ascending on pedicels of 3–4 lines.—Margins of pools and streams in the higher Sierra from Sonora Pass northward; also in Humboldt Co., *Marshall*.

* * *Stems from elongated or rounded and tuberous perennial rootstocks; the radical leaves often simple and those of the stem few.*

4. **C. cuneata,** Greene, Bull. Calif. Acad. i. 74 (1885). Rather slender, 1 ft. high, glabrous: radical leaves 3–4 in. long, two-thirds as wide, 5–7-foliolate; leaflets ovate, toothed or lobed, $1/2$ in. long or more, tapering to slender petiolules of greater length, some of these with a pair of secondary leaflets at base; cauline of 5–9 linear-cuneiform entire leaflets: fl. large, white, changing to rose. In dry ground under oaks etc. near Jolon, Monterey Co., *Greene.*

5. **C. integrifolia,** Greene, Bull. Calif. Acad. ii. 389 (1887); Nutt. in T. & G. Fl. i. 88 (1838), under *Dentaria; C. paucisecta,* Benth. Pl. Hartw. 297 (1849), partly. Rather robust, 1 ft. high, glabrous, somewhat fleshy: radical leaves 1–5-foliolate, the leaflets usually rounded and more or less cordate and nearly or quite entire, 1–$2^{1}/_{2}$ in. broad; upper deeply lobed, or pinnately 3–5-foliolate, the segments linear or linear-oblong, entire: corolla large, white, nodding, the petals only campanulately spreading: pod conspicuously beaked. Common in wet meadows, in open ground.

6. **C. Californica,** Greene. Nutt. l. c. under *Dentaria: C. paucisecta,* Benth. partly. Near the last, but slender, tall, less fleshy; the leaves, both radical and cauline, with broad and ample repandly and mucronulately denticulate leaflets which are of a rich purple beneath: fl. smaller, rose-color. Very common in rich woods, or dry shady banks of the Coast Range; doubtless to be retained as distinct from n. 5, on account of the strictly sylvan habitat, and the differences in form and texture of foliage, though Bentham and other closet botanists have confounded them.

7. **C. cardiophylla.** Stoutish, 1 ft. high or less, glabrous: radical leaves undivided, round-reniform to broadly cordate, slightly and somewhat angularly 5-lobed and mucronately denticulate, 1 in. wide or more; cauline nearly as large, broadly cordate, acute, mucronate-denticulate, tapering from within the broad sinus to a petiole $3/4$ in. long: fl. rather small, white: pods narrowly linear, slender-beaked. In Weldon Cañon of the Vaca Mountains, Solano Co., 1 March, 1885, *Jepson;* an exceedingly well marked new species, with cauline leaves quite like those of *Viola glabella* in outline. The rootstocks are more elongated than in the allied species, and are barely an inch below the surface of the ground; the petioles of the radical leaves very slender and 2–3 in. long.

8. **C. Nuttallii,** Greene. Bull. Calif. Acad. ii. 389 (1887). *Dentaria tenella,* Pursh, Fl. ii. 439 (1814). Rootstock elongated, somewhat jointed and scaly: stem 6–10 in. high, naked below, but with 1–3 (usually 2) palmately or pinnately parted small leaves below the inflorescence; segments narrow-oblong or linear, $1/2$–1 in. long, obtuse, often mucronate, usually entire: radical leaf said to be simple, roundish, 5-lobed: fl. 1–$1^{1}/_{2}$ in. long, in a small terminal cluster, white or rose-color: pod unknown. Plumas Co., *Mrs. Ames,* and northward.

15. NASTURTIUM, *Pliny*. Branching herbs growing in water or wet places, glabrous, or hirsute with simple hairs: the roots annual, biennial or perennial. Leaves usually lyrate-toothed or pinnatifid: the petioles often auriculate-dilated at base. Flowers small, yellow or white. Sepals equal at base, often yellowish, spreading. Petals spreading, obovate or cuneiform, sometimes 0. Pods usually oblong or linear, sometimes much shorter and even subglobose, rather turgid, the strongly convex valves without midvein. Seeds very small, usually numerous and forming 2 rows in each cell, rounded, somewhat flattened, impressed-punctate; cotyledons accumbent.

* *Petals white, exceeding the calyx; stems rooting at the decumbent base; roots fibrous.* Genus CARDAMINUM, Mœnch.

1. N. OFFICINALE, R. Br. Hort. Kew. 2d ed. iv. 110 (1812). *Cardaminum nasturtium*, Mœnch, Meth. 262 (1794). *Sisymbrium nasturtium*, Linn. Sp. Pl. ii. 657 (1753). *Nasturtium aquaticum*, Trag. Hist. 82 (1552). (WATERCRESS). Aquatic, decumbent or procumbent, rooting at the lower joints, the branches 1_2–5 ft. long, stoutish and hollow: leaves pinnate, with rounded or elongated obtusely sinuate leaflets, the terminal one largest: petals $1 \frac{1}{2}$–2 lines long: pods 1_2 in. or more, acute at each end, equalling the spreading pedicels: valves faintly nerved: style short, thick. Abundant in shallow ponds and pools and along streamlets: naturalized from the Old World, where from time immemorial it has been used as a culinary and medicinal herb. It is of very rank growth in California; stems five or six feet high having been observed.

* * *Petals white; stems stout and, with the large leaves, erect from a large perpendicular simple or branching root.* Genus ARMORACIA, Ruppins.

2. N. ARMORACIA, Fries, ex Gray Man. 65 (HORSE RADISH). Earlier radical leaves pinnatifid; later ones very large, oblong, crenate: stem 2–3 ft. high: pods globular (seldom formed): style very short. Escaped from cultivation and naturalized in moist lands along the lower San Joaquin.

* * * *Sepals and petals very small, spreading, yellow or greenish; pod often short; root biennial or annual (except in n. 3).*
Genus RADICULA, Dillenius.

3. **N. sinuatum,** Nutt.: T. & G. Fl. i. 73 (1838). Ascending or decumbent, from slender perennial rootstocks, 3–10 in. high; sparingly villous: leaves of oblong-lanceolate outline, 1–3 in. long, all alike sinuate-pinnatifid with mostly entire lobes: fl. light yellow, 2 lines long, on pedicels twice as long: sepals and petals disposed to be persistent: silicle oval or oblong, $1\frac{1}{2}$–4 lines long: style one-third the length of the silicle. In the Sierra from Lake Tahoe northward and eastward. Our plant is as here described; but much longer and even curved pods are

ascribed to the species by Nuttall, whose type was from the Columbia River.

4. **N. curvisiliqua** (Hook.), Nutt., var. **lyratum,** Wats.: *N. lyratum,* Nutt.; T. & G. Fl. i. 73. Annual or biennial, with several erect or decumbent branching stems seldom 1 ft. long; glabrous, or under a lens sparsely hispidulous: leaves mostly narrowly oblong or oblanceolate and rather regularly pinnatifid into divaricate linear or oblong-lanceolate entire segments: fl. very small, greenish yellow: pods $\frac{1}{2}$ - $\frac{3}{4}$ in. long, linear, more or less curved, pointed with a prominent style, the valves smooth; pedicels half as long as the pod; seeds in 2 rows.—By streamlets, in the Coast Range and the Sierra; common in Marin Co. and southward. Nothing quite answering to the true *N. curvisiliqua* is found in middle California; neither indeed is the typical *N. lyratum* here, and very possibly our plant may prove distinct from both. The northern types of both are of different aspect.

5. **N. occidentale.** Annual, erect, $\frac{1}{2}$ - 1 ft. high, sparingly branched above, or rarely from near the base, glabrous (sparsely and retrorsely hispidulous under a lens), leafy at base: leaves rather broadly oblanceolate, coarsely toothed or somewhat pinnatifid: fl. minute: fruiting raceme elongated and rather dense: pods $\frac{1}{2}$ - $\frac{3}{4}$ in. long, linear, straight or slightly curved, abruptly tipped with a short style, obcompressed, $\frac{3}{4}$ line wide, the thin partition less than $\frac{1}{2}$ line: seeds round-oval, but almost cordate by a deep broad notch at the hilum.—Very common on moist low plains bordering the upper Sacramento, and in the foothills adjacent; a well marked species, most related to the next, though generally labelled "*N. curvisiliqua*" in the herbaria. The long pods, flattened contrary to the partition, are very characteristic.

6. **N. palustre,** DC. Syst. ii. 191 (1821); Mœnch, Meth. 263 (1794), under *Radicula. Roripa nasturtioides,* Spach, Phaner. vi. 506. Stoutish, erect, 2 - 3 ft. high, branching above, glabrous: leaves oblong-lanceolate in outline, coarsely and irregularly toothed or pinnatifid, 2 - 6 in. long: fl. a line long: pods linear-oblong, 3 - 5 lines long, on slender pedicels; style very short; valves thin, smooth and nerveless: seeds in 2 rows, nearly orbicular, scarcely notched at the hilum.- Either uncommon with us, or very rarely collected. Our only specimen which is clearly of this species comes from Humboldt Co., *Chesnut & Drew*. The allied *N. hispidum*, which is hispid-hairy and has globular pods, is in Nevada, near us.

7. **N. dictyotum.** Habit and foliage of the last, 2 -4 ft. high, sparingly hirsute: racemes elongated, rather dense: fl. unknown; pods 3 lines long, of ovate-lanceolate outline; valves of very firm texture, usually with a strong rather tortuous midvein and some anastomosing veinlets; partition rather thick, somewhat favose-reticulate by the impressions of the angular seeds.—On Grand Island, of the lower Sacra-

mento, Sept. 1891, *Jepson*; an uncommonly well marked species in the characters of its fruit. The name has a double application, being suggested by the reticulation of the valves as well as by that of the partition.

16. BARBAREA, *Dodonæus* (WINTER CRESS). Erect branching glabrous biennials or perennials of rather low stature, with angular stems and more or less distinctly lyrate or pinnatifid leaves. Flowers rather small, bright yellow. Sepals equal at base, erect. Pods linear, either somewhat flattened, or more distinctly quadrangular, pointed; valves more or less carinate. Seeds in 1 row, oblong, turgid, marginless; cotyledons oblique.

1. **B. vulgaris,** R. Br. Hort. Kew. iv. 109 (1812). *Erysimum Barbarea*, Linn. Sp. Pl. ii. 660 (1753); Crantz, Austr. 54 (1792), under *Sisymbrium*. Stoutish, 1–3 ft. high; herbage bright green and glossy: leaves mostly radical, the very lowest sometimes simple, oftener with 1 or more pairs of relatively small lobes below a very large terminal one; cauline either simple and toothed, or pinnately parted: fl. 2–3 lines long: pods 1–2 in., erect, ascending, or even arcuate-spreading, flattened, or more or less 4-angled: pedicels always short and stout. Common in moist open ground, or in shady places along streams; varying excessively in foliage and fruit, and perhaps embracing several good varieties or subspecies.

17. ERYSIMUM, *Dioscorides*. Biennials or perennials, ours stout, simple or with few branches. Leaves narrow, entire or runcinately toothed, not clasping. Flowers large, yellowish or orange. Sepals erect, one pair strongly gibbous at base. Petals with long claw and flat blade. Anthers sagittate. Pod 4-angled or flattened, and the valves merely nerved. Seeds in 1 row, not margined; cotyledons incumbent or oblique. It is conceded on all sides that there is no valid distinction between *Erysimum* and *Cheiranthus*, and each of our two species has at one time or another been referred to both genera.

1. **E. asperum,** DC. Syst. ii. 505 (1821); Nutt. Gen. ii. 69 (1818), under *Cheiranthus*. Canescent with short straight closely appressed hairs: stems solitary, rarely with a few branches above, 1–3 ft. high, angular: leaves narrowly spatulate or oblanceolate, entire or runcinate-toothed, 1–3 in. long: fl. large, fragrant: sepals narrow, 4–6 lines long: petals from light yellow to deep orange, 8–12 lines long: pods slender, spreading, quadrangular, commonly 3–4 in. long, 1 line wide, beaked with a stout style. Common in the mountains almost everywhere, but chiefly at some distance from the seaboard; usually with orange-red petals in the Sierra; a less tall form, with pale corollas, abounds in the Mt. Diablo Range, and may prove distinct. This is probably the *E. grandiflorum*, Nutt., referred to the next in the Bot. State Survey.

2. **E. capitatum,** Greene. Dougl.; Hook. Fl. i. 38 (1829), under

Cheiranthus. Stout and low, $1\frac{1}{2}$–$4\frac{1}{2}$ ft. high, sparingly pubescent with appressed bifid or 2-parted hairs; leaves narrow, entire, or sinuately or angularly toothed or lobed: fl. large, cream-color or yellowish, in a depressed terminal corymb, scarcely fragrant: pods in a short raceme; valves nearly flat, with a strong midvein, $1\frac{1}{2}$ lines wide, the whole $1\frac{1}{2}$–$2\frac{1}{2}$ in. long, abruptly and stoutly short-pointed; seeds flattened. Among the sandy or rocky hills of the seaboard only, from Monterey northward to Mendocino Co.; easily mistaken for *E. asperum*, but we have not seen it with even yellow, much less orange-colored flowers. The petals are broader than in that species, but at San Francisco they are invariably almost white. Feb.–May.

18. **BRASSICA,** *Pliny.* Large annuals or biennials, with erect often widely branching stems, lyrate or pinnatifid lower leaves, and yellow flowers. Sepals equal at base. Petals unguiculate; limb obovate. Pods linear or oblong, terete or quadrangular, with a stout 1-seeded or seedless beak; valves 1–5-nerved. Seeds in 1 row, globose; cotyledons incumbent. An Old World genus, of which a few species, cultivated everywhere, have become more or less thoroughly naturalized with us.

* *Sepals erect, enfolding the claws of the petals.*— BRASSICA proper.

1. B. CAMPESTRIS, Linn. Sp. Pl. ii. 666 (1753). Glabrous, glaucous, 2–3 ft. high; lower leaves somewhat rough-hairy, lyrate with large terminal lobe; cauline oblong or lanceolate with a broad auriculate-clasping base: fl. 3–4 lines long: pods nearly terete, 2 in. long or more, ascending, on spreading pedicels; the stout beak 8–10 lines long. Abundant in fields, flowering in the late winter and early spring months; commonly but erroneously called Mustard, it is the Turnip of fields and gardens run wild and become naturalized.

* * *Sepals spreading, releasing the claws of the petals.*
Genus SINAPIS, Tourn.

2. B. NIGRA, Koch; Richl. Deutsch. Fl. iv. 713 (1833); Linn. Sp. Pl. ii. 668 (1753), under *Sinapis*. (BLACK MUSTARD). Not glaucous but dark green, roughish with scattered stiff hairs, stout, 3–12 ft. high; leaves all petiolate; the lower lyrate, with a very large and lobed terminal lobe; the uppermost lobed or toothed or entire: petals 3–4 lines long, twice the length of the yellowish sepals: pods closely appressed to the rachis of the raceme, 4-angled, $\frac{1}{2}$–$\frac{3}{4}$ in. long, sharply beaked with the long style. Common as the preceding, but taking more exclusive possession of fence corners and rich waste lands; flowering only in early summer; the root strictly annual.

3. B. SINAPISTRUM, Boiss. *Sinapis arvensis*, Linn. Sp. Pl. ii. 668 (1753). (CHARLOCK). Annual, the herbage light green, rough with spreading hairs, 2–5 ft. high; lower leaves usually with a large coarsely

toothed terminal lobe and smaller ones of angular outline on the rachis: fl. 4 6 lines long; pods 1 1½ in. long, ascending, nearly cylindrical, with a stout somewhat 2-edged beak a third as long as the prominently nerved valves, often containing a seed, the seeds under each valve 3 8. Common by waysides in the vicinity of Berkeley and Oakland; flowering later than *B. campestris*, but earlier than *B. nigra*.

19. SISYMBRIUM, *Dioscorides*. Erect and rather slender annuals. Leaves not clasping, lyrate-pinnatifid, or (in our species) finely dissected. Flowers small, yellow. Sepals scarcely gibbous at base. Petals unguiculate. Anthers mostly linear-oblong, sagittate. Pods linear or oblong-linear, terete or nearly so, obtuse or short-pointed; valves slightly 1 3-nerved. Seeds usually numerous, small, oblong and teretish; cotyledons incumbent.

* *Seeds in 2 rows; leaves finely dissected.*

1. **S. canescens**, Nutt. Gen. ii. 68 (1818); Gray, Gen. Ill. i. 152. t. 64. Simple or with few branches, ½—2½ ft. high, canescent with short branching hairs: leaves 1 -2 pinnate, the segments more or less deeply toothed or pinnatifid: petals 1 line long or less, about equalling the sepals: pods oblong to linear, or subclavate, ¼ - ½ in. long, on slender spreading pedicels of equal or greater length, acute at each end, and beaked with a very short style: seeds ovate-oblong, ⅓ line long. Plains near Livermore, thence southward throughout the State.

* * *Seeds in 1 row; leaves less dissected.*

2. **S. incisum**, Engelm.; Gray, Pl. Fendl. 8 (1849). Glabrous or glandular-hairy, 1 -2 ft. high, rather freely branching: leaves pinnately divided, the segments lanceolate or linear-lanceolate and incisely serrate: petals lanceolate-spatulate, surpassing the sepals: fruiting racemes elongated, the capillary spreading pedicels about as long as the linear pods, the valves of which are faintly 1-nerved. In the Sierra Nevada, at 6,000 to 10,000 ft. elevation. June Sept.

3. **S. Hartwegianum**, Fourn. Sisymb. 66 (1865); *S. incisum*, var. *Hartwegianum*, Wats. Bot. Calif. i. 41 (1876). Size of the preceding, or smaller, cinereous-puberulent, the racemes more regularly panicled: segments of the pinnately parted leaves 5—7, oblong, obtuse, often 3-lobed: fruiting pedicels and very short acute pods erect and appressed to the rachis more or less closely. In the Sierra, at rather higher elevations than the last, and less common; also in the Colorado Rocky Mountains, and far northward. This apparently very good species has also been referred to *S. Sophia* of Europe by American botanists.

* * *Seeds in 1 row; leaves pinnatifid or entire.*

4. S. OFFICINALE, Scop. Fl. Carniol. ii. 26 (1772); Linn. Sp. Pl. ii. 660

(1753), under *Erysimum*. Rigid, erect, sparingly and divaricately branching above, somewhat hirsute: lowest leaves depressed and rosulate, lyrately and somewhat runcinately pinnatifid, 3–6 in. long: pods terete, ½ in. long, tapering from base to summit, nearly sessile, closely appressed to the rachis in a long slender raceme.—Abundant by waysides and in waste grounds; native of Europe; commonly called Hedge Mustard.

5. **S. acutangulum,** DC. Fl. Fr. iv. 670 (1804). Hirsute with scattered simple hairs, 1–2 ft. high, with ascending branches: leaves 2–6 in. long, mucinate-pinnatifid: pods terete, 1–1½ in. long, less than a line wide, erect or ascending on very short pedicels. Also from Europe, but not as common as the last.

20. SUBULARIA, *Linnæus* (AWLWORT). Aquatic, dwarf, acaulescent, with tufted subulate leaves, and a simple scape, bearing a few minute white flowers. Sepals equal at base (in one foreign species the sepals are united at base, and the stamens perigynous), spreading. Petals without distinct claws. Filaments without wings or appendages. Pods ovoid, slightly obcompressed; style 0. Seeds few, wingless; cotyledons incumbent.

1. **S. aquatica,** Linn. Sp. Pl. ii. 642 (1753). Scapes 1–3 in. high; leaves not so long: fl. scattered, less than a line long; petals not exserted; pods 1½ lines long, about equalling the pedicels, obtuse.—In stagnant pools, on the upper Tuolumne, *Bolander;* plant usually submersed altogether, and therefore easily overlooked. It ought not to be either local or rare in the higher mountains.

21. THLASPI, *Dillenius*. Low glabrous herbs with simple stems. Lower leaves rosulate, entire or toothed: cauline oblong, auricled and clasping. Flowers rather small, white or pinkish. Pods cuneate-oblong or -obcordate, obcompressed but not strongly so; valves acutely carinate or winged: style rather long. Seeds somewhat turgid, wingless; cotyledons accumbent.

1. **T. alpestre,** Linn. Sp. Pl. 2 ed. ii. 903 (1763). Stems several, from a perennial branching rootstock, ½–1 ft. high or more: radical leaves 1 in. long, including the slender petiole, obovate to oblanceolate, entire or few-toothed: cauline obovate to oblong, entire, obtuse or acutish: fl. white, 2–3 lines long: pods obovate to cuneate-oblong, 3–4 lines long, emarginate or truncate at the summit, tipped by a style 1 line long, spreading horizontally on the short pedicels.—Woods of Humboldt Co., *Marshall*, and northward.

2. **T. californicum,** Wats. Proc. Am. Acad. xvii. 365 (1882). Biennial, 6–8 in. high: lower leaves oblanceolate, attenuate to a slender petiole, few-toothed: cauline oblong-lanceolate, narrower toward the base, and usually narrowly auriculate: fl. 2 lines long: pods oblanceolate, acute, 4–5 lines long.—At Kneeland Prairie, Humboldt Co., *Rattan*.

22. BURSA, *Siegesbeck.* Slender nearly glabrous annuals, with simple or pinnate leaves, and small white flowers. Pods oblong or obcordate, more or less obcompressed, ∝ -seeded; valves carinate, 1-nerved. Seeds not winged; cotyledons incumbent.

1. B. PASTORIS, Wigg. Prim. Fl. Hols. 47 (1780). *Thlaspi Bursa pastoris,* Linn. Sp. Pl. ii. 647 (1753). *Capsella Bursa pastoris,* Mœnch, Meth. 271 (1794). Usually hirsute at base, otherwise glabrous, erect, ½–2 ft. high, the stems racemose almost from the base, simple or with few branches: radical leaves usually in a depressed rosulate tuft, runcinate-pinnatifid, or oblanceolate with coarse teeth; cauline sagittate, entire or toothed: pods cuneate-triangular, retuse at summit, 1–2 lines long and broad, on rather long spreading pedicels; seeds minute.— A thoroughly cosmopolitan weed, but native of the Old World; commonly called Shepherd's Purse; flourishing with us at all seasons of the year.

2. B. divaricata, O. Ktze. Rev. Gen. 21 (1891). *Hymenolobus divaricatus,* Nutt. (1838). *Capsella divaricata,* Walp. Slender, often diffusely branching and decumbent or procumbent, 3–8 in. high: lowest leaves sinuate-pinnatifid; cauline entire or nearly so: petals minute, barely equalling the sepals: pod oblong or ovoid, little flattened, 2 lines long or less; obtuse, the valves rather thin; pedicels slender, longer than the pods.— Common everywhere, along the borders of salt marshes: in appearance very unlike *B. pastoris,* and perhaps better received as a distinct generic type. Mar.– May.

23. LEPIDIUM, *Dioscorides.* Low herbs, with pinnatifid or toothed leaves, and small white or apetalous and greenish flowers. Stamens only 4, or even 2. Pod orbicular or ovate, strongly obcompressed, emarginately 2-winged at summit; valves acutely carinate: cells 1-seeded. Seeds not winged; cotyledons usually incumbent, rarely accumbent.

* *Annuals; pedicels flattened.*
← *Pods reticulated.*

1. L. latipes, Hook. Ic. t. 41 (1836). Branching from the base, the short branches stout and depressed, far surpassed by the leaves: these several inches long, irregularly and coarsely pinnatifid, the segments linear, entire or lobed; pubescence scant on the leaves, more dense on the branches, hispidulous: racemes short, dense: pedicels 1–2 lines long; sepals very unequal: petals broadly spatulate, ciliate, greenish, exceeding the sepals: pod broadly oval, 2 lines broad, sparingly pubescent, strongly reticulate, the broad acute wings nearly as long as the body of the pod. In saline soils at Martinez, Alameda, Monterey, etc., and in the interior along the lower San Joaquin. Mar., Apr.

2. L. dictyotum, Gray, Proc. Am. Acad. vii. 329 (1868). Habit and

pubescence of the preceding, but much smaller, the branches at length ascending: leaves narrowly linear, entire or with a few narrow divaricate linear lobes: petals little exceeding the sepals or wanting: pods rounded, 1½ lines broad, emarginate, with short acute wings, finely reticulated and pubescent, exceeding the thick erect pedicels. Originally found in western Nevada, in alkaline soil, but common in the Livermore Valley, also along the borders of marshes at Alameda.

3. **L. oxycarpum**, T. & G. Fl. i. 116 (1838). Very slender, the elongated and loosely racemose branches decumbent or assurgent, nearly glabrous: leaves linear, with a few linear segments or entire: sepals caducous: petals 0: stamens 2: pods on slender deflexed pedicels, glabrous, rounded, 1½ lines broad, the terminal wings tooth-like, short, acute, divergent. Borders of salt marshes at Alameda, Vallejo, *Greene;* also in subsaline soils east of Wild Cat Creek in the Berkeley Hills.

4. **L. Oreganum**, Howell (1887): *L. oxycarpum*, var. (?) *strictum*, Wats. Erect, simple or with few ascending branches, 3–6 in. high, ostensibly glabrous (more or less hispidulous under a lens): leaves linear, with few linear segments or entire: sepals and petals less fugacious: stamens 4: pods round-ovoid, 2 lines broad, the terminal teeth more or less prominent and divergent, the body somewhat hispidulous or glabrate. Plentiful in subsaline soil in the Livermore Valley; also at San Diego, *Cleveland*, and San Bernardino, *Parish*, beyond our limits. The type is from southern Oregon, and has pods less distinctly reticulate, with shorter and less prominent wings; but this and the Californian plants are not specifically different.

++ *Pods faintly or not at all reticulate.*

5. **L. nitidum**, Nutt.; T. & G. Fl. i. 116 (1838): *L. leiocarpum*, H. & A. (1840). Erect and with few ascending branches, or more diffusely branching from the base, ½–1½ ft. high, rather slender, almost glabrous, or the branches distinctly hirsutulous; these racemose almost throughout: lower leaves loosely pinnatifid, segments linear; cauline often entire: petals often present, white: stamens 2 or 4: pods rounded, glabrous and shining, often of a dark purple, or with minute purple dots, 1½ lines broad, with a small abrupt sinus between the short terminal teeth. Var. **insigne**. Stoutish and mostly simple, 4–8 in. high, the mostly solitary fruiting raceme shorter and denser: pods twice as large, round-obovoid.—Very common, especially toward the seaboard, in the middle and southern parts of the State. The earliest flowers, in the Bay region, appearing in January, have conspicuous petals, but are often sterile; the later and fertile are mostly apetalous. Southward the petals are always present. The variety is of the Mt. Diablo Range, east of Livermore, *Greene;* and toward the base of the Sierra, *Sanford*.

6. **L. Menziesii,** DC. Syst. ii. 539 (1821); *L. Californicum,* Nutt.; T. & G. Fl. i. 115 (1838). Low and diffuse, herbage light green, hispid-puberulent or glabrate; branches 3–6 in. long; racemes numerous, rather narrow and dense; leaves of oblong outline, pinnatifid, the segments usually 3-cleft or -toothed; petals 0; pods rounded, 1–1½ lines broad, glabrous, or around the margin more or less hispidulous, faintly reticulate; teeth at the summit very short and obtuse; pedicels short, ascending or spreading, often very little flattened. Common, especially by waysides and in hard clayey soil; late flowering, *i. e.,* Apr.– June.

* * *Stouter and taller; pedicels terete.*

7. **L. intermedium,** Gray, Pl. Wright. ii. 15 (1853). Erect, branching above the middle, 1½–1⅔ ft. high, puberulent or glabrous; lower leaves 1–2 in. long, toothed or pinnatifid; upper entire or only sparingly toothed, oblanceolate or linear; petals 0; pods glabrous, rounded, 1–1½ lines broad, very shortly winged, the obtuse teeth slightly divergent; pedicels 2 lines long. Only occasionally met with in middle western California; more common east of the Sierra; differing from the Atlantic coast *L. Virginicum* in being more slender, and having incumbent cotyledons.

8. L. DRABA, Linn. Sp. Pl. ii. 645 (1753). Biennial or perennial, erect, a foot high or taller, the several stems corymbosely branched at summit; herbage canescently pubescent; lower leaves oblong-obovate, 1–3 in. long, sparingly serrate or entire; cauline narrower, sagittate and clasping; petals white, conspicuous; pods cordate, not winged, turgid, acutish, tipped with a slender short style. Native of Europe, representing a type of Old World species widely different in appearance from all our native kinds, and probably not congeneric with them. It is met with occasionally in old fields at Berkeley and elsewhere in California.

24. THYSANOCARPUS, *Hooker* (LACE-POD). Erect and slender sparingly branched annuals, with minute or rose-colored flowers, in slender elongated racemes. Petals cuneate-obovate, or linear-oblong. Stamens 6, tetradynamous, or sometimes 4 only. Pistil a compressed rounded uniovulate ovary, short slender style, and small obtuse stigma; becoming a plano-convex or concavo-convex samara; the hard substance of the body of the fruit branching into several (12 to 16) radiating lines with diaphanous spaces or even complete rounded perforations between them, the whole forming a crenate wing. Seed solitary, somewhat compressed, wingless. Genus almost too near *Tauscheria* of Asia, which differs in having a beak to the fruit, into which the otherwise involute wing tapers; and in some of our forms the wing is involute.

1. **T. curvipes,** Hook. Fl. i. 69. t. 18 (1829); *T. runcinatus,* Hook.; Don. Dict. i. 196 (1831). A foot high or more, with few and rather strict racemose branches, or smaller and simple-stemmed; radical leaves in a

rosulate tuft, pinnatifid, with short obtuse lobes or subentire, hirsute; cauline oblong- or linear-lanceolate, entire, sagittate-clasping: fr. obovate, seldom 2 lines wide, strongly concavo-convex, glabrous or slightly tomentose, the marginal rays broad, dilated above, rather crowded, with narrow diaphanous spots (rarely a few perforations) between them. Var. (1) **involutus.** Taller and more strict: fr. elliptical, only a line wide; rays nearly obsolete, the purplish subscarious margin closely involute all around; style (rather prominent in fl.) deciduous. Var. (2) **pulchellus.** *T. pulchellus*, F. & M. (1835). Radical leaves merely toothed: pods densely tomentose; the wing rather broader.—The type of this species has not been found south of Mt. Shasta, except in Humboldt Co., *Marshall, Miss Bush*. The first variety is from Sonoma Co., *Bioletti*, and this may not improbably be found distinct. Var. 2 is our most common form in middle California.

2. **T. elegans,** F. & M. l. c. Hook. Ic. 39. Rather stouter, with fewer racemose branches: lower leaves ascending, repand-toothed: fr. 3—4 lines broad, of more rounded outline, nearly plane, the body densely tomentose, the rays separated by regularly ovoid perforations, and joined together beyond them into a very distinct diaphanous nearly entire margin.— Common on low hills of the interior valley in the neighborhood of alkaline or subsaline plains; also beyond the Sierra. Certainly a most distinct species in the character of its fruit, and of peculiar habitat, not being found at all in the Coast Range or toward the sea.

3. **T. laciniatus,** Nutt.; T. & G. Fl. i. 118 (1838): *T. ramosus*, Greene, Bull. Calif. Acad. ii. 390 (1887). Small and simple, or larger with many decumbent branches from the base, glabrous throughout and glaucous: leaves linear, entire, or with a few incised or opposite and divaricate narrow segments: fr. from elliptical with narrow margin, to almost orbicular with broad evenly crenate border, scarcely plano-convex, 1½–2½ lines broad, imperforate, or with irregular deep sinuses between the rays, or rarely with a few perforations, glabrous and very distinctly reticulate-venulose. From Mt. Diablo southward throughout the State, and eastward to New Mexico; very distinct in its vegetative characters, but the silicle variable.

4. **T. radians,** Benth. Pl. Hartw. 297 (1849). Glabrous, 1 ft. high: lower leaves runcinate-pinnatifid; cauline ovate-lanceolate, auriculate-clasping: silicle round-obovate, almost plane, 4–5 lines wide, tomentose, the rays narrowly linear, ending abruptly near the edge of the broad and otherwise diaphanous margin. Common in the interior valleys of Sonoma and Solano counties and northward.

25. **CORONOPUS,** *Ruellius*. Diffuse prostrate heavy-scented annuals, with pinnatifid leaves, and the general aspect of some species of *Lepidium*.

Flowers minute, greenish. Stamens often 2 only. Pods small, short, didymous, 2-celled; cells indehiscent, subglobose, when ripe separating from the persistent linear axis, strongly rugose, 1-seeded.

1. C. DIDYMUS, Smith, Fl. Britt. ii. 691 (1800); Linn. Mant. 92 (1767), under *Lepidium;* Pers. Syn. ii. 185 (1807), under *Senebiera.* Stems diffuse, 1_2 1^1_2 ft. long; the heavy-scented somewhat aromatic herbage more or less hirsute; leaves with small narrow segments; pod a line broad or more, emarginate at base and at summit, strongly reticulate.— Plentiful on bluffs overhanging the sea at Point Lobos; occasional at Berkeley, etc.

26. **RAPHANUS,** *Pliny* (RADISH). Coarse annuals, with large somewhat fleshy lyrate lower leaves, and loose racemes of purple or yellowish large flowers. Sepals erect, the two outer gibbous at base. Petals entire or emarginate, unguiculate. Pod indehiscent, elongated, somewhat moniliform or at least constricted between the seeds, long-beaked. Cotyledons enfolding the radicle.

1. R. SATIVUS, Linn. Sp. ii. 669 (1753). More or less hispid with scattered stiff hairs: fl. 8—10 lines long: petals purplish, with veinlets of darker color, rarely white or yellowish: pod thick, fleshy when young, spongy in maturity, 1 $2\frac{1}{2}$ in. long, 2-5-seeded. —The common Radish; long since become one of the prevalent and troublesome weeds in Californian fields everywhere; flowering and fruiting throughout the year.

27. **CAKILE,** *Serapius* (SEA ROCKET). Glabrous very succulent seaside annuals, with simple leaves and short racemes of smallish purple flowers. Sepals suberect, the two outer gibbous at base. Petals entire, unguiculate. Pod of 2 unequal joints, each 1-seeded, the upper and larger joint deciduous from the other. Seeds in the upper cell erect; in the lower pendulous; cotyledons usually accumbent.

1. **C. edentula,** Hook. Fl. i. 59 (1830); Bigel. Fl. Bost. 157 (1814), under *Bunias: C. Americana,* Nutt. Gen. ii. 62 (1818); Gray, Gen. Ill. i. 170. t. 74. A foot high or more, the stout stem and few ascending branches somewhat flexuous: leaves obovate, sinuately toothed: lower joint of silicle oblong, 3—4 lines long; upper twice as large, ovate, compressed and emarginate at apex.— Common along sandy beaches about the Bay of San Francisco at West Berkeley, Alameda, etc., also at Half Moon Bay; doubtless not rare on the coast and probably indigenous; but from the analogies of plant distribution in America where Old World genera are concerned, we should have expected the other species, *C. maritima* of Europe, to recur on the Pacific coast, rather than that the Atlantic American species should have found place here.

28. **TROPIDOCARPUM,** *Hooker.* Annuals, with light green pubes-

cent herbage, pinnatifid leaves, and loose leafy-bracted racemes of middle-sized yellow flowers. Sepals concave, spreading, equal at base. Petals spatulate-obovate. Stamens tetradynamous; anthers rounded. Silique sessile, elongated, more or less obcompressed, flat or inflated, without partition, indehiscent or the valves (2–4!) opening from above.—A remarkable genus, peculiar to California, and so closely linking this family to Capparideæ that it might quite as well be placed under that order, especially in view of the second species.

1. **T. gracile**, Hook. Ic. t. 43 (1836). Erect, very slender, usually only a few inches high, nearly glabrous: leaves linear, with opposite pairs of linear segments, the floral similar but reduced: stamens very unequal, all exceeding the short pistil: silique linear, 2 in. long, glabrous, flat, indehiscent: seeds in 2 rows. Var. **scabriusculum**. *T. scabriusculum*, Hook. l. c. 52. Much larger, with many decumbent branches, and roughish-pubescent throughout, even to the pods.—Foothills of both ranges of mountains; and on the plains.

2. **T. capparideum**, Greene, 217 (1888). Usually erect, less than a foot high, simple, or with few ascending branches, the stem stoutish but hollow: pods 1½–¾ in. long, linear-oblong, inflated, 2 lines wide, slightly obcompressed (the cross section transversely elliptical), conspicuously 6-nerved: valves 4, 2 deciduous and 2 persistent, the dehiscence beginning at the apex: seeds in 4 rows, *i. e.*, 1 row along either margin of each of the 2 persistent valves.—Abundant in low alkaline soil about the Byron Springs, and near Bethany, at the eastern base of Mt. Diablo.

Order XLIV. FUMARIACEÆ.

De Candolle Systema Naturale Regni Vegetabilis, ii. 105 (1821).

Tender glabrous often glaucous herbs, with a bland watery juice, alternate pinnately or ternately divided or dissected leaves without stipules, and racemose white purple or yellow flowers. Sepals 2, small, deciduous. Petals 4, in 2 dissimilar pairs; one or both of the outer ones saccate at base; inner pair cohering by the callous apex and enclosing the anthers and stigma. Stamens 6, hypogynous; filaments in 2 parcels placed opposite the outer petals, usually diadelphous; anther of the middle stamen in each parcel 2-celled, those of the lateral 1-celled. Ovary of 2 united carpels, 1-celled, with 2 parietal placentæ; style filiform. Fruit a several-seeded siliquose 2-valved 1-celled capsule, or indehiscent. A small family, in some points closely analogous to Cruciferæ; but more related to the Papaveraceæ, from which their irregular flowers not very sufficiently distinguish them.

1. **CAPNORCHIS**, *Boerhaave*. Perennials, with tuberiferous or gran-

ular or scaly subterranean stem or crown, fibrous rootlets, ternately or pinnately compound leaves, and racemose or paniculate flowers; the corolla often persistent over the mature fruit. Corolla flattened and cordate; the 2 outer petals larger, saccate or spurred at base.

* *Flowers on a scape; seeds shining.*

1. **C. formosa,** O. Ktze. Rev. Gen. 15 (1891); Andr. Bot. Rep. vi. t. 393 (1800), under *Fumaria;* Pursh, Fl. ii. 462 (1814), under *Corydalis;* DC. Syst. ii. 109 (1821), under *Dielytra.* Rootstock rather large, creeping, nearly naked; leaves and scapes 2 ft. high, the former twice or thrice pinnately compound, the final divisions incisely pinnatifid; fl. compound-racemose at summit of the naked scape: corolla rose-purple, ovate-cordate, with short spreading tips to the larger petals. Common in woods of the Coast Range and Sierra Nevada, in middle and northern parts of the State, and far northward.

2. **C. uniflora,** O. Ktze. l. c.; Kell. Proc. Calif. Acad. iv. 141 (1871), under *Dicentra;* Greene, Pitt. i. 187 (1888), under *Dielytra.* Roots fleshy and fasciculate, the cluster surmounted by bulb-like fleshy grains, and sending up leaves and a scape 3–5 in. high: blade of leaves of ovate outline, ternately or pinnately divided, the few segments pinnatifid, into few linear-oblong or spatulate lobes: fl. mostly solitary at summit of scape, narrowly oblong-cordate, flesh-color, the 2 outer petals tapering above, at length reflexed. At rather high elevations in the Sierra, near Cisco, etc.

3. **C. pauciflora,** Greene. Wats. Bot. Calif. ii. 429 (1880), under *Dicentra;* Greene, Pitt. l. c., under *Dielytra.* Rootstock running and tuberiferous: scape and leaves very slender, 4–8 in. high, the latter biternate with very narrow segments: fl. 1–3, rose-color; spurs of outer petals stout, straight, not divergent; tips of same spreading or reflexed. — A species of the high mountains west of Mt. Shasta; to be expected in the Trinity Mts.

* * *Flowers panicled or thyrsoidly arranged on a rigid leafy stem; seeds dull.*

4. **C. chrysantha,** Planch. Fl. Serr. viii. 193. t. 820 (1853); H. & A. Bot. Beech. 320. t. 73 (1840), under *Dielytra;* B. & W. Bot. Calif. i. 24 (1876), under *Dicentra.* Pale and glaucous, 2–5 ft. high: leaves bipinnate, the larger 1 ft. long or more; the divisions cleft into few narrow lobes: racemose panicle 1–2 ft. long: corolla linear-oblong, only slightly cordate, golden-yellow: capsule oblong-ovate.— On dry hills from Lake Co., and through the Mt. Diablo Range, to the southern part of the State.

5. **C. ochroleuca,** Greene. Engelm. Bot. Gaz. vi. 223 (1881), under *Dicentra;* Greene, Pitt. i. 187 (1888), under *Dielytra.* Lower and rather stouter than the last: inflorescence thyrsoidly condensed; the somewhat

larger corollas cream-color. — A southern species, but occurring north of Santa Barbara.

2. **CAPNODES,** *Moehring.* Plants differing from *Capnorchis* only by an inequality of the 2 outer petals, only one of which is spurred or saccate at base. Our species are leafy-stemmed, not scapose.

1. **C. Caseanum,** Greene. Gray, Proc. Am. Acad. x. 69 (1874), under *Corydalis.* Perennial, glaucescent, 2–3 ft. high, branching: leaves twice or thrice pinnate; leaflets obovate or oblong, $\frac{1}{2}$ in. long, subsessile, some of them more or less confluent: racemes erect, dense, 3–5 in. long: corolla white or cream-color, with bluish tips to the petals; the straight spur $\frac{1}{2}$ in. long, horizontal or ascending, very obtuse, exceeding the rest of the flower: capsule oval or oblong, turgid, tipped with a slender style: seeds shining. In the Sierra Nevada from near Truckee northward.

2. **C. Bidwellianum,** Greene. Wats. Bot. Calif. ii. 429 (1880), under *Corydalis.* Resembling the last in size and habit; leaflets smaller, acute or acuminate: spur slender, slightly curved. — Same range as the last; first found above Cisco, by *Mrs. Gen. Bidwell.*

Order XLV. PAPAVERACEÆ.

Jussieu, Genera. 236 (1789). Papavera, Adanson, Fam. ii. 425 (1763).

Herbs (*Dendromecon* shrubby) with a colored or milky narcotic juice, commonly glaucous foliage, and mostly solitary showy 4-merous or 6-merous flowers. Sepals 1, 2 or 3, caducous. Petals 4–6, crumpled in the bud. Stamens 6–∞, usually hypogynous (in *Eschscholtzia* mostly epipetalous); anthers innate. Pistil compound and the ovary becoming a capsule, or the carpels nearly distinct, maturing as almost follicular pods. Seeds ∞; albumen fleshy or oily; embryo minute, straight. — A not very large order, but of importance as yielding the opium of commerce; many of the species valued in cultivation as ornamental plants.

1. **PAPAVER,** *Pliny* (Poppy). Glaucescent more or less hispid herbs (ours all annual), with milky juice, alternate lobed or dissected leaves, solitary long-peduncled flowers nodding in the bud. Sepals 2. Petals 4. Ovary with 4 or more intruded placentæ which partially divide the interior of the obovoid or subglobose capsule; this opening by short roundish or triangular apertures near the summit between the parietal ribs: stigma 4–8-lobed, sessile and the lobes radiating over the summit of the ovary and capsule, or raised on a short style and the lobes capitate-recurved. Seeds ∞, small, scrobiculate or reticulate.

1. **P. Californicum,** Gray, Proc. Am. Acad. xxii. 313 (1887); Greene.

FLORA FRANCISCANA.

PART III.

INDEX OF ORDERS.

	PAGE.		PAGE.
PAPAVERACEÆ, - - -	280	UMBELLIFERÆ, - -	313
NYMPHÆEÆ, - - -	287	CORNEÆ, - - - -	335
SARRACENEÆ, - - -	289	ELÆAGNEÆ, - - -	338
DROSEREÆ, - - -	289	DAPHNOIDEÆ, - - -	338
LAURINEÆ, - - -	290	SANTALACEÆ, - -	339
BERBERIDEÆ, - -	291	LORANTHEÆ, - - -	340
RANUNCULACEÆ, - -	293	CAPRIFOLIACEÆ, - -	342
SARMENTOSÆ, - -	311	RUBIACEÆ, - - -	348
ARALIACEÆ, - - -	312	VALERIANEÆ, - -	352

[Issued 1st April, 1892.]

Price, Fifty Cents.

DOXEY & Co., San Francisco : WILLIAM WESLEY & SON, London.
CUBERY & COMPANY, San Francisco.

AN ATTEMPT

TO CLASSIFY AND DESCRIBE THE VASCULAR PLANTS

OF MIDDLE CALIFORNIA.

BY

EDWARD L. GREENE,

Associate Professor of Botany, in the University of California.

SAN FRANCISCO:
CUBERY & CO., PRINTERS, 587 MISSION STREET, BELOW SECOND.
1892.

Copyright, 1891,

By EDWARD L. GREENE.

Pitt. i. 16. Sparsely pilose-pubescent, 1- 2½ ft. high, leafy below; leaves pinnately parted or divided into acutish toothed or 3-lobed or entire segments: peduncles elongated: corolla 2 in. broad; petals brick-red, with a green spot at the base bordered with rose-red: capsule ½ in. long or more, clavate-turbinate, 6- 11-nerved; stigmas sessile and radiating, forming a flat cap to the pod; the short valvular openings somewhat quadrate: seeds coarsely and faintly reticulate.—Summit of the Santa Inez Mts. and northward.

2. **P. Lemmoni,** Greene, Pittonia, i. 168 (1888). Near the preceding, but larger, 1-3 ft. high: corolla 1-3 in. broad, apparently of a deeper red, the base of the petals green: capsule broader and merely obovate; stigmas 7-10, their lower half sessile and radiant upon the pod, the upper half coherent and forming a conical apiculation. Mountains of San Luis Obispo Co.; also said to occur in Marin Co. north of Mt. Tamalpais.

3. **P. heterophyllum,** Greene, l. c.; Benth. Trans. Hort. Soc. 2 ser. i. 408 (1834), under *Meconopsis*. Aspect and size of the two preceding, but the segments of the pinnately divided leaves singularly variable upon the same leaf, some linear, others in close juxtaposition oval; stigmas capitate at summit of a distinct and slender style. Common in middle California, on wooded slopes, and bearing large nodding flowers; or in open fields among growing grain, with small erect flowers; or else these forms represent two species, *M. heterophylla* and *crassifolia*, of Bentham; but from his descriptions no one can decide to which plant belongs either name; moreover, we seem not to have any Californian plant with such capsule as is attributed to *M. heterophylla* in Hooker's plate (Ic. Pl. t. 272). In both our forms the pod is as broadly obovoid, and the openings as small and pore-like as in *P. Californicum* itself.

2. **ARGEMONE,** *Tournefort.* Stout prickly herbs, with sinuate-pinnatifid prickly-toothed leaves, and large short-peduncled white or yellow flowers. Sepals 2 or 3, spinosely beaked. Petals 4-6. Stamens ∞ ; filaments filiform; anthers linear. Ovary oblong, with 3-6 nervelike placentae; stigmas nearly sessile, radiating. Capsule oblong, prickly, 1-celled, opening at summit by 3-6 valves separating from the parietal ribs. Seeds ovoid-globose, pitted.

1. **A. munita,** Dur. & Hilg. Pac. R. Rep. v. 5. t. 1 (1855): *A. hispida*, B. & W. Bot. Calif. i. 21, not of Gray. Perennial, stout, erect, 2½ ft. high, very glaucous, glabrous under the dense armature of straight, spreading or retrorse white prickles: leaves elongated-oblong, cordate-amplexicaul, sinuately lobed: fl. few, terminal, white, 3-4 in. broad; sepals 3: petals 6: ovary densely prickly, the prickles erect. Widely dispersed in the mountain districts, but absent from the Bay region; quite distinct from *A. hispida* of the Great Basin, which has a divided

and subdivided foliage, besides an abundant short-setose pubescence under the armature.

3. PLATYSTEMON, *Bentham.* Annual glaucescent glabrous or hirsute herbs, with entire leaves; the cauline opposite or verticillate. Flowers rather small, white or cream-colored, on slender peduncles. Sepals 3, caducous. Petals 6. Stamens 6 ∞; filaments filiform or flattened; anthers oblong to linear. Carpels 3 ∞, in maturity variously more or less united, or quite distinct. Seeds smooth and shining.- As here received this is a good genus as to agreement of the species in habit and floral characters; but the gynœcium is extremely different in the different species, and even within the limits of the same species; so that no constant characters for these genera are found; and *Meconella* has better claims to generic rank than *Platystigma*.

* *Carpels 6—25, torulose, jointed between the seeds, usually distinct, but not rarely coherent and forming a central seed-bearing cavity which is open at top; stigmas linear.*—PLATYSTEMON proper.

1. **P. Californicus,** Benth. Trans. Hort. Soc. n. ser. i. 405 (1834); Lindl. Bot. Reg. t. 1679; Hook. Bot. Mag. t. 3579. Branching from the base, 6 -12 in. high, sparingly hirsute: lowest leaves alternate; cauline opposite, all linear, entire, 2—4 in. long, sessile or clasping, obtuse: flower-buds ovoid; sepals hirsute: petals $\frac{1}{2}$ in. long or more, pale yellow with a deep greenish-yellow spot at base, sometimes reddish-tinged on the outside; stamens ∞; filaments flattened and ligulate: torulose carpels breaking transversely into 1-seeded indehiscent joints. Var. **crinitus.** *P. crinitus*, Greene, Pitt. ii. 13 (1889). Less branching, nearly acaulescent, the whole plant even to the calyx densely crinite-hirsute: flower-buds globose: petals of a deep greenish yellow almost throughout: torulose pods very short, scarcely longer than the stigmas.—Throughout the western parts of the State, and extremely variable as to size, color and duration of the petals, and as to fruit-structure. Far southward the pods, few and wholly distinct, are, like the flower buds, nodding. In the middle sections the fruit is always erect, and the follicular jointed pods sometimes cohere to form a capsular cavity which itself bears a few naked seeds. This form was first detected by the author in Contra Costa Co., where it is prevalent. The variety is of very singular aspect; but under cultivation at Berkeley, far from its home (mountains of Kern Co.) it scarcely retains the characters I had assigned it as specific.

* * *Carpels 3 only, partly united and forming a 3-lobed 1-celled ovoid capsule open at top; stigmas ovate.*—Genus PLATYSTIGMA, Benth.

2. **P. linearis,** Curran, Proc. Calif. Acad. 2 ser. i. 242 (1888); Benth. Trans. Hort. Soc. n. ser. i. 407 (1834), under *Platystigma.* Acaulescent, 3 -12 in. high, sparsely hirsute: leaves narrowly linear, 1—3 in. long,

acutish: peduncles scapiform, very slender: fl. ½ 1 in. broad: petals as in the last: stamens ∞ ; filaments filiform or flattened: capsule ovate-triquetrous, ½ in. long or more. Less common than the preceding, preferring gravelly hills toward the sea, or bluffs of rivers in the interior: the larger forms showing the ligulate-dilated filaments of n. 1, from which species only the acaulescent habit and the strictly capsular fruit separate it. Mar.- May.

* * * *Carpels 3 only, united, forming a slender elongated and twisted 1-celled capsule; stamens few and definite; stigmas linear.*

Genus MECONELLA, Nutt.

3. **P. Torreyi.** *Meconella Californica*, Torr.; Frem. 2d Rep. 312 (1845). *Platystigma Californicum*, Boland. Cat. 4 (1870). Erect, slender, dichotomous from the base, 3- 8 in. high, glabrous: lowest leaves ovate-spatulate or oblanceolate; upper linear, acute, entire, ½ 1 in. long: fl. ½-1 in. broad, white: stamens (usually 12) in 2 circles; filaments dilated upwards, those of the outer circle conspicuously shorter than those of the inner: capsule narrowly linear, ¾- 1½ in. long. –The plant of the Sierra foothills (Mokelumne Hill, *Bigelow;* Rose Springs, *Mrs. Gates*) is much more slender than that of the San Francisco region, and has flowers only half as large. This, however, can not be referred to the Oregonian type of *Meconella*, for that has but a single set of stamens. In our plant, whether of the Sierra or of the Coast Range, these, though sometimes only 8 or 10, are always in two sets, the outer circle with short, the inner with long filaments.—Mar. -May.

4. **P. denticulatus,** Greene; Bull. Calif. Acad. ii. 59 (1886), under *Meconella;* Bull. Torr. Club, xiii. 218, under *Platystigma*. More slender, 3 10 in. high, simple below, ternately branching above the middle: lowest leaves in a depressed whorl, the rhombic-ovate acute blade shorter than the linear petiolar basal part: cauline in whorls of 3 or more, spatulate or linear, obtuse, remotely and saliently dentate: corolla ¼- ½ in. long: stamens 6 9, in one set and all equal; filaments flattened, but broader below. A very distinct species, common southward beyond our limits, but reaching Monterey Co., *Hickman*. At the time of its publication, Dr. Gray wrote that he had been about to publish it as *Platystigma Clevelandi*. Mar., Apr.

4. **DENDROMECON,** *Bentham*. Shrubs with alternate coriaceous entire leaves, and solitary rather large yellow flowers. Sepals 2. Petals 4. Stamens ∞ ; filaments filiform, short: anthers linear. Ovary linear; style short; stigmas 2, short and erect. Capsule linear, many-nerved, 1-celled, 2-valved, the valves dehiscent somewhat elastically from base to apex. Seeds ∞ , oblong or globose, "finely pitted" (smooth in the extra-limital insular species, *D. flexile*).

1. **D. rigidum**, Benth. Trans. Hort. Soc. 2 ser. i. 407 (1834); Hook. Ic. t. 37; Torr. Bot. Mex. Bound. t. 3. Shrub 2 - 8 ft. high, with many rigid ascending branches and slender branchlets; bark whitish: leaves ovate-to linear-lanceolate, 1—3 in. long, very acute or mucronate, vertical, the very short petiole being twisted, the margin scabrous-denticulate: fl. 1 3 in. broad, the petals nearly rotate-spreading: capsules slightly arcuate, $1\frac{1}{3}$ $2\frac{1}{2}$ in. long: seeds $1\frac{1}{2}$ lines long. - In clayey or gravelly soil among the foothills of both mountain ranges, but more common in the Coast Range. Mar. June.

5. **ESCHSCHOLTZIA**, *Chamisso*. Glabrous and more or less glaucous flaccid herbs, with colorless bitter juice (that of the roots reddish), and alternate dissected leaves. Flowers solitary, yellow or orange-colored. Calyx a synsepalous oblong or conical mitre-like organ deciduous from the more or less funnelform-dilated and variously rimmed torus which bears the 4 petals. Stamens mostly ∞; filaments very short, more or less firmly attached to the base of the petals; anthers linear or oblong, usually longer than the filaments. Ovary linear; style very short; stigmas 2 or more, subulate-filiform. Capsule 10-nerved, 1-celled, ∞-seeded, 2-valved, the valves elastically dehiscent from base to apex, forcibly ejecting the seeds; these spherical, reticulate, tuberculate, or rarely pitted; cotyledons either linear and entire, or deeply bifid into narrowly linear segments.

* *Torus with an erect hyaline inner, and a spreading outer and herbaceous rim; cotyledons deeply bifid, their linear segments divergent.*

← *Perennials.*

1. **E. Californica**, Cham. Hor. Phys. Berol. 73. t. 15 (1820); Linnæa, i. 554 (1826); Raf. Fl. Tell. ii. 92 (1836), under *Omonoia;* T. & G. Fl. i. 63 (1838), under *Chryseis*. Glabrous, glaucescent, the stems decumbent or at length procumbent, 1- 2 ft. long, regularly dichotomous below, above bearing a flower opposite each leaf: leaves ternately dissected, the ultimate segments linear, obtuse: calyx oblong or ovoid, abruptly short-pointed; torus-rim broad: petals about $\frac{3}{4}$ in. long, bright yellow with an orange spot at base: pods small for the size of the plant (2 in. long): seeds conspicuously reticulate.—This, the type of the genus, is found only along the seaboard, in sandy soil, about San Francisco, Monterey, etc.; flowering almost all the year round. We doubt if it has ever been in cultivation; and it is certain that all or nearly all of the so-called *E. Californica* of the seedsmen's and even of the botanists' catalogues belongs to the next.

2. **E. Douglasii**, Walp. Rep. i. 116 (1843); H. & A. Bot. Beech. 320 (1840), under *Chryseis; E. Californica*, Lindl. Bot. Reg. t. 1168 (1828), not Cham. Tufted stems decumbent, 1 ft. high or more, obscurely if at all dichotomous; peduncles more elongated, less regularly opposite the

leaves: calyx ovate-acuminate; outer rim of torus narrow, not exceeding the erect inner one, in age closely deflexed: petals 1 in. long or more, deep yellow, shading into orange at base.—Originally from the plains of the Columbia, this perfectly distinct species is common in Humboldt Co., perhaps even farther southward on the western side. It is not found at all on the plains, but is in the Sierra as far south as Auburn, *Miss Harrison*. Late in summer the flowers, reduced to half their spring size, are apt to be wholly orange-colored, when the species will be easily confounded with the next.

3. **E. crocea,** Benth. Trans. Hort. Soc. 2 ser. i. 407, also Lindl. Bot. Reg. t. 1677 (1834). Stouter than the last, erect or decumbent, the herbage of a deeper green and scarcely glaucescent: fl. fewer, often strictly terminal: calyx large (1 in. long or more), long-conical; outer rim of torus very broad, more or less undulate: petals $1\frac{1}{2}$–2 in. long, deep orange throughout. The most common middle Californian species; abundant in valleys, open plains, and on hillsides in the western parts of the State, in March and April often coloring the landscape for miles with its profusion of orange-colored bloom. The later and reduced flowers are paler than those of spring, whereas in the preceding these are of a deeper color. The buds also of the later and starved specimens of *E. crocea* lose much of their conical shape and approach those of *E. Douglasii;* but the broad torus-rim of the present species is a constant character; and while much more abundant than the other in general, the two do not encroach upon each other's territory.

4. **E. glauca,** Greene, Pittonia, i. 45 (1887). Erect or decumbent, very slender, 2–4 ft. high, very glaucous: leaf-segments linear, less divergent than in the preceding: calyx slender-conical, varying in late specimens to ovate with slender acumination; outer torus-rim narrow, spreading: petals 1—$1\frac{1}{2}$ in. long, light yellow with a very distinctly rhombic deep orange spot at base.—Species exceedingly well marked by a certain not well definable grace of its very beautiful white-glaucous foliage: discovered by the author on Santa Cruz Island, but found on the mainland in the mountains near Santa Cruz, by *Dr. Parry*. The dry-season flowers in this species become reduced in size, but undergo no change of color.

++ *Annuals.*

5. **E. compacta,** Walp. Rep. i. 116 (1843); Lindl. Bot. Reg. t. 1948 (1837), under *Chryseis: E. tenuisecta,* Greene, Pitt. i. 169 (1888). Annual, erect, 1–2 ft. high, glabrous, light green, more or less glaucescent: leaves finely dissected, the ultimate segments linear-cuneiform, 3-toothed or -cleft at the broad apex: calyx very thin and partly diaphanous, slender-conical; outer torus-rim broad, thin: petals $\frac{3}{4}$–$1\frac{1}{2}$ in. long, light yellow, shading into orange below the middle: segments of coty-

ledons almost filiform.—From Butte Co. southward on the plains of the interior, and also in the hill country about San Bernardino. A strictly annual species, the early state of which appears with one or more almost scapiform long peduncles rising from a compact tuft of radical leaves. Lindley's figure of *E. compacta* may be conceived to represent the maturer state of a plant which I in 1888 described as new, but the plate does not indicate the peculiarly thin and delicate texture, nor the very slenderly attenuate form of the calyx as displayed in my type; but specimens from the plains of the lower San Joaquin are more at agreement with the figure. Mar.–May.

* * *Annuals; the torus with no spreading rim (except in n. 6); cotyledons entire.*

+— *Branching and leafy above; peduncles subterete, 8-striate.*

6. **E. ambigua.** Slender, branching from the base, decumbent, glaucous and scabrous-puberulent throughout, 1 ft. high or less: leaves small, ternately dissected, the ultimate segments short, approximate in threes: calyx ovate-acuminate, about 4 lines long, or 5: torus small, but with ample rim: petals deep yellow, 1 in. long or less: pods 1 in. long or more: mature seeds unknown.—In fields near Cholame, San Luis Obispo Co., *Mr. & Mrs. Lemmon*. Species connecting the two very natural sections of the genus. The general aspect is that of *E. hypecoides;* the conspicuously rimmed torus forbids its being joined to that species, and would bring it nearer to Lower Californian *E. peninsularis*, to which, however, it can by no means be referred.

7. **E. hypecoides,** Benth. Trans. Hort. Soc. l. c. (1834): *E. Austinæ*, Greene, Bull. Calif. Acad. i. 69 (1885). Scabrous- or even hirsute-pubescent below, glabrous above, glaucescent: branches many from the annual root, decumbent at base, 1 ft. high or less, slender, sparingly leafy: leaves small; segments rather few, linear-cuneiform: calyx oblong-conical, $\frac{1}{2}$ in. long, abruptly slender-pointed: torus short-tubular, $1\frac{1}{2}$ lines deep; outer margin a mere herbaceous ring, the inner erect, hyaline: petals 1 in. long: seeds with a faint irregular reticulation.—Widely dispersed along the foothills from southern Oregon, *Howell*, to Santa Cruz and Kern counties in this State. In cultivation at Berkeley, this is greatly admired on account of the rich profusion of small flowers. It was at first mistaken by me for a perennial. The early flowering state exhibits few and somewhat scapiform long peduncles. Apr.–June.

8. **E. minutiflora,** Wats. Proc. Am. Acad. xi. 122 (1876). Size and habit of the preceding, though more slender, more freely branching and with shorter peduncles: fl. very small (only $\frac{1}{4}$ in. broad): capsule very slender, $1\frac{1}{2}$ in. long: seeds nearly smooth, scarcely $\frac{1}{2}$ line thick. Of the desert regions along the eastern base of the Sierra; probably in Mono and Inyo counties.

+— +— *Subacaulescent; scapose peduncles quadrangular.*

9. **E. rhombipetala,** Greene, Bull. Calif. Acad. i. 71 (1885); Gray, Proc. Am. Acad. xxii. 273 (1887). Glaucous and tuberculate-scabrous throughout; stemless or the stems stout, depressed, very leafy, the stout 4-angled peduncles little exceeding the subradical leaves: torus subcylindrical, with 2 minute approximate scarious margins: petals 1½ in. long, rhombic-ovate, fugacious: capsules very large for the plant (3–4 in. long): seeds large, very distinctly and regularly favose-reticulate. A very common plant in grain-fields along the eastern foothills of the Vaca Mts. and the Mt. Diablo Range. Feb.—May.

10. **E. Lemmoni,** Greene, W. Am. Sc. iii. 157 (1887). Habit of the last, or rather more slender, the scapose peduncles longer (6–12 in. high), the whole plant, even to calyx and capsules, almost hoary-pubescent: torus urceolate, 3–4 lines long, nearly glabrous, constricted just below the narrow double hyaline border: calyx ovate, long-acuminate: petals orange, about 1 in. long. -Fields near Cholame, San Luis Obispo Co., *Mr. & Mrs. Lemmon.*

11. **E. cæspitosa,** Benth. Trans. Hort. Soc. l. c. 408 (1834): *E. tenuifolia,* Benth. l. c.; Hook. Bot. Mag. t. 412; Greene, Bull. Calif. Acad. l. c. Very slender, stemless, more or less scabrous or hirsutulous, at least near the base: earliest leaves simple and very narrowly linear, the succeeding ones merely bifid or trifid, the later more compound: scapes 3–8 in. high: corolla 1 in. broad, rotate-spreading: seeds more or less densely muricate-squamose and bur-like. Common in the foothills of the Sierra from Butte Co. southward to Fresno. A small but beautiful plant with the best of specific characters; although the very different *E. hypecoides* was strangely confused with it by Dr. Gray in his latest dissertation on these plants.

Order XLVI. **NYMPHÆEÆ.**

Salisbury, in Koenig & Sim's Annals of Botany, ii. 70 (1805).

Aquatic perennial herbs, often with milky juice. Leaves alternate, peltate or cordate, involute from both margins in the bud, commonly floating when mature, and, like the naked 1-flowered peduncles, arising from a submersed stem or stout horizontal rootstock. Sepals and petals distinct (in our genera hypogynous). Stamens few or many. Carpels distinct, or united and forming a capsular fruit.

1. **NYMPHÆA,** *Theophrastus* (Yellow Pond-Lily. Spatterdock). Rootstocks stout, creeping in the muddy beds of lakes or sluggish streams. Leaves large, leathery, cordate. Sepals 6—12, imbricated, rounded and concave, yellowish or reddish within. Petals and stamens numerous,

short, hypogynous, crowded around the base of the ovary; filaments short; anthers truncate, opening toward the ovary by linear slits. Ovary oblong or ovate, 12—20-celled; the sessile broad flat stigma with as many radiating striæ. Seeds without an aril.

1. **N. advena,** Ait. Kew. ii. 226 (1789), and 2 ed. iii. 295 (1811), under *Nuphar*. Rhizome several feet long, 2 in. thick, creeping, rooting from beneath, the upper side marked with scars of former petioles: leaves cordate, with narrow or closed sinus, $\frac{1}{2}$- 1 ft. long, $\frac{1}{2}$- $\frac{3}{4}$ ft. wide, floating or slightly raised above the water on the stout usually semiterete petioles: sepals 6, the 3 outer less than 1 in. wide, nearly orbicular, greenish, the 3 inner more than twice as large, narrowed at base, yellow: petals about 15, small, concealed beneath the many stamens: ovary ovoid; stigma sessile, $\frac{1}{2}$—$\frac{3}{4}$ in. broad, 15-rayed: seeds oval, light brownish, smooth and shining. In deep sloughs about Stockton, *Sanford;* the flowers often fully five inches in diameter. May, June.

2. **N. polysepala,** Greene, Bull. Torr. Club, xv. 84 (1888); Lawson, Trans. Roy. Soc. Canad. iv. 120; Engelm. Trans. St. Louis Acad. ii. 282 (1865), under *Nuphar*. Near the preceding, but leaves of broader and more rounded outline: sepals 9—12, all but the outer of a rich brownish red: rays of stigma 15—21, the margin somewhat crenate. In the typical form, or near it, at Eureka, Humboldt Co., and in our northern mountain lakes generally; but common toward the coast in the vicinity of San Francisco, etc. with fewer and less highly colored sepals, thus approaching *N. advena* rather too closely.

2. **BRASENIA,** *Schreber* (WATER-SHIELD). Roots fibrous. Submersed stems, petioles, etc. coated with a colorless transparent jelly. Leaves centrally peltate, floating. Flowers pedunculate, purplish. Sepals 3 or 4. Petals 3 or 4. Stamens 18—36. Ovaries 6—18, becoming oblong-ovate indehiscent and somewhat drupaceous 1-seeded fruits.

1. **B. peltata,** Pursh, Fl. ii. 389 (1814); Thunb. "Acta. Ups. vii. 142. t. 14," under *Menyanthes;* F. v. Muell. Pl. vict. 15 (1874), under *Cabomba*. *Hydropeltis purpurea*, Michx. Fl. i. 324. t. 29 (1803). Stems of variable length, 1—10 ft.: leaves of a dark purplish green, oval or elliptical, 2—5 in. long, glabrous and shining above, gelatinous beneath, on petioles 6—15 in. long: fl. brownish purple, scarcely 1 in. broad; petals much like the sepals but a trifle longer and thinner: stamens dark purple: carpels oblong, acuminate and tipped with the persistent style.—A curious aquatic, widely dispersed in the northern hemisphere; occurring also in Australia; not common on the Pacific slope of N. America, but plentiful in deep sluggish waters about Stockton, *Sanford*, and also detected in Clear Lake, Lake Co., *Bolander*, as well as in the northern Sierra, *Brewer*.

Order XLVII. SARRACENEÆ.

Dumortier, Analyse des Familles des Plantes, 53 (1829).

A singular family of acaulescent herbs with hollow pitcher-like leaves; here represented by the monotypical genus

1. CHRYSAMPHORA, *Greene* (DARLINGTONIA). Scapose 1-flowered stem, and long yellowish trumpet-shaped leaves from horizontal rootstocks. Calyx of 5 narrowly oblong imbricated sepals, persistent. Petals 5, ovate-oblong, erect, with a small ovate tip to the oblong main part. Stamens 12–15, in one series, hypogynous; filaments subulate; anthers oblong, of 2 unequal cells turned edgewise by a twist of the filament. Ovary somewhat turbinate, being dilated towards the truncate or concave summit, exceeding the stamens, 5-celled; the cells opposite the petals; style short, 5-lobed, the lobes recurved; stigmas thickish, introrsely terminal. Capsule loculicidally 5-valved. Seeds ∞, ovate-clavate, thickly beset with short slender projections.

1. **C. Californica,** Greene. Pittonia, ii. 191 (1891); Torr. Sm. Contr. vi. 4. t. 1 (1853), under *Darlingtonia*. Pitcher-like leaves $1\frac{1}{2}$–$2\frac{1}{2}$ ft. high, very gradually dilated upwards, somewhat spirally striate, the uppermost portion reticulate or mottled; the vaulted lid or hood ending in a forked foliaceous appendage; the body of the pitcher also appendaged by a wing 2–4 lines wide running down one (the inner) side; scape exceeding the leaves, scaly-bracted; fl. nodding; the yellowish sepals expanding; the purplish or reddish petals suberect; the whole flower 2–3 in. broad. In bogs, at middle elevations of the Sierra, chiefly northward, about Mt. Shasta, etc. May.

Order XLVIII. DROSEREÆ.

Salisbury, Paradisus Londinensis, 95 (1807). DROSERACEÆ, De Candolle, Théorie Élémentaire, 214 (1813).

An order of small bog herbs, represented here by two species of the typical genus,

1. DROSERA, *Linnæus* (SUNDEW). Low perennials or biennials, the purplish or brownish herbage beset with bristles whose gland-like tips secrete a drop of glistening viscous liquid. Leaves (in ours radical and rosulate-tufted) petiolate, with a villous stipular fringe at base; the blade inflexed or involute in bud. Flowers in an unilateral scorpioid raceme or spike which is bracted, but the flowers not in the axils of the bracts. Calyx 5-parted, persistent, the segments imbricate in bud.

Petals 5, convolute in bud. Stamens 5, hypogynous. Styles mostly 3, each 2-parted, the filiform or clavate forks stigmatose down the inner side. Capsule oblong, 1-celled, with 3 parietal placentae, 3-valved. Seeds ∞, small.

1. **D. rotundifolia**, Linn. Sp. Pl. i. 281 (1753). Leaves spreading; the rounded blade 2–6 lines broad, abruptly narrowed to the slender petiole: scape 3–6 in. high, few-flowered: petals oblong, 2 lines long, little exceeding the oblong sepals: styles very short: capsule included in the calyx: seeds linear, with a loose testa. In cold swamps of the mountains, from Mendocino Co. northward; also in the Sierra.

2. **D. Anglica**, Huds. Fl. Angl. 135 (1798). Leaves ascending, cuneate-oblong, narrowed to the slender petiole: scape sometimes forked at the tip, few-flowered: petals linear-oblanceolate, 3–4 lines long, nearly twice as long as the oblong sepals: capsule exceeding the calyx: seeds linear, with loose testa.—In the Sierra Nevada northward, *Lemmon*.

Order XLIX. LAURINEÆ.

Ventenat, Tableau du Règne Végétal, ii. 245 (1799).

An extensive and important family of aromatic trees and shrubs. We have but one species.

1. **UMBELLULARIA**, *Nuttall* (CALIFORNIA LAUREL). Arborescent, evergreen. Leaves alternate, petiolate, coriaceous, entire, exstipulate, highly odoriferous. Flowers perfect, in peduncled terminal and axillary small capitate umbels: these in bud enclosed within an involucre of about 4 broad caducous bracts. Perianth with no tube; segments 6, the 3 outer enfolding the others, all deciduous. Stamens 9; the outer series (6) spreading, the inner (3) erect and near the pistil; a circle of 6 stout short-stipitate glands intervening between the 2 series; anthers 4-celled, the cells of the outer series introrse, those of the inner extrorse. Fruit drupaceous, inserted on the enlarged and thickened base of the calyx.

1. **U. Californica**, Nutt. Sylv. i. 87 (1842); Mez, Laur. 482 (1889); H. & A. Bot. Beech. 159 (1840), under *Tetranthera*; Nees, Syst. 463 (1836), under *Oreodaphne*. Tree 10–75 ft. high, the growing twigs and the inflorescence very minutely puberulent: leaves oblong-lanceolate, 2–4 in. long, short-petioled, bright green and shining: peduncles 1½–1 in. long: pedicels of the 5–10 flowers 1–5 lines; involucral bracts imbricate: sepals yellowish-green, 1½–2 lines long, oblong: drupes on stout peduncles, ovoid or subglobose, about 1 in. long, dark purple, the pulp and putamen thin.—Common throughout the State, chiefly along streams in the mountains and among the hills; the foliage pungently aromatic.

Order L. **BERBERIDEÆ.**

Ventenat, Tableau du Règne Végétal, iii. 83 (1799).

Shrubs or herbs, with alternate or radical exstipulate leaves (traces of stipules in our species of *Berberis*). Sepals and petals 3 or 6 each (in *Achlys* none), hypogynous. Stamens 6 or 9; anthers opening by valves hinged at top. Pistil 1; style short or 0. Fruit a berry or a 1-celled capsule.

1. **BERBERIS,** *Brunfels* (OREGON GRAPE. BARBERRY). Ours low evergreen shrubs, with yellow inner bark and wood, and pinnate prickly leaves; putting forth, early in the season, clustered terminal and axillary racemes of yellow flowers. Sepals 6, subtended by 3 or more small bracts. Petals 6, opposite the sepals. Stamens 6. Berries globose or oblong, in our species dark blue, covered with a bloom.

1. **B. repens,** Lindl. Bot. Reg. t. 1176 (1828). Less than a foot high: leaflets 3–7, ovate, acute, dull green and glaucescent, not shining: racemes few. Sparsely wooded hills in Lake Co., and far northward and eastward beyond California.

2. **B. dictyota,** Jepson. Bull. Torr. Club, xviii. 319 (1891). Size of the preceding but stouter: leaflets 3–5, rather remote, hard-coriaceous, ovate, undulate, with rigidly spinescent teeth; the lower face pale green, upper bright green and shining, both faces very strongly reticulate-veiny. An imperfectly known, but well marked species; the foliage extremely rigid and strikingly netted-veined. Marysville Buttes, *Jepson*.

3. **B. Aquifolium,** Pursh, Fl. i. 219 (1814); Nutt. Journ. Philad. Acad. vii. 11 (1834), under *Mahonia*. Often 5 or 6 ft. high: leaflets 7–9, the lowest pair distant from the stem, ovate, acute, bright green and glossy, sinuately dentate, the spinose teeth not very prominent: racemes mostly terminal: fruit nearly globose. In the Sierra from Kern Co. northward: credited to Monterey in the "Botany of California;" but we doubt its occurrence there. With the herbarists the species is too often mixed with *B. repens*.

4. **B. pinnata,** Lag. Gen. et Sp. 14 (1816). Habit of the last, but a smaller shrub; leaflets thinner, more prominently prickly, and the lowest pair near the base of the petiole: flowers more profuse, appearing in the axils as well as at the ends of the branches: fruit oblong. Hills of the Coast Range, from Monterey northward to Marin Co.

5. **B. nervosa,** Pursh, Fl. i. 219. t. 5 (1814), as to leaf only. *Mahonia glumacea*, DC. Syst. ii. 21 (1821). Stem simple, 1 ft. high or less, at summit bearing a crown of large leaves, and many dry chaffy persistent

bracts: leaves 1—2 ft. long; leaflets 11—17, ovate, acuminate, somewhat palmately nerved: racemes elongated: berries larger than in other species. - In deep woods near the coast from Monterey northward.

2. VANCOUVERIA, *Morren & Decaisne.* Perennial acaulescent herbs, with leaves 2—3-ternately compound, and scapes bearing a raceme or panicle of small white or yellow nodding flowers. Sepals 6, obovate, reflexed, subtended by 6—9 oblong membranaceous bractlets. Petals 6, deflexed, but with cucullate-incurved tips. Stamens 6, erect, closely appressed to the pistil. Carpel 1; the stigma cup-shaped; ovules 10 or fewer, in 2 rows on the central suture. Capsule dehiscing by a dorsal valve. Seeds oblong, somewhat curved, with a fleshy aril.—Naturally only a hexamerous-flowered group of the genus *Epimedium.*

1. **V. hexandra,** Morr. & Dec. Ann. Sc. Nat. 2 ser. ii. 351 (1834); Hook. Fl. i. 31. t. 13 (1829), under *Epimedium.* More or less villous with brownish hairs, 1—2 ft. high: leaves long-petioled, spreading: leaflets 1—2 in. broad, petiolulate, subcordate, obtusely 3-lobed, emarginate, the color light green, the texture thin, the whole leaf dying at end of summer: scape exceeding the leaves; fl. 8—15 in a nearly simple raceme, yellowish-white or yellow, $\frac{1}{2}$ in. long: ovary glandular-pubescent.—From Humboldt Co., *Chesnut & Drew,* northward, in woods.

2. **V. parviflora,** Greene, Pittonia, ii. 100 (1890); *V. hexandra,* Greene, Bull. Calif. Acad. i. 66, not Morr. & Dec. Half as large as the preceding; leaves dark green or purplish, subcoriaceous, persistent through the year: fl. small and numerous, 25—50 in a panicle, white, or with a tinge of lavender: ovary glabrous. Common on bushy hills of the Coast Range, from Santa Cruz to Marin Co.; perhaps also farther northward. This, although formerly taken to be the real *V. hexandra,* is a most distinct species.

3. ACHLYS, *De Candolle.* Perennial herb; the long-petioled leaves parted into 3 broad fan-shaped leaflets. Flowers small, spicate at summit of a tall slender scape. Sepals and petals 0. Stamens 6—12; filaments filiform, unequal, dilated upwards; anthers didymous, broader than long. Ovary ovate; stigma sessile. Capsule small, spherical, dehiscent by a valve, 1-seeded. Seed affixed to the base of the capsule.

1. **A. triphylla,** DC. Syst. ii. 35 (1821); Smith; Rees Cycl. (1812 ?), under *Leontice.* Rootstock creeping, ascending: leaves 2 or 3, on erect petioles 1 ft. long or more: leaflets 3—5 in. long, palmately nerved, the outer margin irregularly and coarsely sinuate: scape solitary, equalling the leaves; spike 2—4 in. long; fl. small, white, fragrant: fr. 2 lines thick.—From Mendocino Co., *Bolander,* northward, in deep woods.

Order LI. RANUNCULACEÆ.

L. Gerard, Flora Gallo-Provincialis, 378 (1761); A. L. de Jussieu, Mem. Acad. Paris, 214 (1773).

Herbs (our species of *Clematitis* shrubby) with colorless acrid juice. Leaves alternate (opposite in *Clematitis*; the cauline ones whorled in *Anemone*), usually lobed or ternately divided, the petiole dilated at base and more or less clasping the stem, seldom stipulate. Inflorescence various. Sepals usually 5 (3–6), deciduous (in *Paeonia* persistent). Stamens ∞, hypogynous (in *Paeonia* perigynous); anthers adnate, extrorse, opening by slits. Pistils 1–∞, simple, becoming achenes or follicles (in *Actæa* berry-like fruits). Seeds with horny albumen and minute embryo. A varied, rather large, and interesting order, furnishing many ornamental and several important medicinal plants. The affinities of the family seem to be many; it being at several points in close contact with Rosaceæ; at others indistinguishable from Berberideæ except by the rimose dehiscence of the anthers; while the analogies of habit and some essentials of vegetative character link it as closely as possible to the Umbelliferæ; moreover, some authors have found it difficult to draw the line between these plants and some Papaveraceæ.

Hints of the Genera.
Fruit a head or spike of achenes;
    Flowers complete, - - - - - - - - - 3, 4, 5
    " incomplete, the petals wanting;
        Sepals green and inconspicuous, - - - - 13
        " petaloid, - - - - - - - 1, 2
Fruit of 3 or more follicles;
    Flowers irregular, - - - - - - - - - 6, 7
    " regular;
        Petals tubular and spur-like, - - - - - 12
        " rounded and concave, - - - - - - 8
        " narrow and elongated, - - - - - 11
        " none; sepals petaloid, - - - - 9, 10, 14
Fruit many-seeded but berry-like, - - - - - - - 15

1. CLEMATITIS, *Dioscorides.* Half woody, climbing by the tortuous petioles of the compound leaves which are opposite, with clustered or solitary flowers in the axils. Sepals 4, petaloid, valvate in bud. Petals inconspicuous or 0. Pistils ∞; styles persistent, becoming feathery appendages of the large compressed and capitate-clustered achenes.

    * Some of the outer filaments sterile and dilated into petals.
        Old genus Atragene.

1. C. verticillaris, DC. Syst. i. 166 (1818). *Atragene Americana,* Sims, Bot. Mag. t. 887 (1806). Slender, nearly glabrous, leaves ternate; the petiolulate leaflets ovate or subcordate, acute: fl. solitary, nodding, on a peduncle equalling the petiole of the leaves, 2–3 in. broad, bluish-purple: outer stamens sterile and enlarged to narrow spatulate petals.— In mountains from near Cape Mendocino northward and eastward.

＊ ＊ *Petals 0; our species diœcious.* —CLEMATITIS proper.

2. **C. lasiantha**, Nutt.; T. & G. Fl. i. 9 (1838). Silky-pubescent throughout, even to the outside of the sepals: leaflets 3, ovate, somewhat cuneate at base, coarsely toothed or 3-lobed or -parted: fl. large, solitary, erect on 1–2-bracted peduncles; sepals white, $\frac{3}{4}$ in. long.— In the hilly districts, trailing over rocks and shrubs: the flowers often several at each node of the stem, but the peduncles 1-flowered. Apr.

3. **C. ligusticifolia**, Nutt. l. c. Leaves glabrous throughout, or somewhat pubescent, in our district silky-tomentose beneath: stems elongated, often climbing small trees to the height of 30 ft.: leaves 5-foliolate: leaflets broadly ovate to lanceolate, usually 3-lobed: fl. panicled in the axils; sepals scarcely $\frac{1}{2}$ in. long.— In middle Calif. less frequent than the last, but rather common in the Mt. Diablo Range, also occasional in Marin Co.: most showy in autumn, when laden with its abundant heads of white feathery-tailed achenes. Fl. July.

2. **ANEMONE**, *Dioscorides* (WIND FLOWER). Perennial herbs with radical lobed or divided leaves, and a cauline involucral whorl of about three, or these more or less united. Flowers 1 to several, on erect peduncles. Sepals 5 or more, petaloid, imbricate in bud. Petals 0. Stamens and pistils ∞. Fruit a head of compressed pointed (in some species feathery-tailed) achenes.

＊ *Styles long, in fruit becoming plumose tails.*— Old genus PULSATILLA.

1. **A. occidentalis**, Wats. Proc. Am. Acad. xi. 121 (1876): *A. alpina*, Hook. Fl. i. 5, not Linn. Stems clustered, stout, $\frac{1}{2}$–$1\frac{1}{2}$ ft. high, 1-flowered: plant more or less villous: radical leaves long-petioled, biternate, the divisions pinnate, the lateral ones subsessile; the segments pinnatifid with narrow laciniate-toothed lobes: involucral leaves similar, subsessile about midway of the stem: sepals 6 or 7, $\frac{1}{2}$–$\frac{3}{4}$ in. long, pale bluish-purple: head of achenes globose, the silky tails 1 in. long or more. Dry ridges, at high altitudes, on Lassen's Peak; doubtless also in the Trinity Mts., and far northward. July, Aug.

＊ ＊ *Achenes not plumose-tailed.*— ANEMONE proper.

2. **A. Drummondii**, Wats. Bot. Calif. ii. 424 (1880): *A. multifida*, B. & W. Bot. Calif. i. 4, not Poir.: *A. Baldensis*, Hook. Fl. i. 15, not Linn. More or less villous, 3–10 in. high, 1–3-flowered: radical leaves rounded and ternately multifid, on petioles of 1–3 in.; the involucral similar, subsessile: sepals 5–8, white, 4–6 lines long: achenes oblong, 2 lines long, densely woolly, in an ovate head.— Habitat of the preceding, but in moister ground near snow. Aug., Sept.

3. **A. deltoidea,** Hook. Fl. i. 6, t. 3 (1829). Stem solitary, slender, 1 ft. high or less, glabrous: radical leaves ternate, the leaflets deltoid-ovate, sometimes trifid; involucral sessile, rhomboid, incisely serrate: sepals about 5, oval, large, white: achenes ovate, pubescent, in a rounded head. In Humboldt Co. and northward; on the border of woods, at considerable elevations. July, Aug.

4. **A. nemorosa,** Linn. var. **Grayi.** *A. Grayi,* Behr. & Kell. Bull. Calif. Acad. i. 5 (1884): *A. Oregana,* Gray, Proc. Am. Acad. xxii. 308 (1887). Stem very slender, 1 ft. high, solitary, from a horizontal running rootstock: radical leaf remote from the stem, of reniform outline, trifid, the segments serrate; involucral long-petioled, 3-foliolate, the terminal leaflet 3-lobed, each lateral one 2-lobed, all coarsely serrate: sepals about 6, oval, white or bluish: achenes 12–20, oblong, 2 lines long, pubescent, tipped with a hooked beak, forming a globose head, the long pedicel at length curved into a ring.—Moist shady slopes of the higher Coast Range mountains, from Santa Cruz northward; common about Lagunitas on Mt. Tamalpais. The Oregon form, with deep blue sepals, on which Dr. Gray founded his species, is inseparable from our own, in which the sepals are often bluish, as indeed they are in the genuine Old World *A. nemorosa* in certain localities. Mar. May.

3. **MYOSURUS,** *Lobelius* (MOUSETAIL). Small stemless glabrous annuals, with narrow entire leaves, and many slender 1-flowered (sometimes short or obsolete) scapes. Sepals 5, spurred at the base. Petals 5, consisting of an oblong blade with a nectariferous gland or pit at base, and a filiform claw. Stamens 5–15. Pistils ∞, crowded on a long slender receptacle; becoming a spike of small rather thin-walled achenes.

1. **M. minimus,** Linn. Sp. Pl. i. 284 (1753). Scape 1–5 in. high, rather stout, gradually thickened under the fruiting spike, this long-conical, 1–2 in. long: sepals with prominent slender spur: carpels crowded, the more or less distinctly rhomboid top with a manifest costa ending in an appressed straight beak: seeds oval or oblong. Var. (1) **apus,** Greene, Bull. Calif. Acad. i. 277 (1885). Fruiting spike nearly cylindrical, nearly or quite sessile among the leaves. Var. (2) **filiformis,** Greene, l. c. Spike nearly cylindrical, very slender, on a slender scape: carpels in few series, minute, with delicate costa, the slender beak shortly and abruptly recurved.—The Californian plant nearest the Old World type is much larger, with relatively longer scapes and shorter more conical spikes. This is found in the Livermore Valley (State Survey n. 1193), but is more common northward, reaching Vancouver Island. The first variety is on the lower San Joaquin, in subsaline soil, where it grows with the other, and flowers earlier. Var. 2 is not rare, occurring at San Francisco, in the Oakland Hills and beyond them; but its best type is of the far South and insular (Guadalupe Island). Mar. May.

2. **M. apetalus,** Gray, var. **lepturus,** Gray, Bull. Torr. Club, xiii. 2 (1886). Smaller and slender; the spikes (conical in the type) shorter but often very slender: achenes with an ascending or somewhat spreading prominent beak: seeds elongated-oblong. Livermore and Sacramento valleys, and far northward and eastward. The type of this species is of South America. Mar. May.

3. **M. alopecuroides,** Greene, Bull. Calif. Acad. i. 278 (1885); Gray, Bull. Torr. Club, xiii. 3. Scapes 1 2 in. high, stout, thickened upwards, or nearly obsolete, and the conical spikes subsessile: top of carpel oblong, with a wide soft-cellular border, and a conspicuous broad flattened spreading beak: seeds oval, striate-reticulate. - Interior valley; near Antioch, *Mrs. Curran,* and about Vacaville, *Jepson;* the specimens of the latter with nearly or quite sessile flowers and spikes, these less elongated and somewhat oblong-conical. Mar. -May.

4. **KUMLIENIA,** *Greene.* Flaccid perennial, with rounded and lobed mostly radical leaves, and a nearly leafless 1 2-flowered stem. Sepals 5 7, white-petaloid and conspicuous. Petals 5, small and inconspicuous, consisting of an oval fleshy nectariferous-pitted blade and slender claw. Stamens ∞. Pistils ∞; becoming in maturity a rounded head of elongated thin and somewhat utricular 1-seeded carpels, each tipped with a persistent hooked style.

1. **K. hystricula,** Greene, Bull. Calif. Acad. i. 337 (1886); Gray, Proc. Am. Acad. vii. 328 (1868), under *Ranunculus.* Radical leaves long-petioled, round-reniform, with 5 broad but not deep rounded lobes; stems 4 10 in. high, with 1 or 2 leaves (3-lobed), and as many white flowers, the expanded corolla-like calyx ³⁄₄ in. broad: ripe carpels brownish, of papery texture or thinner, lanceolate, 3 lines long including the uncinate style, loosely investing the linear-oblong acute seed.- A rare plant of the eastern slope of the Sierra at middle elevations; wearing the general aspect of a *Ranunculus,* but displaying the flowers of a *Caltha* augmented by the nectary-like petals of *Helleborus,* the utricular fruit peculiar. It has been found at Forest Hill near Newcastle, *Bolander,* and on wet mossy rocks in the Yosemite, *Parry.* Apr.—June.

5. **RANUNCULUS,** *Pliny* (BUTTERCUP. CROWFOOT). Mostly perennial, with a tuft of fibrous or fleshy-fibrous roots, terete stems, which are erect, procumbent, creeping or submersed, and entire or cleft or divided (sometimes submersed and capillary-dissected) leaves. Flowers solitary, or few and scattered, regular, yellow (sometimes white). Sepals 5, commonly reflexed. Petals 5 (rarely 10 or more), spreading, with a nectariferous scale or pit at base within. Stamens and pistils ∞, the latter becoming compressed smooth or tuberculate or even muricate, glabrous or pubescent, usually beaked achenes, disposed in globose or somewhat elongated heads.

* *Mostly land plants, the leaves seldom cut into capillary segments; peduncles not reflexed after flowering; flowers mostly yellow; petals with a scale concealing the nectary; achenes without distinct transverse ridges.*— RANUNCULUS proper.

　+-*Leaves undivided; achenes not strongly compressed.*

1. **R. Bolanderi,** Greene, Bull. Calif. Acad. ii. 58 (1886). Erect, stout, 1½–3 ft. high; peduncles and calyx pubescent: radical leaves few, with long petioles and reduced blade; cauline lanceolate, 3–6 in. long, rather remotely and coarsely callous-denticulate: petals broadly obovate, 3–4 lines long: achenes many, in a large ovate head, the beak subulate, acute, slightly incurved.— Long Valley, Mendocino Co., *Bolander;* apparently rare. The species is much like *R. lingua* of Europe, but very distinct from the next, with which it has been confused.

2. **R. alismæfolius,** Geyer; Benth. Pl. Hartw. 295 (1849); Gray, Proc. Am. Acad. xxi. 368, excl. syn. and var. Erect, 1 ft. high, pilose-pubescent below, usually glabrous above; leaves mostly radical, short-peduncled, suberect, lanceolate, entire: stem branching above, and peduncles elongated: sepals and petals often persistent: achenes with a slender straight beak, and disposed in a depressed-globose head.— Moist meadows of the Sierra Nevada at middle and higher altitudes. June, July.

3. **R. alismellus.** *R. alismæfolius,* var. *alismellus,* Gray, Proc. Am. Acad. vii. 327 (1868), xxi. 368. Dwarf, acaulescent, glabrous: leaves elliptical, entire, acute, thin and flaccid, reclining or fully sustained on the long slender petiole: scapes exceeding the leaves, very slender (often with a leaf), 1-flowered: fl. ½ in. broad: carpels few, the mature ones unknown.— Near the summit of the Sierra, also in the Trinity Mts., at higher elevations than the last, and in very wet ground. It has often been confused, by collectors and amateurs, with the next, to which it is about as much allied as to the preceding.

4. **R. Flammula,** Linn. var. **intermedius,** Hook. Fl. i. 11 (1829). Stems slender or even filiform, rooting at the lower joints, 4–10 in. long: leaves lanceolate, entire: fl. 2–5 lines broad: achenes few, with a very short stout straight beak.— Frequent in both ranges of mountains, along the muddy or sandy margins of pools, or in the smallest forms quite submersed. Californian specimens are more slender and smaller than the British American type of this variety; they are not, however, properly referable to *R. reptans.* That is distinguished by linear leaves and a prominent slender curved beak to the achene.

5. **R. pusillus,** Poir. Encycl. vi. 99 (1804): *R. pusillus,* var. *trachyspermus,* Gray, Proc. Am. Acad. xxi. 367. Annual, erect, slender, 5–10 in. high: leaves lanceolate or linear, the radical on slender petioles, entire;

cauline sometimes toothed: fl. very small: achenes small, papillose-roughened, rather many, in a roundish or oval head.—Rare in California, though common in the southern Atlantic states; found in Napa Valley, *Bigelow*, and in Marin Co., *J. P. Moore*. The achenes are either smooth or rough in even the eastern plant, so that the designating of ours as a variety seems unwarranted. May.

6. **R. Cymbalaria,** Pursh, Fl. ii. 392 (1814). Glabrous, somewhat succulent, low, the crown of the perennial root sending out filiform runners: leaves long-petioled, small, round-ovate, obtuse, crenately lobed: scapes usually naked, erect, 3—6 in. high, 3 5-flowered; petals linear or linear-oblong, few or many (4—8): achenes minute, short-beaked, striate on the sides, crowded in an oblong head.—At Mono Lake, *Bolander*, and in the Kern Co. mountains, *Palmer*; apparently not in western Calif., but common in alkaline soils in many parts of western N. America.

7. **R. Lemmoni,** Gray, Proc. Am. Acad. x. 68 (1874). Subacaulescent, villous-pilose below: leaves lanceolate or linear, entire: peduncles scapiform, 1-flowered; petals small, spatulate-oblong: achenes turgid, villous-pubescent, with an inflexed subulate beak, in a depressed-globose head. - Sierra Valley, Plumas Co., *Lemmon*.

+ + *Leaves mostly ternately lobed, cleft or divided; achenes usually much flattened (except in n. 8).*

8. **R. glaberrimus,** Hook. Fl. i. 12 t. 5. A (1829). Glabrous, flaccid but rather fleshy, 3—6 in. high: leaves all petiolate; radical rounded, 3-lobed or coarsely toothed; cauline subcuneate, trifid or entire: fl. several, large; sepals spreading; petals 3 - 4 lines long, obovoid: achenes plump, smooth, puberulent, with a short curved beak, and disposed in a large globose head. Var. **ellipticus.** *R. ellipticus*, Greene, Pitt. ii. 110 (1890). Radical leaves elliptical, acute, entire: stems shorter; fl. fewer, often apetalous. The type, a plant of the far north and east, reaches our borders on the eastern slope of the Sierra northward. The variety, a plant of different aspect, and with very dissimilar foliage, is found not far from Truckee, *Mr. Sonne*, where it appears as if confluent with the type.

9. **R. oxynotus,** Gray, Proc. Am. Acad. x. 68 (1874). Stout, glabrous, the tufted stems 3—6 in. high, at base encased within the dead petioles of the preceding year: leaves crowded, subreniform, or rounded and with cuneate base, crenately 5- 9-lobed, $\frac{1}{2}$—$\frac{3}{4}$ in. broad; cauline broadly cuneiform, with 3 - 5 oblong lobes: sepals pilose: petals 4 lines long: achenes oblong, smooth, carinate on the back, acuminate with the subulate curved style, disposed in an oblong thick fleshy head.—At great elevations in the Sierra, near snow. July- Sept.

10. **R. repens,** Linn. Sp. Pl. 554 (1753). Pubescent, the stems 1—2

ft. long, trailing, rooting at the lower joints; leaves ternately parted and often subdivided; sepals spreading; petals 5; achenes 1¼ lines long, rather sharply margined, the nearly straight beak about 1½ line long.— Frequent in lawns at Golden Gate Park and elsewhere in the Bay district, but scarcely naturalized. It is common about Eureka, Humboldt Co., *Marshall*, where it may be either native or introduced.

11. **R. maximus,** Greene, Bull. Torr. Club, xiv. 118 (1887): *R. macranthus*, Bot. Calif., not Scheele. Pilose or hirsute, the stems stout, 2–5 ft. long, trailing but not rooting; leaves broad, ternate, the radical on petioles 1 ft. long or more; leaflets laciniately lobed; sepals reflexed; petals 5–8, oblong-obovate, obtuse, 7–10 lines long; achenes thickish; the beak long, straight or slightly incurved; head roundish or broadly ovate. The type of this species is of the Bay region, and is not common. It was formerly abundant in the marsh at Newbury Station, Berkeley, and is in the hills east of Alameda and Oakland; being also credited to Marin Co. In a smaller state, with quite small flowers, though with the same large achenes and broad leaflets, it appears in the middle Sierra.

12. **R. Bloomeri,** Wats. Bot. Calif. ii. 426 (1880). Nearly as large as the last, the usually glabrous but sometimes pilose herbage of a peculiarly light green, the stems ascending; earliest leaves round-cordate, coarsely crenate-toothed or lobed; the later ones 3–9-foliolate, with leaflets 1 in. long and nearly as broad, the coarse teeth rounded and somewhat regular; petals 5, retuse, ¾ in. long; achenes long-beaked, forming a roundish head. Common on low grounds adjacent to San Francisco Bay. A very distinct species, much more frequent than the last. Feb.–May.

13. **R. Californicus,** Benth. Pl. Hartw. 295 (1849): *R. delphinifolius*, T. & G. Fl. i. 659 (1840), not HBK.: *R. dissectus*, H. & A. Bot. Beech. 316 (1840), not Bieb.: *R. Deppei*, Nutt.; T. & G. Fl. i. 21 (1838), under *R. acris*, therefore by implication wrongly described as to calyx. The type deep green, with little or no pubescence, erect or decumbent, 1–1½ ft. high, the stems freely branching and many-flowered; leaves ternately much dissected; sepals reflexed; petals 10–15, obovate-oblong, 4–5 lines long; achenes 1½ lines long, much compressed, the beak short and recurved; head globose. Var. (1) **laetus.** Strictly erect, stoutish and fistulous, rather stiffly hirsute and glaucescent below, the herbage of a light yellowish green; segments of the much dissected leaves broader than in the type; fl. and fr. the same. Var. (2) **canescens.** Stout and low, the basal parts canescently long-villous; leaves scarcely dissected, the 3 cuneate main segments only deeply incised; fl. large, fully an inch wide; fr. as in the above forms. Var. (3) **cuneatus.** Slender, decumbent, the nascent parts silky-pubescent, the plant otherwise glabrous; leaves more or less deeply cleft into 3 cuneate lobes or segments, these incisely toothed; fl. small; achenes very many, in a round-ovoid head. Var. (4) **latilobus,**

Gray, Proc. Am. Acad. xxi. 375 (1886), in part, excl. *R. Ludovicianus*. Size, habit, fl. and fr. of type, but leaves cleft below the middle into 3 broad cuneate-obovate coarsely toothed lobes. The prevailing buttercup of western California, apparently not reaching the foothills of the Sierra eastward, but ranging north and south toward the seaboard almost throughout the State, and running into several very well characterized varieties or subspecies, all of them carrying invariably that one mark, the multiplicity of petals, by which, along with the reflexed sepals, the species is distinguished from its European analogue, *R. acris*. I here take as the typical form the common plant of the Bay region. From the summits of the Oakland Hills down to the bay and the ocean, the unpastured hills and level lands are almost yellow with its bloom in March. Var. 1 is of the interior, about Suisun, and also in San Mateo Co., occupying low meadow lands adjacent to the brackish marshes. Var. 2 belongs to the middle elevations of the Mt. Diablo Range and the valleys among them, from Niles to the hills east of Livermore, thence southward to San Luis Obispo Co. It was a part of my *R. Ludovicianus*. Var. 3 is confined to the wet meadows that lie back of the ocean in San Mateo Co., doubtless also reaching San Francisco Co. In cultivation at Berkeley it behaves very unlike the other forms, is almost annual, *i. e.*, many individuals come to flowering the first year from the seed, and die before the end of the year. Other individuals are of perennial duration. Var. 4 is n. 374 of the State Survey, from Santa Barbara. *R. Ludovicianus*, properly defined, is a very different plant.

14. **R. Ludovicianus,** Greene, Bull. Calif. Acad. ii. 58 (1886), excl. plant of San Luis Obispo Co. Ascending, less than a foot high; stems stout, striate, and, with the foliage, somewhat villous-tomentose or ternate: radical leaves parted into 3 broad coarsely and callously toothed segments; cauline parted into few narrow lobes: petals as in the last: achenes larger and much less compressed, often sparsely strigose-hispid and somewhat papillose.- In mountain meadows of the southeastern parts of the State, from Kern Co., *Mrs. Curran*, to San Bernardino, *Parish*, n. 1890. In the original account of the species I erroneously included with it the var. *canescens* of the last, thus leading Dr. Gray to merge the whole in var. *latilobus* of the preceding.

15. **R. rugulosus,** Greene, Pittonia, ii. 58 (1890). Nearly glabrous, the stems slender and decumbent, or stoutish and reclining, $1\frac{1}{2}$ 3 ft. long: leaves about 5-parted or -divided, the divisions cleft into linear or lanceolate segments: petals 7 11, spatulate-oblong, $\frac{1}{2}$ in. long: achenes barely a line long including the short recurved style, the sides uneven with low rugosities. —This appears as in some sort replacing *R. Californicus* in the southeastern mountains and foothills. The type is from the Chowchilla Mts., a slender suberect plant. A coarser form, with broader

leaflets and a depressed mode of growth, bearing much likeness to *R. repens*, has been collected near Visalia, and east of Stockton.

16. **R. canus,** Benth. Pl. Hartw. 294 (1849): *R. occidentalis*, var. *canus*, Gray, Proc. Am. Acad. viii. 374: *R. Californicus*, var. *canus*, B. & W. Bot. Calif. i. 8. More or less silky-canescent, and most so when young; stems erect, stoutish, 1½ – 2 ft. high; leaves ternately dissected into many narrow acute segments: petals 5 (rarely 7 – 10), round-obovate: achenes large, much compressed, the beak broad at base, short and hooked: head globose. Plains and hills of the interior, especially about Antioch, and northward to Chico. *Mrs. Bidwell;* abundantly floriferous and very showy; apparently intergrading with *R. Californicus* on the one hand, and *R. occidentalis* on the other, but sufficiently distinct. Apr., May.

17. **R. occidentalis,** Nutt. var. **Eiseni,** Gray, Proc. Am. Acad. xxi. 373 (1886): *R. Eiseni*, Kell. Proc. Calif. Acad. vii. 115 (1877). Erect, slender. 1 ft. high, more or less villous or hirsute with long spreading or appressed white hairs: leaves 3-lobed or -parted, the broad cuneiform segments of the radical ones trifid: fl. loosely corymbose-panicled: sepals reflexed: petals 5, obovate-oblong, 3 – 4 lines long: achenes broad and rounded, compressed and thin, glabrous, tipped with a short recurved beak. Var. **Rattani,** Gray, l. c. Radical leaves more deeply parted: stem taller, more freely branching and floriferous: achenes rather smaller, with relatively longer beak, their sides hairy and papillose.- True *R. occidentalis* is probably not within our limits, unless perchance it occurs in Humboldt Co. The var. *Eiseni* is common in the Sierra Nevada at middle elevations, on dry open slopes. Var. *Rattani* is rather of the Coast Range, from Marin Co. northward. Apr.--May.

18. **R. Nelsonii,** Gray, Proc. Am. Acad. viii. 374 (1872): *R. recurvatus*, var. *Nelsonii*, DC. Syst. i. 290 (1818): *R. tenellus*, Nutt.; T. & G. Fl. i. 23 (1838), not Viviani (1830). Densely hirsute, at least below, with short spreading or deflexed brownish hairs: stem 1 ft. high or more, the branches slender, spreading: radical leaves palmately 3 – 5-lobed or -parted, the segments cuneate, trifid: fl. very small and inconspicuous: sepals reflexed: petals elliptic-oblong: achenes in a globose head, small, smooth, glabrous or with short recurved hairs, the beak commonly as long as the body, closely recurved at tip. Var. **tenellus,** Gray, Am. Acad. viii. 373. Nearly glabrous: fl. very small: achenes with a much shorter beak.- The type of this, common in openings among the forests from Oregon to Alaska, is found in Humboldt Co., *Chesnut & Drew;* the variety is of the Sierra Nevada, extending southward to Fresno Co.

19. **R. hebecarpus,** H. & A. Bot. Beech. 316 (1840). Slender, erect, branching and leafy, 5 -15 in. high, pilose-pubescent: radical leaves rounded or reniform, deeply lobed or cleft, the segments 3-lobed; fl.

numerous, minute, on filiform pedicels: achenes rather few, in a globose head, rounded and flattened, papillose and short-hairy, the beak very short, the small tip abruptly recurved. Var. **pusillus**, B. & W. Bot. Calif. i. 9; Greene, Bull. Torr. Club, xiv. 116. Annual, very slender, usually reclining; organs of the fl. few and definite (4 or 5 only in each circle). The type is perennial, extending from the foothills of the Sierra near Chico southward to the southern seaboard. In the Bay region we have only the variety. It is common in the hilly districts from Napa Co. to San Mateo, in the shade of oaks, and along streams. Apr., May.

20. R. MURICATUS, Linn. Sp. Pl. i. 555 (1753). A stoutish and glabrous annual with yellow-green herbage, and round-reniform slightly lobed leaves: fl. small and inconspicuous: achenes very large, with stout ensiform beak and coarsely muricate-prickly sides.— Common about San Francisco, at the Presidio, Mountain Lake, and in the Mission Hills, in wet sandy soil, or about springs; also in Marin Co.; doubtless naturalized from Europe: flowering at all seasons.

21. **R. delphinifolius**, Torr.; Eaton, Man. 2 ed. 395 (1818), not HBK. (1821); Britt. Bull. Torr. Club, xviii. 363: *R. lacustris*, Beck & Tracy, Eaton, Man. 3 ed. 395 (1822); Greene, Pitt. ii. 62: *R. multifidus*, Pursh (1814), not Forsk. (1775): *R. Purshii*, Hook. (1829). Perennial, aquatic; stems a few inches to several feet long, mostly submersed, as are also most of the leaves, these all divided into numerous linear-filiform segments: fl. (and some of the uppermost leaves) emersed and almost floating, 1 in. broad, yellow: sepals spreading: petals 5, broadly obovate: achenes slightly flattened, the sides faintly rugose, the basal part corky-thickened, the rather long and slender beak straight or somewhat incurved.— In mountain lakes; Humboldt Co., *Chesnut & Drew*. A handsome aquatic, more common on the Atlantic side of the continent than with us. Late in the season it may be found in a terrestrial form, with leaves less finely dissected. June, July.

* * *Aquatics, with leaves mostly capillaceous-multifid and submersed; peduncles opposite the leaves, recurved in fruit; petals white, with naked nectariferous pit: achenes little compressed, transversely rugose-striate.* Genus BATRACHIUM, S. F. Gray; Wimmer; Fries.

22. **R. aquatilis**, Dod. Pempt. 576 (1583); Ray, Syn. 3 ed. 249 (1724); Linn. Sp. Pl. i. 556 (1753). Perennial, the emersed and floating leaves when present, roundish, 3-lobed: sepals deciduous: styles subulate: achenes slightly rugose, usually hispidulous, 12–20 in a rather compact globose head. Frequent in ponds and ditches, sometimes in running streams, or, on muddy shores appearing in a dwarf and wholly terrestrial form. The Old World type, with round-reniform emersed foliage, and hispidulous achenes, not known in eastern America, has been found in

Humboldt Co., *Chesnut & Drew*. A state with flowers constantly 4-merous occurs in the sloughs near Stockton. Flowering and fruiting takes place at all seasons except midwinter.

23. **R. Lobbii,** Gray, Proc. Am. Acad. xxi. 364 (1886): *R. hydrocharis Lobbii,* Hiern, Journ. Bot. ix. 66 (1871): *R. aquatilis,* var. *Lobbii,* Wats. Bibl. Ind. 17 (1878). Annual, the floating leaves always present, deeply 3-lobed, the middle lobe usually elliptical and entire, the laterals somewhat larger, oblong, obcordate at summit: sepals persistent, embracing the few (4–6) finely and rather sharply rugose achenes; style filiform, deciduous except the base, which remains upon the achene as a short beak. In early spring only, in winter pools which go dry in summer; frequent in Marin Co.; also in the Berkeley Hills toward San Pablo.

6. **DELPHINIUM,** *Dioscorides* (LARKSPUR). Erect herbs, with petioled palmately divided leaves, and irregular mostly either blue or scarlet flowers disposed in terminal racemes. Sepals 5, colored and petaloid, the upper one produced into a long hollow spur, the others plane. Petals 4 (or 2 at the least), 2 of them developed backwards into a spur which is inserted into the spur of the calyx. Stamens ∞, unequal. Pistils mostly 3 (1–5), becoming many-seeded follicles.

* *Flowers blue, varying to pink or flesh-color (never scarlet).*
+· *Root a cluster of coarse thickish and half-woody fibres.*

1. **D. Californicum,** T. & G. Fl. i. 31 (1838): *D. exaltatum,* H. & A. Bot. Beech. 318 (1840), not Ait. Stems usually several, stout, 3–5 ft. high, leafy up to the raceme, pubescent: leaves ample, deeply 5-cleft, the segments variously lobed: raceme strict, often dense, 1–1½ ft. long: fl. small, either dull greenish or whitish, with a tinge of flesh-color or purple, little expanded, externally rather densely velvety-pubescent: follicles oblong, turgid, erect.—One of our few well-marked species of this intricate genus, inhabiting the Coast Range; preferring moist places on open or sparsely wooded hills. Apr.–June.

2. **D. glaucum,** Wats. Bot. Calif. ii. 427 (1880): *D. scopulorum,* B. & W. l. c. i. 11, not Gray: *D. scopulorum,* var. *glaucum,* Gray, Bot. Gaz. xii. 52 (1887). Size, habit and leafiness of the above, but glabrous and glaucescent: leaves ample, thin, deeply 5-parted; segments laciniate-toothed or -lobed: fl. in a narrow elongated raceme, on short slender pedicels, of a rather pale but very clear blue: sepals narrow, not wide-spread, glabrous or nearly so; spur tapering gradually to the end, which is abruptly curved downward.—In the high Sierra, where it is common, growing in masses, and occupying damp ground bordering marshes and streamlets.

3. **D. hesperium,** Gray, Bot. Gaz. xii. 51 (1887): *D. simplex,* Boland. Cat. 4; B. & W. i. 10, not Dougl. Stem solitary, rather slender and strict,

$1\frac{1}{2}$ — $2\frac{1}{2}$ ft. high; herbage canescent with a short and close, or coarser and spreading pubescence; conspicuously leafy at base of stem only, the leaves becoming sparse and small toward the raceme, all rather small, much dissected, the lobes linear, obtuse: raceme dense, elongated: fl. well expanded, deep blue (except in the albino state, then pinkish): spur stout and straight, about as long as the sepals: follicles erect, pubescent. Var. **Hanseni.** Very slender, with narrow, elongated and rather lax racemes of flowers one half as large as in the type, and of less intense blue. The genuine form of this handsome species is of the Coast Range, where it is noted as our only late-flowering species, appearing in June, after the dry season has set in; the lower leaves at time of flowering having mostly died away. The variety in Amador Co., *Geo. Hansen.*

4. **D. recurvatum,** Greene, Pittonia, i. 285 (1889). Stem solitary, stoutish and hollow, 1 — 2 ft. high; herbage cinereous-pubescent throughout, or nearly glabrous, glaucescent: leaves mostly subradical; petiole elongated; leaf-segments cleft into about 3 linear obtuse lobes: raceme long, occupying more than half the stem, rather open, the lower pedicels more elongated: fl. lavender-color or bluish: sepals linear-oblong, widespread, or at length recurved, the blunt spur curved upwards.—In moist and subsaline ground on the plains of the San Joaquin from near Tulare northward to Byron; also in a taller and more leafy variety near Antioch, *Chesnut & Drew;* Sacramento Valley, *Jepson.* Mar. — May.

5. **D. variegatum,** T. & G. Fl. i. 32 (1838): *D. grandiflorum,* var. *variegatum,* H. & A. Bot. Beech. 317 (1840). Pubescent; stem simple, 1- $1\frac{1}{2}$ ft. high: leaves few, 3-parted into cuneiform segments, these cleft into broad linear lobes: raceme short, lax and few-flowered, the pedicels elongated, ascending: sepals dark blue, obovate-oblong, acutish. $\frac{3}{4}$ in. long; spur short, straight: lower petals round-obovate, 3-lobed, the upper small, white: ovaries appressed-pubescent. Var. **apiculatum.** *D. apiculatum,* Greene, Pitt. i. 285 (1889). Flowers smaller, many, on short suberect pedicels, forming a compact cylindrical raceme: segments of the leaves broader; herbage coarsely and retrorsely pubescent. The type is of open fields and hills along the seaboard from Monterey to San Luis Obispo. The variety belongs to the interior valley of the State from one to two hundred miles farther north.

6. **D. ornatum.** Puberulent; stem simple, stoutish, 1 ft. high: leaves on slender petioles, the blade divided and subdivided into rather few narrowly linear acutish segments: raceme strict: fl. very large, of a rather light blue and white, forming a rather dense short raceme, the pedicels erect: sepals 1 in. long, oblong, obtuse, the inner ones, and also the white petals, with conspicuously crisped margins: ovaries appressed-pubescent. -Related to the preceding, but very distinct in foliage and characters of sepals and petals. It is known only in herbarium specimens

(State Survey n. 409) obtained at Nipoma, San Luis Obispo Co., and which have been confused with those of *D. variegatum*.

7. **D. trollifolium**, Gray, Proc. Am. Acad. viii. 275 (1872). Glabrous or somewhat pubescent; 2 ft. high or more: lower leaves very large, long-petioled, 5–7-lobed; lobes cuneate, usually closely approximate, the sinus closed, the summit laciniately cleft and toothed: raceme loose, the pedicels, especially the lower, elongated and ascending: fl. middle-sized, bright blue (but petals white); the stout spur as long as the sepals or longer, gently curved downwards throughout its length: follicles glabrous, recurved-spreading.—In shady places from Humboldt Co. southward to Monterey, in the Coast Range only, and not seen in the Bay region. The Monterey plant (*McLean*, 1874) is not quite like the type, and may be distinct.

+ + *Roots scarcely woody-fibrous, thick and more or less fleshy, often tuberiform or grumous.*

8. **D. distichum**, Geyer; Hook. Lond. Journ. Bot. vi. 67 (1847): *D. azureum*, Torr. Bot. Wilkes Exp. partly, not Michx. Root fleshy and branching, with fibrous rootlets: stem mostly solitary, strict, 1–2 ft. high, very leafy up to the narrow and dense virgate raceme: leaves light green, glabrous, thickish, the lowest cleft to the middle, or more deeply, into oblong callous-tipped lobes; divisions of the upper successively deeper and narrower, those of the uppermost narrowly linear: fl. smallish, somewhat 2-ranked, in a narrow elongated raceme; spur ¾ in. long, straight, horizontal, twice the length of the sepals: follicles short, thickish, erect. —Plains of Humboldt Co. or Mendocino, *Kellogg*, northward to the British boundary, in rather moist open ground. July, Aug.

9. **D. uliginosum**, Curran, Bull. Calif. Acad. i. 151 (1885). Roots scarcely fleshy, rather woody-fibrous but thick: stem leafy at base only, 1 ft. high or more; herbage deep green, almost glabrous: leaves flabelliform, 3-cleft, the segments about 3-toothed: fl. few, rather large, deep blue, in an open raceme; spur slender, straight, equalling the sepals.— Wet open ground near Epperson's, Lake Co., *Mrs. Curran*. July.

10. **D. Menziesii**, DC. Syst. i. 355 (1818). Root a cluster of short roundish or compressed tubers: stem solitary, rather slender, 1 ft. high or less, leafy below, but leaves few, long-petioled, palmately parted, more or less pubescent: fl. few and large (1½ in. broad), on long ascending pedicels; spur short, stout, straight: follicles short, thick, divergent. From near San Francisco, where it is rare, northward to Mendocino and Humboldt counties, where it is common, as also far northward beyond our borders. Apr.—July.

11. **D. Andersonii**, Gray, Bot. Gaz. xii. 50 (1887): *D. decorum*, var.

Nevadense, Wats. Bot. Calif. i. 11 (1876). Root with more numerous and less fleshy tuberiform branches: stem low, stoutish; herbage somewhat fleshy, glabrous, glaucescent: leaves few, deeply cleft, the segments cuneiform, deeply 3-lobed: raceme long and rather lax: fl. 1 in. broad, deep blue, the long spur mostly strongly uncinate-incurved at tip: follicles short, erect.—In the Sierra Nevada, near Truckee, *Sonne,* etc. May, June.

12. **D. decorum,** F. & M. Ind. Sem. Petr. iii. 33 (1836). Root grumose, the tuberiform branches ending in many coarse fibres: stem solitary, slender, simple, mostly less than 1 ft. high: herbage of a very pale green, pubescent or nearly glabrous: leaves small, parted into 3–5 rather widely sundered segments, these broad-cuneiform, obtusely or acutely 3-lobed in the radical ones, narrow and entire in the few cauline: fl. rather small, in a somewhat open or more condensed raceme, deep blue, except the white uppermost petals; spur straight: follicles glabrous, widely divergent in maturity. Common along the borders of thickets, or in more open stony places, among the hills of the Coast Range, from perhaps Santa Cruz, northward to Humboldt Co. Apr.

13. **D. patens,** Benth. Pl. Hartw. 296 (1849): *D. decorum,* var. *patens,* Gray, Bot. Gaz. xii. 54 (1887). Pale green and glabrous, or deeper green and glandular-pubescent, very slender, 1–2 ft. high: leaves larger than in the last, deeply 5-parted; segments narrowly cuneiform and deeply incised: raceme very lax, the small flowers on almost filiform spreading pedicels: spur longer than the sepals, abruptly narrowed to the uncinate tip: follicles glabrous or glandular-pilose, divergent.—At middle elevations of the Sierra, on the westward slope, where it is the analogue of *D. decorum,* and evidently more than a mere variety of that species.

* * *Scarlet flowered species; the roots not fleshy.*

14. **D. nudicaule,** T. & G. Fl. i. 33 (1838): *D. sarcophyllum,* H. & A. Bot. Beech. 317 (1840). Glabrous or slightly hairy; stem simple, 1–2 ft. high, the leaves all near the base, long-petioled, 3–5-lobed, the segments mucronately 3–7-toothed or lobed: raceme very lax, somewhat pyramidal, the lower pedicels greatly elongated: fl. 1 in. long or more; sepals bright scarlet, not widely expanding, the spur straight; petals yellow: follicles glabrous, divergent at summit, sometimes narrowed at base to a short stipe. Rocky slopes and summits of the Coast and Mt. Diablo ranges of mountains, from Mendocino Co. to Mt. Hamilton and Santa Cruz; also in the Sierra Nevada, according to Brewer & Watson.

15. **D. cardinale,** Hook. Bot. Mag. t. 4887 (1855): *D. coccineum,* Torr. Pac. R. Rep. iv. 62 (1857). Stout, leafy up to the long rather dense raceme, and often 5–8 ft. high: leaves large; segments acuminate: fl. 1 in. broad; sepals widely expanding: follicles (often 5 or 6) erect. –A magnificent species, common in the Coast Range from perhaps near Monterey southward throughout the State. July.

7. **ACONITUM,** *Theophrastus* (MONKS' HOOD. ACONITE). In all respects like *Delphinium* (its separation therefrom a mere conventionality), save that the sepals are never wide-expanded, but rather connivent, and that the uppermost one is arched into a hood or helmet-shaped organ, instead of being prolonged into a spur.

1. **A. Columbianum,** Nutt. T. & G. Fl. i. 34 (1838): *A. nasutum*, Hook. Fl. i. 26 (1829), not Fisch.: *A. Fischeri*, B. & W. i. 12, not Reichenb. Glabrous or sparingly pubescent, 2-5 ft. high; leafy up to the rather long and very loose raceme: leaves 3-5-cleft: segments broadly cuneate, laciniately toothed or cleft: hood narrowly oblong, with a salient acute beak: follicles 3 (rarely 5), glabrous, erect. Frequent at rather high altitudes in the Sierra Nevada; also in the Coast Range from Lake Co., northward, in moist shades, along cold brooklets, etc. June-Sept.

8. **PÆONIA,** *Dioscorides*. Stout perennials with more or less distinctly tuberous and clustered, or at least fleshy and branching roots. Leaves ternately compound. Flowers solitary at the ends of the mostly simple stems. Sepals 5, persistent. Petals 5 or more, rounded, concave, red. Stamens ∞, inserted on a plane disk founded on the united bases of the sepals. Pistils 2-5, becoming large leathery follicles.

1. **P. Brownii,** Dougl. Hook. Fl. i. 27 (1829); Greene, Gard. & Forest, iii. 356. Roots deep-seated, elongated and branching, fleshy but scarcely tuberous: stem 1 ft. high and, with the herbage, very glaucous: leaves of cordate-ovate outline, the numerous segments oblong, obtuse. In the higher Sierra, from the middle, or perhaps even from southerly sections of the State northward to British Columbia. It is also credited to Marin Co. by *Dr. Behr,* whose plant, however, is more likely to be of the next species. June.

2. **P. Californica,** Nutt.; T. & G. Fl. i. 41 (1838); Greene, l. c. Roots rather more distinctly tuberous-thickened: stem and herbage when mature scarcely glaucous: leaves of pedate outline (broader than long), their segments lanceolate or oblong, acute. On hillsides, among bushes, in the southern parts of the State, both along the seaboard and far inland; first discovered near Santa Barbara, by *Nuttall;* readily distinguishable from the preceding, even in the herbarium. March.

9. **CALTHA,** *C. Gesner* (MARSH MARIGOLD). Perennial herbs of very wet ground. Roots coarse-fibrous, fascicled. Leaves mainly or altogether radical, undivided, cordate or subsagittate. Flowers showy, terminal, solitary or several. Sepals 5-12, petaloid. Petals 0. Stamens ∞. Pistils 5—12, becoming small follicles.

1. **C. leptosepala,** DC. Syst. i. 310 (1818); Hook. Fl. i. 22. t. 10. Leaves long-petioled, from oval to reniform, cordate at base, the sinus

narrow or closed, the margin crenate or entire: stems several, leafless, 1-flowered; sepals about 10, linear-oblong, white, or with a tinge of lurid purple on the outside: carpels 8—10, very shortly stipitate, pointed with the short slender style.—In marshy grounds of the higher Sierra.

10. ISOPYRUM, *Linnæus.* Low slender very flaccid perennials, with bi- or triternately compound leaves, and a few smallish white flowers. Sepals 5, petaloid. Petals 0. Stamens 10—40. Pistils 3—6, becoming transversely veined rather few-seeded follicles.

1. **I. occidentale,** H. & A. Bot. Beech. 316 (1840). Roots fascicled, fibrous: stem 6—10 in. high, parted above into few 1-flowered branches: leaflets 4—8 lines long, irregularly 3-lobed: pods short, sessile, obliquely pointed. — Under oaks, or other trees and shrubs, among the foothills on either side of the valley of the Sacramento; Forest Hill, *Bolander*, and Vaca Mts., *Jepson.* Mar., Apr.

2. **I. stipitatum,** Gray, Proc. Am. Acad. xii. 54 (1876): *I. Clarkei,* Kell. Proc. Calif. Acad. vii. 131 (1877). Much smaller, the roots thickened almost to the tuberous: leaflets deeply 3-parted into oblong segments: fl. solitary: follicles broadly oblong, obtuse, distinctly stipitate.—Mendocino Co., *J. H. Clarke*, and northward. Apr., May.

11. COPTIS, *Salisbury* (GOLD-THREAD). Low perennials; the hard tenacious running rootstocks and fibrous roots yellow. Leaves all radical, ternately compound, coriaceous, evergreen. Flowers few, borne umbellately at summit of a naked scape. Sepals 5—7, petaloid, narrow. Petals as many, narrow. Stamens 10—20. Pistils 3—5 or more, becoming stipitate follicles containing several crustaceous shining seeds.

1. **C. laciniata,** Gray, Bot. Gaz. xii. 297 (1887): *C. asplenifolia,* Wats. Bot. Calif., not Salisb. Leaves trifoliolate, the lateral leaflets shortstalked, all ovate, nearly 3-parted, their divisions incised, acute: sepals and petals linear-attenuate: mature pods longer than their stipe: seeds oval. — A rare denizen of deep moist woodlands in Mendocino and Humboldt counties, *G. R. Vasey, C. C. Marshall.*

12. AQUILEGIA, *Tragus* (COLUMBINE). Perennials, branching above. Leaves mostly radical and biternate; the leaflets with rounded lobes; texture membranaceous. Flowers, at the ends of the few branches, large, showy, usually somewhat nodding. Sepals 5, plane, colored like the petals. Petals 5, tubular, projecting like hollow spurs behind the sepals, and ending in a small globular cavity which is filled with honey. Pistils 5, becoming follicles, each with many black shining seeds.

1. **A. truncata,** F. & M. Ind. Sem. Petr. Suppl. 8 (1843); C. A. Mey. Sert. Petr. t. 11: *A. Californica,* Lindl. Gard. Chron. 836 (1854): *A. eximia*

RANUNCULACEÆ. 309

Van Houtte, Fl. Serr. t. 1188 (1857). Glabrous, or in southerly stations notably puberulent, 1 3 ft. high: fl. 1½ -2 in. broad, red tinged with yellow: sepals widely spreading or reflexed: petals truncate, the limb very short: spurs ½ ¾ in. long, thick and blunt. Common in shady ravines and on banks of streams. Apr., May, on the seaboard; June, July, in the Sierra.

2. **A. leptocera,** Nutt. Journ. Philad. Acad. vii. 9 (1834): *A. cærulea*, B. & W. Bot. Calif. i. 10, not James: *A. macrantha*, H. & A. Bot. Beech. t. 72 (1840). Neither as tall nor as branching as the above: fl. very large, pale yellow; sepals spreading, 2 3 in.; spurs slender, 2 3 in. long; pistils in the central earliest flowers commonly 7 or 8.—Woods of the higher Sierra; equally related to *A. cærulea* of the Rocky Mts., and to *A. chrysantha* of Mexico, but distinct from both. June, July.

13. **THALICTRUM,** *Dioscorides* (MEADOW-RUE). Tall perennials, often heavy-scented, with fibrous roots, hollow stems, bi- or triternately compound leaves, and many panicled greenish imperfect (diœcious) flowers. Sepals 4 7, small, deciduous. Petals 0. Stamens ∞, with slender linear anthers on rather long almost capillary filaments. Pistils few or many, becoming ribbed or veined achenes which are tipped with the persistent beak-like style, and disposed in roundish heads.

1. **T. polycarpum,** Wats. Bot. Calif. ii. 424 (1880), partly. Stout, 3—4 ft. high, glabrous, not glaucous, aromatic-scented: leaves short-petioled or sessile; leaflets with acute or acuminate lobes: sepals lanceolate, not scarious: achenes very many in the head, broadly obovoid, short-stipitate, compressed, turgid, the style abruptly curved, the sides marked with low more or less anastomosing veins.—In open places near streams, chiefly in the Coast Range above and below San Francisco. Easily recognized by the peculiar aromatic odor, sharp-pointed leaf-lobes, and crowded heads of large turgid achenes. May, June.

2. **T. cæsium.** Tall as the last, but less robust, not at all aromatic, glaucous throughout, even to the achenes: sepals oblong, obtuse, thin and somewhat scarious-margined: achenes from nearly orbicular to broadly lanceolate, veined as in the last, but not turgid. Foothills of the Sierra, from Calaveras Co. northward; common near Chico: also in Lake Co., *Mr. Whitmore*. Apparently confused with *T. polycarpum* hitherto, but occupying a different range, and very clearly distinct.

3. **T. hesperium,** Greene, Pittonia, ii. 24 (1889): *T. dioicum*, Boland. Cat. 3 (1870), not Linn.: *T. Fendleri*, B. & W. Bot. Calif. i. 4 (1876) partly, not Engelm.: *T. Fendleri*, var. *platycarpum*, Trel. Proc. Bost. Soc. xxiii. 304 (1886). Tall, scentless, glabrous except the growing parts and the lower face of the leaves, which have a sparse minutely gland-tipped

pubescence: lobes of the leaflets rounded: sepals 5, lanceolate, not scarious: achenes obliquely oval or semi-obovate, substipitate, the ribs or veins distinct and parallel. Of more general distribution than any of the above, inhabiting both the inner Coast mountains and the Sierra Nevada; common along streams in the Oakland Hills, but not reported from west of San Francisco Bay, where *T. polycarpum* replaces it. The species seems almost or quite confluent with *T. Fendleri* of the southern Rocky Mountains.

4. **T. occidentale,** Gray, Proc. Am. Acad. viii. 372 (1872). Rather slender, 2 ft. high: leaflets thin, sparingly glandular-puberulent beneath: achenes long ($1\frac{1}{4}$ – $1\frac{1}{2}$ in.), lanceolate, not oblique, substipitate, tapering above to a slender beak, the side parallel-ribbed.—A species of the distant north and north-east, but frequent in extreme northern Calif., reaching our limits in Sierra Co., at Gold Lake, *C. A. Ramm.* June–Aug.

5. **T. sparsiflorum,** Turcz.; Ind. Sem. Petr. i. 40 (1835): *T. Richardsonii*, Gray, Am. Journ. Sci. xlii. 17 (1842). Tall and rather slender, thin-leaved, strongly rue-scented: heads of achenes nodding: achenes very oblique, much flattened, tipped with a short incurved style, the sides with low nerves. In the higher Sierra, from Donner Lake northward.

14. **TRAUTVETTERIA,** *Fischer & Meyer.* Perennial herbs, with palmately lobed leaves both radical and cauline. Flowers small, white, terminal, in a corymbose panicle. Sepals 3–5, concave, petaloid, caducous. Petals 0. Pistils ∞, becoming membranaceous 4-angled somewhat bladdery 1-seeded fruits disposed in heads. Seed ascending.

1. **T. grandis,** Nutt.; T. & G. Fl. i. 37 (1838); Wats. Bot. Calif. ii. 425. Slender, 1–3 ft. high: leaves few, the radical long-stalked, all of thin texture, deeply 5–7-lobed, the lobes acuminate, laciniate-toothed, beneath showing a sparse curled pubescence: achenes little more than a line long, broadly gibbous at base, disposed in globose heads.—A plant of the far North, found in Plumas Co., on Mill Creek, *Mrs. Austin.*

15. **ACTÆA,** *Linnæus* (HERB-CHRISTOPHER, BANE-BERRY). Perennials, with roots somewhat knotted and tuberous. Leaves ample, ternately compound. Flowers small, white, in a terminal raceme. Sepals about 4, caducous. Petals 1 or more. Stamens ∞. Pistil 1; stigma sessile, 2-lobed. Fruit indehiscent, berry-like, the fleshy pericarp, with a false line of dehiscence on one side, enclosing 2 closely packed vertical rows of flattened semiorbicular seeds.—A small genus, seeming to form one of the connecting links between this family and the Berberideæ.

1. **A. arguta,** Nutt.; T. & G. Fl. i. 35 (1838): *A. spicata,* var. *arguta,* Torr. Pac. R. Rep. iv. 63 (1857): *A. rubra,* var. *arguta,* Greene, Pitt. ii. 108 (1890). Stem 2–3 ft. high, bearing one or more large stalked leaves;

leaflets acute, coarsely and incisely serrate; raceme oblong, 1—2 in. long, often with one or more short branches at base: sepals obovate, concave: petals with rhombic-ovate acute limb and nearly filiform claw: stamens 25 or 30: filaments filiform, or slightly enlarged under the minute roundish anthers: berries rather obliquely oval, as large as peas, deep cherry, red, or occasionally snow-white.—Common on wooded northward slopes, under hazel bushes, etc., in the Berkeley Hills; also in the woods of the Coast Range from Santa Cruz Co. northward. Specimens from the mountains of Fresno Co. seem doubtfully referable to this. Distinct enough from the Old World *A. spicata*, which has emarginate petals, black berries, etc.; more nearly allied to the East American *A. rubra*. Feb.

Order LII. SARMENTOSÆ.

L. Gerard, Flora Gallo-Provincialis, 378 (1761). SARMENTACEÆ, Vent. Tabl. iii. 167 (1799). VINIFERÆ, J. St.-Hilaire, Exp. Fam. Nat. ii. t. 79 (1805). AMPELIDEÆ, HBK. (1815). VITACEÆ, Lindl. Intr. 2 ed. 30 (1836). VITES, Juss. (1789).

A small family, important as containing the Grape; closely allied to the Araliaceæ, and connecting with Rhamneæ in some aspects of flower and fruit; also with Ranunculaceæ, through *Clematitis*, in vegetative characters, habit, etc. We have but one native species.

1. VITIS, *Varro* (GRAPE). Shrubs with watery juice, climbing by branching tendrils placed opposite the leaves. Flowers small, greenish, very numerous, in thyrsiform clusters opposite the leaves. Calyx minute, cup-like, with or without traces of 4 or 5 teeth. Petals 4 or 5, distinct and spreading, or distinct at base only and united at apex, then falling off like a calyptra. Stamens as many as the petals and opposite them, inserted on a perigynous disk or elevation of the torus: filaments slender; anthers introrse. Pistil with a short style or none; stigma slightly 2-lobed. Fruit baccate, 1—4-seeded. Seeds bony, rather large, grooved on one side; embryo small, in a hard albumen.

1. **V. Californica,** Benth. Bot. Sulph. 10 (1844). Stem often 1—2 in. thick below, climbing trees to the height of 20—50 ft.: leaves 3 in. long, nearly as broad, round-cordate with deep and narrow sinus, obtuse, rather coarsely serrate, sometimes 3-lobed, canescently tomentose beneath, and when young more or less so on both faces: fr. 4 lines thick, in large clusters, purple, glaucous: seeds broad.—Along streams almost throughout the State, except in the higher mountains; but also absent from the immediate seaboard, especially in the northwestern districts.

2. **V. VINIFERA,** Linn., the wine grape, native of the Old World, has escaped from cultivation, and will occasionally be seen in a wild state.

Order LIII. ARALIACEÆ.

A. Richard; Dictionaire Classique d'Histoire Naturelle, i. 506 (1822). ARALIÆ, Juss. (1789). Tribe of UMBELLIFERÆ, Baillon (1880).

Herbs, shrubs or trees, with mostly stout hollow stems, and alternate lobed or compound leaves. Flowers small, in simple but often panicled or racemosely arranged umbels. Calyx joined to the ovary, entire or toothed. Petals 5--10, deciduous. Stamens as many or twice as many as the petals, inserted around the border of the calyx outside of an epigynous disk. Ovary more than 2-celled; styles as many as the cells, sometimes connate. Fruit berry-like. Seeds pendulous; embryo minute; albumen fleshy. An order closely allied to the Grape Family; but so near to Umbelliferæ as to be scarcely separable therefrom by its fleshy and more than bicarpellary fruits.

1. **ARALIA,** *Vaillant* (SPIKENARD). Our species a very coarse perennial herb, with ternately compound leaves and large serrate leaflets. Calyx 5-toothed or entire. Petals 5, ovate, slightly imbricate. Stamens 5. Disk depressed or 0. Fruit laterally compressed, becoming 3--5-angled, fleshy externally; endocarp chartaceous.

1. **A. Californica,** Wats. Proc. Am. Acad. xi. 144 (1876). Herbaceous, unarmed, 6--10 ft. high, from a large thick perennial root: leaves bipinnate, or the upper pinnate with only 1--2 pairs of leaflets: leaflets cordate-ovate, 4--10 in. long, abruptly acuminate, simply or doubly serrate with short acute teeth: umbels in loose terminal and axillary compound or simple racemose panicles which are 1--2 ft. long, each umbel subtended by several linear bractlets: fl. 2 lines long; disk and style-base (stylopodium) obsolete; styles united for half their length: fr. 2 lines long.—In shaded and moist ravines of the Coast Range.

2. **HEDERA,** *Pliny* (IVY). Shrubby, climbing by aerial roots. Leaves coriaceous, evergreen, simple, lobed. Flowers in a terminal panicle of umbels. Calyx 5-toothed. Petals 5. Stamens 5. Styles united into a single very short one. Cells of the ovary 5 or 10. Berry smooth and black, with 2--5 seeds.

1. H. HELIX, Gerard, Herb. Em. 857 (1633); Park. Theatr. 679 (1640); Linn. Sp. Pl. i. 292 (1753). Leaves ovate, angularly 3--5-lobed, those of the sterile and young shoots more deeply so than those of the flowering branches; these bushy, erect, projecting a foot or more from the climbing main stem: umbels globose: fl. yellowish-green.- The English Ivy, common on trees in parks, and on buildings, and well adapted to our climate, fruits freely here, and will often be met with wild, as an escape from cultivation.

Order LIV. UMBELLIFERÆ.

Morison, Plantarum Umbelliferarum Distributio Nova, 5 (1672); Ray, Methodus Plantarum, 47 (1703); Van Royen, Flora Leydensis Prodromus, 91 (1740); Haller, Hortus et Ager Gottingensis, 171 (1753); L. Gerard, Flora Gallo-Provincialis, 230 (1761); Crantz, Inst. ii. 113 (1766); Juss. Gen. 218 (1789).

Herbs with mostly hollow, often striate, angled or fluted stems, mostly compound leaves which are prevailingly alternate; the petiole dilated or even sheathing at base. Flowers small, in simple or compound umbels (sometimes sessile and therefore capitate). Calyx almost wholly adnate to the 2-celled ovary. Petals 5, mostly valvate in bud, usually inflexed at apex in flower. Stamens 5, epigynous, alternate with the petals; anthers ovate, subdidymous. Styles 2, simple, more or less dilated at base into a *stylopodium*. Fruit of 2 closely approximated and often ribbed, sometimes winged, always 1-seeded carpels; the intervals between the ribs usually occupied by one or more oil-tubes or *vittæ*. The face by which the two carpels meet or partly cohere is called the *commissure*. A slender prolongation of the axis between these faces is called a *carpophore*, which, in maturity, is apt to split into 2 branches, with a carpel suspended from each. An extensive and very natural family, of considerable economic importance, on account of the wholesome fleshy roots of some species, the aromatic seeds of others, and the medical properties inherent in many. The green herbage in many is acrid and poisonous; the rootstocks, tuberous roots, etc., of the half-aquatic species are dangerous to cattle and horses that are apt to feed on them in early spring.

Hints of the Genera.

Umbels simple, or imperfectly or irregularly compound;
    Leaves simple, neither spinosely nor setaceously toothed, - - 1, 2
    "   spinosely toothed, or lobed or parted, - - - - 3, 4
Umbels regularly compound; leaves compound, often finely dissected;
    Ribs of the carpels with barbed or hooked prickles, - - 28, 29
    Fruit more or less flattened laterally, broadly ovate or subglobose or elliptic-oblong, not broadly winged;
        Oblong or rounded; ribs filiform or prominent; oil-tubes 2 or 3 in the intervals, - - - - - 5
        Broadly ovate; ribs prominent, obtuse; oil-tubes none, 6
        Ovate or oblong; ribs prominent, corky, oil-tubes 1–3, 7, 8
        Very small; ribs not prominent; oil-tubes 1 to the interval, - - - - - - - - 10, 11
        Ovate or oblong; ribs filiform; oil-tubes 1 or 2 to the interval, - - - - - - · - - 12, 13
        Ovate, with broad commissure; ribs rather prominent; oil-tubes 1–3, - - - - - - - 14–16
        Linear or linear-oblong, rather large, not winged, - 22, 25–27
    Fruit not compressed; ribs corky, rounded, - - - - - 9
    "   somewhat compressed dorsally; some of the ribs narrowly winged, - - - - , - - - - - 17, 18

Fruit much compressed dorsally and winged;
Lateral wings broad, distinct, the dorsal ones often less
prominent, - - - - - - - - 19–21
Lateral wings thin, coherent until maturity; dorsal
ribs filiform, - - - - - - - - 22—24

1. **HYDROCOTYLE,** *Tournefort* (MARSH PENNYWORT). Low glabrous herbs, growing in or near water, with creeping stems. Leaves rounded, toothed or lobed, sometimes peltate; stipules scale-like. Flowers inconspicuous, in simple umbels, or in whorls one above another, on a scapiform erect peduncle. Calyx-teeth obsolete. Petals entire, acute. Fruit flattened laterally, suborbicular, acutely margined, and with 2 or more less prominent ribs or nerves on each side; oil-tubes 0: carpels coherent.

1. **H. prolifera,** Kellogg, Proc. Calif. Acad. i. 15 (1854). Herbage light green and flaccid: leaves about 1 in. broad, peltate, emarginate at base, simply crenate, on petioles 1–4 in. long: peduncles equalling or exceeding the leaves: fl. in 1–4 whorls, each 4- 12-flowered, with many bractlets; pedicels 1—6 lines long: fr. 1 line wide, emarginate at base; ribs 2 on each side, prominent.—Said to occur near San Francisco, a locality which we have not been able to confirm; but it is common in the interior, near Sacramento, *Drew;* also in the Suisun marshes, where it exhibits dichotomously branched peduncles. June—Aug.

2. **H. ranunculoides,** Linn. f. Suppl. 177 (1781). Herbage dark green, of firm texture: leaves (sometimes floating) 1- 2 in. broad, round-reniform, 3–7-cleft, the lobes crenate; petioles 2–10 in. long; peduncles much shorter ($\frac{1}{2}$ 3 in.), reflexed in fruit: fl. 5 10 in a capitate umbel: fr. very shortly pedicellate, 1 -1$\frac{1}{2}$ lines broad, with thickened but scarcely angled margins, rather obscurely nerved on each side, longer than the pedicels.—Abundant in shallow ponds, margins of lakes, etc., along the seaboard.

2. **BOWLESIA,** *Ruiz & Pavon.* Slender very flaccid herbs, with sparse stellate pubescence, and opposite simple leaves with scarious lacerate stipules. Flowers minute, white, in simple few-flowered umbels on axillary peduncles. Calyx-teeth rather prominent. Petals elliptical, obtusish. Fruit broadly ovate, with narrow commissure, turgid, becoming depressed on the back, without ribs or oil-tubes.

1. **B. lobata,** R. & P. Fl. Peruv. iii. 28 (1802). Annual, the slender stems more or less dichotomous, 2 in. to 1 ft. long: leaves round-reniform or cordate, $\frac{1}{2}$ 1$\frac{1}{2}$ in. broad, shorter than the slender petioles, deeply 5-lobed; lobes acutish, entire or few-toothed: umbels short-peduncled, 1—4-flowered: fr. 1 line long, sessile or nearly so, pubescent, the inflated calyx not adherent to the carpels. -Among rocks, under trees etc., on hillsides from the valley of the Sacramento southward; frequent in the Coast hills south of San Francisco, and in Napa Valley. Apr., May.

UMBELLIFERÆ. 315

3. ERYNGIUM, *Nicander* (BUTTON SNAKEROOT). Perennial herbs with rigid coriaceous spinosely toothed or divided leaves, and white or blue flowers sessile in dense heads which are encircled by a series of bracts forming an involucre; each flower also subtended by a rigid bract. Calyx-teeth manifest, rigid, persistent. Fruit ovoid or obovoid, scarcely compressed, covered with hyaline scales or vesicles; ribs obsolete; oil-tubes 0; carpels and seeds semiterete.

1. **E. armatum,** C. & R. Bot. Gaz. xiii. 141 (1888). Diffusely branching, 1 ft. high or more: radical leaves oblanceolate, serrately or spinosely dentate or incised, attenuate to a margined petiole; cauline narrower, sessile: heads peduncled, globose, $\frac{1}{2}$ in. thick; bracts of involucre triangular-lanceolate, entire, thick-margined, 1 in. long and much exceeding the head; bractlets similar and as prominent: fr. with lanceolate-acuminate calyx-lobes longer than the styles. Common in low ground, on the plains and among the foothills, almost throughout the State.

2. **E. Vaseyi,** C. & R. l. c. 142. Smaller, branching above: leaves oblanceolate, irregularly spinulose-serrate, attenuate at base: involucral bracts narrow, rigid, spinescent at tip and spinose-toothed, 1 in. long or less; bractlets similar: fr. with lanceolate acuminate-cuspidate calyx-lobes exceeding the short styles. From near Mt. Shasta southward to San Luis Obispo Co.

3. **E. petiolatum,** Hook. Fl. i. 259 (1833). Erect, 1–5 ft. high, branching above: radical leaves oblanceolate, irregularly spinose-serrate, narrowed to an elongated fistulous petiole, or the very lowest (submersed when young) reduced to a long terete petiole; cauline mostly sessile: heads peduncled, globose, $\frac{1}{2}$ in. high; involucral bracts linear-lanceolate, spinosely tipped and toothed, often 1 in. long; bractlets lanceolate, cuspidate-tipped, little exceeding the flowers, scarious-winged below: fr. with calyx-lobes like the bractlets but smaller, shorter than the long styles. Var. **minimum,** C. & R. l. c. Only 1–3 in. high, all the parts correspondingly reduced; bracts of the involucre equalling the heads.— In marshes, but less common than the above; the variety only in the Sierra, at Donner Lake, *Sonne.*

4. **E. articulatum,** Hook. Lond. Journ. Bot. vi. 232 (1845). More or less branching, erect, decumbent or rarely prostrate: radical and lower leaves consisting of a long articulated petiole with or without a small lanceolate entire or laciniate blade; cauline sessile: bracts of involucre $\frac{1}{2}$ in. long, exceeding the heads, linear, cuspidate, spinosely toothed; bractlets tricuspidate, little exceeding the flowers, the central cusp largest: calyx-lobes lanceolate, cuspidate, little exceeding the styles. Var. **microcephalum,** C. & R. Revis. Umb. 99 (1888). Very small and slender: bracts ovate-acuminate, little surpassing the heads, these only 2–3 lines

long: calyx-lobes short-mucronate. In swamps and wet meadows from San Luis Obispo Co. to Plumas.

3. **E. Harknessii,** Curran, Bull. Calif. Acad. i. 153 (1885). Slender, not rigid, dichotomously branching, 2–4 ft. high: leaves much as in the last, but blade of the lowest with perfectly entire and unarmed margin; cauline petiolate, sparingly soft-spinulose on the margin: heads round-ovate, $\frac{3}{4}$ in. high, blue; bracts of the involucre longer than the head but deflexed: calyx-segments subulate, pungently mucronate, equalling the long styles. In the Suisun Marsh, *Bolander* (1864), *Greene, Curran* (1883); named by the latter, in compliment to Dr. Harkness.

4. SANICULA, Brunfels (SANICLE).

Glabrous perennials (n. 1 biennial), with chiefly radical leaves, these mostly palmately divided and sometimes subdivided. Flowers unisexual, in irregularly compound few-rayed umbels; these involucrate with sessile leaf-like usually toothed bracts; the bracts of the involucels usually small and entire. Calyx-teeth somewhat foliaceous, persistent. Fruit subglobose or obovoid, densely uncinate-prickly or tuberculate; ribs obsolete; oil-tubes many. Seed hemispherical.

* *Mature fruit pedicelled; leaves palmately lobed or divided.*

1. **S. Menziesii,** H. & A.: Hook. Fl. i. 258. t. 90 (1833); Bot. Beech. 141 and 347 (1840). Biennial: stem solitary, erect, branching loosely above, 2–5 ft. high: leaves 2–3 in. broad, of rounded outline, but with deep broad lobes and cordate base, the shining surface delicately rugose; the 3–5 lobes sharply toothed, the teeth setaceously tipped; cauline leaves parted or divided into about 3 narrow segments: involucre small, of 2 or 3 narrow leaflets; the involucels of 6–8 lanceolate entire bracts a line long: sterile fl. nearly sessile: fruits 4–8 in each head, becoming distinctly pedicellate and divergent, obovate, a line long or more, covered with hooked prickles. Abundant in moist open woods, and along streams in shade of thickets, throughout middle California toward the seaboard, and far northward. May, June.

2. **S. arctopoides,** H. & A.; Hook. l. c. t. 91 (1833); Bot. Beech. 141 (1840). The whole herbage of a greenish yellow, and with an offensive odor: main stem simple, very short; the many scape-like flowering branches at first depressed, later becoming elongated and divergent, 3–6 in. long, each bearing an umbel of 1–3 elongated rays: leaves deeply 3-parted, the lanceolate segments once or twice laciniately cleft: involucre of 1 or 2 leaflets; heads large, $\frac{1}{2}$ in. broad, encircled by 8 or 10 oblanceolate mostly entire bracts which are yellow and resemble the rays of a composite: fr. $1\frac{1}{2}$ lines long, naked at base, strongly armed above. — Plentiful on bleak hills near the sea, at San Francisco and far northward; also here and there in the interior of California. Feb.–Apr.

* * *Mature fruit sessile; leaves palmately divided (except in n. 5).*

3. **S. nudicaulis,** H. & A. Bot. Beech. 317 (1840); *S. laciniata,* H. & A. l. c. Stems several, slender, erect, 1 ft. high or more: leaves long-petioled, of cordate outline, 3-parted; divisions laciniately once or twice pinnatifid, the segments with widely spreading acute often spinosely pointed teeth; fl. yellow, in many small heads disposed in compound umbels terminating sparingly leafy branches: fr. naked at base, uncinate-bristly above. Wooded hills, among bushes, along borders of thickets etc., from Humboldt Co. to Monterey, towards the sea. It is not improbable that *S. nudicaulis* and *laciniata* may be proven distinct; but while they are treated as one, to the former name must be conceded the priority which belongs to it. Mar. May.

4. **S. Nevadensis,** Wats. Proc. Am. Acad. xi. 139 (1876). Stout and low, the numerous branches ascending and almost scapiform, 1–6 in. long; leaves ternate, the divisions oblong-ovate, 3–5-lobed, the segments lobed or toothed: rays about 5, sometimes branched: involucre of pinnatifid bracts: fl. greenish or yellowish, the sterile ones equalling the pedicels: fr. covered with stout forked prickles. Foothills of both ranges of mountains, but chiefly northward.

5. **S. maritima,** Kellogg; Bot. Calif. ii. 451 (1880); Greene, Pitt. i. 269 (1889). Stoutish, 1 ft. high, rather fleshy: radical leaves long-petioled, the lowest oblong-cordate, not lobed, but crenate-dentate; some of the later more or less deeply 3-lobed, 2–4 in. long: involucre of large leaf-like lobed or parted bracts: umbel of about 3 elongated rays: fl. yellow, the sterile ones short-pedicellate: fr. nearly naked below, prickly above, 2 lines long. —In moist lowlands adjacent to salt marshes about San Francisco Bay near Alameda, San Francisco, etc. Mar. May.

* * * *Fruit sessile; leaves pinnately divided and subdivided.*

6. **S. bipinnatifida,** Dougl.; Hook. Fl. i. 258 (1833). Stoutish, slightly fleshy, 1–2 ft. high, herbage of a peculiarly dark green: leaves mostly radical, but an opposite pair on the stem near the base, with 1–3 above these, all pinnately 3–7-parted, the divisions incisely toothed or lobed, decurrent on the toothed rachis, the teeth acutely or somewhat setaceously pointed: umbel of 3 or 4 greatly elongated rays: fl. of a very dark purplish red: fr. 1½ lines long, prickly. Very common on hillsides and open grounds generally. Mar. May.

7. **S. bipinnata,** H. & A. Bot. Beech. 347 (1840). Like the last in size: segments of the bipinnate leaves remote, not decurrent, narrowly obovate, cuneate, incisely mucronate-dentate: umbel compound: fl. yellow: fr. naked at base, echinate above. Apparently of the interior only, from Kern Co. to Butte. Feb. Apr.

8. **S. tuberosa,** Torr. Pac. R. Rep. iv. 91 (1857). Very slender, the

solitary erect freely branching stem 6–18 in. high, from a small roundish not deeply seated tuberous root: leaves small, finely twice or thrice pinnate, the ultimate segments small: umbels 1 4-rayed, small: fl. yellow, the sterile ones long pedicelled: fr. broader than long, tuberculate. -Rocky hills, in sterile clayey soil, in both the Sierra and the Coast Range. Mar.–May.

5. ARRACACIA, *Bancroft*. Perennials, glabrous or pubescent. Roots thick, elongated, yellow, fragrant. Leaves mostly radical, pinnately or ternately compound. Involucre sometimes wanting. Involucels conspicuous. Flowers yellow. Calyx-teeth obsolete or prominent. Fruit somewhat flattened laterally, with prominent equal filiform ribs, and thin pericarp. Oil-tubes conspicuous, 3–6 in the intervals, 4–10 on the commissural side.

1. **A. Hartwegi,** Wats. Proc. Am. Acad. xxii. 415 (1887); Gray, Proc. Am. Acad. vii. 342 (1868), under *Deweya;* C. & R. Rev. Umb. 121 (1888), under *Velæa*. Sub.acaulescent, light green, the petioles and veins somewhat scabrous: leaves biternate and quinate; leaflets obovate or oval-oblong, 1–2 in. long, mostly confluent, coarsely and deeply mucronate-serrate: peduncles 1–2 ft. high; umbel 16–20-rayed, usually without involucre, but the umbellets subtended by linear-oblong reflexed bractlets; rays 2½–4 in. long; pedicels short: fr. nearly orbicular, smooth, 3–4 lines long, 2½–3 lines broad, sharply ribbed.—Foothills of the Sierra, from Butte Co. southward; also near San Francisco.

2. **A. Kelloggii,** Wats. l. c.; Gray, l. c. 343, under *Deweya;* C. & R. l. c., under *Velæa*. More slender than the last, mostly puberulent: leaves triternate: leaflets ovate, ½–¾ in. long, usually 3-lobed: umbel 8–16-rayed, mostly without involucre, the involucels of small linear bractlets; rays 1–3 in. long: fr. 1–2 lines long, nearly as broad, retuse at base, the ribs filiform.—Hills of the Coast Range, in wooded or open ground.

3. **A. Parishii,** Greene. C. & R. Rev. Umb. 121 (1888), under *Velæa*. Nearly acaulescent, 1 ft. high, glabrous, somewhat fleshy: leaves ternately pinnatifid; segments ovate, irregularly lobed and cuspidate-toothed, the margins revolute: umbel about 20-rayed, with no involucre, but involucels of a few setaceous bractlets; rays 2 in. long or more; pedicels about 4 lines: calyx-teeth prominent: fr. glabrous, 3 lines long, with prominent ribs. Of the South chiefly, but reaching our limits in Tulare Co., at 8,000 ft. in the mountains.

4. **A. vestita,** Wats. Proc. Am. Acad. xxii. 415 (1887); l. c. xvii. 374 (1882), under *Deweya;* C. & R. l. c. 122 (1888), under *Velæa*. Stemless, 2–4 in. high, hoary with a short hirsutulous pubescence: ternately compound leaves only 1–2 in. long; segments crowded and confluent, only 1–2 lines long; involucre 0; rays many, 4–8 lines long, the involucels

of several short lanceolate bractlets: fr. sessile, pubescent, $2\frac{1}{2}$ lines long, with inconspicuous ribs. Mountains of Tulare Co., *Palmer*, southward to San Bernardino, at great elevations.

6. **CONIUM,** *Linnæus* (POISON HEMLOCK). Tall glabrous biennial, with large ternately-dissected thin leaves, and compound umbels of small white flowers terminating the paniculate branches. Calyx-teeth obsolete. Fruit broadly ovate, laterally compressed; carpels with 5 prominent obtuse often undulate or crenulate ribs, and no oil-tubes.

1. C. MACULATUM, Linn. Sp. Pl. i. 243 (1753). Root fusiform: stem stout, fistulous, 3–7 ft. high, glaucescent, spotted with purple: leaves a foot long or more, two-thirds as broad: segments $\frac{1}{2}$ in. long, pinnatifid, the lobes acute: umbels 12–20-rayed: rays 1–$1\frac{1}{2}$ in. long: fr. $1\frac{1}{2}$ lines long, shorter than the pedicels. Waste grounds, in shady places; rather rare in California, but of rank growth.

7. **SIUM,** *Dioscorides* (WATER PARSNIP). Glabrous perennial aquatics, with angled stems, pinnate leaves with leaflets pinnatifid or serrate, and white flowers; the involucres and involucels of several bracts. Calyx-teeth minute. Fruit oblong, ovate or nearly globose; ribs prominent or obscure; oil-tubes few or many in the intervals.

* *Fruit with corky ribs; oil-tubes between them.*—SIUM proper.

1. **S. heterophyllum,** Greene, Pittonia, ii. 102 (1890): *S. cicutæfolium*, Bot. Calif. i. 261 partly, not Gmel. Stem stoutish and brittle, strongly angular and somewhat flexuous, 3 ft. high, from a cluster of fleshy fibrous roots, these thickened below the middle: lowest leaves simple, 2–10 in. long, rhombic-lanceolate, serrate or laciniate, on a stout fistulous petiole which is still longer and usually submersed; the later radical 3-lobed or -parted, thus passing to the cauline which are truly pinnate, but mostly with only 2 or 3 pairs of leaflets, these broadly lanceolate, acute, serrate: bracts of involucre broadly lanceolate, acute at each end: fr. $1\frac{1}{2}$ lines long, broadly ovoid; oil-tubes broad, solitary between the ribs, 2 on the commissural side: cross-section of seed angular.—Common in brackish swamps, under the influence of tide-water, at Suisun, Stockton, etc.

2. **S. cicutæfolium,** Gmel. Syst. ii. 482 (1791): *S. lineare*, Michx. Fl. i. 167 (1803). Taller, more slender, less branching, not flexuous: leaves all pinnate, the leaflets of the earliest often pinnatifid or even dissected into filiform subdivisions, those of the later in 6–8 pairs, oblong-lanceolate to linear, 2–4 in. long, acuminate, sharply serrate: involucre and involucels of 6–8 linear bracts: fr. oblong, $1\frac{1}{2}$ lines long: oil-tubes narrower, 2 or 3 in each interval, 3 or more on the broad side.—On the eastern slope of the Sierra northward, in Plumas Co., etc.

** *Fruit with angled corky covering; oil-tubes beneath this.* –
Genus BERULA, Koch.

3. **S. erectum,** Huds. Fl. Angl. 103 (1762): *S. angustifolium,* Linn. Sp. Pl. 2 ed. ii. 1672 (1763). *Berula angustifolia,* Koch; Mert. & Koch, Deutsch. Fl. ii. 455 (1826). Stem angular, 1–3 ft. high, from a stoloniferous crown, usually erect, corymbosely branching above: leaflets about 6 pairs, ovate-oblong to linear, $1\frac{1}{2}$–2 in. long, often laciniate at base, the upper ones usually more or less deeply incised: peduncles 1–2 in. long; rays 1 in. or less; involucre and involucels of 6–8 linear entire lanceolate bracts: fr. $\frac{2}{3}$ line long, less compressed than in the above: oil-tubes small, in twos and threes, concealed beneath the corky covering (confluent ribs). Sierra Co., *Lemmon,* and near Tehachapi, *Greene;* usually in shallow but cold water, about mountain springs, etc.: apparently not in western California.

8. **CICUTA,** *Bessler* (WATER HEMLOCK). Glabrous tall branching perennials of marshes and stream banks. Rootstocks short and erect, or horizontal and rooting from beneath. Leaves pinnately or ternately compound. Umbels of white flowers many-rayed; involucre small or 0; involucels of several small bractlets. Calyx-teeth small, acute. Stylopodium depressed. Fruit broadly ovate or rounded, slightly compressed laterally, but the commissure narrow; ribs broad, obtuse, corky; oil-tubes solitary in the intervals. Seed subterete.

* *Rootstock short, erect; roots fascicled, fleshy.*

1. **C. Bolanderi,** Wats. Proc. Am. Acad. xi. 139 (1876); Greene, Pitt. ii. 6. Roots numerous, very coarse, 4–7 in. long, whorled around the base of a short-conical strictly erect axis: stem stout, erect, 4–9 ft. high, purplish below and very glaucous, paniculate from below the middle: radical leaves on petioles 2 ft. long or more, the blade twice or thrice pinnate: leaflets narrowly lanceolate-acuminate, 2–4 in. long, closely and sharply serrate, the setaceous tips of the teeth somewhat spreading. Marshes about Suisun Bay, near Benicia, Martinez, Suisun, etc.; also in similar situations (always within reach of tide-water) near Napa.

2. **C. occidentalis,** Greene, Pittonia, ii. 7 (1889): *C. maculata,* Bot. Calif., not Linn. Roots few, at the base of a more slender often somewhat ascending axis, 3–5 in. long, fusiform, often $\frac{3}{4}$ in. thick above the middle: stem stout, 3–6 ft. high, green, scarcely glaucous, paniculate from toward the base: leaves bipinnate: leaflets 2–3 in. long, narrowly lanceolate, coarsely serrate.— In the Sierra Nevada, the type on the eastward slope mainly or wholly. Farther westward, at Tehachapi, etc., in a taller coarse leaved form, which may be a variety or a distinct species. This is known only in flower. June.

* * *Rootstock horizontal, only partly or not at all subterranean, emitting roots from beneath only; roots fleshy-fibrous, cylindrical.*

3. **C. Californica,** Gray, Proc. Am. Acad. vii. 344 (1868); Greene, Pitt.

ii. 10: *C. vivosa*, var. *Californica*, C. & R. Rev. Umb. 130. Rhizome freely branching, the branches 1/4 — 1 ft. long, the older portion slender (1/2 in. thick or more) with long internodes, upper end abruptly clavate-enlarged and short-jointed: stem erect, 3—6 ft. high: lowest leaves bipinnate, the upper simply pinnate; leaflets ovate-lanceolate: involucre nearly obsolete: seed sometimes with 2 oil-tubes in the intervals.— In eddies and along the margins of swift-flowing mountain streams of the Coast Range only, from near Santa Cruz to the Oakland Hills; the naked branching claviform rhizomes conspicuous, growing among bowlders and barely above water.

9. ŒNANTHE, *Dioscorides.* Aquatic perennials, with glabrous decompound leaves and involucrate umbels. Calyx-teeth rather prominent, acute. Stylopodium short-conical: styles elongated in age. Fruit oblong, not compressed, with broad commissure, rounded corky ribs, and oil-tubes solitary in the narrow intervals. Seed compressed dorsally, flat on the face.

1. **Œ. Californica,** Wats. Proc. Am. Acad. xi. 139 (1870); H. & A. Bot. Beech. 142 (1840), under *Helosciadium. Cicuta Californica,* Greene. Pitt. i. 271, not Gray. Rootstocks erect or ascending, 1—2 in. long, 3/4 in. thick, solid: stem solitary, decumbent or procumbent, rooting at the lower joints, erect above and with one or more umbelliferous branches: leaves ternate and bipinnate (or the upper ones simply pinnate), the pinnæ nearly sessile: leaflets approximate, ovate, acutish, toothed, at base often lobed, 1/2—1 in. long: umbels with few linear bracts or none: fr. 1 1/2 lines long, oblong, obtuse at each end, tipped with the long spreading styles; ribs and commissure very corky: seed usually angled; oil-tubes at the angles. Very common, forming dense masses covering shallow pools, or stretches of muddy shore back of the salt marshes and among the hills throughout the Bay region and northward. The rather succulent herbage appears to be innocuous, and is said to be eaten by cattle without causing poisoning. It is therefore an exception among aquatic umbellifers. Apr. Nov.

10. APIUM, *Brunfels.* Glabrous biennial, with pinnately or ternately compound leaves, and nearly naked umbels of small whitish flowers. Calyx-teeth obsolete. Stylopodium depressed or 0. Fruit ovate or broader; the carpels straight, obtusely ribbed; oil-tubes solitary in the intervals. Seed nearly terete.

1. A. GRAVEOLENS, Linn. Sp. Pl. i. 264 (1753). (CELERY). Biennial, with fibrous roots: stem erect, 2—3 ft. high, branching freely: leaves pinnate; leaflets in 1 or 2 pairs, cuneate-obovate or rhomboidal, sparingly toothed, 1—2 in. long, those of the uppermost leaves 3 only, oblanceolate, nearly entire: umbels sessile or short-peduncled; rays 6—12, slender, 1 in.

long: fr. ⅔ line long. Very common in marshy grounds throughout the Bay region, where its habits are quite those of an indigenous plant, but it is assumed to have established itself in the first place as an escape from the gardens.

11. APIASTRUM, *Nuttall.* A small and rather delicate branching annual, with leaves dissected into linear segments. Umbels sessile in the forks, or opposite the leaves, naked, few-rayed. Calyx-teeth obsolete. Petals ovate, concave, obtuse. Stylopodium depressed; styles short. Fruit cordate, laterally compressed, the commissure narrow; ripe carpels incurved, with 5 often obscure rugulose ribs; oil-tubes broad and solitary in the intervals, with a narrow one under each rib.

1. **A. angustifolium,** Nutt.; T. & G. Fl. i. 644 (1840); Torr. Bot. Mex. Bound. t. 28. A few inches to nearly a foot high; branches more or less dichotomous: leaves 1–2 in. long, biternately or triternately dissected into narrowly linear or almost filiform segments: rays of umbel very unequal: fr. 1½ line long, somewhat broader, the 5 primary ribs occasionally supplemented by 4 less prominent intervening ones. Common in early spring, from Mendocino Co. southward.

12. CARUM, *Turner (Dioscorides?).* Glabrous erect rather slender herbs, our species perennial, with tuberous or fusiform or coarse-fibrous usually fascicled roots, pinnately ternate leaves with few linear leaflets, and involucrate umbels of white flowers. Calyx-teeth small. Fruit ovate to linear-oblong; pericarp thin, with obtuse often filiform ribs; oil-tubes solitary in the intervals.

1. **C. Kelloggii,** Gray, Proc. Am. Acad. vii. 344 (1868); Greene, Pitt. i. 273 (1889), under *Atænia*. Stems several, 3–6 ft. high, from a strong tuft of coarse hard fibrous roots: lower leaves ternate, the pinnate divisions with linear segments 1–3 in. long or more: involucre and involucels prominent, somewhat scarious: calyx-teeth subulate, conspicuous: fr. oblong, 1½–2½ lines long; stylopodium prominent, styles as long: seed sulcate beneath the large oil-tubes. Very common on open plains and hillsides about San Francisco Bay. July–Oct.

2. **C. Gairdneri,** Gray, Proc. Am. Acad. vii. 344 (1868); H. & A. Bot. Beech. 349 (1840), under *Atænia;* T. & G. Fl. i. 612, under *Edosmia*. Stem solitary, 1–4 ft. high, from a fascicle of fusiform tuberous roots: leaves mostly simply pinnate, with 3–7 linear or almost filiform leaflets 2–6 in. long, the lowest rarely themselves pinnately divided, the uppermost cauline usually simple: involucre of few bracts or 0: involucels of linear-acuminate bractlets: fr. ovate, ½–1 line long, with long styles: seed terete. Var. **latifolium,** Gray, l. c. Smaller and very slender; leaflets broader, linear-lanceolate, ½ in. wide.—Throughout the State; the variety in the mountains northward and eastward. July—Oct.

3. **C. Oreganum,** Wats. Proc. Am. Acad. xx. 368 (1885): Nutt. herb., under *Edosmia;* Greene, Pitt. i. 274 (1889), under *Ataenia.* Aspect of the last, but with lower leaves more finely divided, the short lobes linear; fr. oblong, $1\frac{1}{2}$—2 lines long: seed sulcate beneath the oil-tubes, slightly concave on the face, with central ridge. In the mountains along our northeastern borders.

13. **EULOPHUS,** *Nuttall.* Our species at agreement with those of the preceding genus in habit, general aspect, vegetative characters, flowers, and even fruit, save that here the pericarp is thinner, the ribs less prominent, the oil-tubes usually several in the intervals, and the face of the seed more concave. It is likely that a more natural classification of these plants would be to merge them in *Carum,* or to receive the entire series as one geographical genus under the name *Ataenia.*

1. **E. Californicus,** C. & R. Rev. Umb. 114 (1888); Gray, Proc. Am. Acad. vii. 346 (1868), under *Podosciadium;* Torr. Pac. R. Rep. iv. 93 (1857), under *Chaerophyllum.* Root unknown: stem 3—4 ft. high, nearly simple: lowest leaves 1 ft. long, triternately dissected into many linear entire or toothed segments; uppermost simple: involucre and involucels of many scarious lanceolate long-acuminate bracts: fr. linear-oblong, 4 lines long; oil-tubes solitary in the intervals, 4 on the face: seed-face deeply concave. Stanislaus Co., near Knight's Ferry, thence southward to the Santa Lucia Mts. May.

2. **E. Parishii,** C. & R. l. c.; also Bot. Gaz. xii. 157 (1887), under *Pimpinella.* Root deep-seated, tuberiform: stem solitary, erect, 1—2 ft. high: lower leaves ternate, with linear-lanceolate leaflets 1—3 in. long, 2—5 lines wide, the terminal leaflet remote from the others; uppermost simple: umbel 8—10-rayed, with few or no bracts; involucels of 2—6 narrowly lanceolate bractlets: fr. ovate or oblong, $1\frac{1}{2}$—2 lines long; carpel with 5 slender ribs; oil-tubes 2—4 in the intervals, 6 on the face. Foothills of the Sierra, from Placer Co. southward.

3. **E. Bolanderi,** C. & R. Rev. Umb. 112 (1888); Gray, Proc. Am. Acad. vii. 346 (1868), under *Podosciadium.* Stem 2 ft. high: leaves pinnate, the segments narrowly linear: umbels many-rayed, the rays 5—9 lines long; bractlets of the involucels scarious, exceeding the pedicels: petals with the inflexed tip very long-acuminate: fr. oblong, $1\frac{1}{2}$ lines long, the narrow ribs becoming elevated and undulate; oil-tubes 2—5 in the intervals but small and often obscure, 6 on the face: seed much compressed dorsally.—In the Sierra, at higher elevations than the last, and ranging northward.

4. **E. Pringlei,** C. & R. l. c. 113. Slender, 1—2 ft. high, the pinnately compound leaves with broad inflated midrib, the divisions cut into narrowly linear segments: umbel 3—8-rayed; involucre scant; involucels

of many scarious lanceolate bractlets a third as long as the pedicels: fr. oblong, 2–2½ lines long; oil-tubes 3–5 in the intervals, 8 on the face of the carpel. Var. **simplex**, C. & R. l. c. Leaflets linear-lanceolate, entire: oil-tubes 2 or 3 in the intervals, 4 on the face. The type is southerly, in San Luis Obispo and Kern counties; the variety in Sierra Co.

14. **PIMPINELLA**, *Brunfels*. Perennials with decompound foliage and nearly naked umbels. Calyx-teeth obsolete. Fruit ovate or broader than long, laterally compressed but with broad commissure; carpels 5-angled, with distant usually slender ribs, and several oil-tubes in the intervals. Seed somewhat flattened dorsally, with plane or slightly convex face.

1. **P. apiodora**, Gray, Proc. Am. Acad. vii. 345 (1868). Stoutish, erect, glabrous, 2–3 ft. high, sweet-scented: leaves mostly radical, 2–3-ternate: leaflets cuneate-ovate, laciniately pinnatifid and toothed, 1 in. long: umbels long-peduncled, 6–15-rayed; rays 1–2 in. long, hispidulous-puberulent: fl. white or pinkish: fr. broadly ovate (not known in its mature state), 1½ lines long: oil-tubes 4–6 in the intervals, 8 or more on the face. — From the Bay region eastward to the borders of Nevada.

15. **PODISTERA**, *S. Watson*. Acaulescent, dwarf and cespitose, with pinnately parted leaves, no involucre, involucels of 3—5-cleft bractlets, and pinkish flowers. Calyx-teeth prominent. Fruit elliptic-ovate, glabrous; carpels with filiform ribs, the cross-section oblong-pentagonal; oil-tubes 2 or 3 in the intervals, 6 on the face.

1. **P. Nevadensis**, Wats. Proc. Am. Acad. xxii. 475 (1887); Gray, Proc. Am. Acad. vi. 536 (1865), under *Cymopterus*. Obscurely puberulent: leaves 3–4 lines long, thickish, the 3–7 lanceolate segments acute, entire: peduncles very short; umbels of 3—5 umbellets which are sessile, with very short pedicels, and equalled by the involucels: fr. little more than a line long, nearly sessile. — A low densely matted herb found among rocks near the summit of Mt. Dana, at an elevation of about 13,000 ft.

16. **FŒNICULUM**, *Pliny* (FENNEL). Perennial, erect and tall, with dark green striate stem, and equally dark sweet-scented and -flavored leaves dissected into countless linear-setaceous leaflets. Flowers yellow, in umbels destitute of bracts and bractlets. Calyx with turgid border and no teeth. Fruit oblong; carpels 5-ribbed; oil-tubes solitary in the intervals, 2 on the face.

1. F. VULGARE, Gerarde, Herb. Em. 1032 (1633); Park. Theatr. 884 (1640); Ray, Syn. 2 ed. 111 (1696), 3 ed. 217 (1724); Gærtn. Fr. et Sem. i. 105 (1788). *Anethum Fœniculum*, Linn. Sp. Pl. i. 263 (1753). Cultivated from ancient times, and formerly in high repute as a medicinal and culinary herb; naturalized in many parts both of the Old World and the

New, and common in central and southern California, attaining the height of 3–6 ft.; readily known by its dark green finely dissected foliage and large umbels of greenish-yellow small flowers. May–Sept.

17. **LIGUSTICUM**, *Dioscorides*. Stoutish and rather tall perennials, with ternately decompound leaves, and white flowers in many-rayed umbels. Calyx-teeth obsolete. Stylopodium mostly conical, the margin at base undulate. Fruit ovate or oblong, somewhat compressed dorsally, the commissure broad; ribs somewhat prominent or even wing-like, the lateral ones usually broadest; oil-tubes 3–5 in the intervals, 6–10 on the face. Seed with rounded or angular back, plane or concave on the face.

1. **L. apiifolium,** Gray, Proc. Am. Acad. vii. 347 (1868); Nutt.; T. & G. Fl. i. 641 (1840), under *Cynapium*. Stems erect, 2–4 ft. high, the inflorescence puberulent; leaves mostly radical, ternate or biternate, then once or twice pinnate, the ultimate segments $\frac{3}{4}$–$1\frac{1}{2}$ in. long, ovate, laciniate-pinnatifid; umbel many-rayed; involucels of several narrowly linear long bractlets: fr. oval, $1\frac{1}{4}$–2 lines long; stylopodium short-conical; ribs narrow, acute; oil-tubes 3–5 in the intervals, 4–8 on the face; seed with rounded back and concave face.—Neighborhood of Yosemite, and thence far northward.

2. **L. Grayi,** C. & R. Rev. Umb. 88 (1888): *L. apiifolium*, var. *minus*, Gray, Bot. Calif. i. 264 and ii. 451. Stem 1–2 ft. high, the inflorescence glabrous; leaves ternate, then pinnate, the segments ovate, laciniate-pinnatifid; fr. narrowly oblong, 2–$2\frac{1}{2}$ lines long; ribs almost wing-like; oil-tubes 3–5 in the intervals, 8 on the face; seed with angled back and slightly concave face.—Habitat of the preceding.

18. **SELINUM,** *Theophrastus*. A genus of precarious status, the species easily referable either to the preceding or to the next, at agreement with both, as they with each other, in habit; differing from *Ligusticum* in having the carpel more decidedly winged, and the oil-tubes usually only one in each interval.

* *Involucels conspicuous; pedicels glabrous; wings of carpel thin.*

1. **S. Pacificum,** Wats. Proc. Am. Acad. xi. 140 (1876). Leaves ternately bipinnate; segments ovate, acutish, 1 in. long, laciniately toothed and lobed; peduncles stout, the umbel about 15-rayed; bracts of involucre 1 in. long, equalling the rays, lobed and toothed; involucels of several linear entire or 3-toothed bractlets: fr. oblong, 3–4 lines long; wings rather narrow; oil-tubes conspicuous, rarely 2 in the intervals; seed channelled under the dorsal oil-tubes.—Near Sausalito, and in the Mission Hills.

* * *Umbels naked; rays and pedicels hoary-pubescent; wings of carpel corky.*

2. **S. capitellatum,** Wats. Bot. King Exp. 126 (1871); Gray, Proc.

Am. Acad. vi. 537 (1865), under *Sphaenosciadium*. Stout, 2—5 ft. high, glabrous except the tomentose inflorescence: leaves large, bipinnate, with dilated petioles; leaflets few, 1—2 in. long, oblong to linear-lanceolate, laciniately lobed or coarsely toothed: umbels 6—8-rayed; umbellets capitate and globose, 3—6 lines thick: fr. cuneate-obovate, $1\frac{1}{4}$ in. long; carpels strongly ribbed, the lateral wings broader than the 3 dorsal: oil-tubes solitary, the seed with corresponding shallow grooves.—Banks of streams in the higher Sierra, from Mono Co. northward to Donner Lake.

3. **S. erygniifolium,** Greene, Pittonia, ii. 102 (1890). Stoutish, $1\frac{1}{2}$ ft. high: stem and bladdery-dilated petioles glabrous, the leaves roughish-pubescent and the inflorescence white-tomentose: leaves bipinnate, rather small; leaflets ovate, acute, $\frac{1}{2}$ in. long or less, with rather stiffly setaceous recurved tips and teeth: fr. unknown.— In the vicinity of the Yosemite, 1889. *Elmer Drew*.

19. **ANGELICA,** *Braunschweig*. Perennials, stout and tall. Segments of the large pinnately or ternately compound leaves broad, toothed; petioles dilated. Umbels many-rayed, nearly or quite naked. Flowers white or purple. Calyx-teeth minute or obsolete. Fruit ovate or oblong, strongly flattened dorsally, with a very broad commissure, margined by a broad somewhat scarious wing; dorsal ribs prominent, more narrowly winged; oil-tubes solitary or in pairs in the intervals. Face of seed plane or slightly concave.

1. **A. tomentosa,** Wats. Proc. Am. Acad. xi. 141 (1876). Hoary-tomentose, or the stem in age glabrate: leaves quinately bipinnate; leaflets firm, ovate, acute, very oblique at base, 2—4 in. long, the lower sometimes lobed, serrate with unequal acute teeth; umbels naked, often dense; rays 1—3 in. long: fr. 3 lines long, broadly elliptical, the lateral wings thin, the dorsal acutish: seed thin, plane on the face, channelled on the back by the impressed dorsal oil-tubes.—Rocky banks of streamlets among the hills from Mendocino and Napa counties southward.

2. **A. Breweri,** Gray, Proc. Am. Acad. vii. 348 (1868). Glabrous or finely puberulent, 3—4 ft. high: leaves ternate or quinate and pinnate; leaflets lanceolate or oblong-lanceolate, 2—3 in. long, sharply serrate, the teeth cuspidate, the lower somewhat lobed: peduncles long, sometimes bearing 1 or 2 dilated subscarious bracts; umbels naked; rays 2 in. long: fr. pubescent, oblong, 4 lines long, the lateral wings narrow, corky, the dorsal obtuse, not prominent; oil-tubes usually 6 (the dorsal or lateral in pairs), besides 2—4 on the face: seed more or less concave on the face, the oil-tubes forming deep channels on the back.—In the Sierra Nevada, from Mariposa Co. to Plumas.

3. **A. lineariloba,** Gray, l. c. 347. Glabrous, stout, 2—3 ft. high: leaves twice or thrice quinate; leaflets 1—2 in. long, linear, cuspidately

acuminate, entire, or the lowest 3-parted, with their decurrent sometimes coarsely toothed lobes divaricate: umbels naked; rays 1—2 in. long: fr. glabrous, 4 lines long by 2 broad: lateral wings narrow, somewhat corky: oil-tubes solitary, or the lateral in pairs; seed nearly plane on the face, channelled under the dorsal oil-tubes.—At Mono Pass in the Sierra, thence southward.

20. CYMOPTERUS, *Rafinesque*. Low perennials, often subacaulescent, from a fleshy tuberous or fusiform root. Leaves pinnately decompound, with small narrow segments. Umbels usually both involucrate and involucellate. Flowers white or yellow. Calyx-teeth minute or distinct. Fruit ovate or elliptical, obtuse or retuse, somewhat flattened dorsally, the lateral ribs and some or all of the others with broad thin wings. Oil-tubes narrow, one or several in the intervals. Seed dorsally flattened, more or less concave on the face.

1. **C. terebinthinus,** T. & G. Fl. i. 624 (1840); Hook. Fl. i. 266. t. 95 (1833), under *Selinum*. Glabrous, $1\frac{1}{2}$—$11\frac{1}{2}$ ft. high, leafy at base: leaves rather rigid, tripinnate; leaflets linear-oblong, acute, entire or 1—2-toothed, barely a line long: rays unequal, $1\frac{1}{2}$—2 in. long; involucre of a single linear leaflet or none: involucels of several short linear bractlets; pedicels 1—2 lines long: fl. yellow: fr. 3—4 lines long, 2 or 3 broad; calyx-teeth evident; wings rather thin-corky, a line broad; oil-tubes 2—4 in the intervals, 4—10 on the face.—In the Sierra Nevada, at great elevations.

2. **C. cinerarius,** Gray, Proc. Am. Acad. vi. 535 (1865). Stemless, from a subterranean creeping rootstock: peduncles (2—3 in. high) and petioles glabrous: leaves of cordate outline, bipinnate, the segments toothed, glaucous-cinereous with a fine roughish pubescence: umbellets few, subsessile; involucels of many somewhat united subscarious long-acuminate bractlets: fl. white: fr. $1\frac{1}{4}$ in. long; wings narrow, undulate; oil-tubes 3 in the intervals, several on the face: seed narrow, strongly incurved, showing a deeply concave face.—Habitat of the preceding, but more restricted; Sonora Pass, and above Lake Mono.

21. OROGENIA, *S. Watson*. Small subacaulescent perennials, with fusiform or tuberous roots, ternate leaves with linear segments, and white flowers in naked umbels: rays few, very unequal. Calyx-teeth minute. Fruit oblong, slightly flattened laterally. Carpels compressed dorsally, with filiform dorsal and intermediate ribs, the laterals excessively corky-thickened and involute, forming a cavity which is partitioned by a projection from the face of each carpel; oil-tubes 3 in the intervals, 2—4 on the face.

1. **O. fusiformis,** Wats. Proc. Am. Acad. xxii. 474 (1887). The stoutish stem 3—6 in. high, from a long fusiform root: leaves 2—3-ternate; ter-

minal leaflets often 3-parted, all 1 in. long or less: umbel 6 10-rayed: fr. 1¼ in. long, 1½ lines broad. Eastward slope of the Sierra, in Plumas and Nevada counties.

22. LEPTOTÆNIA, *Nuttall.* Stout and tall (except n. 4) glabrous subacaulescent perennials, with thick often very large fusiform roots, pinnately decompound leaves, and yellow or purple flowers. Fruit strongly compressed dorsally, oblong or elliptical, with thick corky lateral wings, the dorsal and intermediate ribs filiform or obscure: oil-tubes 3–6 in the intervals, 4–6 on the face, mostly small, sometimes obsolete. Seed thin, with plane or slightly concave face.

* *Oil-tubes obsolete or very obscure.*

1. **L. dissecta,** Nutt.; T. & G. Fl. i. 630 (1840); Gray, Proc. Am. Acad. vii. 348 (1868), under *Ferula.* Leafy at base, 1–3 ft. high: leaves broad, 1 ft. long, ternate and thrice pinnate; segments ovate or oblong, ½–1 in. long, laciniate-pinnatifid and toothed, puberulent on the veins beneath and along the margins: umbel 8–20-rayed, involucrate with few linear bracts, the bractlets of the involucels more numerous: fl. yellow or purplish: fr. sessile or nearly so, 5–9 lines long, about 3 lines broad: seed face plane. –Throughout the State, both east and west.

2. **L. multifida,** Nutt. l. c. Near the preceding, but the leaves more finely divided: umbels without involucre: carpels 4–6 lines long: seed-face concave. Only on the eastward slope of the Sierra.

3. **L. anomala,** C. & R. Rev. Umb. 53 (1888). Acaulescent, glabrous; scape slender, ½–1 ft. high, bearing a 3–6-rayed naked umbel: leaves pinnate, with few distant narrowly linear divisions: rays of umbel 1–3 in. long; bractlets of involucels veiny, scarious-margined, more or less united: fr. small, oblong, 4 lines long, 2 lines broad, the lateral ribs considerably thicker than the body of the carpel, the dorsal slender-filiform or obsolete: calyx-teeth occasionally manifest.

* * *Oil-tubes 3 or 4 in the intervals, 6 on the face.*

4. **L. Californica,** Nutt.; T. & G. Fl. i. 630 (1840); Gray, Proc. Am. Acad. vii. 348 (1868), under *Ferula.* About 1–2 ft. high: leaves ternate and pinnate or twice ternate; leaflets cuneate-obovate, 1–2 in. long, usually 3-lobed, coarsely toothed above: stem with 1 or 2 leaves: umbel 15–20-rayed, naked or with 1 or 2 narrow bracts; bractlets 0; fr. 5–7 lines long, 3–4 lines broad; margin of carpels thinnish, the dorsal ribs indistinct. Hills of Kern Co. and northward along the Sierra.

23. PEUCEDANUM, *Theophrastus.* Perennials of diverse habit, ours mostly low and subacaulescent, with fusiform root. Leaves ternately or pinnately dissected. Involucre 0; involucels usually present. Flowers white or yellow. Calyx-teeth obsolete or manifest. Fruit strongly flattened dorsally, oblong to suborbicular, glabrous or tomentose; carpel with dorsal and intermediate ribs filiform and approximate, the lateral ones developed into a broad thin wing which until maturity is coherent with that of its companion carpel, forming a broad scarious wing to the fruit as a whole. Oil-tubes 1—8 in the intervals, 2—10 on the commissural side. Seed flat, with plane or concave face.

* *Stout; leaves finely dissected; fruit-wings broad; oil-tubes 1—3 in the intervals; fl. white (purplish in n. 4).*

1. **P. eurycarpum,** C. & R. Rev. Umb. 61 (1888): *P. macrocarpum,* var. *eurycarpum,* Gray, Proc. Am. Acad. viii. 385 (1870). Root tuberous-enlarged: stem 1—2 ft. high, branching, pubescent: leaves subdivided into countless small linear cuspidate segments: umbel 3—12-rayed, with involucels of lanceolate acuminate often united bractlets; rays $1\frac{1}{2}$—4 in. long; pedicels 1—5 lines: fr. glabrous, 5—9 lines long, broadly elliptical, the wings as broad as the body or broader, the ribs filiform; oil-tubes large, solitary in the intervals, 2 on the face.—Plains and hills of the interior, from the Sacramento northward.

2. **P. dasycarpum,** T. & G. Fl. i. 628 (1840): *P. Pringlei,* C. & R. Bot. Gaz. xiii. 209 (1888). Subacaulescent from a fusiform root, tomentose-pubescent: leaves rather small, with countless short linear segments: peduncles stout, $\frac{1}{2}$—1 ft. high; umbel 6—12-rayed, with involucels of linear-lanceolate more or less tomentose bractlets; rays 1—3 in., pedicels 3—5 lines long: fr. nearly orbicular, 4—7 lines long, nearly glabrous or coarsely pubescent, the thin scarious wings broader than the body: oil-tubes large, usually solitary in the intervals, 4 on the face: seed deeply sulcate under the oil-tubes, plane on the face.—In the interior almost throughout the State.

3. **P. tomentosum,** Benth. Pl. Hartw. 312 (1849). Subacaulescent, more or less densely villous-tomentose and purplish: leaves cut into very small filiform or very narrow segments: peduncles 1 ft. high or more: umbel of 4—8 equal rays 1—3 in. long; involucels of linear-lanceolate or ovate-acuminate bractlets: calyx-teeth manifest: fr. ovate to orbicular, 5—9 lines long, densely tomentose; wings rather thick, from somewhat narrower to even broader than the body, the prominent ribs concealed by the tomentum: oil-tubes mostly 3 in the intervals, 4 on the face: seed somewhat concave. Common on bushy hills and open plains.

4. **P. Austinæ,** C. & R. Bot. Gaz. xiii. 208 (1888). Acaulescent, less than a foot high, apparently glabrous (minutely and sparsely scabrous under a lens): leaf-segments ovate, pinnately lobed or incised: umbel few-rayed, with small involucels: fl. white, or with a tinge of flesh-color: fr. elliptical, 3—4 lines long; wings not as broad as the body; oil-tubes

solitary in the dorsal intervals, 2 in the lateral, 4 on the face: seed concave on the face, with a central ridge. Plumas Co., *Mrs. Austin*, and northward.

5. **P. Vaseyi,** C. & R. l. c. 144 (1888). Subacaulescent, 6 8 in. high, pubescent: leaves small (1 2 in. long); petioles inflated; blade bipinnate, the small ovate segments irregularly 3 5-lobed: umbel with 2 5 equal rays; involucels of obovate petiolulate toothed bractlets: fl. perhaps yellowish: fr. broadly oblong, $\frac{1}{2}$ in. long, emarginate, glabrous; wings twice as broad as the body; ribs prominent; oil-tubes solitary in the intervals, 4 on the face. Among the foothills east of Sacramento and southward.

* * *More slender, leaves much dissected; fl. yellow (except in n. 8).*

6. **P. utriculatum,** Nutt.; T. & G. Fl. i. 628 (1840). Rather slender, usually erect and branching, 1 ft. high or more, glabrous or puberulent: petioles short, their margins greatly dilated and forming a membranous saccate cavity; ultimate segments of the decompound leaves narrowly linear, $1\frac{1}{2}$ in. long or less: umbel 5 20-rayed, with involucels of dilated obovate often toothed petiolulate bractlets: fr. glabrous, broadly elliptical, 2 5 lines long; wings thin, as broad as the body; oil-tubes large and solitary in the intervals (or sometimes with 1 or 2 short accessory ones), 4 6 on the face; seed somewhat concave on the face. Common throughout the State, at least in the western portion, and on plains of the interior.

7. **P. carnifolium,** T. & G. Fl. i. 628 (1840). Herbage and general aspect of the last, but acaulescent or nearly so; petioles without bladdery dilatation; leaf-segments $1\frac{1}{2}$ 2 in. long; bractlets of involucels often lanceolate: fr. 3 4 lines long: wings narrow and thickish; ribs obsolete: oil-tubes indistinct, 2 or 3 in the intervals, none on the face. Common in central parts of the State.

8. **P. Parishii,** C. & R. Bot. Gaz. xiii. 209 (1888). Caulescent, 4 12 in. high, pallid and pubescent: leaves pinnate, with pinnatifid or entire leaflets, the ultimate segments oblong-linear or linear, cuspidate, toothed or entire: umbel unequally 4 12-rayed, with involucels of small linear-lanceolate acuminate bractlets; rays 1 5 in. long; pedicels 2 4 lines: fl. white: fr. obovoid, 4 6 lines long, glabrous, broadly or narrowly winged; ribs filiform or obsolete; oil-tubes small, often obscure, 6 8 in the intervals, 8 to 10 on the face. From Colusa Co., southward to San Bernardino.

* * * *Leaves not finely dissected, the leaflets with broad or elongated segments; fl. yellow.*

9. **P. triternatum,** Nutt.; T. & G. Fl. i. 626 (1840); Pursh, Fl. i. 197

(1814), and Hook. Fl. i. 264. t. 94, under *Seseli*. Herbage pale green, puberulent; stem mostly leafless, 1–2½ ft. high, simple; leaves bi- or triternate; leaflets 2–4 in. long, narrowly linear to linear-lanceolate; umbel unequally 5–18-rayed, with no involucre, but involucels of lanceolate or setaceous bractlets; fr. narrowly oblong, 1/4–1/3 in. long, glabrous or pubescent; wings narrow; ribs prominent; oil-tubes very large and broad, solitary in the intervals, 2 on the face: seed-face slightly concave. Var. **alatum**, C. & R. Rev. Umb. 70, has very narrowly linear leaflets, and a broad wing to the fruit. Eastward and northward along the Sierra.

10. **P. leiocarpum**, Nutt. l. c. (1840); Hook. l. c. 263. t. 93, under *Seseli*. Habit of the preceding but stouter, 1/2–1½ ft. high; leaflets thick, ovate to narrowly lanceolate, 1–2 in. long, acute, or in the broader forms sharply toothed at the broad apex; umbel few-rayed, naked, as are also the umbellets; fr. 4–5 lines long, 2 lines broad, narrowed below; ribs rather prominent; wings half as wide as the body; oil-tubes 1 or 2 in the intervals, 4 on the face. From Sierra Co. northward.

11. **P. parvifolium**, T. & G. Fl. i. 628 (1840); H. & A. Bot. Beech. 348 (1840), under *Ferula*; *P. Californicum*, C. & R. Bot. Gaz. xiii. 143 (1888). Short-caulescent, slender. 6–10 in. high: leaves bipinnate but with confluent upper leaflets; leaflets broad, obtuse, truncate or emarginate at the very apex, irregularly incised and with broad cuspidate teeth; umbel 8–10-rayed; involucels of linear or lanceolate-acuminate bractlets; fr. broadly elliptical to orbicular, 3 lines long; wings broader than body; ribs prominent; oil-tubes solitary in the intervals, 2–4 on the face.— Near Monterey, and southward in the Coast Ranges.

* * * * *Stem stout, tall, angular, leafy; leaves pinnate; fl. yellow.—*
Genus PASTINACA, Tourn.

12. P. SATIVUM, Wats. Bot. King Exp. 128 (1871); Linn. Sp. Pl. i. 262 (1753), under *Pastinaca*. *Peucedanum Pastinaca*, Baill. Hist. vii. 96 (1880). Biennial, branching, 2–4 ft. high; stem angular or fluted; herbage nearly glabrous, of a somewhat yellowish green: leaflets of the pinnate leaves large, ovate or oblong, incisely toothed; involucre and involucels small or 0; fr. oval, 2–3 lines long, broadly winged, prominently ribbed; oil-tubes solitary in the intervals. The *Parsnip* of the farms and gardens, native of Europe, is spontaneous here and there by waysides and in waste lands.

24. **HERACLEUM**, *Linnæus* (COW PARSNIP). Perennial or biennial, with stout hollow fluted stem, ample lobed or compound leaves, and very large umbels of white flowers. Calyx-teeth small or obsolete. Fruit round-obovate, very much flattened dorsally, somewhat pubescent. Carpel with dorsal ribs filiform, the margin winged, the wings coherent when young, strongly nerved toward the margin: oil-tubes solitary in the

intervals, obclavate, extending from the apex downward to or below the middle of the carpel, 2 on the face: seed thin and flat.

1. **H. lanatum,** Michx. Fl. i. 166 (1803). Stem 3–8 ft. high: leaves ternate, 1–2 ft. long, the stout petioles and veins hirsute beneath, the base of the petiole much dilated; leaflets 4–10 in. long, rounded and subcordate, the lobes somewhat palmately arranged, acuminate, toothed; rays many, 3–6 in. long: fl. large, white, irregular, the outer petals being larger: fr. broadly obovate, 4–6 lines long, slightly pubescent. In wet open ground, or in moist thickets, from the seaboard to the Sierra.

25. MYRRHIS, *Morison* (SWEET CICELY). Perennials with thick aromatic roots, rather slender stems not tall, ternately-compound mostly radical leaves: involucres and involucels reduced or obsolete. Flowers white. Calyx-teeth obsolete. Fruit linear to linear-oblong, more or less attenuate at base, acute at summit, glabrous or bristly along the ribs. Carpel nearly pentagonal in section, flattened dorsally if at all. Oil-tubes obsolete in mature fruit. Seed-face slightly concave to deeply sulcate.

* *Involucre and involucels scant or 0; fruit nearly glabrous, not attenuate to a slender base.*—Genus GLYCOSMA, Nutt.

1. **M. occidentalis,** B. & H. f. Gen. Pl. i. 897 (1865); Gray, Proc. Am. Acad. vii. 346 (1868); Nutt.; T. & G. Fl. i. 639 (1840), under *Glycosma*; C. & R. Rev. Umb. 119 (1888), under *Osmorrhiza*. Stoutish, puberulent or pubescent: leaflets oblong, 1½–4 in. long, acute, coarsely serrate, rarely incised: umbel 5–12-rayed. naked or with 1 or 2 bracts; rays 1–5 in. long, mostly erect; pedicels 1–3 lines: fr. 7–12 lines long, 1½ lines wide, obtuse at base, glabrous, with prominent acute ribs; the mostly conical stylopodium together with the style ½–1 line long.—Dry woods in the foothills and at middle elevations.

2. **M. Bolanderi,** Gray, l. c. *Osmorrhiza occidentalis,* var. *Bolanderi,* C. & R. l. c. (1888). Leaflets ovate, incised: pedicels shorter: ribs of carpel obtuse; otherwise as in the preceding, of which it may be a variety only.—Woods of Mendocino Co., *Bolander;* apparently rare.

3. **M. ambigua,** Greene. Gray, Proc. Am. Acad. viii. 386 (1872), under *Glycosma*. Tall, glabrous or hairy at the nodes, the petioles and leaf-veins beneath pilose: leaflets ovate-oblong, acute, 2- or 3-cleft and incisely toothed: fruiting umbel with spreading rays: fr. 6–7 lines long, linear-oblong, acutish at each end, the ribs more or less setulose toward the base.—Woods near Cahto, Mendocino Co.

* * *Involucre and involucels caducous or 0; fruit attenuate at base; ribs bristly.* Genus OSMORRHIZA, Rafinesque.

4. **M. brachypoda,** Greene. Torr.; Dur. Pl. Pratt. 89 (1855), under *Osmorrhiza*. Stout, pubescent or glabrous: leaflets 1 in. long or more,

acute, laciniately lobed or toothed; umbel only 1–4-rayed, with involucre and involucels of linear bracts and bractlets, the latter exceeding the flowers: fr. 6–8 lines long, 3 lines wide, abruptly attenuate at base, rough-bristly on the very prominent ribs. — From near Monterey to Nevada Co. and southward.

5. **M. nuda,** Greene. Torr. Pac. R. Rep. iv. 93 (1857), under *Osmorrhiza*. Slender, 2–3 ft. high, more or less pilose-pubescent: leaves twice ternate: leaflets 1–2 in. long, ovate, acute or obtusish, rather deeply cleft and toothed: umbel long-peduncled, 3–5-rayed, naked or with small caducous bracts and bractlets; pedicels $\frac{1}{4}$–$\frac{3}{4}$ in. long: fr. slender, 3–7 lines long, with slenderly attenuate base; carpels acutely ribbed; stylopodium very short. — Common in shady woods along streams almost throughout the State.

26. CHÆROPHYLLUM, *Columna*. Rather slender annuals with ternately compound leaves, and small white flowers in almost naked umbels. Calyx-teeth obsolete. Fruit lanceolate, or ovate-oblong and beaked at summit, the beak not as long as the body; ribs of carpel equal; oil-tubes present.

1. C. ANTHRISCUS, Lam. Encycl. i. 685 (1783); Linn. Sp. Pl. i. 256 (1753), under *Scandix*. *Anthriscus vulgaris*, Pers. Syn. i. 320 (1805). Weak and often half reclining; small umbels opposite the leaves, about 3-rayed: fr. about 2 lines long including the short beak, roughened with short rigid incurved bristles. — In sandy soil at Alameda, *Dr. W. P. Gibbons*; naturalized from Europe.

27. SCANDIX, *Theophrastus*. Annual, with pinnately decompound leaves cut into countless slender segments. Flower and fruit much as in *Chærophyllum*, except that the beak of the carpel far exceeds the body.

1. S. PECTEN VENERIS, Dod. Pempt. 689 (1583); Linn. Sp. Pl. i. 256 (1753); Crantz, Austr. 189 (1769), under *Chærophyllum*; All. Fl. Pedem. ii. 29 (1785), under *Myrrhis*. Erect, 1 ft. high more or less, leafy throughout, but radical leaves ample, of oblong outline, cut into many short ligulate acuminate lobes: bractlets of involucels many: fr. 1½–3 in. long including the beak which is the conspicuous part of it, the body and the margins of the beak with tubercles ending in short prickles. — A weed in fields and by waysides in Napa, Contra Costa and Alameda counties; first detected in Napa Valley, by *Mr. Sonne*, in 1888; introduced from Europe.

28. DAUCUS, *Galen*. More or less hispid annuals and biennials, with pinnately decompound leaves, involucres and involucels of lobed or divided bracts, and white flowers. Outer rays of umbel longest and in fruit connivent over the inner, giving a concave top to the umbel as a whole. Calyx 5-toothed. Fruit ovate or oblong; carpels semiterete or dorsally flattened; primary ribs filiform and bristly, the secondary more prominent, winged with a row of more or less united barbed prickles. Oil-tubes solitary under the secondary ribs. Seed nearly flat on the face.

1. **D. pusillus,** Michx. Fl. i. 164 (1803): *D. microphyllus,* Presl.; DC. Prodr. iv. 213 (1830). Annual, erect, or the branches short and almost prostrate, 1½ -2 ft. high, retrorsely hispid: leaves bipinnate, the segments pinnatifid, with short narrowly linear lobes: rays 2 6 lines long, nearly equal; involucre bipinnatifid, equalling the umbel; involucels equalling the greenish white flowers: fr. 1½ 2 lines long, short-pedicellate, the prickles usually equalling or exceeding the width of the body: seed slightly concave on the face. Found in nearly all parts of the State; on bluffs and hills near the sea often depressed and condensed. The herbage has a reputation as an antidote to the poison of rattlesnakes.

2. D. CAROTA, Linn. Sp. Pl. ii. 242 (1753). Biennial, stout, 2- 3 ft. high, hispid: leaves 1 ft. long or less: involucre of many pinnatifid bracts equalling the large umbel; bractlets scarious, with an herbaceous midrib: fl. white, but the central one of each umbellet abortive and dark purple: fr. oblong-ovoid, the spines as long as its diameter: fruiting umbel deeply concave, resembling a bird's nest. - The common *Carrot* of the gardens; already becoming a wayside weed in middle California.

29. **CAUCALIS,** *Theophrastus.* Genus in nature scarcely distinct from *Daucus,* but fruit more compressed laterally, and the seed-face deeply channelled.

1. C. NODOSA, Huds. Fl. Angl. 114 (1798): Linn. Sp. Pl. i. 240 (1753), under *Tordylium;* Gaertn. Fr. et Sem. i. 82. t. 20 (1788), under *Torilis.* Branching at base, the long branches reclining, leafy throughout and retrorsely hispid: leaves pinnate, with pinnatifid divisions: umbels small, naked, subsessile opposite the leaves: carpels unequal, the larger one a line long: surface tuberculate and prickly, the prickles barbed or incurved at summit.—An obscure weed, common in many parts of the State; native of Europe.

2. **C. microcarpa,** H. & A. Bot. Beech. 348 (1840). Erect, slender, 6 -15 in. high, nearly glabrous: leaves much dissected, hispidulous: umbels terminal and at the ends of the branches, subtended by two or more foliaceous dissected bracts, 3 -6-rayed; rays slender, 1 3 in. long; umbellets few-flowered, the pedicels unequal; involucels of short entire bractlets: fr. oblong-ovoid, 2 lines long, armed with uncinate prickles.— Very common, but slender and obscure. Like *Daucus pusillus* it is regarded as efficacious against the venom of the rattlesnake.

Order LV. CORNEÆ.

De Candolle, Prodromus, iv. 271 (1830). Cornaceæ, Lindl. Intr. 2 ed. 49 (1836).

Trees, shrubs or undershrubs, with opposite exstipulate leaves, and naked or involucrate cymose or capitate or amentaceous inflorescence. Calyx-tube coherent with the ovary; limb 4-lobed or obsolete. Petals 4, epigynous, valvate in bud. Stamens 4, alternate with the petals; anthers 2-celled. Style filiform; stigma simple. Fruit drupaceous, 1–2-seeded. Seed pendulous; embryo minute; albumen fleshy.

1. CORNUS, *Pliny* (Dogwood). Deciduous shrubs, or low semi-herbaceous plants. Flowers perfect, not in aments. Drupe globose, ovoid or oblong; putamen 2-celled, 2-seeded.

* *Flowers white, not involucrate, cymose, appearing later than the leaves; drupe small, subglobose.* —Cornus proper.

1. **C. glabrata,** Benth. Bot. Sulph. 18 (1844); C. & E. Bot. Gaz. xv. 89. Shrub 5–12 ft. high, with gray bark, and nearly or quite glabrous twigs and foliage: leaves oblong to narrowly ovate, acute at each end, or acuminate at apex, 1–2 in. long, green alike on both faces; petioles short, slender: fl. in many small open flat-topped cymes: fr. globose, white; stone little compressed, not furrowed, broader than high, the breadth 2 lines or more.—In the Coast and Mt. Diablo Ranges from Monterey Co. northward, also in the foothills of the Sierra.

2. **C. Torreyi,** Wats. Proc. Am. Acad. xi. 145 (1876); C. & E. l. c. 34. Size and habit unknown: leaves obovate or oblanceolate, abruptly acute or short-acuminate, on rather long and slender petioles, lower face paler and somewhat pubescent with loose silky hairs: cyme loose and spreading: fr. white; stone obovoid, $2\frac{1}{2}$–$3\frac{1}{3}$ lines long, somewhat flattened, acute at base, ridged on the edges, tubercled at summit, higher than broad.—Yosemite Valley, *Dr. Torrey*.

3. **C. Greenei,** C. & E. Bot. Gaz. xv. 36 (1890). Size and habit unknown: twigs and inflorescence appressed-pubescent: leaves ovate, obovate or oval, acutish or rounded at base, acute or acuminate at apex, appressed-pubescent or glabrate above, beneath scarcely lighter but with a sparse appressed pubescence of stiffish hairs of which some are straight, others curved: fl. large, in loose paniculate cymes: calyx-teeth triangular: styles with enlarged greenish tips: fr. dark blue or purple; stone globular, not furrowed, slightly ridged.—Probably of middle California, but the specimens, as found in the old herbarium of the University, were without a label, and neither the special locality nor the collector's name can now be guessed.

4. **C. stolonifera,** Michx. Fl. i. 92 (1803): C. & E. l. c. 86. Stems numerous, clustered, decumbent, forming a low thicket; twigs nearly glabrous, red-purple: inflorescence appressed-pubescent: leaves mostly oval or oblong and rather abruptly acuminate or only acute, at base usually obtuse, both faces very minutely and sparsely appressed-pubescent, but the lower very pale and appearing as if glaucescent: cymes small, flat-topped: calyx-teeth minute; fr. globose, white; stone little or much compressed, furrowed on the edges.—An eastern species, crossing the continent however in northerly districts, and reaching California, in the northern counties; Trinity Mts., *C. C. Marshall.*

5. **C. pubescens,** Nutt. Sylv. iii. 54 (1842); C. & E. l. c. 37. Shrub 6–15 ft. high, with smooth reddish branches: leaves ovate, acute, 2–4 in. long, paler and more or less pubescent beneath: fl. in convex cymes: fr. white, subglobose, 2 lines broad: stone somewhat flattened, mostly oblique, with a more or less prominently furrowed edge, the sides more or less prominently ridged. Var. **Californica,** C. & R. l. c. *C. Californica,* C. A. Mey. Pubescence said to be loose and spreading; leaves more rounded and broader; stone smaller, etc.—Throughout the State, in the variety chiefly; the type being mostly of a more northerly habitat. Our common form has a vernal and also an autumnal season of flowering and fruiting.

* * *Flowers greenish, involucrate with 4 small caducous bracts, umbellate, appearing with the leaves; drupe elongated.—*
Subgenus MACROCARPIUM, Spach.

7. **C. sessilis,** Torr.; Dur. Pl. Pratt. 89 (1855), and Bot. Mex. Bound. 94. t. 7: C. & E. l. c. 33. Shrub 6–15 ft. high: bark of twigs greenish: leaves rather crowded, ovate, short-acuminate, pale and appressed silky-pubescent beneath: umbel terminal, nearly sessile, becoming lateral by the development of the twig; pedicels many, slender, silky, 3–4 lines long: involucre nearly as long, very thin, deciduous: petals narrow, acuminate: fr. oblong, $\frac{1}{2}$ in. long or less.—By streams at middle elevations of the Sierra, from Placer Co. northward.

* * * *Plant low, semiherbaceous; fl. greenish, in a dense cyme, subtended by 4 petaloid white bracts.—*Subgenus CORNION, Spach.

8. **C. Canadensis,** Linn. Sp. Pl. i. 118 (1753). Flowering stem simple, erect from a subterranean creeping rootstock, 4–10 in. high: leaves mostly in an apparent whorl of 6 at the summit, ovate to oblong, 1–2$\frac{1}{2}$ in. long, acute at each end, subsessile: peduncle solitary, 1 in. long: involucral bracts ovate, 4–8 lines long: ovary silky: fr. globular, 2 lines thick.—In moist woods of the Coast Range from Mendocino Co. northward.

* * * * *Arborescent; fl. greenish, sessile on a thick convex receptacle, subtended by 4–6 large white petaloid bracts.—*
Genus BENTHAMIDEA, Spach.

9. **C. Nuttallii,** Audubon, Birds, 467 (1838); T. & G. Fl. i. 652. Tree 15–70 ft. high, with ascending or widely spreading branches and smooth bark: leaves 3–5 in. long, obovate, acute at each end, pubescent: bracts of involucre usually 6, obovate to oblong, $1\frac{1}{2}$–3 in. long, abruptly acute to acuminate, white, often tinged with red: head $\frac{1}{2}$–1 in. broad, very dense: fr. 5–6 lines long, scarlet.—In the Coast Range from Monterey northward. May–July.

2. GARRYA, *Douglas.* Evergreen shrubs with greenish bark, and opposite entire coriaceous leaves. Flowers diœcious, in axillary pendulous aments, solitary or in threes between the decussately connate bracts. Petals 0. Calyx of sterile flowers 4-parted, with linear valvate segments. Stamens 4; filaments distinct; disk and rudimentary ovary 0. Calyx of fertile flower with a shortly 2-lobed or obsolete limb; disk and rudimentary stamens 0; ovary 1-celled, with 2 pendulous ovules; styles 2, stigmatic on the inner side, persistent. Berry ovoid, 1–2-seeded, dark blue or purple when ripe.

1. **G. elliptica,** Dougl.; Bot. Reg. t. 1686 (1835); Greene, Gard. and Forest, iii. 198. Shrub usually clustered, 5–15 ft. high: leaves $1\frac{1}{2}$–3 in. long, dark green, elliptical, rounded or acute and mucronate at apex, truncate or rounded at base, the margin undulate, glabrous above, tomentose beneath: aments solitary or several; the sterile 4–10 in. long, their silky bracts truncate or acute; calyx-segments cohering at tip: fertile aments stouter, 2–6 in. long; bracts acute or acuminate; ovary sessile, densely silky-tomentose: fr. globose, 4 lines thick.—In rich shady places along streams, from Monterey northward. Feb.

2. **G. Fremonti,** Torr. Pac. R. Rep. iv. 136 (1857). Shrub 5–10 ft. high, glabrate: leaves light green, ovate or oblong, not undulate, $1\frac{1}{2}$–$2\frac{1}{2}$ in. long, on petioles of $\frac{1}{2}$ in. or less: aments solitary, 2–3 in. long, with acute somewhat silky bracts: ovaries nearly glabrous: fr. globose, 2 lines or more in thickness, short-pedicellate.—From Mt. Hamilton northward, on dry slopes and summits.

3. **G. buxifolia,** Gray, Proc. Am. Acad. vii. 349 (1868). Shrub 2–5 ft. high: leaves oblong-elliptical, 1–$1\frac{1}{2}$ in. long, 4–8 lines broad, acute at each end, glabrous above, densely white-silky beneath; petioles 1–3 lines long: fertile aments 1 in. long, the short bracts acute, more or less silky: fr. glabrous, globose, subsessile, $2\frac{1}{2}$–3 lines thick.—Apparently local on Red Mountain, Mendocino Co.

DIVISION II. SYMPETALÆ.

Corolla of petals which are more or less united and forming a tubular part; or corolla sometimes wanting and replaced by a colored and corolla-like synsepalous calyx.

ORDER LVI. ELÆAGNEÆ.

A. Richard, Mém. de la Soc. d'Hist. Nat. de Paris, i. 375 (1823).

A small family, as nearly allied to Salsolaceæ as to any other, and here represented by a single species of

1. **LEPARGYRÆA**, *Rafinesque*. Shrubs with branchlets more or less spinescent, opposite leaves which are entire and with a scurfy indument. Flowers diœcious, small, clustered in the axils of the branchlets. Calyx of sterile flower deeply 4-parted. Stamens 8, with as many alternating glands, subsessile, shorter than the limb of the calyx. Calyx of fertile flower with ovoid tube and spreading 4-lobed limb, the throat beset with 8 contiguous glands. Pistil simple; ovary enclosed within the calyx-tube and appearing inferior, 1-ovuled: style elongated, acute, stigmatose up and down one side. Fruit appearing drupaceous by the enclosure of the stone-like seed within the fleshy persistent calyx-tube.

1. **L. argentea**, Greene, Pittonia, ii. 122 (1890); Nutt. in Fras. Cat. (1813), under *Elæagnus*, and Gen. ii. 240 (1818), under *Shepherdia*. Shrub 5 -18 ft. high, little spinescent: leaves silvery-scurfy on both sides, mostly oblong, obtuse, cuneate at base, 1 - 1½ in. long: staminate fl. 1½ lines long; pistillate 1 line: fr. ovoid, 2½ lines long, subsessile, scarlet, acidulous and edible.--From the Mono Lake region northward, on the eastward slope of the Sierra only. Common in the Great Basin, and called Buffalo Berry.

ORDER LVII. DAPHNOIDEÆ.

Ventenat, Tableau du Règne Végétal, ii. 235 (1799).

Small family, intermediate between the preceding and the Caprifoliaceæ. We have but one member, of the genus

1. **DIRCA**, *Linnæus* (LEATHERWOOD). Branching deciduous shrubs, with smooth and very tenacious brown bark; the wood also very tough and flexible. Flowers in fascicles of about 3, appearing before the

leaves, but from the same buds, and these of yellowish or whitish very silky caducous scales, which appear as an involucre to the flowers. Calyx 0. Corolla tubular, but slightly oblique, yellowish, nodding, 4-lobed. Stamens 8, inserted at base of the corolla-tube, exserted; filaments filiform; anthers small, oblong. Ovary sessile, 1-celled; style longer than the stamens. Corolla deciduous from the growing ovary, this becoming a somewhat drupaceous small fruit.

1. **D. occidentalis,** Gray, Proc. Am. Acad. viii. 631 (1873): *D. palustris*, Torr. Pac. R. Rep. 77 (1857), not Linn. Shrub 4–7 ft. high: bud-scales densely white-villous: leaves oval with rounded base, 1–3 in. long: fl. canary-yellow, subsessile, 3–4 lines long, rather deeply 4-lobed, the lobes nearly truncate, somewhat connivent, rendering the upper and broader part of the organ slightly urceolate. On moist well shaded northward slopes of the Oakland and Berkeley Hills. Feb., March.

ORDER LVIII. SANTALACEÆ.

Robert Brown, Prodromus Floræ Novæ-Hollandiæ, 350 (1810).

Represented by two species of the genus

1. **COMANDRA,** *Nuttall*. Glabrous pale and glaucous low perennials, the erect stems from subterranean rootstocks. Leaves alternate, subsessile, entire. Flowers greenish-white, in small terminal and axillary umbels. Perianth urceolate or campanulate, with a 5-toothed persistent limb. Ovary surmounted by a 5-lobed disk which is free from the perianth. Stamens short, their linear filaments attached by a tuft of basal hairs to the base of the perianth-lobes. Style filiform; ovary coherent with the perianth-tube, about 3-ovuled, becoming a drupe-like 1-seeded fruit.

1. **C. umbellata,** Nutt. Gen. i. 157 (1818); Linn. Sp. Pl. i. 208 (1753), under *Thesium*. Stem 6–15 in. high: leaves oblong, obtuse or acute, 1½–1½ in. long, umbels few-flowered, corymbosely clustered at summit of stem: fl. 1½–2 lines long, on slender pedicels, the oblong slightly spreading lobes white, about equalling the green tube, which is conspicuously continued above the ovary: style slender: fr. not very fleshy, globose, 2–3 lines thick, on slender pedicels 2–3 lines long. –Western slope of the Sierra, from Fresno Co. northward.

2. **C. pallida,** A. DC. Prodr. xiv. 636 (1857). Aspect of the preceding, but with narrower and acute leaves: fr. ovoid, 3–4 lines long, sessile, or the pedicels very short and stout.—Perhaps not within our limits; but found east of the Sierra northward, and in Oregon.

ORDER LIX. **LORANTHEÆ.**

Jussieu & Richard, Annales du Museum, xii. 292 (1808).

Evergreens, half-shrubby and parasitic on trees and shrubs; color yellowish-green or yellow. Branches dichotomous; the joints swollen. Leaves opposite, either coriaceous, or reduced to more or less distinctly connate scales. Flowers (diœcious in our genera) of 2–5 sepals coherent at base and valvate in æstivation, no petals; anthers as many as the calyx-segments and (in ours) sessile upon them; ovary inferior, 1-celled, 1-ovuled, becoming a 1-seeded berry with glutinous epicarp.

1. **PHORADENDRON**, *Nuttall* (MISTLETOE). Flowers globose, imbedded in the rachis of jointed spikes. Calyx 3- (rarely 2- or 4-) lobed. Anthers sessile on the base of the lobes, 2-celled, opening by a pore or slit; pollen-grains smooth. Stigma sessile, obtuse, entire or more or less distinctly 2-lobed. Berry globose, pulpy, translucent, crowned with the persistent calyx-lobes. Embryo with foliaceous cotyledons.

* *Leafy species; leaves dilated upwards.*

1. **P. flavescens**, Nutt. Pl. Gamb. 185 (1848); Engelm. Pl. Fendl. 58 (1849); Pursh, Fl. i. 114 (1814), under *Viscum*. *V. leucocarpum*, Raf. Fl. Ludov. 79 (1817). Branches 1 ft. long or more, terete, pubescent when young, usually of a light or yellowish green as also the foliage: leaves oblanceolate to obovate and nearly orbicular, $\frac{1}{2}$–2 in. long, obtuse, 3-nerved, in age glabrous: bracts of the inflorescence connate into a short truncate cup: fl. depressed-globose, the calyx-lobes ciliate: staminate spikes opposite or verticillate, usually shorter than the leaves, 3–7-jointed, the many fl. in 4–6 rows on each side, fragrant with the odor of pond-lilies; anthers transverse, opening by 2 pores: pistillate spikes mostly opposite, shorter than the staminate: berries white, 2 lines thick. Var. **villosum**, Engelm. l. c. *P. villosum*, Nutt. l. c. Leaves smaller, spatulate to orbicular, permanently pubescent: spikes smaller. Parasitic on various exogenous trees in the interior of the State.

2. **P. Bolleanum**, Engelm. Bot. Calif. ii. 105 (1880); Seem. Bot. Herald, 295. t. 63 (1856), under *Viscum*. Branches $\frac{1}{2}$ ft. long or less, terete, puberulent when young: leaves very thick, spatulate to linear, obtusish, nerveless, $\frac{1}{2}$–1 in. long: spikes opposite or in fours, with connate minutely ciliate bracts: the staminate of two 6–12-flowered joints, the fertile of a single 2-flowered joint: anthers transverse, opening by pores: berry white, $1\frac{1}{2}$ lines thick.—From Placer and Lake counties southward, chiefly on firs and junipers.

* * *Leaves reduced to short mostly connate scales.*

3. **P. juniperinum**, Engelm.; Gray, Pl. Fendl. 58 (1849). Branches

1_2 1 ft. long, very numerous (forming a rounded rather dense mass), terete (the ultimate twigs quadrangular), glabrous: scales broadly triangular, obtusish, connate or distinct, ciliate: staminate spikes of a single 6 8-flowered joint (rarely 2): anthers transverse, opening by pores: pistillate spikes 2-flowered: berry globose, 1½ lines thick, whitish or light red. Var. **Libocedri**, Engelm. Bot. Calif. ii. 105. Branches longer and more slender, the ultimate twigs more sharply quadrangular. From Yuba Co. southward; the type on junipers, the variety on cedar.

2. **RAZOUMOFSKYA,** *Hoffmann.* Small yellow or greenish leafless parasites upon coniferous trees: leaves represented by connate scales. Flowers axillary and terminal, solitary or several from the same axil. Staminate fl. mostly 3-parted, compressed, or the terminal ones globose: anthers sessile on the lobes, orbicular and 1-celled, dehiscent by a circular aperture at base; pollen-grains spinulose. Pistillate fl. ovate, compressed, 2-toothed, subsessile, the pedicel in fruit elongated and recurved. Fruit compressed, elastically dehiscent at the circumscissile base, forcibly ejecting the seed. Cotyledons rudimentary, indicated by a notch in the axis of the embryo.

* *Staminate fl. on peduncle-like joints, all or nearly all terminal.*

1. **R. Americana,** O. Ktze. Rev. Gen. ii. 587 (1891); Nutt. in Pl. Lindh. ii. 214 (1850), under *Arceuthobium.* Slender, dichotomously or verticillately branching, greenish-yellow; staminate plants often 3 4 in. long; fertile much smaller; staminate fl. 1 line broad or more, with round-ovate acutish lobes; the pistillate ½—1 line long; fr. 2 lines long. Parasitic on *Pinus contorta;* flowering in autumn, fruit maturing a year later.

** *Staminate fl. sessile, mostly axillary.*

2. **R. Douglasii,** O. Ktze. l. c.; Engelm. in Wheeler's Rep. 253 (1878), under *Arceuthobium.* Slender, greenish yellow, ¼—1 in. high, much branched, but not verticillately, the accessory branchlets behind (not beside) the primary ones: spikes short, mostly 5-flowered: staminate fl. less than a line wide, with round-ovate acutish lobes: fr. 2½ lines long. Var. **abietinum** (Engelm. Bot. Calif., under *Arceuthobium*). Fertile plant larger (1 3 in. high), the sterile smaller, with spreading or even recurved branchlets: fr. smaller (scarcely 2 lines long). The type on *Pseudotsuga taxifolia,* east of the Sierra southward, not for a certainty within our limits; the variety, or probably distinct species, on *Abies concolor* farther northward, in Sierra Co., *Lemmon*.

3. **R. occidentalis,** O. Ktze. l. c.; Engelm. Bot. Calif. ii. 107 (1880), under *Arceuthobium.* Stout, 2 5 in. high, paniculately much branched: sterile plants brownish-yellow, smaller; fertile commonly darker, olive-brown: staminate fl. in long dense spikes, often 9 -17 on a single axis,

their buds ventricose with the upper edge curved outward; calyx 3–5 (usually 4-) -parted, 1½–2 lines wide; anthers sessile below the middle of the lanceolate acuminate lobes: fr. 2½ lines long. On various conifers of both mountain ranges, from middle parts of the State northward.

Order LX. **CAPRIFOLIACEÆ.**
Richard, Dict. Class. d'Hist. Nat. iii. 172 (1826).

Shrubs often trailing or climbing. Leaves opposite, mostly exstipulate. Flowers terminal and cymose or subspicate, or solitary or in pairs in the leaf-axils, regular or irregular. Calyx-tube coherent with the ovary; limb 5-toothed or obsolete. Corolla 4–5-lobed or -cleft; the lobes imbricate in bud. Stamens distinct. Ovary 2–5-celled, or by abortion 1-celled after flowering. Fruit a berry or drupe.

Corolla rotate or broad-campanulate, regular; style short or 0, - - - 1, 2
" campanulate to tubular, more or less irregular; style elongated, - 3—5

1. SAMBUCUS, *Pliny* (ELDER). Shrubs or small trees, with stout thick and very pithy shoots and branches, and pinnate foliage; leaflets 5–11, serrate: young shoots and foliage heavy-scented. Flowers small, white or cream-color, very many, in compound cymes at the ends of terminal and lateral shoots. Calyx with 5 minute teeth. Corolla rotate, 5-lobed. Stamens 5. Stigmas and ovary-cells 3—5. Fruits of the nature of drupelets, though berry-like, each with 3 (rarely 4 or 5) separate seed-like nutlets, each with a single seed.

1. S. callicarpa. *S. racemosa*, Gray, Bot. Calif. i. 278, not Linn. Arborescent, 10–25 ft. high, clustered, and each of the several trunks often a foot in diameter; bark light brown, more flaky than fissured; pith of shoots white: young twigs and foliage pubescent with sparse stiff short somewhat retrorse hairs: young leaves with free ligulate callous-tipped stipules 1–3 lines long: leaflets 2–5 pairs, often with conspicuous false stipellæ, or the later leaves on vigorous shoots completely bipinnate, the ordinary leaflets from oval to oblong-lanceolate, abruptly acuminate, closely and rather deeply serrate, thin: cymes rather small but flat-topped: corolla rotate, white: fr. bright red.—Very common by streams, and even in moist lowlands near the sea along the Coast Range; perhaps also in the Sierra. The arborescent habit, stipulate and often bipinnate leaves, but more than all the broad and flat rather than thyrsoid inflorescence and fruit-clusters, mark this as a species very distinct from the Old World *S. racemosa*, in which latter the corolla-lobes, moreover, are closely reflexed against the pedicel. The eastern shrub, *S. pubens*, is easily distinguishable from both by a character not hitherto mentioned, *i. e.*, the large and rounded very conspicuous winter-buds. The red-berried elder of the northern woods, from Oregon to Alaska, is not *S. racemosa*, for it has, like our species, very ample and almost flat-topped cymes; but neither am I confident of its identity with *S. callicarpa*. Our tree has small winter-buds, and is hardly in flower before April, putting forth its leaves in March.

2. **S. melanocarpa**, Gray, Proc. Am. Acad. xix. 76 (1883). Shrubby, only 6–8 ft. high: leaves never bipinnate; stipules 0: leaflets 5–7, oblong-lanceolate, abruptly acuminate: cyme rather ample, broad but convex: fr. black, without bloom. - Common in the northern Rocky Mountains, reaching the Sierra Nevada, according to Gray. The species is nearest to *S. pubens*, having rather conspicuous winter-buds, and convex or even somewhat pyramidal inflorescence.

3. **S. velutina**, Dur. & Hilg. Pac. R. Rep. v. 8 (1855); *S. Mexicana*, Presl.; DC. Prodr. iv. 322 ? (1830). Shrub 5–6 ft. high, velvety-tomentose, the upper face of the leaves glabrous: leaflets 5–11, obliquely ovate-lanceolate, acute, sharply serrulate, subcoriaceous: corymb small, flattish: fr. dark purple, of an agreeable flavor. The type of this is from the plains of Kern Co.; and to the species may probably be referred an almost herbaceous elder common in the Sacramento Valley wheat fields. The shoots of this are simple, 5 or 6 ft. high, and bear an ample terminal cyme. But the leaves in this scarcely shrubby plant are commonly altogether bipinnate. Its fruit is unknown, for the plants, springing up in the fields after the spring plowing and as if from rootstocks, are cut down by the reapers while in flower. It seems unlikely that *S. Mexicana* can be this species.

4. **S. glauca**, Nutt.; T. & G. Fl. ii. 13 (1841). Arborescent, often 30 ft. high at southerly stations, and the solitary trunk not rarely more than a foot thick, covered with a dark close very distinctly and rather finely fissured bark: twigs long and slender; leaves exstipulate, coriaceous, glabrous; leaflets 3–5 pairs, lanceolate, acuminate, sharply serrulate, seldom or never divided: cymes large, flat: fl. white: fr. blue with a dense bloom but black beneath it. In rather dry and sparsely wooded ravines and open grounds throughout the middle and southern parts of the State; flowering and fruiting at intervals throughout the long season from March to December. Fruit acidulous, and when cooked not unpalatable. In San Diego Co. the author once measured a healthy tree of this species with trunk three feet thick.

2. **VIBURNUM**, *Pliny*. Shrubs or small trees, with tough and flexible (not pithy) branches, simple leaves, and terminal flattened cymes of white flowers. Corolla rotate or open-campanulate. Ovary 1-celled, 1-ovuled, becoming a drupe with a single more or less flattened stone. Embryo minute.

1. **V. ellipticum,** Hook. Fl. i. 280 (1833); Gray, Syn. Fl. i. 10. Stems 2–5 ft. high, with glabrous pale brown bark: leaves from round-oval to elliptic-oblong, rounded at both ends, dentate above the middle, 2 in. long, subcoriaceous, 3–5-nerved from the base, the nerves ascending or parallel: corolla 4–5 lines wide: stone of the bluish black fr. deeply and broadly sulcate on both faces, the furrow of one face divided by a median ridge.—In mountain woods, from Placer and Mendocino counties northward.

3. **OBOLARIA,** *Siegesbeck* (Twin-Flower. Linnæa). Trailing evergreen shrub, with erect long and slender 2-flowered peduncles. Calyx with limb 5-parted into subulate-lanceolate lobes, constricted above the globular tube, deciduous from the fruit. Corolla campanulate-funnelform, not gibbous at base, almost equally 4-lobed. Stamens 4, in unequal pairs, included. Ovary 3-celled, becoming a dry indehiscent 1-seeded fruit.

1. **O. borealis,** O. Ktze. Rev. Gen. i. 275 (1891); Linn. Sp. Pl. ii. 631 (1753), under *Linnæa*. Prostrate branches slender, pubescent: leaves obovate, crenately few-toothed, ½–1 in. long, on very short petioles: filiform peduncles 2–6 in. high, with 2 bracts near the summit, and from the axil of each a rose-colored flower nodding on a filiform pedicel; pedicels bibracteolate near the summit: corolla ½ in. long or more, very fragrant. In woods of pine and spruce, from Humboldt and Plumas counties northward. July.

4. **SYMPHORICARPOS,** *Dillenius* (Snowberry). Low branching shrubs, erect or spreading, never climbing. Leaves small, membranaceous, mostly entire. Flowers small, nearly or quite regular, axillary and terminal, solitary or in dense spicate clusters, white or pinkish. Calyx with globular or oblong tube and 4–5-toothed persistent limb. Corolla either short-campanulate and slightly gibbous, or salverform and regular, 4–5-lobed. Stamens inserted on the throat of the corolla and as many as its lobes. Ovary 4-celled; 2 cells containing a few sterile ovules, the other 2 each with a single suspended ovule. Fruit globose, berry-like, containing two seed-like smooth 1-seeded nutlets.

* *Flower short, urceolate- or open-campanulate, only 2–3 lines long.*

1. **S. racemosus,** Michx. Fl. i. 107 (1803); Bot. Mag. t. 2211; Bart. Fl. Am. Sept. i. t. 19. Usually 3–4 ft. high, slender, with spreading branches: leaves round-oval to oblong, 1 in. long, glabrous above, pubescent along the veins beneath: axillary clusters mostly few-flowered, the lowest 1-flowered: corolla reddish or pinkish, 2 lines long, slightly gibbous, moderately villous within, cleft above the middle: fr. $\frac{1}{3}$–$\frac{1}{2}$ in. thick, subglobose, snow-white. On banks of streams in shady places almost everywhere in the Coast Range; more slender than the eastern type,

with a more graceful habit, thinner foliage and smaller fruit: very possibly a distinct species. Fl. May, fr. Oct.

2. **S. ciliatus,** Nutt.; T. & G. Fl. i. 4. ? Low and diffuse, seldom 1 ft. high, with many very slender but rather rigid leafy branches, and few-flowered clusters: leaves oval, obtuse, $1\frac{1}{2}-3$ in. long, glabrous above, pubescent along the veins beneath, the margin rather densely ciliate: corolla rose-red, 2 lines long, slightly gibbous, cleft to the middle or more deeply, scarcely villous within: fr. small, globose, snow-white.—Common in the Oakland Hills on northward slopes, and answering well to Nuttall's description of *S. ciliatus* as to leaf and flower, though he says nothing of the diminutive size, and even compares his Santa Barbara shrub with *S. vulgaris*. We would offer for our plant the provisional name *S. nanus*, in case it should prove distinct. It is earlier in its flowering, and much earlier in maturing its fruit, than the preceding species. Fr. July.

3. **S. mollis,** Nutt. l. c. (1841); Gray, Syn. Fl. i. 14. Stems more surculose and straggling, but several feet long: leaves orbicular or broadly oval, $1\frac{1}{2}-1$ in. long, soft-pubescent beneath or sometimes on both faces, and even almost tomentose: fl. solitary or in short clusters: corolla open-campanulate, 5-lobed nearly to the middle, pubescent within: stamens and style included: berries rather small, white. Var. **acutus,** Gray, l. c. Leaves soft-tomentulose, oblong or oblong-lanceolate, acute at both ends or acuminate, sometimes irregularly and acutely toothed.—Wooded hills of both ranges of mountains; the variety (probably a species, but little known) on Lassen's Peak.

* * *Corolla with more elongated and narrow tube; lobes short.*

4. **S. rotundifolius,** Gray, Pl. Wright. ii. 66 (1853), and Journ. Linn. Soc. xiv. 11. Low, rather stout and rigid: leaves very pale and glaucescent, subcoriaceous and more or less densely soft-pubescent: corolla elongated-campanulate, 3–4 lines long; its tube pubescent within below the stamens, twice or thrice the length of the lobes: fr. elliptical, small; nutlets oval, broad and obtuse at both ends.— Chiefly on the eastward slope of the Sierra; but near Tehachapi, *Greene;* also in Fresno Co.

5. **S. oreophilus,** Gray, Journ. Linn. Soc. l. c. 12 (1875). Erect with rather slender but short spreading or ascending branches: leaves oval or oblong, rather thin, not pale, glabrous or pubescent: corolla tubular-funnelform, 4–6 lines long, yellowish-white, with or without a reddish tinge; tube nearly glabrous within, 4 or 5 times the length of the slightly spreading lobes: nutlets of the elongated drupe oblong, pointed at base.—East of the Sierra only.

5. **CAPRIFOLIUM,** *Brunfels* (HONEYSUCKLE). Erect or trailing or

climbing shrubs, with leaves either membranous or subcoriaceous, occasionally stipulate. Flowers larger and showy, in pairs on an axillary peduncle, or verticillate-spicate at the ends of the branches. Calyx-limb small and 5-toothed, or obsolete. Corolla more or less gibbous at base, or bilabiate, or both. Stamens 5, on the tube of the corolla. Ovary 2–3-celled, becoming a few-seeded purple or red or yellow berry.—A rather too heterogeneous assemblage, doubtless better received as several distinct genera, according to the views of the earlier sytematists.

* *Stems erect, never twining or climbing; leaves all distinct; fl. axillary, 2 on each slender peduncle, their ovaries distinct or connate, the corollas $1\frac{1}{2}$ in. long; calyx-limb minute or obsolete.* —
Genus XYLOSTEON of many authors.

1. **C. cœruleum**, Lam. Fl. Fr. ii. 366 (1778); Linn. Sp. Pl. i. 174 (1753), under *Lonicera*. Low (1–2 ft.), pubescent or glabrate: leaves thin, pale or glaucescent, ovate-oblong, obtuse, entire, 1 in. long or more: peduncles shorter than the flowers: corolla ochroleucous, gibbous at base, narrowly funnelform, scarcely at all bilabiate: bracts subulate or linear, commonly larger than the ovaries; these completely united, forming a roundish or ovoid 2-eyed sweetish berry which is black but glaucous. —From Mariposa Co. northward, in the Sierra Nevada.

2. **C. conjugiale**, O. Ktze. Rev. Gen. i. 274 (1891); Kell. Proc. Calif. Acad. 67 (1862), under *Lonicera*. *L. Breweri*, Gray, Proc. Am. Acad. vi. 537 (1865). Shrub 3—5 ft. high, freely branching and bushy: leaves thinnish, bright green, pubescent when young, ovate or oval, acute or acuminate, 1—$2\frac{1}{2}$ in. long, short-petioled: peduncles slender, 3–5 times the length of the dark maroon flowers: bracts subulate, caducous; calyx-teeth subulate: corolla 4–5 lines long, gibbous-campanulate, the upper lip crenately 4-lobed, throat and lower part of filaments and style very hirsute: berries red, almost wholly connate. —Woods of the higher Sierra.

3. **C. involucratum**, O. Ktze. l. c.; Banks in Richardson, App. Frank. Voy. 5 (1823); Spreng. Syst. i. 759 (1825), under *Lonicera;* Bot. Reg. t. 1179. Erect, 1–3 ft. high: leaves large, thin, ovate-elliptical, acuminate, short-petiolate, 2–3 in. long, pubescent on the veins beneath and on the margin: peduncles short, the fl. subtended by a pair of large ovate foliaceous bracts: calyx-limb obsolete: corolla yellowish, gibbous at base, narrowly funnelform, with scarcely spreading lobes; these commonly acutish: berries distinct, black. A species of the Rocky Mountain and far northeastern regions, perchance reaching our borders in the Sierra eastward. The next has needlessly been confused with it, being clearly distinct.

4. **C. Ledebourii**, Greene. Esch. Mem. Acad. Petrop. x. 284 (1826), under *Lonicera*. *L. Mociniana*, DC. Prodr. iv. 336 (1830): *L. intermedia*,

Kell. Proc. Calif. Acad. ii. 154. ? Stouter, 5 15 ft. high, often with the very long sarmentose branches reclining on or half climbing over other shrubs or small trees: leaves of firmer texture than in the last, and more hairy: corolla more strongly gibbous at base, strictly salverform above the gibbosity, the short rounded lobes spreading abruptly, the whole almost scarlet without, yellow within. –Very common along streams almost throughout western California, ranging far northward, and also extending into the mountains of Arizona; the habit of the shrub, and more especially its very different corolla, marking it as distinct from *C. involucratum*. Feb.–May.

* * *Stems usually more or less twining; upper leaves often connate-perfoliate; fl. sessile in spiked whorls at the ends of the branches; corolla 1 in. long, mostly bilabiate.*–CAPRIFOLIUM proper.

5. **C. ciliosum**, Pursh. Fl. i. 160 (1814); Poir, Encycl. v. 612 (1804), under *Lonicera: C. occidentale*, Lindl. Bot. Reg. t. 1457. Usually depressed, only a foot or two high, and almost prostrate: leaves ovate or oval, glaucous beneath, usually ciliate, otherwise glabrous, the uppermost one or two pairs connate into an orbicular or elliptical disk: whorl of flowers usually 1 only, rarely 2 or 3: corolla glabrous or sparingly pilose-pubescent, crimson-scarlet without, yellow within, ventricose-gibbous below; the limb slightly bilabiate; lower lobe 3–4 lines long.–In the Sierra at middle altitudes or lower, and far northward and eastward. A small but very beautiful species. June–Aug.

6. **C. hispidulum**, Lindl. Bot. Reg. t. 1761 (1835), var. **Californicum.** *Lonicera Californica*, T. & G. Fl. ii. 7 (1841): *L. hispidula*, var. *vacillans*, Gray. Twining, 10–25 ft. high, the ultimate branches often a yard or two in length and drooping, hispidulous and somewhat glandular as to the upper portion and about the inflorescence; leaves ovate-oblong or elliptical, acutish, 1–3 in. long, the lower pairs without stipules, the intermediate with broadly ovate stipular appendages often ½ in. long and as broad, the one or two floral pairs connate, all very glaucous beneath, pale and glaucescent above, thickish but hardly subcoriaceous: spikes 1–5, each with 3–6 whorls of pink flowers; corolla hispidulous, ½–¾ in. long; anthers exserted, narrowly linear, 2½ lines long: berries crimson.–Common in moist ravines and on shady banks, climbing over small trees, along the seaboard only; very beautiful in flower. The Oregonian type of the species is much smaller, seldom or never twining, and with strongly ciliate leaves, even the uppermost small and distinct.

7. **C. interruptum**, Greene. Benth. Pl. Hartw. 313 (1849), under *Lonicera. L. hispidula*, var. *interrupta*, Gray. Stoutish, erect and bushy, 4–7 ft. high, less disposed to twine or climb; bark of branches white and almost shining, glabrous: leaves of a very pallid hue, white-glaucous beneath, glaucescent above, 1 in. or more in breadth, mostly

orbicular or round-obovate, never stipulate, several of the uppermost pairs connate; fl. numerous, in several interrupted spikes; corolla ⅓ in. long, yellow, glabrous. Common on dry bushy hills of the inner Coast Ranges, and in similar places among the Sierras northward. May, June.

8. **C. subspicatum,** Greene. H. & A. Bot. Beech. 349 (1841), and Torr. Bot. Mex. Bound. t. 23, under *Lonicera. L. hispidula,* var. *subspicata,* Gray. Size of the last, bushy but more straggling, the many ultimate branches and branchlets short and all floriferous; almost all the plant, except the upper face of the leaves, densely glandular-pubescent: leaves small, coriaceous, narrowly oblong, obtuse, tapering to a distinct petiole, very veiny on both faces, deep green above, apparently white-tomentose between the veins beneath, none of them either stipulate or connate: spikes numerous, almost panicled, but the whorls of flowers few, each whorl with a pair of bracts at base: corolla ½ in. long, yellowish. In the Coast Ranges, but southward only. How this and the preceding can have passed for varieties of *C. hispidulum* passes our comprehension. Probably no country can exhibit three species of this group which are more pronouncedly distinct.

Order LXI. RUBIACEÆ.

L. Gerard, Flora Gallo-Provincialis, 224 (1761): Jussieu, Gen. 196 (1789).

Our species herbs (*Cephalanthus* and some species of *Galium* shrubby) with opposite or verticillate mostly exstipulate entire leaves, and 4-merous perfect (rarely diœcious) flowers. Calyx-limb obsolete, or of 4 teeth. Stamens distinct, alternate with the corolla-lobes and inserted on its throat or tube. Ovary 2 — 4-celled, with a solitary ovule in each cell. Fruit indehiscent, dry or baccate.

1. **CEPHALANTHUS,** *Linnæus* (BUTTON-BUSH). Shrubs with opposite or ternate leaves, and flowers in dense globose terminal and axillary peduncled heads. Calyx inverse-pyramidal, 4-toothed. Corolla with long slender tube and small 4-cleft limb. Stamens 4, short, on the throat of the corolla. Style slender, long-exserted; stigma capitate; ovary 2-celled. Fruit achene-like, 1 - 2-seeded.

1. **C. occidentalis,** Linn. Sp. Pl. i. 95 (1753). Shrub or small tree, with ovate-lanceolate leaves 3 -5 in. long, rather glossy above, often more or less pubescent: fl. white, in heads 1 in. thick, these solitary or few or several toward the ends of the branches. River banks of the interior, especially of the Sacramento from Shasta Co. to Solano; also on the lower San Joaquin. June—Aug.

2. **KELLOGGIA,** *Torrey.* Slender perennial, with opposite leaves, and loosely cymose-panicled small pinkish flowers. Calyx-tube obovoid, somewhat flattened laterally, beset with bristles; teeth 4, very small, subulate, persistent. Corolla funnelform; lobes valvate. Stamens 4, in the throat of the corolla; filaments short; anthers linear. Style slender; ovary 2-celled; ovules erect. Fruit small, oblong, coriaceous, uncinate-hispid, splitting into two 1-seeded carpels which are indehiscent, 1-seeded. Embryo large, straight, in fleshy albumen.

1. **K. galioides,** Torr. Bot. Wilkes Exp. t. 6 (1862); Gray, Proc. Am. Acad. vi. 539 (1865). About 1 ft. high, glabrous or minutely pubescent: leaves lanceolate, sessile, the stipules small and scarious: fl. small, in a loose dichotomous cyme; the long pedicels thickened above and articulated with the flower: corolla white, with a tinge of rose-purple, 3–5 lines long, pubescent on the outside. Foothills of the Sierra Nevada, from Kern Co. northward, in shady woods, and also at higher altitudes. A neat but not showy plant, not very different from *Galium* except in the elongated corolla.

3. **SHERARDIA,** *Dillenius.* Annual, slender, rough, with angular stem, and exstipulate leaves in verticels of 6. Flowers umbellate. Calyx-limb of 4–6 accrescent teeth. Corolla salverform, with a slender tube and 4-cleft limb. Stamens 4. Fruit didymous, of 2 dry indehiscent 1-seeded carpels, crowned by the calyx-teeth, and separating from each other when ripe.

1. **S. arvensis,** Linn. Sp. Pl. i. 102 (1753). About 3–6 in. high, hispidulous-roughened or nearly glabrous: leaves obovate-lanceolate, acute: fl. in small subsessile umbellate cymes: corolla bluish. Vicinity of Berkeley; first found by *Mr. C. T. Blake,* in 1889; observed also in 1891, on the university grounds, by *Mr. Bioletti;* naturalized from Europe.

4. **GALIUM,** *Dioscorides* (BEDSTRAW. CLEAVERS). Herbaceous (rarely suffrutescent), with slender angular stems, verticillate leaves without stipules (or the smaller leaves of the whorls to be considered as stipular organs ?), and small usually cymose flowers. Calyx-limb obsolete. Corolla rotate, 4-parted (sometimes 3- or 5-parted). Stamens as many as the corolla-lobes, short. Styles 2, short; stigmas capitate: ovary 2-lobed, 2-celled, 2-ovuled. Fruit didymous (biglobular), dry or fleshy, separating into 2 closed 1-seeded carpels which are indehiscent, and glabrous, hispid, echinate or hirsute.

* *Fruit dry when ripe.*—GALIUM proper.
+– *Annuals.*

1. **G. bifolium,** Wats. Bot. King Exp. 134. t. 14, f. 8 (1871). Simple, or at length with few branches, erect, 3–6 in. high, glabrous: leaves in

pairs, or in larger specimens 4 in the whorl, lanceolate, the alternate pair (answering to stipules?) much smaller: peduncles solitary, lateral and terminal, naked, 1-flowered, when in fruit about equalling the leaves, spreading: corolla minute, white: fr. recurved, minutely uncinate-hispid.—In moist shades of the higher Sierra.

2. G. SPURIUM, Linn. Sp. Pl. i. 106 (1753): *G. Vaillantii*, DC. (1805); *G. Aparine*, Gray, Bot. Calif. in part, not Linn. Branching chiefly from the base, diffuse, 1–2 ft. high, glabrous except the retrorsely scabrous angles of the stem and veins and margins of the leaves: leaves 6–8 in the whorl, linear-oblanceolate, cuspidate: fl. 3–9 in axillary umbellate cymes; corolla pale green, the segments acuminate: pedicels recurved after flowering: fruit large, coarsely tuberculate, more or less uncinate-hispid. Mostly in the mountains back from the seaboard; less common than the next.

3. G. APARINE, Linn. l. c. 108. Taller and more slender, 3–5 ft. high (or often only a few inches), climbing by the retrorse prickliness of the angles and leaf-margins: corolla minute, white: pedicels straight in fruit: surface of carpel smooth but densely uncinate-hispid. Very common in shady or open places in woods and along the salt marshes; readily distinguished by the minute white corollas, straight pedicels, and smaller and more prickly carpels. Though both these species are as much at home in our woods and thickets as any indigenous plants, it is probable that they came hither from the Old World within the last two centuries. The villous pubescence at the nodes, which is a part of the ascribed character of this species, is with us seldom noticeable; and yet, in a specimen from the Marysville Buttes, collected by Mr. Jepson, the whole stem is villous at the angles, without a trace of the usual retrorse prickles. Mar. May.

+ + *Perennials.*

++ *Stem wholly herbaceous; fruit not hirsute.*

4. G. asperrimum, Gray, Pl. Fendl. 60 (1859). Diffusely branching, 1–2 ft. high, the numerous slender branches and the leaf-margins more or less ciliolate-scabrous, scarcely spinulose: leaves in sixes, lanceolate or oblanceolate: fl. many, in naked cymes; corolla white (turning dark in drying): fr. granulate and rather densely setulose.—At middle altitudes in the Sierra, from Mariposa Co. northward, but not very common.

5. G. triflorum, Michx. Fl. i. 80 (1803). Stem flaccid, 1 ft. long or more, reclining or at least decumbent, retrorsely aculeate-scabrous on the angles, or smoothish: leaves in sixes, thin, elliptic-lanceolate, acute at both ends, or cuspidate-acuminate, the margins and often the midrib beneath beset with very short usually retrorse and hooked prickles:

peduncles few, once or twice 3-forked; pedicels divergent: corolla greenish: fr. hirsute with slender hooked bristles, or when ripe merely roughened. In woods of both the Sierra and the Coast Range.

6. **G. trifidum,** Linn. Sp. Pl. i. 105 (1753). Erect or reclining, rather slender, 5–20 in. high, glabrous, except the retrorsely scabrous angles of the stem, and the more hispidulous but sparse roughness of the margins of the leaves and the midrib beneath: leaves (in our forms) usually in fours or fives, linear or oblanceolate, or lanceolate-oblong, obtuse, 4–7 lines long: peduncles slender, scattered, 1–several-flowered: fl. minute, white, often 3-merous: fr. small, smooth, glabrous. In wet grounds, toward the seaboard in a very large form, and in the Sierra in the smallest states.

++ ++ *Stem suffrutescent; fruit very bristly or hirsute.*

7. **G. angustifolium,** Nutt.; T. & G. Fl. ii. 22 (1841). Rigid, much branched, shrubby at base, 1½–4 ft. high, smooth and glabrous, or minutely pruinose-puberulent: leaves in fours, narrowly linear, ½–1 in. long, or on branches shorter: fl. diœcious, in many rather densely panicled cymes: corolla small, light-green: fr. with white bristles about as long as the body. In Kern and Santa Barbara counties and southward.

8. **G. Matthewsii,** Gray. Proc. Am. Acad. xix. 80 (1883). Shrubby at base, smooth and glabrous, paniculately much branched: leaves in fours, oblong- to ovate-lanceolate, some of the upper cuspidate-acute, all 2–3 lines long, with stout midrib and no veins: fl. in naked panicles: bristles of unripe fruit rigid, not longer than the body.—In Inyo Co.

9. **G. multiflorum,** Kellogg, Proc. Calif. Acad. ii. 97, f. 26 (1863): *G. Bloomeri,* Gray, Proc. Am. Acad. vi. 538 (1865). A few in. to 1 ft. high, from a suffrutescent base, the erect stems tufted and little branching, glabrous, pruinose-puberulent or pubescent: leaves in fours, from broadly ovate to ovate-lanceolate, mucronate-apiculate or abruptly acuminate, thickish, 4–7 lines long, with 2, or sometimes 4, lateral nerves from the base: uppermost leaves often in pairs only: fl. thyrsoid-paniculate: corolla greenish: fr. densely white-hirsute, the hairs longer than the body. In the higher mountains, chiefly on the eastern slope of the Sierra.

* * *Perennial or shrubby; leaves in fours, 1-nerved; fruit baccate.* —
Genus RELBUNIUM, Endl.

10. **G. pubens,** Gray, Proc. Am. Acad. vii. 350 (1868). Herbaceous, somewhat cinereous with a fine pubescence partly soft and partly scabrous: stems diffuse, 1–2 ft. long: leaves thickish, roundish-oval to oblong, mostly pointless, ½ in. long, the margins often hispidulous-scabrous: sterile fl. in small loose cymes, the fertile more scattered: young fr. smooth and glabrous, probably fleshy when ripe. In and near the Yosemite Valley.

11. **G. Californicum,** H. & A. Bot. Beech. 349 (1840). Herbaceous from slender creeping rootstocks, in low tufts, or diffuse with slender stems a foot long, hispid or hirsute, rarely glabrate in age: leaves thinnish, ovate or oval, apiculate-acuminate, ¼—1½ in. long, margins and midrib hispid-ciliolate: fr. blackish, glabrous, on recurved pedicels.—Common in shady places, of the Coast Range chiefly.

12. **G. Nuttallii,** Gray, Pl. Wright. i. 80 (1852): *G. suffruticosum*, Nutt. in T. & G. Fl. ii. 21 (1841), not H. & A. (1840). Suffrutescent, tall and climbing, often 3–4 ft. high, mostly glabrous, except the minutely aculeolate-hispidulous angles of stems and margins of leaves, these also sometimes naked: leaves small, oval to linear-oblong, mucronate, mucronulate, or obtuse: fr. smooth and glabrous, purple. —In thickets of the Coast Range.

13. **G. Bolanderi,** Gray, Proc. Am. Acad. vii. 350 (1868): *G. margaricoccum*, Gray, l. c. xiii. 371. Less shrubby, 1—2 ft. high, glabrous, sometimes pubescent; angles of stem hardly scabrous: leaves oblong-linear or lanceolate, acutish, ½ in. long, thickish, with margins and midrib naked or sparsely hispidulous: corolla dull purplish: berry large, smooth, white. On the western slope of the Sierra, from the Yosemite northward, and in Humboldt Co.

14. **G. Andrewsii,** Gray, Proc. Am. Acad. vi. 537 (1865). Small and densely matted; nearly or quite glabrous, the herbage bright green and shining: leaves crowded, acerose-subulate, either naked or sparsely spinulose-ciliate, 2—4 lines long: fl. diœcious, the sterile in few-flowered terminal cymes; fertile solitary, subtended by a whorl of leaves which are longer than the deflexed fruiting pedicel: berry smooth, blackish.— Dry summits of the Coast Ranges, from Lake Co. southward.

ORDER LXII. **VALERIANEÆ.**

Dufresne, Histoire Naturelle et Médicale de la Famille des Valerianées (1811). VALERIANACEÆ, Lindl. Synops. 137 (1829).

Herbs with opposite leaves, no stipules, and mostly complete flowers, in a cymose or thyrsoid inflorescence. Calyx-tube coherent with the ovary; limb either obsolete, or composed of teeth which develop as a pappus or feathery crown upon the fruit. Corolla more or less irregular; the limb bilabiate, the lobes imbricate in bud. Stamens 1–3, epipetalous. Filaments and style filiform; stigma undivided and truncate, or minutely 3-cleft. Fruit an achene; seed pendulous. -A small family, of some culinary and medicinal value; early referred to Umbelliferæ, to which the plants bear some considerable analogy.

FLORA FRANCISCANA.

PART IV.

Issued August 5th, 1897.

Price, One Dollar.

PAYOT, UPHAM & Co., San Francisco: WILLIAM WESLEY & SON, London.

FLORA FRANCISCANA.

AN ATTEMPT

TO CLASSIFY AND DESCRIBE THE VASCULAR PLANTS

OF MIDDLE CALIFORNIA.

BY

EDWARD L. GREENE,

Professor of Botany, in the Catholic University of America,
Washington, D. C.

SAN FRANCISCO:
CUBERY & CO., PRINTERS, 587 MISSION STREET, BELOW SECOND.
1897.

Copyright, 1891,
By Edward L. Greene.

1. **VALERIANA,** Tournefort (VALERIAN). Perennials, with roots more or less fleshy, and with a peculiar unpleasant odor. Flowers in terminal cymes. Calyx of 5–15 setiform lobes, which are inrolled and scarcely seen in the flower, in fruit unrolling and appearing like a plumose pappus to the fruit. Corolla salverform, the tube not spurred. Stamens 3. Ovary ripening into a flattened achene, which is mostly 1-nerved on one face, 3-nerved on the other, and with a more or less distinct nerve on each margin.

1. **V. sylvatica,** Banks; Richards. App. Frankl. Journ. 2 ed. 2 (1823). Minutely pubescent, or almost glabrous; stem 8–30 in. high, from short ascending rootstocks: radical leaves mostly simple and ovate to oblong, or some 3–5-foliolate; cauline more or less distinctly petioled, 3–11-foliolate or -parted, the divisions entire or few-toothed: fruiting cymes open, thyrsoid-paniculate: corolla 3 lines long, the tube short.—In subalpine moist places of the Sierra Nevada.

2. **VALERIANELLA,** Tournefort. Low and dichotomous, or taller and simpler annuals, with cymose inflorescence. Corolla more or less bilabiate, spurred or gibbous at base. Calyx-limb none, therefore no pappus to the variously winged often meniscoid glabrous or pubescent fruit.

* *Stems dichotomous ; fruit not obviously winged or meniscoid.—* VALERIANELLA proper.

1. V. OLITORIA, Poll. Palat. i. 30 (1776). Only 3–6 in. high, slightly pubescent: fl. very small, the corolla-limb barely a line wide, pale blue: achenes obliquely obovoid, with midnerve more or less distinct on both faces.—Waste lands about San Francisco; infrequent.

* * *Not dichotomous ; cymes thyrsoidly congested at summit of the almost simple stem ; achenes usually winged laterally, often appearing meniscoid.*—Genus PLECTRITIS, DC.

2. **V. macrocera,** Gray, Proc. Am. Acad. xix. 83 (1883); T. & G. Fl. ii. 50 (1841), under *Plectritis.* Corolla white, only a line long, with stoutish spur sometimes as long as the body, sometimes shorter; limb somewhat equally spreading, hardly bilabiate, or equally 4-lobed and the posterior lobe emarginate-bifid: fr. glabrous or puberulent, obtuse or lightly lineate-sulcate on the dorsal angle, the broad wing, circumscribing the ventral face of the achene, spreading or incurved.--On hillsides. April–June.

3. **V. ciliosa,** Greene, l. c. Slender, erect, simple, seldom a foot high: corolla small, deep pink, very distinctly bilabiate, the slender and tapering

spur much longer than the body: fr. of roundish outline but with a stout and prominent apiculation, the otherwise glabrous back provided with a broad ribbon-like keel, this densely ciliate on both margins, the subrostriform apiculation bearing similar hairs, the turgid, abruptly inflexed wings leaving on the ventral face an elliptic acute opening.—Plentiful on the northward slopes of low hills west of Napa Valley ; growing with the preceding, and quite as distinct from it in form and coloring of corolla as in the striking characters of the fruit. April.

4. V. congesta, Lindl. Bot. Reg. t. 1094 (1827). Corolla rose-purple, 3—4 lines long, with obviously bilabiate 5-cleft limb, the lobes oblong, obtuse; tube very gibbous, spurred at base, the spur short, arcuate, obtuse: fr. pubescent, the keel prominent, obtuse, the circumscribing ventral-face wing broad, involute.—An Oregonian species, attributed also to California by some, but not positively known to us as Californian.

5. V. magna, Greene, Proc. Philad. Acad., 1895, p. 548. Glabrous, the stout stems sharply angular, 3—5 ft. long, tortuous, half-reclining among shrubs, or on fences, and with rather numerous small branches: corolla white, bilabiate, with ample funnelform tube, and a short thick spur produced beyond the base: fr. glabrous externally, triquetrous-ovoid, the ventral concavity formed by the quite simple wings almost closed below, open above, the wings strongly hispid-ciliate within.— Collected only by the author, in Knight's Valley, Sonoma Co., June, 1891. Species remarkable for its great size, and half-climbing habit ; the fruit showing affinity with *V. aphanoptera*.

6. V. samolifolia, Gray, l. c.; DC. Prodr. iv. 642 (1830), under *Betckea*. Corolla a line long, obscurely bilabiate, with short obconic-saccate spur: fr. triquetrous, wholly destitute of wing, glabrous or a little pubescent.—Near the coast, from Sonoma Co. northward.

ORDER LXIII. DIPSACEÆ.

Coulter, Mem. in Act. Genev. ii. 13 (1823). Juss. Gen. 194 (1789) in part.

Herbs with opposite leaves, and flowers in dense involucrate peduncled heads ; each flower in the head enclosed within a tubular involucel and subtended by a bract. Calyx-tube adherent to the ovary ; limb entire, or toothed, or with bristle-like segments that persist upon the fruit. Corolla inserted at summit of calyx-tube, 4- or 5-lobed. Stamens 4, epipetalous, alternate with the corolla lobes. Style filiform ; stigma simple, longitudinal or subcapitate. Fruit achene-like, crowned with the calyx-limb, 1-seeded. Seed pendulous; albumen fleshy.

1. **DIPSACUS,** *Tournefort* (TEASEL). Tall coarse biennials with muricate or prickly stem and foliage ; the cauline leaves connate. Involucre of rigid spreading unequal bracts ; bracts of receptacle rigid, acuminate. Involucel sessile, 4-angled, 8-ribbed, terminated by 4 short teeth. Calyx-limb cup-shaped, quadrate or 4-lobed. Corolla funnelform, 4-cleft.

1. D. FULLONUM, Mill. Dict. (1768.) Stout, erect, very rough with short prickles, 4–6 ft. high ; radical leaves 8–12 in. long, elliptic-lanceolate, arcuate ; cauline connate-perfoliate: heads large, ovoid or oblong, on stout naked peduncles: bracts of receptacle rigid, recurved at the tips, as long as the flesh-colored corollas: stamens exserted.—Very common coarse weed in low and rich waste lands.

2. **SCABIOSA,** *Brunfels*. Soft unarmed plants, with peduncled globose or hemispherical heads, the flowers of the outer circle often larger than the others. Receptacle bearing hairs or soft scales among the flowers. Calyx-limb a cup-shaped border with 4 or more teeth or bristles. Corolla funnelform or salverform, often slightly irregular.

1. S ATROPURPUREA, Linn. Sp. Pl. 100 (1753). Suffrutescent, freely branching, 2–3 ft. high: radical leaves lyrate ; cauline pinnate, the segments oblong, toothed or incised: heads low hemispherical, in fr. ovate: corollas dark maroon to rose-purple, flesh-color, and white, the outer circle of them larger and exceeding the involucre ; calyx-limb pedicellate, in fruit bearing 5 pappus-like bristles.—An escape from the gardens of old-fashioned flowers, and become a luxuriant street and wayside weed in many places.

ORDER LXIV. **COMPOSITÆ.**

Vaillant, Act. Acad. Paris, 143 (1718); Adans. Fam. ii. 103 (1763). CORYMBIFERÆ of Jussein (1789), and of many earlier authors.

Herbs or shrubs with watery or resinous (never milky) juice, foliage various, the individual flowers small, in dense closely involucrate heads, the head often resembling a simple flower. Calyx wholly or partially adherent to the 1-celled, 1-ovuled ovary ; the limb represented, if at all, by one or more scales, awns or bristles called the pappus. Corollas tubular, palmatifid or ligulate ; the tubular ones 4—5-toothed or -cleft, often called disk-corollas ; the ligulate commonly toothed at apex, known as the ray-corollas. Stamens mostly 5, syngenesious, their anthers thus forming a tube around the style. Pollen-grains globose, echinate. Style in all fertile flowers 2-cleft at summit (except in one suborder), stigmatose on the margin, the upper portion of the forks usually not stigmatose, often variously hairy or appendaged. Fruit 1-seeded, indehiscent, commonly crowned by its pappus of capillary or plumose bristles, or of scarious scales; at the insertion on the common receptacle often subtended by a bract; this commonly called the chaff: the receptacle described as naked when the chaff is wanting: the surface of the receptacle being diagnosed as alveolate, foveolate, or merely areolate, according as the insertion of the achenes forms deeper or shallower depressions; or fimbrillate when the receptacle around these scars rises in teeth, or awns. — Our largest natural order, so-called, of flowering plants; the genera and species most conveniently considered under subordinal, natural, by not easily definable groups.

Rays none; style branches elongated, usually clavate-thickened upward and obtuse; stigmatic only below the middle..................1. EUPATORIACEÆ.

Rays usually present; anthers not caudate; style-branches of perfect flowers flattened, and with a distinct terminal appendage..................2. ASTERACEÆ.

Rays none; anthers caudate; style-branches of perfect flowers with no appendage; the stigmatic lines reaching almost to the naked truncate or obtuse summit.
..................3. GNAPHALIACEÆ.

Rays none; fertile fl. apetalous or nearly so; the staminate involucres forming a raceme above the axillary pistillate one; pappus none.....4. AMBROSIACEÆ.

Rays seldom wanting; anthers not caudate; involucre not scarious; receptacle chaffy; pappus never of capillary bristles..................5. HELIANTHACEÆ.

Rays present, fertile, the achenes of each more or less enfolded by its involucral bract; receptacle chaffy, style-branches subulate, hispid........6. MADIACEÆ.

Rays present; receptacle naked, or merely fimbrillate; pappus paleaceous or aristiform, or when bristly rigid..................7. HELENIOIDEÆ.

Anthers not caudate; bracts of involucre more or less scarious; style-branches truncate; pappus a scarious crown, or a circle of small scales, or wanting.
..................8. ANTHEMIDEÆ.

Anthers not caudate; receptacle naked; involucres not imbricated, mostly cylindrical, the bracts not scarious; pappus of many soft-capillary bristles.
..................9. SENECIONIDEÆ.

Rays none; anthers caudate; style-branches united, stigmatic to the obtuse summit, smooth and naked, but often with a pubescent node below; receptacle densely setose.
..................10. CYNAROCEPHALÆ.

Flowers bilabiate, the marginal ones with lower lip elongated and ray-like; styles with long truncate branches..................11. MUTISIACEÆ.

Suborder 1. EUPATORIACEÆ.

Heads rayless. Corollas all tubular and regular, never yellow, though sometimes cream-color. Anthers without tails. Style-branches elongated, usually clavate, minutely papillose or puberulent, the stigmatic lines only near the base.

Hints of the Genera.

Achenes 4-angled; pappus partly squamellate, - - - - - - - 1
Achenes 5-angled; pappus bristly, - - - - - - - - - - - 2
Achenes 10-striate, - - - - - - - - - - - - - 3

1. **TRICHOCORONIS**, *Gray*. Weak and flaccid fibrous-rooted perennial of muddy shores. Leaves opposite or attenuate, sessile. Heads few, peduncled, terminating somewhat corymbose branches. Flowers flesh-color. Style-branches scarcely clavate; rather linear and flattish. Pappus of small awns and intervening paleæ.

1. **T. riparia**, Greene, Eryth. i. 42 (1893); Pitt. ii. 216 (1891), under *Bioletlia*. Stems assurgent, hardly a foot high, sparsely pubescent: leaves linear-lanceolate, remotely serrate, slightly auricled at base: heads $2\frac{1}{2}$ lines broad: achenes $\frac{3}{4}$ line long, sharply 4-angled, the sides dark brown, the angles hispid-ciliolate toward the summit; pappus of 4 barbellate bristles and as many intervening minute fimbriate-lacerate scales. Banks of the lower San Joaquin.

2. **EUPATORIUM**, *Tournefort*. Our species a suffrutescent herb, with leaves mostly alternate. Achenes 5-angled, with no intervening ribs or striæ. Pappus of many rather rigid scabrous bristles.

1. **E. occidentale**, Hook. Fl. i. 305 (1833). Nearly glabrous, slightly glandular, 1—2 ft. high, the somewhat tufted stems erect and simple from a woody base: leaves ovate, truncate or subcordate at base, sparingly dentate, 1—2 in. long, in very short petioles: cymes somewhat panicled at and near the summit of the stem: involucral bracts about 15, in 2 series, lanceolate, nearly nerveless: corollas pinkish.—Crevices of rocks in the higher Sierra, and more plentiful eastward in Nevada. Aug.-Oct.

3. **COLEOSANTHUS**, *Cassini*. Perennial, often suffrutescent. Inflorescence of terminal and subterminal short clusters of narrow heads. Involucre of striate-nerved scales, the outer shorter. Corollas slender, 5-toothed. Style bulbous at base. Achenes 10-striate or -ribbed. Pappus of numerous but uniserial scabrous or barbellate bristles.

* *Heads small, numerous, paniculately disposed.*

1. **C. Californicus,** O. Ktze. Rev. Gen. i. 328 (1891); T. & G. Fl. ii. 79 (1841) under *Bulbostylis*. Shrubby at base, 2–3 ft. high, paniculately branching: leaves alternate, broadly ovate or triangular, irregularly crenate-toothed, about 1 in. long, 3-ribbed and roughish, and, with the whole plant, somewhat glandular-puberulent, heads spicate or racemose along the leafy branches, mostly nodding, ½ in. long, 10–15-flowered: scales of involucre with mostly obtuse straight tips.—Usually along stream banks, in gravelly places, and chiefly in the inner Coast Ranges. Sept.—Dec.

2. **C. multiflorus,** O. Ktze. l. c.; Kell. Proc. Calif. Acad. vii. 49 (1876), under *Brickellia*. Habit of the preceding, but more compactly branching, and more profusely flowering: leaves 2 in. long, ovate-lanceolate, entire, strongly 3-nerved, glabrous and glutinous: very numerous heads ⅓ inch long, 3—5-flowered: achenes sparsely hairy.—On rocks, in King's River Cañon; collected by *Dr. Kellogg,* in 1866.

3. **C. longifolius,** O. Ktze. l. c.; Wats. Am. Nat. viii. 301 (1873), under *Brickellia*. More woody than the preceding, also more slender and loosely branching, with fewer heads: leaves 2—5 in. long, lanceolate-linear, entire, or the larger somewhat sinuate-toothed, 3 nerved, slightly scabrous, the margins more decidedly so: heads in terminal clusters on the branchlets, 3 lines long, 3—5-flowered: bracts of involucre obtuse, conspicuously striate: achenes minutely hairy on the angles.—Near Owen's Lake, and elsewhere in Inyo Co., *Coville*.

* * *Heads few, larger, solitary at the ends of branchlets, or somewhat corymbose.*

4. **C. linifolius,** O. Ktze. l. c.; Eaton, Bot. King Exp. 137 (1871), under *Brickellia*. Glandular-puberulent; stems many, 12—16 in. high, from a woody base, corymbose at summit: leaves sessile, elliptic-lanceolate, entire, 1 in. long: heads large, solitary, on elongated leafy-bracted peduncles, 40–50-flowered: achenes with a double row of minute bristles along the striate: pappus almost plumose.—Inyo Co., *Coville*.

5. **C. Greenei,** O. Ktze. l. c.; Gray, Proc. Am. Acad. xii. 58 (1876), under *Brickellia*. Scarcely a foot high, very viscid: leaves ovate, obtuse, somewhat serrate, sessile or short-petioled: heads terminal, ¾ in. high, on short leafy peduncles, and leafy-bracted under the involucre, the proper bracts of which are lanceolate or linear, acuminate and glabrous: achenes with a few short bristles on the ribs, and some hirtellous hairs at summit.—High mountains from Tehama Co., northward.

6 **C. microphyllus,** O. Ktze. l. c.; Nutt. Trans. Am. Phil. Soc. vii. 286 (1840), under *Bulbostylis*. Glandular-pubescent and viscid, 1—2 ft.

high, tufted from a woody base, and paniculately branched, the short leafy branchlets terminated by 1—3 rather large heads: leaves subcordate or ovate, scabrous in age, sparingly denticulate or almost entire, the larger 1½ in. long, those of the branchlets only 1 or 2 lines: heads nearly ½ in. long, about 15-flowered: inner bracts of involucre firm, the outer with greenish spreading tips, the outermost passing into the small leaves of the branchlets.—Eastern slope of the Sierra, near Lake Tahoe, etc.

7. **C. grandiflorus**, O. Ktze. l. c.; Hook. Fl. ii. 26 (1834), under *Eupatorium*. Puberulent or almost glabrous: stem 2–3 ft. high, herbaceous from a fleshy perennial root: leaves deltoid-cordate, or the upper deltoid-lanceolate, 2–4 in. long, coarsely dentate-serrate and with an entire acuminate apex: heads terminal, nodding, in a cymose panicle: involucre ½—¾ in. long, about 40-flowered, its bracts thinnish, scarious-margined, obtuse or acutish, or some of the outer and shorter with a subulate acumination: pappus white, merely scabrous, somewhat deciduous.—Yosemite Valley and northward, in the Sierra Nevada.

Suborder 2. ASTERACEÆ.

Plants with a watery (never balsamic) juice, destitute of aromatic and bitter qualities, the leaves and heads only, in some, resinous. Leaves mostly alternate. Receptacle seldom chaffy. Anthers obtuse and entire, or only emarginate at base. Style-branches not clavate, often with filiform, or shorter and broader, papillose or hispid appendage. Pappus in most of ours of rather firm scabrous bristles. Disk-flowers yellow, changing to red or purple in some of the genera having cyanic rays.

Hints of the Genera.

* *Flowers of both ray and disk permanently yellow.*

Pappus of several short scales, - - - - - - - - - - 4, 5
Pappus of a few stout deciduous awns, - - - - - - - - - 6
Pappus of few and persistent slender bristles, - - - - - - 7
Pappus of many and persistent slender bristles;
   Heads with ligulate ray flowers;
      Heads few, ½ in. high, spicate or racemose, - - - - 10, 18
      Heads many, panicled or cymose;
         Ray-achenes with no pappus, - - - - - - 8
         All the achenes pappose;
            Pappus simple, - - - - - - 11
            Pappus double, - - - - - - - 9

Heads solitary;
    Peduncles scape-like, - - - - - - - 10, 11
    Branched shrubs, - - - - - - - - - 11, 12
Heads panicled or thyrsoid-congested; style-branches long,
    slender, - - - - - - - - - - 12, 13
    Style-branches shorter;
        Corolla ventricose, - - - - - - - 15
        Corolla tubular, - - - - - - - - 14
Heads small, in a flat-topped cluster, - - - - - - 16
Heads small, in a panicled or thyrsiform terminal inflorescence, 17
Rays palmatifid, - - - - - - - - - 19

    * * *Ray-corollas when present not yellow, except in* n. 18.
      +— *Pappus not excessively copious or accrescent.*

Pappus of 3–5 awns or bristles, - - - - - - - - 7
Pappus 0; herb acaulescent, - - - - - - - - - 27
Pappus of many reddish or tawny bristles;
    Ray-corollas palmatfiid, - - - - - - - - - 19
    Ray-corollas ligulate;
        Style-tips very bristly, - - - - - - 20
        Style-tips not bristly;
            Suffrutescent, - - - - - - 18
            Acaulescent herbs, - - - - - - 26
            Branching herbs. - - - - 21
Pappus of white, or merely yellowish or brownish bristles;
    Corollas all permanently white, or at least pale;
        Rays many, very small, scarcely spreading; heads small, panicled or
            racemose, - - - - - - - - - - 26
        Rays few, broader, spreading; heads cymose-corymbose, - 22, 24
Disk-corollas yellow, changing to red, brown or purple;
    Rays 15–40, broad-linear; involucre imbricated;
        Stems simple; head solitary, - - - - - - - - 25
        Stems branching; heads many, - - - - - - - - 23
    Rays 50–100 or more, very narrow; involucre scarcely imbricate, - - 28, 29

        +— +— *Pappus copious, accrescent.*

Annual herbs; rays minute, - - - - - - - - - - - 30
Perennial or shrubby; rays none, - - - - - - - - - 31

4. **GUTIERREZIA,** *La Gasca.* Nearly glabrous, somewhat resiniferous freely branching herbaceous or suffrutescent plants. Leaves alternate, narrow, entire. Heads small, spherical, hemispherical, or narrower, usually corymbosely arranged at summit of stem and branches. Invol-

ucral bracts coriaceous, the outer successively shorter, often with greenish but usually appressed tips. Flowers of both ray and disk permanently yellow. Style appendages slender. Achenes angled or striate, mostly silky. Pappus paleaceous.

1. **G. Californica,** T. & G. Fl. ii. 193 (1842); DC. Prodr. v. 313 (1836) under *Brachyris*. Stems tufted, ascending from a woody base, 1½ ft. high, loosely paniculate: leaves linear, acute, scabrous: heads few, solitary or in pairs or threes at the ends of the branchlets, turbinate or obovate, 3 lines high; fl. of disk and ray each 8—10: achenes densely silky: pappus of about 12 unequal acutish scales, none longer than the achene.—Dry hills of the Coast Range. June—Sept.

2. **G. lucida.** *Xanthocephalum lucidum*, Greene, Pitt. ii. 282 (1892). Size and habit of the preceding, but of a light green and very resinous: inflorescence densely cymose-panicled and heads very narrow, little more than a line long, mostly with only 2 flowers, 1 of the ray and 1 of the disk: bracts of involucre closely appressed: pappus of 4 or 5 acute scales.- Dry hills far southward, bordering the Mohave Desert.

5. **EASTWOODIA,** *Brandegee.* Nearly allied to *Gutierrezia*, but with few and very large heads. Achenes enfolded by a complicate receptacular bract as in many Helianthem. Rays wanting.

1. **E. elegans,** Brandg. Zoe, iv. 397 (1894). Shrub 1½ ft. high; branches with a white bark: leaves small and scattered, oblanceolate-linear, entire, sparingly scabrous: heads solitary, terminating leafy branchlets.—Dry hills east of Bakersfield, Cariso Plains, etc.

6. **GRINDELIA,** *Willdenow.* Coarse herbs or suffrutescent plants, with sessile rigid most serrate leaves, and rather large hemispherical heads terminating corymbose branches. Bracts of involucre imbricated in many series, with usually narrow herbaceous squarrose-recurved tips. Flowers of both disk and ray very numerous, permanently yellow. Style-appendages lanceolate or linear. Achenes short, thick, compressed or turgid, truncate, glabrous. Pappus of 2—8 deciduous stout awns or bristles.

* *Herbaceous perennials, flowering in early summer.*

1. **G. camporum.** Greene, Man. 171 (1894). Stems white and shining, tufted from a perennial root, 2 ft. high, glabrous, very leafy up to the loosely corymbose heads, even the branches of the corymb conspicuously leafy-bracted; radical leaves almost wanting, cauline oblanceolate-spatulate, sessile and clasping, 2 in. long, saliently serrate-toothed;

bracts of flowering branches nearly entire, spreading; involucres ½—¾ in. wide, their bracts with long linear recurved tips : ray-achenes obscurely triquetrous, with 3 or more pappus-awns ; disk-achenes compressed, obliquely biauriculate or unidentate at summit, and with pappus of 2 bristles.—Common on rich plains of the interior. June Sept.

2. **G. hirsutula**, H. &. A. Bot. Beech. 147 (1833) : *G. rubricaulis*, DC. Prodr. v. 316 (1836); Greene, Man. 171. Rather slender, ascending, 2 ft. high, stems from brownish to dull red, herbage scarcely glutinous, roughish-pubescent or even somewhat hirsute : radical leaves numerous, tufted, oblanceolate, coarsely serrate; cauline reduced, few and remote : heads solitary or few, nodding in the bud : inner bracts of involucre closely imbricated and very glutinous, without spreading tips : achenes mostly thin and flat, with obcordate summit and only 2 pappus-awns.—Open glades among the wooded hills.

3. **G. patens**, Greene, Pittonia, ii. 290 (1892). Foliage and pubescence of the preceding, nearly, but stem stouter, erect, the flowering branches at no stage nodding at summit ; heads larger, ½—1 in. broad ; bracts of involucre mostly linear or lanceolate-foliaceous, straight and widely spreading, some of the inner with shorter and recurved tips: disk-achenes with obcordate summit and only 2 awns.—Hillsides and plains about San Francisco Bay.

4. **G. robusta**, Nutt. Trans. Am. Phil. Soc. vii. 314 (1840). Stems stout, ascending, 1½ ft. high : leaves broadly cordate-oblong, obtuse, coarsely serrate, 1½ in. long, often 1 in. broad, subcoriaceous, glabrous, on the margins pubescent : heads very few and large, corymbosely disposed ; outer bracts of involucre rather leafy, the others narrow and squarrose : pappus-awns 2.—Open ground near the sea, at Monterey and southward.

* * *Late-æstival and autumnal species.*
+— *Herbaceous perennial.*

5. **G. procera**, Greene, Man. 172 (1894). Strictly erect, 5—7 ft. high, simple up to the corymbose-paniculate summit, the stout white stem scabro-puberulent, plant otherwise glabrous, slightly glutinous : lower leaves unknown ; upper cauline lanceolate, attenuate-acute, entire, 2—3 in. long ; involucres small, low-hemispherical ; bracts with appressed base and short slender recurved tips : rays short : pappus-awns 2.— Bottom lands of the lower San Joaquin, in places inundated in spring and early summer.

← ← *Suffrutescent species.*

6. **G. paludosa,** Greene, l. c. About 5 ft. high, sterile leafy shoots a foot high, or more, surviving the winter, the plant otherwise herbaceous; herbage glabrous except the scabrous-ciliolate leaf-margins; only the involucres glutinous : leaves slightly fleshy, oblong-lanceolate to spatulate-oblong, 2—3 in. long, conspicuously serrate, at least those of sterile shoots with a broad cordate-clasping base, the lobes surrounding the stem : involucre squarrose : achenes with prominent turgid angles, those of the ray triquetrous, of the disk compressed : awns 2 only, even in the ray, stout, strongly flattened.—Brackish marshes of Suisun Bay.

7. **G. cuneifolia,** Nutt. Trans. Am. Phil. Soc. vii. 315 (1840). Bushy, 2—4 ft. high, glabrous : leaves thickish and rather fleshy, 3—4 in. long, cuneate-spatulate to linear-oblong, entire or sharply denticulate, clasping though not auricled at the broad base ; involucre ½ in. high, glutinous, the bracts all with squarrose green tips : pappus-awns usually several, compressed, barbellulate.—Borders of salt marshes and along tidal sloughs about S. F. Bay and southward along the coast.

Two other *Grindelia* species have been attributed to our district, one of which is certainly fictitious, the other probably so. *G. humilis*, H. & A., is based on an abnormal twig broken off from a shrubby species and described as a small herb. *G. Pacifica*, M. E. Jones, obtained at Santa Cruz, and described as 6 inches high, and as having filiform root-leaves, oblanceolate stem-leaves, and lax involucral scales, may be an abnormal state of some herbaceous species.

7. **PENTACHÆTA,** *Nuttall.* Slender almost glabrous small vernal annuals. Leaves alternate, linear, entire. Involucres solitary, hemispherical or campanulate, of thin scarious-margined appressed mucronulate bracts in 2 series. Rays white, yellow, or wanting. Disk-corollas yellow, very slender. Style-appendages filiform-subulate, hispid. Achenes pubescent. Pappus of 3—5 slender bristles.

* *Flowers of both disk and ray yellow.*

1. **P. aurea,** Nutt. Trans. Am. Phil. Soc. vii. 336 (1840). Diffusely branching, 4—12 in. high : heads rather large and many-flowered, the rays often 40 or more : involucral bracts broadly lanceolate, setaceously acuminate, with green middle portion and scarious margins : achenes somewhat villous-pubescent ; pappus-bristles 5—8.—Common in the extreme southern part of the State ; occurring as far northward as San Luis Obispo Co.

* * *Ray-flowers white or purplish, or none.*

2. **P. bellidiflora**, Greene, Bull. Calif. Acad. i. 86 (1885). Sparingly branching, the peduncles somewhat scapiform : involucre hemispherical, many-flowered : rays 8—14, white or reddish ; achenes oblong-turbinate, villous : pappus-bristles 5 or none.—Open hills and sterile slopes in Marin and San Mateo counties ; not common. April, May.

3. **P. aphantochæta**, Greene, Bot. Gaz. viii. 256 (1883). Very slender, only 2—3 in. high, usually simple and monocephalous : whole plant purplish, the peduncle white-villous under the small head : outer series of corollas rose-red, claviform-urceolate, i. e., widening upwards, the throat abruptly contracted under the minute teeth : pappus of 3—5 short bristles or cusps, or obsolete.—Frequent on open hills in San Mateo and Contra Costa counties, thence northward. April, May.

4. **P. alsinoides**, Greene, Bull. Torr. Club. ix. 109 (1882). Dichotomously branching, only 2—5 in. high : involucre turbinate, of 5—7 bracts and 3—7-flowered : rays 0 : disk-corollas filiform, not deeply cleft : achenes obovate-clavate ; pappus-bristles 3, very slender.—An obscure hillside plant, but not rare in middle portions of the State.

8. **HETEROTHECA**, *Cassini*. Tall hairy herbs, with alternate leaves, and a terminal corymbose panicle of middle-sized heads. Involucres ovate ; their bracts closely imbricated in many series, without spreading tips. Flowers yellow ; those of the ray pistillate, of the disk perfect, the later with ovate or lanceolate style-appendages. Achenes compressed, pubescent, those of the ray thin-triquetrous with caducous pappus or none ; pappus of disk achenes of an outer series of sparse short bristles, and an inner, more copious series of longer ones.

1. **H. grandiflora**, Nutt. Trans. Am. Phil. Soc vii. 315 (1840). Annual or biennial, 3—6 ft. high, hirsute, the inflorescence viscid and strong-scented by a coat of short gland-tipped hairs : cauline leaves oval or oblong, coarsely toothed, partly vertical by a twist in the petiole, this at base bearing 2 stipuliform lobes : involucre ½ in. high : ray achenes without pappus, those of the disk with but faint traces of the outer and shorter bristles.—Frequent along railways in Contra Costa Co.; an immigrant from more southerly portions of the State, where it is common. July—Dec.

9. **CHRYSOPSIS**, *Elliott*. Perennials, leafy-stemmed and of rather low growth. Leaves sessile, entire or nearly so. Heads middle-sized, terminating corymbose or fastigiate branches. Involucres ovate or broader, of narrow regularly imbricated bracts in several series. Style-appendages linear-filiform to slender-subulate. Achenes compressed,

obovate to linear-fusiform; pappus fuscous, of many capillary scabrous bristles, with or without an outer series of short bristles or paleæ.

* *Heads radiate; outer pappus setose-squamellate.*—CHRYSOPSIS proper.

1. **C. hispida,** Nutt. Trans. Am. Phil. Soc. vii. 316 (1840); Hook. Fl. ii. 22 (1834), under *Diplopappus.* Stems tufted, decumbent, 1 ft. long or less, simple up to the terminal fastigiate corymb of heads : pubescence all short, stiff and spreading, not dense : leaves 1 inch long or more, spatulate-oblanceolate, spreading, or even somewhat deflexed on the stem, and with fascicles of smaller ones in the axils : heads ½ in. high.—Near Summit Soda Springs, Placer Co.; not otherwise known in California; but found by the railway near Verdi, in western Nevada, by *Mr. Sonne,* where it may have been introduced from the Rocky Mountain region, where it is common. July - Sept.

2. **C. sessiliflora,** Nutt, l. c. Slender, sparsely pilose-hispid, viscid-glandular : leaves oblanceolate, sharply pointed : corymbose branches ending in about 3 subsessile heads ½ in. high, leafy-bracted at base : bracts of involucre not pubescent but very viscid-glandular : achenes slender-fusiform, silky-pubescent ; outer pappus slenderly squamellate. —Santa Clara Co., and southward. June.

3. **C. Bolanderi,** Gray, Proc. Am. Acad. vi. 543 (1865). Stoutish, ½ —1 ft. high ; pubescence long-silky : heads few and subsessile : bracts of involucre not glandular, silky-villous : outer pappus of narrow paleæ nearly half as long as the achene.—On stony hilltops toward the sea ; flowering in summer.

4. **C. echioides,** Benth. Bot. Sulph. 25 (1844). Rigid, brittle, 2—3 ft. high, often suffrutescent, hoary with a dense hirsute and hispid pubescence : leaves rigidulous, small : heads less than ½ in. high, in short fastigiate corymbs; bracts hirsutulous : achenes silky but the hairs not appressed : setulose outer pappus not conspicuous.—Sandy plains, and banks of streams, from Solano Co. southward, east of the mountains. Aug.—Oct.

* * *Rays none; outer pappus obsolete.*—Genus AMMODIA, Nutt.

5. **C. rudis,** Greene, Man. 174 (1894), and Eryth. ii. 106. Erect, 1—3 ft. high, rigid, brittle, rough-hairy but not hoary, glandular, heavy-scented: involucres in a narrow leafy panicle: bracts of involucre acute, midrib prominent, margin scarious: achenes oblong, pubescent ; pappus copious, slender, scabrous, seldom a trace of the short outer series.— Common along stream banks ; heretofore referred to *C. (Ammodia) Oregana,* from which it is altogether distinct. July—Oct.

10. PYRROCOMA, *Hooker*. Rigid perennial herbs, with coriaceous mostly radical leaves from a fusiform root. Stems leafy-bracted, bearing racemose or panicled middle-sized heads. Bracts of hemispherical involucre many, rigid, with herbaceous more or less squarrose tips. Flowers yellow; those of the ray rather numerous, short, pistillate ; of the disk tubular, slightly dilated upwards. Style-appendages subulate-linear, pubescent. Achenes more or less flattened and striate, glabrous or pubescent. Pappus of copious reddish or brownish slender but rigid unequal bristles.

1. **P. elata**, Greene, Man. 173 (1894). Stout, erect, 1—3 ft. high, glabrous: radical leaves long-petioled, 6—8 in. long, lanceolate, entire; cauline 1—3 in., sessile, ascending, rigidly ciliolate: heads 1½ in. high and as broad, disposed in an interrupted spike or narrow panicle: involucral bracts rigid, imbricated in several series, the green tips acute, spreading: achenes flattened, closely costate, pubescent.—A somewhat rare plant of subsaline soils at Calistoga and near San Jose. July—Oct.

2. **P. paniculata**, T. & G. Fl. ii. 244 (1842); Nutt. Trans. Am. Phil. Soc. vii. 331 (1840), under *Homopappus*. Stoutish, erect, 1—2 ft. high, glabrous : radical leaves oblong-lanceolate, obscurely and remotely serrulate; cauline shorter and broader, sessile and clasping: heads (panicled in the Oregonian type) differing according to the varieties. Var. **virgata** (Gray). Heads few, small, but broad, forming a virgate spike. Var. **stenocephala** (Gray). Heads few, narrow and cylindric, forming a raceme or panicle.—The type not within our limits: the two varieties on the eastern slope of the Sierra.

3. **P. apargioides**, Greene, Eryth. ii. 70 (1894); Gray, Proc. Am. Acad. vii. 354 (1868), under *Aplopappus*. Low-growing, and the texture not rigid : radical leaves broadly lanceolate, from serrate to laciniate-toothed and pinnatifid: scapiform flowering stems several or many, decumbent or ascending, a few inches high, or seldom nearly a foot high, bearing one or several middle-sized heads: involucre hemispherical, ½ in. high; bracts lanceolate, loosely imbricated in few ranks: rays 20 or more: pappus rather soft.—Eastern slope of the Sierra Nevada.

11. STENOTUS, *Nuttall*. Glabrous evergreen suffruticose or shrubby plants. Leaves alternate, narrow, entire. Heads solitary on scapiform peduncles, or at the ends of the branches. Involucre hemispherical; bracts in two or three series, membranaceous, scarious-margined, closely appressed. Flowers yellow; rays few ; disk-corollas dilated above, deeply 5-toothed. Style-appendages filiform, flattened, puberulent : achenes oblong, somewhat compressed, densely villous; pappus very slender, permanently white.

1. **S. acaulis,** Nutt. Trans. Am. Phil. Soc. vii. 334 (1840); Journ. Philad. Acad. vii. 33 (1834), under *Chrysopsis*. Woody caudex cespitose, very leafy; leaves oblanceolate, or narrowly spatulate, entire, mucronate-acute, rigid, pale, very minutely scabrous: flowering stem only a few inches high, scape-like and naked, or leafy only below, bearing a single head: bracts of the hemispherical involucre few, in 2 or 3 series, ovate, acute, with scarious margins: rays 9–12: achenes silky-pubescent; pappus rather scanty. —Near the summit of the Sierra.

2. **S. linearifolius,** T. & G., Fl. ii. 238 (1842); DC. Prodr. v. 347 (1836), under *Aplopappus*. Very leafy, 1–1¼ ft. high; leaves 1 in. long, linear, acute, spreading, punctate and resiniferous, 1-nerved: head about 1 in. broad, on a peduncle: rays 12–14: achenes densely white-villous; pappus copious, fragile or deciduous. —Mt. Diablo towards the summit, and southward. May–July.

12. **MACRONEMA,** *Nuttall.* Low and many-stemmed suffrutescent plants, with thinnish sessile entire foliage either glandular or puberulent, but not resinous-punctate, and middle-sized or rather large heads solitary or glomerate at the ends of the branches. Bracts of the campanulate involucre in few ranks and not very unequal: innermost thin-chartaceous or partly scarious; outer loose and foliaceous, or with leafy tips. Rays usually present. Style-appendages long and filiform. Achenes slender, compressed, few-nerved, soft-pubescent.

1. **M. suffruticosum,** Nutt. Trans. Am. Phil. Soc. vii. 322 (1840). Stems glandular-pubescent or puberulent, 6–8 in. high; branches very leafy, the leaves broadly oblanceolate: the few heads ⅓–¾ in. high; involucral bracts loose and mostly foliaceous-tipped: rays 2–5.—Alpine or subalpine in the Sierra Nevada from Mariposa Co. northward.

2. **M. molle,** Greene, Eryth. ii. 73 (1894); Gray, Proc. Am. Acad. xvi. 80 (1880), under *Aplopappus*. Stems more slender and elongated, and with the leaves pale with a scanty soft-tomentose pubescence: leaves linear lanceolate: bracts of the involucre (except the linear scarious inner ones) short-lanceolate, with erect herbaceous tips.—About the headwaters of the Sacramento, *Pringle.*

3. **M. Greenei,** Greene, l. c.; Gray, l. c., under *Aplopappus*. Stems a foot high, glabrous, or glandular-pubescent toward the summit: leaves spatulate-oblong to oblanceolate, obtuse or mucronate: heads solitary or few, ½ in. high; bracts of the involucre in about 3 series, lanceolate to linear, the green tips of all but the innermost elongated-subulate, spreading. —Range of the preceding.

4. **M. discoideum,** Nutt. Trans. Am. Phil. Soc. vii. 322 (1840). *Aplopappus Macronema,* Gray. Seldom a foot high, rigidly erect, spar-

ingly leafy, the branches white with a dense tomentum: leaves viscidly glandular, not woolly: heads mostly solitary, nearly an inch high, rayless; involucre of few bracts.—Subalpine or alpine in the Sierra Nevada, and far eastward. Aug., Sept.

13. **CHRYSOTHAMNUS,** *Nuttall.* (RABBIT BUSH). Shrubby or half-shrubby plants, with very narrow entire subcoriaceous foliage, both stem and leaves sometimes white with a pannose tomentum. Heads numerous and narrow, seldom with rays, the bracts of the involucre narrow, commonly in 4 or 5 vertical ranks. Corollas slender, somewhat funnelform or claviform, with spreading or suberect teeth. Style-tips lanceolate-subulate to filiform. Achenes narrow, not compressed. Pappus of subequal bristles.

* *Destitute of tomentum, sometimes puberulent; heads and achenes rather short; style-tips not exceeding the stigmatic portion.*

1. **C. puberulus,** Greene, Eryth. iii. 93 (1895). Erect and with short fastigiate leafy branches, ½ 1 ft. high, the foliage and twigs from scabro-puberulent to hispidulous: leaves from spatulate-linear to linear, 1-nerved, ¼—1 in. long, often serrulate-scabrous on the margin: heads in small clusters at the ends of the many branchlets; involucre barely 3 lines high, its bracts obtusish, mucronulate.—Eastern base of the Sierra, toward the borders of Nevada. July–Sept.

2. **C. stenophyllus,** Greene, l. c. 94. More densely tufted than the last, more leafy, the cymes larger and fastigiate: bark of branches very white and shining: leaves narrowly linear, acute, 1-nerved, their margins serrulate-scabrous, otherwise glabrous: involucre narrow, its bracts more distinctly 4-ranked, seldom or never mucronate.—Habitat of the last. July–Sept.

3. **C. tortifolius.** *Bigeloria Douglasii,* var. *tortifolia,* Gray. Shrub 2—4 ft. high, the stout leafy branches with a smooth very white bark and terminating in a broad compound fastigiate-cymose cluster of heads: leaves oblong-linear or lanceolate, very acute, about 2 in. long, distinctly 3-nerved, often twisted, the margin serrulate-scabrous, otherwise usually glabrous: heads about 4 lines high, the linear-oblong obtuse bracts distinctly 4-ranked.—Hills along the Truckee River. *Sonne,* and plains of Lassen Co., *Mrs. Austin.* Sept.

4. **C. humilis,** Greene, Pittonia, iii. 24 (1895). Depressed, only 6—8 in. high, much branched: cinereous-puberulent throughout: leaves rather sparse, suberect, 1 in. long, narrowly oblanceolate, acute: heads in small terminal clusters, subsessile, 4 lines high; bracts about 3 in each vertical rank, lanceolate, with greenish obtuse delicately ciliolate tips:

flowers cream color: achenes nearly linear, angular and pubescent.—Plains of the Truckee River, near Truckee, *Mr. Sonne.* Sept.

* * *Stem, and often foliage also pannose-tomentose; heads and achenes narrow and elongated; style-tips very long and filiform.*

+ *Ray-flowers wanting.*

5. **C. speciosus,** Nutt. Trans. Am. Phil. Soc. vii. 323 (1840). Stout shrub 2–4 ft. high, the numerous leafy branches ending in a broad cymose corymb: leaves narrowly linear, 2–3 in. long, and with the branchlets of the inflorescence minutely white-tomentose: heads 5 lines high; bracts of involucre firm, acutish, not ciliate, tomentose on the back, or the inner ones glabrous except near the tip, all in vertical ranks of 3 or 4: corollas with slender almost glabrous tube longer than the subcylindric rather deeply 5-toothed limb.—Plains of Plumas Co. and northward, east of main Sierra.

6. **C. Californicus,** Greene, Eryth. iii. 111 (1895). Seldom 2 ft. high, stout; flowering branches densely white-tomentose, the foliage more loosely and flocculently so: leaves broad for this group, spatulate-linear and oblanceolate, acute, barely 2 in. long: heads in an ample pyramidal panicle: involucre 1/2 in. high or more, glandular-puberulent, the bracts only 3 in each vertical rank and very unequal, the inner ones lanceolate, all acute: corolla with short tube not as long as the sub-cylindric rather deeply toothed limb: achenes with a dense rather coarse pubescence.—Common at considerable elevations in the Sierras of Nevada and Placer counties. Sept., Oct.

7. **C. occidentalis.** *C. Californicus,* var. *occidentalis,* Greene, l. c. in part. Taller and more slender, in no part loosely or flocculently tomentose; leaves narrowly linear, the flowering branches somewhat numerous and reedy, bearing the heads in smallish and dense clusters: bracts of the involucre 4 in each rank, ovate- to linear-lanceolate, cuspidately acutish.—In Kern and Santa Barbara counties. Sept.

8. **C. Mohavensis.** Greene, l. c. 113. Stout, 3–5 ft. high, the coarse and often flexuous ultimate branches usually glabrate, glutinous and nearly leafless; leaves when present sparse, 1 in. long: cymose heads only 4–5 lines high; bracts rather thin, viscid-puberulent, narrow but obtuse, about 5 in each rank: achenes appressed-villous.—Of the Mohave Desert, but now found along the railway in the Tehachapi region of Kern Co. Oct.

9. **C. Nevadensis,** Greene, l. c. 114; Gray, Syn. Fl. 136 (1884), under *Bigelovia.* Only 1–2 ft. high, forming a dense reedy suffrutescent tuft;

cinereous-tomentulose oblanceolate leaves recurved at tip and mucronate: heads few and elongated (¾ in. high), somewhat thyrsoidly or even racemosely disposed; bracts of the involucre tomentulose, ending in a long rigid recurved acumination: corollas merely ochroleucous.—Valley of the Truckee River, and in Plumas Co., always east of the Sierra Nevada.

10. **C. Bolanderi**, Greene, l. c.; Gray, Proc. Am. Acad. vii. 354 (1868), under *Linosyris*. Stout and low, only a few inches high, the stem woolly but leaves green, glabrate and viscid, oblanceolate, obscurely 3-nerved: heads few and large (¾ in. long); bracts of the involucre rather few and thin, lanceolate, with a soft acuminate apex, some of the outer and shorter herbaceous-tipped.—Subalpine or alpine in the higher Sierra, about Mono Pass, etc.

11. **C. ceruminosus**, Greene, l. c. 94; Dur. & Hilg. Pac. R. Rep. v. 9, t. 6 (1855), under *Linosyris*. Early glabrate (or perhaps never tomentose) and balsamic-viscid, 2 ft. high or more: leaves scattered on the slender fastigiate branches, spreading or recurved, almost filiform: heads cymose-fascicled, about 5 lines high; viscid involucral bracts, narrowly lanceolate, abruptly ending in a spreading setiform tip: corolla deeply cleft.—Apparently a very local species of the region of the Tejon Pass; perhaps of the more typical group of the genus.

+ + *Heads with some ligulate ray-corollas.*

12. **C. Bloomeri**, Greene, l. c. 115; Gray, Proc. Am. Acad. vi. 541 (1865), under *Aplopappus*. Usually glabrous and resinous, sometimes tomentulose, 1–2 ft. high, with erect rigid virgate branches abundantly leafy: leaves spatulate-linear to almost filiform, 1–2 in. long: heads thyrsoidly or almost racemosely disposed, ⅜ in. high; inner bracts of the involucre oblong-lanceolate or linear, with thin-scarious erose-ciliate margins, some of the outer with filiform herbaceous tips: rays 1—4: achenes sparsely pubescent.—Eastern foothills of the Sierra Nevada, from Kern Co. northward.

14. **CHRYSOMA**, *Nuttall*. Evergreen shrubs of low stature, with narrow entire mostly subterete punctate leaves, and terminal cymose or corymbose clusters of small heads. Involucre turbinate; bracts mostly lanceolate, very regularly imbricated, margins subscarious. Flowers yellow, none turning red or brown. Disk-corollas slender-tubular with subcampanulate throat and deeply cleft limb. Style-appendages filiform, acuminate, hirsutulous. Achenes more or less distinctly prismatic. Pappus of scabrous slender bristles dull-white or yellowish, becoming reddish.

* *Leaves broad or narrow, but plane.*

1. **C. cuneata**, Greene, Eryth. iii. 11 (1895); Gray, Proc. Am. Acad. viii. 635 (1873), under *Aplopappus*. Freely branching and spreading, 1 ft. high or more, balsamic-glutinous: leaves coriaceous, cuneate- or spatulate-obovate, retuse, ½ in. long more or less, resinous-punctate: heads ½ in. high, in a terminal fasciculate corymb; bracts lanceolate or almost linear, obtusish: rays 1-5, or 0: style-tips slender-subulate: achenes pubescent.—In crevices of rocks of the Sierra Nevada, at middle elevations, from Placer Co. southward. Aug., Sept.

2. **C. arborescens**, Greene, Eryth. iii. 10 (1895); Gray, Bot. Mex. Bound. 79 (1859), under *Linosyris*, and Proc. Am. Acad. viii. 640 (1873), under *Bigelovia;* Greene, Man. 175, under *Ericameria*. Erect, fastigiately branching, 3-10 ft. high, densely clothed with very narrow-linear leaves 1½—3 in. long, 1 line wide: heads in a terminal-fastigiate corymb, 20—25-flowered: turbinate involucre scarcely 3 lines high; bracts lanceolate, acute: rays seldom present: achenes short, apparently quadrangular, silky-pubescent.—At considerable elevations in the mountains of Sonoma, Marin and Contra Costa counties, and in the middle Sierra Nevada. Sept.—Dec.

* * *Leaves very narrow and subterete; shrubs of heath-like aspect.*

3. **C. ericoides**, Greene, l. c. 11; Less. Linnæa, vi. 117 (1831), under *Diplopappus;* H. & A. Bot. Beech. 146 (1833), under *Aplopappus*. Diffusely branching, 1½—1½ ft. high, the branches fastigiate-corymbose, very leafy throughout: leaves linear, terete, those of the branches ½—¾ in. long, deflexed, bearing in their axils very short branchlets hidden by two-ranked closely imbricated shorter ones: involucres ¼ in. high; bracts tomentose-ciliolate: rays about 5, short: achenes subcylindrical, striate, glabrous.—Sandy hills and beaches from Bolinas Bay southward, chiefly or altogether near the sea. Aug.—Dec.

4. **C. teretifolia**, Greene, l. c. 12; Dur. & Hilg. Pac. R. Rep. v. 9, t. 7 (1855), under *Linosyris;* Gray, Proc. Am. Acad. viii. 644 (1873), under *Bigelovia*. Rigidly and fastigiately branching: leaves 1 in. long, involute-filiform, thus at least in appearance terete: heads open-paniculate, 4 or 5 lines high; bracts narrowly oblong to broadly linear, obtuse, all but the innermost tipped with a green gland-like spot: achenes pubescent.—Mountains along the western border of the Mohave Desert; seldom collected, and not well known.

15. **ISOCOMA**, *Nuttall*. Rather rigid tufted erect suffrutescent plants, with thick slightly succulent leaves, and a corymbose terminal cluster of smallish rayless heads. Bracts of several-flowered involucre

coriaceous, closely imbricated, the tips herbaceous but appressed, obtuse or acutish. Corollas permanently yellow; tube slender; limb ventricose, the segments being more or less strongly connivent about the style, the pubescent appendages of which are ovate or somewhat narrower. Achenes short, compressed or subterete, silky-pubescent. Pappus-bristles numerous, unequal, the inner longest and often perceptibly flattened and awn-like, hardly scabrous.

1. **I. vernonioides,** Nutt. Trans. Am. Phil. Soc. vii. 320 (1840). *Bigeloria Menziesii,* Gray. Glabrous or loosely pubescent, 2–4 ft. high, erect: leaves oblanceolate, more or less serrate, 1–2 in. long, often with many fascicled ones in the axils: heads 4 lines high, campanulate; bracts of involucre obtusish: pappus-bristles stout, none very perceptibly flattened.—Common shrub of S. Calif., found at Black Point, San Francisco, where it may have been introduced accidentally.

2. **I. arguta,** Greene, Man. 175 (1894). Branches 6–10 in. high, more or less pubescent or hirsute below, glabrous above, leafy throughout; leaves diminishing upwards, the lowest 1 in. long, all broadly oblanceolate, of coriaceous texture, with saliently spreading coarse and acute or mucronate teeth: heads 1/3 in. high, turbinate, 12–15-flowered: inner pappus-bristles distinctly flattened and tapering very gradually from base to apex. Subsaline plains east of the Vaca Mts., in Solano Co., *Jepson.*

3. **I. acradenia,** Greene, Eryth. ii. 111 (1894); Bull. Torr. Club. x. 126 (1883), under *Bigeloria.* Very many slender stems 1–2 ft. high forming tufts from a woody base: leaves spatulate-linear, entire: heads glomerate-cymose, 4 lines high, 10–20-flowered: involucre campanulate; the obtuse apex of the oblong bracts with a protuberant rounded resiniferous gland: bristles of the pappus rigid, very unequal.—Western borders of the Mohave Desert.

16. **EUTHAMIA,** *Cassini.* Erect glabrous perennials, very leafy, the branching more or less distinctly corymbose. Leaves nearly linear, entire, pellucid punctate. Heads small, clustered at the ends of the branches. Involucral bracts firm, imbricated, glutinous. Flowers permanently yellow; those of the ray about twice as many as those of the disk. Achenes short, turbinate, villous-pubescent.

1. **E. occidentalis,** Nutt. Trans. Am. Phil. Soc. vii. 326 (1840); T. & G. Fl. ii. 226 (1842), under *Solidago.* Somewhat paniculately branching, 3–6 ft. high: leaves lanceolate-linear, obscurely 3-nerved: bracts of involucre linear-lanceolate, acute: rays 16–30; disk-flowers 8–14, their

style-tips obtuse.—Common in low grounds along rivers and on the borders of marshes. Aug.—Oct.

17. SOLIDAGO, *Vaillant* (GOLDEN ROD). Strict simple-stemmed perennials, with alternate more or less serrate leaves. Inflorescence a terminal cluster of many small heads, usually disposed in scorpioid racemes and forming a panicle; otherwise forming a thyrsus. Involucre narrow; bracts in two or more series, neither herbaceous tipped or glutinous. Flowers all permanently yellow; the outer and ligulate short, the inner narrow-funnelform. Style-appendages flattened, lanceolate. Achenes terete or prismatic, 5—10-nerved, glabrous or pubescent. Pappus a series of unequal scabrous permanently white bristles.

* *Heads numerous, small, in a more or less pyramidal panicle of secured racemes.*

1. **S. sempervirens,** Linn. Sp. Pl. ii. 878 (1753). Bright green and glabrous, leafy throughout, 2—8 ft. high: leaves rather fleshy, lanceolate to linear, the upper acute, lower obtuse, all entire: panicle narrow, dense, virgate: heads 3—4 lines long: bracts of involucre lanceolate, scabrous-ciliolate: rays 8—10, rather large, golden yellow: achenes minutely pubescent.—Marshes about San Francisco, at Laguna Honda, and in similar situations southward. Aug.—Nov.

2. **S. Guiradonis,** Gray, Proc. Am. Acad. vi. 543 (1865). Glabrous, slender, 2—3 ft. high: leaves bright green, thickish, entire; the lowest 6 in. long, less than ½ in. wide, lanceolate or oblanceolate, tapering gradually into a long narrow base or margined petiole, somewhat 3-ribbed: heads few, in a narrow virgate panicle: involucral bracts lanceolate-subulate: rays 8 or 9, small: disk-flowers 10 or 12: achenes nearly glabrous.—Along streams in the mountains of Fresno Co. and southward, *Guirado, Rothrock.*

3. **S. spectabilis,** Gray, Proc. Am. Acad. xvii. 193 (1882). Two feet high: heads many, in a long and narrow compound thyrsus: herbage light green, seeming glabrous, but with a scattered short rough pubescence under a lens: lower and radical leaves elongated, oblanceolate, narrowed to a petiole, often 1 inch wide and 3-nerved, entire or with a few serratures; cauline lanceolate, or the small upper ones linear, acute: involucral bracts mostly obtuse.—Plumas Co., *Mrs. Austin,* and elsewhere along the eastern slope of the Sierra Nevada.

4. **S. elongata,** Nutt. Trans. Am. Phil. Soc. vii. 327 (1840). Puberulent, 1 to 2 ft. high equally leafy up to the long panicle: leaves thinnish, lanceolate, acute, sparingly serrate, 2—3 in. long: branches of panicle scarcely secund, ascending; heads small; bracts of involucre linear,

acutish or obtuse: rays 10—16, narrow: achenes pubescent.—Shady or moist grounds, from middle California northward. July—Oct.

5. **S. Californica**, Nutt. Trans. Am. Phil. Soc. vii. 328 (1840). Roughish with an almost cinereous short pubescence; commonly 2—4 ft. high: leaves ampler and more numerous below, passing from obovate to oblong-lanceolate and lanceolate, and from obtuse to acute, the lower and broader more or less serrate: panicle usually virgate but loose, 4—12 in. long, the racemiform clusters secund but seldom recurved: heads 3 lines high; bracts lanceolate-oblong or oblong-linear, obtusish, pubescent, rays 7—12, pale yellow: achenes pubescent. Var. **Nevadensis**, Gray. Inflorescence more secund: heads smaller: involucre mostly glabrous.—Very common in dry soil. The variety peculiar to the Sierra Nevada. July—Oct.

* * *Heads fewer and larger, somewhat thyrsoidly congested.*

6. **S. spathulata**, DC, Prodr. v. 339 (1836). Glabrous, slightly glutinous, with the odor of Grindelia, 1—2 ft. high: stems decumbent and even suffrutescent at base: lower leaves spatulate, 2—4 in. long, rounded at apex, serrate: heads 4 lines high, almost as broad, about 25-flowered, disposed in short racemes thyrsoidly crowded at and near the summit of the stem: bracts of involucre oblong or broadly linear, all but the inmost series obtuse and green-herbaceous almost throughout, the inner acutish and with a green midvein: rays short: achenes pubescent.—On bluffs near the sea at Point Lobos; also in the Mission Hills, and southward to Monterey. Aug., Sept.

18. **HAZARDIA**, *Greene*. Low shrubs or suffrutescent plants, with subcoriaceous more or less persistent toothed or serrated leaves, and spicate or somewhat thyrsoidly congested heads. Involucre oblong-obovid or obconic, its numerous bracts in many series, closely imbricated, often with more or less distinctly squarrose-spreading tips. Heads 20—40-flowered. Rays yellow, or none. Disk-corollas narrow, merely 5-toothed, yellow, changing to red or brown. Style-tips linear-subulate. Achenes fusiform, slightly compressed, few-nerved. Pappus reddish.

1. **H. squarrosa**, Greene, Eryth. ii. 112 (1894); H. & A. Bot. Beech. 146 (1833), under *Aplopappus*. Suffruticose, erect, 2—3 ft. high, glandular and glutinous: leaves oblong, 1 in. long, spinulose-dentate: heads many, spicately thyrsoid toward the summit of the branches, ½ in. long: bracts of the involucre rigid, appressed, multiserial, with abruptly spreading tips: rays none: achenes sparsely pubescent.—Hills along the coast from Monterey southward.

2. **H. obtusa.** Size and habit of the preceding, but stouter, the much larger heads scarcely more than simply spicate toward the ends of the branches: leaves round-obovate, about 1 in. long, sessile by a broad clasping base, rather closely spinescent-toothed: heads rather more than ³⁄₄ in. high; bracts of the involucre closely imbricated in many series, without squarrose tips, and all very obtuse, even almost truncate, but with a short cusp: rays none: achenes wholly glabrous; pappus of a light reddish brown.—A very distinct species, known only by specimens obtained in San Emidio Cañon, Kern Co., in 1894, by Miss Eastwood.

3. **H. Whitneyi,** Greene, Pittonia, iii. 43 (1896); Gray, Proc. Am. Acad. vii. 353 (1868), under *Aplopappus*. Suffrutescent, 1–2 ft. high, glandular-scabrous, the upper part of the stem sometimes tomentosely pubescent: leaves 1–2 in. long, oblanceolate, acute, serrate, sessile by a half-clasping base: heads ½ in. high or more, subpaniculate on short terminal and subterminal branches, or subsessile and glomerate: bracts of the involucre narrowly lanceolate, acute, the tips often slightly recurved, not squarrose: rays 6–8, short, golden-yellow: achenes glabrous; pappus deep reddish-brown.—Subalpine in the Sierra Nevada, and the inner Coast Range northward.

19. **LESSINGIA,** *Chamisso.* More or less floccose-woolly annuals with alternate more or less serrate leaves and small cymosely panicled heads of yellow, whitish or purplish flowers, these all perfect. Corollas with slender tube and long narrow lobes; those of the marginal row more deeply cleft on one side and imitating a palmatifid ligule. Involucre campanulate or turbinate; bracts much imbricated and appressed, herbaceous-tipped. Anthers with slender-subulate appendages. Appendages of style-branches obtuse or truncate, densely hispid, often with a setiform cusp amid the hairs. Achenes turbinate or cuneiform, silky-villous. Pappus-bristles rigid, scabrous, red or brownish.

* *Yellow-flowered species.*

1. **L. Germanorum,** Cham. in Linnæa, iv. 203 (1829). Low, slender, branching and spreading from the base; branchlets at length glabrate, purple: lower leaves sinuate-pinnatifid, those of the branches narrowly oblanceolate: involucre hemispherical, its bracts more or less green-herbaceous not glandular.—San Francisco and southward in sandy soil near the sea.

2. **L. glandulifera,** Gray, Proc. Am. Acad. xvii. 207 (1882). Erect, stoutish, diffusely branched above: leaves more irregularly and deeply toothed or cleft, those of the stem more numerous, ovate or oblanceolate, and of the branchlets minute and almost crowded, rigid, beset

along the margin with yellowish large glands: involucre campanulate to turbinate, its bracts more or less glanduliferous.—Plains of the lower San Joaquin and southward.

3. **L. pectinata**, Greene, Proc. Philad. Acad., 1895, p. 548. Slender and very diffuse, the ascending or more widely spreading and almost prostrate branches a foot long or more; herbage scarcely tomentose even when young, green, sparingly stipitate-glandular and scabrous throughout; only the lowest leaves woolly, these and also those of stem and branches pectinate-pinnatifid, the segments pungently spinescent-tipped: involucral bracts narrow, acute, glandular-puberulent. —Sandy soil near the sea at Monterey, *Hartweg*, *Parry*, *Tidestrom* and others.

4. **L. parvula.** Low and slender, branched from the base, the spreading branches only 3–6 in. long; herbage more or less white-woolly and also glandular, some or all of the glands stipitate: radical leaves somewhat lyrately pinnatifid, the cauline 5–7-lobed, the lobes spinescent; involucres small and almost campanulate, only $1\frac{1}{2}$ lines high and almost as broad, the short bracts numerous and imbricated, the short outer ones often woolly, the inner granular-glandular and also with a few large glands, all abruptly acute: bristles of the pappus very fine and only light-brownish.—Common in the interior of Monterey and San Luis Obispo counties, *Hickman*, *Lemmon*. It seems to have formed a part of Gray's *L. ramulosa tenuis*, that author not having known that the flowers are yellow.

* * *Flowers lilac or purplish.*

← *Bracts of the involucre without cartilaginous tips.*

5. **L. ramulosa**, Gray, in Benth. Pl. Hartw. 314 (1849). Erect, 1–2 ft. high, very loosely branching, the glabrate branchlets and upper leaves more or less hirtellous and glandular: leaves oblong to lanceolate, the lower spatulate, entire or toothed, the small ones of the branchlets with partly clasping base: involucre campanulate or turbinate, 10–20-flowered: corollas short, purple: style-appendages with minute setiform tip.— Dry hills from middle sections of the State northward.

6. **L. leptoclada**, Gray, Proc. Am. Acad. vii. 351 (1868). Taller and more slender, with almost filiform branchlets bearing few or solitary 5–20-flowered heads: upward leaves somewhat sagittately adnate to the branches at base: involucre turbinate; its bracts in many ranks: corollas elongated: style-appendages with a conspicuous subulate tip.—Same range as the preceding.

7. **L. adenophora**, Greene, Bull. Calif. Acad. i. 190 (1885). Erect, 1–2 ft. high, loosely branched from near the base: stem-leaves $\frac{1}{2}$–$1\frac{1}{4}$ in.

long, ovate and round-ovate, pungently acute, sessile and somewhat cordate-clasping, densely soft-woolly above, glabrate beneath, the margins closely beset with small stipitate glands: bracts of the narrow-campanulate involucres very acute, suberect, more or less stipitate-glandular: corollas red-purple: bristles of the pappus more or less completely united into 5 paleæ.—Mountains of Lake and Colusa counties.

8. **L. nemaclada,** Greene, l. c. Size of the last, but the branching more paniculate: leaves lightly floccose above, glandular-scabrous beneath: branchlets very slender and heads small (3—5-flowered); bracts of the involucre with spreading tips; pappus of few or many awn-like bristles which are sometimes by cohesion reduced to 5 broad awns or paleæ as in the last.— Foothills of the mountains on both sides of the Sacramento, in El Dorado and Colusa counties.

9. **L. virgata,** Gray, in Benth. Pl. Hartw. 315 (1849). More densely woolly: stem and virgate branches rigid: upper leaves appressed, concave, carinate-nerved: heads spicately sessile in the axils of the leaves: involucre cylindrical, woolly, 5 7-flowered : fl. nearly white : style-branches with a conspicuous subulate tip.—Sandy plains of the Sacramento and San Joaquin. Sept.

10. **L. hololeuca.** With the habit of *L. ramulosa*, but stouter, the branches rigidly ascending, the whole plant even to the involucre white-tomentose: leaves all entire, the oblong and ovate-oblong cauline ones sessile and cordate-clasping, ending in a short spinescent tip: heads broadly turbinate ; involucral bracts short and in only a few series, straight and spinescent at tip: corollas red-purple: pappus of rufous bristles.— Low hills east of the Santa Rosa Valley, Sonoma Co.; collected by the author, 15 Sept. 1888.

+ + *Inner bracts of the involucre cartilaginous-aristate.*

11. **L. Parryi,** Greene, Bull. Calif. Acad. i. 192 (1885). Woolly throughout, 2—10 in. high, erect, not stout, sparingly branched: heads few and spicate, or some solitary at the ends of the branches: corollas pink: pappus reddish: style-appendages bristly-hairy but not cuspidate. —Mountains of Kern Co., at Keene Station, etc.

12. **L. nana,** Gray, in Benth. Pl. Hartw. 315 (1849). Stout short branches prostrate, only 2—4 in. long, or even obsolete and the heads sessile among the radical leaves, the whole plant white-tomentose except the long white cartilaginous awns of the inner bracts of the involucre: heads large, sessile: corollas red: pappus dark-red.— Sandy plains and hills in the valleys of the Sacramento and San Joaquin. Aug., Sept.

20. CORETHROGYNE, *DeCandolle.* Genus very nearly allied to *Lessingia;* distinguished chiefly by the numerous and altogether ligulate violet ray-corollas, thus more resembling *Aster*, but herbage hoary or woolly, and the flowering season spring and early summer. Involucres from hemispherical to turbinate. Style-appendages bristly but not cuspidate.

* *Heads solitary or few, large; involucres hemispherical.*

1. **C. obovata,** Benth. Bot. Sulph. 22 (1844); *C. spathulata*, Gray, Proc. Am. Acad. vii. 351 (1868). Stems decumbent, 1–2 ft. long; cauline leaves 1 in. long or more, obovate, bullate-rugose, toothed above the middle; peduncles and involucres minutely glandular, the plant otherwise hoary-tomentose: rays about 20, linear.— Hills along the seaboard, from Marin Co. to Humboldt. May, June.

2. **C. Californica,** DC. Prodr. v. 215 (1836). Stems tufted, ascending or erect, 1–2 ft. high, at summit bearing several corymbosely disposed long-peduncled large heads: leaves from spatulate-lanceolate to linear, 2 in. long or less, acute, entire, or the larger with a few teeth below the apex: peduncles and bracts glandular: rays 25 or more, oblong-linear, dark-violet.— Hills of the Coast Range from Alameda Co. (Arroyo del Valle, *Greene*) southward, chiefly toward the sea.

3. **C. caespitosa.** *C. Californica*, Greene, Man. 178, not of DC. Suffrutescent, diffuse and depressed, the numerous brittle lignescent branches forming a mat, only the pedunculiform ultimate branchlets suberect, each with a single large head; stems and foliage white with a persistent cottony wool: leaves small, numerous, oblanceolate, acute, serrate: involucres and rays much as in the last.—Obtained by the author, at Crystal Springs, San Mateo Co., 22 June, 1886, and not otherwise known.

* * *Heads numerous, smaller, panicled; the involucres turbinate (except in n. 4).*

4. **C. viscidula.** Tall and slender, loosely corymbose-panicled, only the young plant in any degree hoary or flocculent, the stem and branches at flowering time purplish and glandular-scabrous: leaves narrowly oblanceolate, acute, serrulate, reticulate-venulose, glandular-scabrous on both faces, the peduncles and their small bracts very scabrous and with a few short-stipitate larger glands; the small nearly hemispherical involucres of more numerous bracts and more strongly imbricated than in the foregoing species, these also viscid-glandular: rays only 12—18: pappus only light-brownish.—Well marked species known to me only in specimens obtained at Monterey by *Dr. Parry* in 1888.

5. **C. filaginifolia,** Nutt. Trans. Am. Phil. Soc. vii. 290 (1840); H. & A. Bot. Beech. 146 (1833), under *Aster*. Suffrutescent, erect, 2 ft. high or more, paniculately branched above, the branches and foliage woolly: leaves scattered, spatulate-oblong, mucronate, the lowest serrate toward the apex; heads small; bracts of the turbinate involucre merely granular, not at all woolly or glandular: rays few, rather short.—Hills of Monterey Co. to Mariposa, thence southward.

6. **C. virgata,** Benth. Bot. Sulph. 23 (1844). Taller than the last, often 3 ft. high; heads far more numerous, forming a virgate panicle: leaves from narrow-obovate to oblanceolate, serrate: pedicels and involucres densely beset with viscid short-stalked glands.—Abundant along the seaboard southward; flowering later than *C. filaginifolia*.

21. **MACHÆRANTHERA,** *Nees.* Annuals, biennials or short-lived perennials, glabrous or with more or less often glandular pubescence, the leaves usually incised or pinnatifid. Involucres mostly somewhat turbinate; the firm bracts imbricated, numerous, glandular, and with recurved or squarrose herbaceous tips. Rays numerous, purple, or sometimes wanting. Disk-corollas yellow, slender-tubular, with 5 short erect teeth. Style-tips subulate or narrower, connivent. Achenes little compressed, strongly striate, pubescent. Pappus bristly or capillary, brownish, scabrous.—Genus more related to *Corethrogyne* than to *Aster*.

1. **M. Shastensis,** Gray, Proc. Am. Acad. vi. 539 (1865). Stems several from a perennial root, erect, 6--10 in. high, cymose-paniculate; herbage cinereous-puberulent, the inflorescence granular: lowest leaves oblanceolate, petiolate, acute, the cauline oblong, obtuse, all entire: bracts of the turbinate involucre rather few, coriaceous, erect, only the outer ones with short herbaceous tips, these scarcely spreading: rays few, red-purple: achenes long and narrow.—Mountains of Plumas Co., *Mrs. Austin*, and northward.

2. **M. montana,** Greene, Pitt. iii. 60 (1896). Habit of the preceding, but larger, more densely puberulent; leaves narrower, strongly serrate-toothed; heads more racemose; turbinate involucre of more numerous and closely imbricated bracts, all with subulate herbaceous squarrose-recurved tips: rays blue or purplish. From the vicinity of Mono Lake eastward to the Rocky Mts.

3. **M. leucanthemifolia,** Greene, l. c. 61; also Eryth. iii. 119 (1895), under *Aster*. Apparently only biennial, a foot high, loosely and almost divaricately branched; herbage pale but puberulent: lowest leaves spatulate-obovate, coarsely and deeply toothed, the teeth spinescent, the cauline similar but small: bracts of the turbinate involucre with

spinescent recurved tips: rays about 20, blue, broad and short.— Species of western Nevada, but apparently reaching our borders in Mono Co.

4. **M. incana,** Greene, l. c. 62; Lindl. Bot. Reg. t. 1693 (1834). under *Diplopappus.* Biennial, tall and loosely branching, commonly 2 ft. high or more; herbage pale and only sparingly puberulent: lowest leaves narrowly lanceolate, acute, entire, or with a few salient teeth; cauline linear, obtuse, entire: bracts of the hemispherical involucre with long slender acute spreading or recurved herbaceous tips: rays very numerous and narrow, violet: achenes appressed-pubescent.—Foothills and plains of Kern Co.

22. **SERICOCARPUS,** Nees. Perennials with upright leafy stems, and small heads of white flowers in a terminal corymbose cyme. Leaves entire, sessile. Involucre narrow, of imbricated and appressed coriaceous bracts with green-herbaceous squarrose tips. Rays few (5 or 6). Style-tips lanceolate-subulate. Achenes narrow, little compressed, 2-nerved, silky-pubescent.

S. Californicus, Durand, Pl. Pratt. 90 (1855). Commonly 3 ft. high, stout, sparsely woolly-hairy below, scabrous above: leaves thinuish, lanceolate, 2 in. long or more, 1-nerved: heads in sessile glomerate clusters at the ends of the corymbose branches; bracts of the involucre fewer and less squarrose than in the eastern and typical species.— Foothills of the Sierra, from Nevada Co., *Pratten,* to Amador, *Bioletti,* also near Donner Lake on the eastern slope, *Sonne.* A rare and most distinct species, strangely referred to the Oregonian *S. rigidus* by recent writers.

23. **ASTER,** *Tournefort.* Leafy-stemmed autumnal herbs with panicled or somewhat corymbose heads. Involucre hemispherical to campanulate, of several series of unequal imbricate bracts with herbaceous tips. Rays many, not very narrow, white, pinkish, or bluish. Disk-corollas yellow changing to red-purple; tube slender; limb funnelform. Style-appendages from triangular-lanceolate to slender-subulate. Achenes compressed. Pappus copious, dull white, or rarely more deeply colored, scabrous.

* *Erect perennials, leafy-stemmed; heads in a somewhat flat-topped terminal cluster.*

1. **A. radulinus,** Gray, Proc. Am. Acad. viii. 388 (1872). Stoutish, roughish-pubescent, 1 ft. high or more, the mostly solitary stem usually bearing an open corymb of middle-sized heads: leaves rigid, obovate-oblong, acute, sharply serrate above, tapering to a narrow entire base, scabrous both sides: involucre obconic, 4—5 lines long; bracts rigid,

appressed, acutish or mucronate, the tips green: rays white; disk-corollas becoming red: achenes minutely pubescent; pappus rather rigid.—Borders of woods and thickets; early-flowering. July—Sept.

2. **A. Torreyi,** Porter, Bull. Torr. Club. xvii. 37, t. 100. Near the preceding, but taller, not as stout, less pubescent; stems clustered, the rootstocks freely branching: leaves spatulate-lanceolate, acute, serrate, 2–3 in. long: heads numerous, smaller than in the last: involucral bracts oblong- to spatulate-linear, slightly woolly-ciliate, the green appressed tips acutish: rays few, pinkish.—At Blue Cañon, and in Summit Valley, *Torrey, Greene;* referred by Dr. Gray to the preceding, but apparently a fair species or subspecies. Aug.- Oct.

* * *Perennials, stems less leafy, often decumbent; inflorescence more paniculate or racemose.*

3. **A. Menziesii,** Lindl.; Hook. Fl. ii. 12 (1834). Strictly erect, 2 ft. high, usually simple and very leafy up to the mostly simply racemose or racemose-paniculate inflorescence, the whole plant cinereously and roughly pubescent: leaves oblong-lanceolate, acute, 1–3 in. long, remotely serrate or entire, sessile by a broad auriculate-clasping base: involucre broadly turbinate, ½ in. high; bracts somewhat spatulate, well imbricated, the broad green tips obtuse: rays light violet, rather short. Vaca Mountains, *Jepson,* and southward. Sept.- Dec.

4. **A. invenustus,** Greene, Man. 179 (1894). Stout stems 2 ft. long or more, ascending from a decumbent base; herbage cinereous with scabrous and short-hirsute pubescence: lower cauline leaves lanceolate-spatulate, 2–3 in. long, with remote and slight serratures: heads very numerous in an ample cymose panicle; involucres nearly hemispherical, ¼ in. high, the almost wholly green-herbaceous very obtuse spatulate-linear bracts in rather few ranks; rays dull pale purplish.—Napa Co., to Amador, and southward. Aug., Sept.

5. **A. Chilensis,** Nees, Ast. 123 (1832): *A. Chamissonis,* Gray; Torr. Bot. Wilkes Exp. 341 (1874). Erect, stoutish, 2–4 ft. high, glabrous or somewhat hirsute, the stem occasionally with strongly hirsute lines: leaves lanceolate, acute, entire, 2—5 in. long, entire, or obscurely serrate, the whole margin scabrous: heads ½ in. high, in a more or less ample panicle of short loose leafy racemes; bracts of campanulate or broadly obconic involucre much imbricated, linear or linear-spatulate, with short and rounded green tips: rays 25—30, purple or violet. ½ in. long.--Toward the seaboard only; common and variable; some forms very showy. Aug.—Oct.

6. **A. Sonomensis,** Greene, Man. 180 (1894). Slender, decumbent at base, 1 –1½ ft. high, glabrous, only the leaf-margins scabrous-ciliolate: radical leaves, narrowly lanceolate, very regularly, though remotely and slightly serrate-toothed, tapering to a long petiole, this with a dilated and strongly ciliate basal part: heads rather few in a terminal corymbose panicle; involucres ¼ in. high, broad-campanulate to broad-obconic, the well imbricated bracts narrowly oblanceolate, acute; rays purplish, rather narrow, ½ in. long.—In open plains of the Sonoma Valley, in low subsaline ground, *Greene;* also at Elk Grove, near the Sacramento River, *Drew.* Sept.

7. **A. denudatus,** Nutt. Trans. Am. Phil. Soc. vii. 292 (1840). Erect, 1 ft. high or more, scarcely decumbent, glabrous, glaucous, only the peduncles and sometimes the involucres pubescent: leaves chiefly radical, subcoriaceous, lanceolate or spatulate, entire, the margins strongly ciliolate-scabrous with short incurved hairs: inflorescence a terminal corymbose panicle: heads rather small; involucral bracts regularly imbricated, erect, the outer obtuse, with conspicuous green tips, the inner acute: rays light-blue: achenes nearly glabrous, substipitate. Common open meadow species of the Great Basin, reaching our borders in the valley of the Truckee River. Aug., Sept.

8. **A. frondeus,** Greene, Proc. Philad. Acad., 1895, p. 551. Stems decumbent at the base, thence erect, often 2 ft. high, simple and sparingly leafy with large leaves up to the summit, which bears one or more very large hemispherical heads on short peduncles; herbage green and nearly glabrous: radical leaves obovate-oblong, petiolate, 4 or 5 in. long, the lower cauline nearly as long but spatulate and sessile; the uppermost auriculate-clasping: heads 1½ in. high and very broad: bracts of involucre nearly equal, the outer broader and more foliaceous, but even the spatulate inner ones mainly herbaceous: rays many, ½ in. long, purplish.—Mountains of Plumas Co.; differing from the Rocky Mountain type of the species in having a thinner foliage.

9. **A. Oreganus,** Nutt. in T. & G. Fl. ii. 163 (1841); Trans. Am. Phil. Soc. vii. 296 (1840), under *Tripolium.* Often 3 ft. high or more, erect, leafy, contractedly or amply paniculate; herbage merely somewhat scabrous, the mostly narrowly lanceolate leaves from 3 or 4 in. long on the main stem to 1 or 2 in. long on the branches, all sessile and slightly auriculate clasping, those of the branchlets mucronate: heads hemispherical, 4 or 5 lines high; bracts in 2 or 3 series, all with short linear coriaceous base and ample spreading broader herbaceous tips: rays many, flesh-color to purplish. Frequent on moist stream banks east of the Sierra, and along the Washoe foothills east of Truckee, thence to Plumas and Modoc counties and northward. Sept.

10. **A. Yosemitanus.** *A. adscendens*, Lindl. var. *Yosemitanus*, Gray. Twice as tall as the last, erect, far more leafy, the leaves thin and deep green, pubescence scanty or none: leaves from spatulate- to linear-lanceolate, 1½—3 in. long, acute, entire or with a few obscure serratures in the middle: heads of middle size, loosely paniculate, or more frequently subcorymbose; bracts of the involucre oblong-linear or sub-spatulate, more or distinctly imbricated, the green tips scarcely spreading: rays violet.—Foothills of the eastern slope of the Sierra, and up to almost subalpine elevations, in dry wooded or open ground. Very distinct from the last, under which (as *A. adscendens*) it was placed by Dr. Gray. It is also variable, some specimens by *Geo. Hansen*, from Amador Co., being racemose and much too like what a glabrous *A. Menziesii* might be.

11. **A. lentus,** Greene, Man. 180 (1891). Erect, slender, 4–6 ft. high, slightly succulent, glabrous except a slight pubescence under the heads, and a delicately serrulate-scabrous margin to the leaves: lowest leaves 3–5 in. long, lanceolate-linear, slightly falcate, those of the flowering branches straight and half-clasping at the sessile base: panicle loose and ample, often a yard long, the branches loosely racemose: heads 4—5 lines high; involucres oblong; bracts linear, acute, appressed, green-herbaceous and somewhat succulent almost throughout: rays many, ¾ in. long, light purple.—Plentiful along tidal streams in western part of the Suisun marsh; the largest and most showy of Californian species Oct., Nov.

* * * *Stems ascending, from a perennial root; heads few, glomerate or interruptedly thyrsoid.*

12. **A. integrifolius,** Nutt. Trans. Am. Phil. Soc. vii. 291 (1840). Stem simple, often a foot high or more, stout, ascending, purplish, sparingly leafy: leaves firm, oblong-spatulate, the larger 3–6 in. long, the smaller lanceolate, all entire: heads few, large, terminal and axillary; the peduncles and imbricated herbaceous bracts of the involucre viscidly glandular: heads ½ in. high, not broad; involucral bracts few-ranked, not squarrose: rays 15—25, bluish-purple: achenes compressed-fusiform 5-nerved, sparsely pubescent. Subalpine near the summits of the northern and middle Sierra. A peculiar species, not intimately related to any other.

* * * * *Tall annual, open-paniculate, with many small heads.*

13. **A. exilis,** Ell. Sk. ii. 344 (1824). Stem erect, 2–4 ft. high, stout below, above paniculately parted into many slender but rather rigid branches: lowest leaves lanceolate, the upper linear, mostly entire: heads

3 lines high, narrow; bracts of the involucre linear-subulate: rays 15—40, bluish-purple. Common in moist subsaline or alkaline soil. Aug.—Oct.

24. LEUCOSYRIS. Glabrous pale glaucescent perennial, woody at base, the intricately branched reedy stems nearly leafless, the branchlets terminating each in a small head of white flowers. Involucre campanulate or turbinate, of lanceolate acute chartaceous bracts regularly imbricated in about 3 series. Rays wanting. Style-appendages linear-subulate. Achenes terete, silky-pubescent.

1. **L. carnosa.** *Linosyris carnosa*, Gray, Pl. Wright. ii. 80 (1853); *Bigelovia intricata*, Gray, Proc. Am. Acad. xvii. 208 (1882); *Aster carnosus*, Gray, Syn. Fl. 202 (1884). Tufted reedy pale-green stems 2—3 ft. high; the sparse lower leaves linear, very fleshy, entire, 1 in. long or less, those of the branches reduced to subulate scales: heads scattered, 3–4 lines high.—In subsaline moist ground from Tulare Co., *Congdon*, and Kern Co., *Greene*, southward and eastward.

25. LEUCELENE, *Greene.* Low perennials, the tufted leafy stems (in our species monocephalous) from a lignescent base or rootstock. Leaves small, ascending, oblong-linear or lanceolate, rigid, entire. Involucres turbinate, or campanulate, imbricated, the bracts narrow, nearly plane, with narrow scarious margins and without herbaceous tips. Rays white or purple. Disk-corollas tubular-funnelform, 5-toothed. Style-appendages (in our species) subulate-linear. Achenes compressed, pubescent. Pappus double, an inner series of bristles and an inconspicuous outer one of squamellæ.

1. **L. alpina.** *Chrysopsis alpina*, Nutt. Journ. Acad. Philad. vii. 34, t. 3, fig. 2 (1834). *Aster scopulorum*, Gray, Proc. Am. Acad. xvi. 98 (1880). Stems tufted, rigid, only a few inches high, bearing a single peduncled head: leaves $\frac{1}{2}$ in. long or more, scabro-puberulent, callous-margined and mucronate: rays $\frac{1}{2}$ in. long, deep violet.—At middle altitudes of the eastern slope of the Sierra, in dry rocky soil.

26. OREASTRUM, *Greene.* Stemless perennials, with narrow sub-coriaceous leaves and scapiform monocephalous branches from a somewhat fusiform or branched tap-root. Bracts of involucre narrow, subequal, in about 2 series. Rays purple, rather numerous. Disk-corollas yellow, tubular-funnelform, with 5 erect teeth. Style-branches filiform, hirsutulous. Achenes subterete, 5–8-costate. Pappus uniserial, bristly, barbellate scabrous, brownish.

1. **O. Andersonii,** Greene, Pitt. iii. 147 (1896); Gray, Proc. Am. Acad. vi. 540 (1865), under *Erigeron*, and l. c. vii. 352 (1868), under *Aster* Long radical leaves linear, acute, 3—5-nerved, narrowed at base; those of the scapes much reduced, the uppermost mere subulate bracts: the large hemispherical involucre somewhat woolly, its bracts linear-lanceolate, herbaceous, biserial: ligules uniserial, broadly linear: achenes pubescent. A beautiful rather dwarf aster-like perennial in moist subalpine places of the middle Sierra.

2. **O. elatum,** Greene, l. c. Glabrous, even to the involucre; scapiform branches decumbent at base, 12—20 in. high, leafy-bracted; lower cauline leaves linear, acute, 4 6 in. long: heads more than 1 in. broad, low-hemispherical; bracts in about 3 series, all with broad-linear subcartilaginous base, and triangular-subulate herbaceous long tip: rays deep violet: ovaries glabrous. On Mt. Dyer, Plumas Co., July, 1879, *Mrs. Austin.*

27. **BELLIS,** *Tournefort.* (DAISY.) Low herbs. Involucres broad, many-flowered; bracts of nearly equal length Rays many, white or reddish. Disk-corollas yellow. Style appendages short, triangular. Achenes obovate, compressed, nerved on the margins. Pappus none.

1. B. PERENNIS, Linn. Sp. Pl. ii. 886 (1753). Perennial, acaulescent: leaves obovate: scapes several, each with a single head. Escaped from gardens, and naturalized in the western counties from at least Marin and Alameda northward.

28. **CONYZELLA,** *Dillenius.* (FLEABANE.) Annuals with panicled small heads of whitish very inconspicuous flowers. Involucres ovoid, of 2 or 3 series of imbricated subulate bracts. Rays few and small, short and suberect, or none. Disk-corollas all slender and 4 5 toothed, or some of the outer filiform and truncate. Style-appendages short, obtuse. Achenes compressed, with marginal ribs and a sparse strigose pubescence. Pappus scanty and simple, the scabrous bristles short and fragile, or longer, firmer and more enduring.

1. C. CANADENSIS, Rupr. in Mem. Acad. Petersb. Ser. 7. xiv. n. 4 (1869); Linn. Sp. Pl. ii. 863 (1753), under *Erigeron.* Sparsely hispid or nearly glabrous; stem stout, erect, 1 6 ft. high, with countless small subcylindric heads in a rather dense panicle: lowest leaves spatulate, upper linear; heads only 2 lines high: rays white, very short. A weed in cultivated lands, or by waysides; less common in California, where it is doubtless barely naturalized, than in the Atlantic States, where it is thought to be indigenous. Sept. Nov.

2. **C. linifolia.** *Erigeron linifolius*, Willd. Sp. iii. 1955 (1803). Not as tall as the last, seldom 2 ft. high, more strict and less paniculate, in the early state of flowering scarcely more than racemose: herbage somewhat canescently hirsute and also scabrous with a minute appressed pubescence: lowest and linear-spatulate leaves incisely toothed, the narrowly linear upper ones entire: involucre ¼ in. high, broadly ovoid, cinereous-pubescent: rays very small: pappus slightly accrescent, firm and reddish or brownish. —Established on sandy banks of the Sacramento River opposite Sacramento, where it was collected by the author in 1895. Native of tropical America. July – Sept.

3. **C. Coulteri.** *Conyza Coulteri*, Gray, Proc. Am. Acad. vii. 355 (1868). *Erigeron discoideus*, Kell. Proc. Calif. Acad. v. 55 (1872). Size of the last, the heads quite as large, but closely panicled as in n. 1; herbage viscidly pubescent, and also with some longer and hirsute hairs: leaves spatulate-oblong and oblanceolate, from coarsely toothed to laciniate-pinnatifid: involucre rather softly hirsute: corollas white, none ligulate, the female ones consisting of a short truncate tube, and these extremely numerous, only a few of the central ones of the head bisexual: pappus very fine and white, somewhat accrescent. – Not infrequent in the interior of the State, yet nowhere common, and probably not native. Aug., Sept.

29. **ERIGERON**, *Linnæus*, in part. Genus of perennials and biennials, very diverse in habit, the more typical species readily distinguished from *Aster* by long narrow subequal involucral bracts, and excessively numerous almost filiform ray-flowers. Other groups quite like *Aster* in these particulars; others wholly destitute of rays; but all differing from the genuine Asters in a far more simple inflorescence, a coarser and more scanty pubescence, more scanty pappus, and especially by short and rounded or blunt style-tips.

* *Perennials with radical leaves largest, the cauline diminished upwards; heads solitary or few; pappus simple.*

← *Heads subracemose; rays very many, short, filiform, suberect.* —Genus TRIMORPHA, Cass.

1. **E. racemosus**, Nutt. Trans. Am. Phil. Soc. vii. 312 (1840): *E. armeriæfolius*, Gray, not of Turcz. Stems commonly several from the root, slender, erect, 1 ft. high more or less, glabrous or with a few bristly hairs; the oblanceolate and linear leaves hirsute-ciliate below the middle, all entire: heads simply racemose, or somewhat corymbose-panicled: involucre 3—4 lines high, broad-campanulate or subhemispherical; filiform rays purplish; pappus somewhat accrescent, dull-white

or reddish-brown. In moist ground, at higher than middle elevations of the Sierra, and probably on the eastern declivity only.

+ + *Heads solitary or sub-corymbose; rays longer, broader and spreading.*

2. **E. Philadelphicus**, Linn. Sp. Pl. ii. 863 (1753). Hirsute, 1—3 ft. high: radical leaves obovate or spatulate, the scattered cauline ones oblong or oblong-lanceolate, with broad clasping base, all irregularly toothed: heads less than 1 in. broad, in an ample loose terminal cymose-corymb: rays very many and narrow, flesh-color to bright pink. — Along streamlets and the borders of boggy places, in both Coast Range and Sierra.

3. **E. glaucus**, Ker. Bot. Reg. i. t. 10 (1815). *Aster Californicus*, Less. Linnaea, vi. 121 (1826). Monocephalous and somewhat leafy branches several, from a stoutish branching and very leafy caudex; plant commonly pale (hardly glaucous) and more or less villous or hirsute: leaves obovate to spatulate-oblong, 2—4 in. long, entire, or with a few serratures: those of the branches gradually reduced to bracts: heads $1\frac{1}{2}$ in. broad including the numerous though not very narrow lilac to light-violet rays. — Along the seaboard only, either in sandy soil, or on cliffs overhanging the sea; flowering at almost all seasons.

4. **E. sanctarum**, Wats. Proc. Am. Acad. xxiv. 83 (1889). Only 4—8 in. high, erect, slender and rather leafy, from slender branching rootstocks, the stem naked and monocephalous at summit: leaves entire, the lowest oblanceolate, the upper linear, pubescent: heads of middle size: involucre of linear acuminate bracts in 2 series, the outer hispidulous: rays very numerous, purple. Santa Inez Mountains, and also toward the sea in the same region.

5. **E. simplex.** *E. uniflorus* of American authors, not of Linnaeus. Stem solitary, simple, 2—10 in. high, monocephalous, nearly glabrous below, more or less hairy above, and the involucre densely villous-hirsute: lowest leaves spatulate, obtuse, the cauline oblanceolate to linear, acutish: head nearly an inch broad including the broad spreading white or pinkish rays — Alpine on the high summits of the Sierra northward. Common in the Rocky Mountains, and hitherto confused with the Old World *E. uniflorus*, which has twice as many very narrow and erect purple rays.

6. **E. frondens.** Stoutish, erect, $1\frac{1}{2}$—2 ft. high, with a copious tuft of radical leaves, and free propagation by subterranean slender offshoots: herbage green but scabrous-hirsutulous, the short hairs appressed: basal leaves lanceolate or oblong lanceolate, 3—5 in. long including the

petiole, remotely serrate-toothed: cauline many, nearly as large, sessile and half-clasping: heads 3—8, long-peduncled, 1¼ in. broad including the numerous and narrow white rays: bracts of the involucre linear, slenderly acuminate, subequal.—Collected by the author, at Summit Station, and at foot of the Washoe Mts. south of Truckee, in the summer of 1895. Species remarkably combining the characters of *E. Philadelphicus* and *E. Coulteri*.

7. **E. Coulteri**, Porter, Fl. Colo. 61 (1874). More slender, the monocephalous stem and one or two radical leaves from a slender ascending rootstock, this in no degree proliferous; herbage nearly glabrous, never scabrous; the lanceolate basal leaves very acute, the few serratures also acute, the slender petioles longer than the blade: heads 2 in. broad including the numerous narrow white rays; bracts of the involucre broader than in the last and less acuminate.—Subalpine woods of the Sierra from above Donner Lake northwards, but more plentiful in the Rocky Mts. of S. Colorado.

8. **E. salsuginosus**, Gray, Proc. Am. Acad. xvi. 93 (1880); Richards. App. Frankl. 2 ed. 32 (1823), under *Aster*. Rootstocks short, thickish, branching: stems more or less decumbent, 12—20 in. high, the summit or peduncles lanate-pubescent: leaves mostly glabrous except on the margin; the radical spatulate or almost obovate, attenuate to a marginal petiole; cauline gradually smaller, relatively broader and sessile, all entire or nearly so, obtusish or acute: head usually solitary; bracts of involucre in several series, linear-subulate with spreading tips, viscidulous: rays about 40, broad, purplish or violet: disk ½ in. broad. Var. **angustifolius**, Gray. Leaves all narrower, the radical oblanceolate, merely scabrous on the margin, the upper cauline linear-lanceolate, all acute: heads commonly 3.—Type frequent in Plumas Co. and northward; the variety very common at Summit Station in the Sierra Nevada. July—Oct.

9. **E. Sonnei**, Greene, Pittonia, i. 218 (1888). Stems slender, 6—10 in. high, apparently from a horizontal rootstock: herbage canescently strigulose: leaves mostly basal, 2—3 in. long, lanceolate, narrowed to a petiole: pedunculiform upper part of stem leafy-bracted, usually monocephalous: involucre campanulate, 4—5 lines high; bracts subequal, in about 2 series: rays 9—12, broad, purplish or nearly white.—Western slope of Washoe Mts., Nevada, but barely beyond the limits of Nevada Co., California. *Mr. Sonne*. Evidently forming a part of the *E. Nevadensis*, Gray, and should possibly be considered the type of that species.

10. **E. Nevadensis**, Gray, Proc. Am. Acad. viii. 649 (1873), in part. Tufted leaves, and subscapiform monocephalous stems often numerous,

from a branched or simple crown of a long tap-root; herbage strigose-pubescent: leaves lanceolate to linear, mostly 1-nerved, the radical 3–5 in. long: heads ½ in. high, solitary at summit of the leafy-bracted stems, these 4–12 in. high, usually decumbent: bracts of involucre equal: rays few and broad, white or purplish: achenes large; pappus coarse. High mountains above Donner Lake, etc., *Sonne*, and more abundant, as well as larger, in the mountains of W. Nevada.

11. **E. corymbosus**, Nutt. Trans. Am. Phil. Soc. vii. 308 (1840). Stems and leaves from a long tap-root with slightly branching crown or short caudex; herbage soft-cinereous: radical leaves narrowly oblanceolate, acute, entire, mostly distinctly 3-nerved, 4–8 in. long; cauline many, narrower and linear: stems 4–10 in. high: heads 3–8, peduncled and forming a corymb: involucre nearly hemispherical, ½ in. high: rays 40–50, white or bluish. Common in Lassen Co., thence far northward.

12. **E. Lassenianus.** Habit of the last, but only half as large, slender, the tap-root seldom or never with branching crown, the numerous stems strongly decumbent; pubescence fine but somewhat spreading: spatulate-linear radical leaves only 2–3 in. long, 1-nerved: heads numerous, often subracemose, small; the broad white or purplish rays only about 12–18. --Mt. Dyer, and elsewhere, at considerable elevations among the mountains bordering Lassen and Plumas counties, *Mrs. Austin.*

13. **E. tener**, Gray, Proc. Am. Acad. xvi. 91 (1880). Silvery-whitish with a fine and close pubescence : radical leaves ovate to lanceolate, 1½–2 in. long, on slender petioles: stems several, very slender, ascending, bearing 2 or 3 heads: bracts of the involucre in 2 or 3 ranks: rays 25–30, bluish.—Crevices of rocks, in subalpine situations about the sources of the Sacramento; more common in W. Nevada. July, Aug.

14. **E. concinnus**, T. & G. Fl. ii. 174 (1841). Stems few, or many and densely tufted, from an entirely herbaceous tap-root, mostly very erect, 1 ft. high or less, usually branching above and with several heads: leaves oblanceolate, thin, hirsutulous, but the stems more notably and almost hispidly hirsute: heads hemispherical; disk ½ in. broad; involucre hispid; rays 50—80, long and narrow, white, purplish or blue; pappus double, the outer of distinct subulate or even oblong paleæ.- In Plumas Co. and northward, east of the Sierra only. July, Aug.

15. **E. aphanactis.** *E. concinnus*, var. *aphanactis*, Gray, Proc. Am. Acad. vi. 540. Foliage and pubescence of the last, but only half as tall and more condensed; involucre low-hemispherical, rays none; disk-corollas permanently yellow, never brownish in age: exterior pappus not paleaceous, but consisting of a few short bristles.—Eastern slope of

the Sierra from Truckee River southward. Very distinct from *E. concinnus*, not growing with it, but of decidedly its own and a more southerly geographic range. July, Aug.

16. **E. filifolius**, Nutt. Trans. Am. Phil. Soc. vii. 328 (1840). Canescent or cinereous with very fine close pubescence, without hirsute hairs: stems often 1 ft. high, the base subligneous: leaves long and linear-filiform or quite filiform, the lowest dilated upward: involucre canescent: rays 50—80, white to violet: pappus simple.—In Plumas Co., *Lemmon*, and far northward.

17. **E. peucephyllus**, Gray, Proc. Am. Acad. xvi. 89 (1880). Tufted stems branching, very leafy at base, the subscapiform peduncles with only 2 or 3 heads, or even monocephalous: leaves almost filiform, cinereous-pubescent: involucre hirsute-pubescent: rays 30—40, pale blue or purplish, rather short: pappus conspicuously double, the outer squamellate. In half desert regions east of the Sierra, on the northeastern borders of our district.

* * *Low tufted perennials; peduncles slender, nearly leafless, scape-like and monocephalous.*

+ *Leaves narrow and entire.*

18. **E. pygmaeus.** *E. Nevadensis*, var. *pygmaeus*. Gray. Dwarf and cespitose, the long tap-root surmounted by a stout much branched and half subterranean caudex: the tufted leaves cinereously strigose, less than ½ in. long, oblanceolate: scapes slender, scarcely 1 in. long: bracts of involucre unequal, not numerous, nor more pubescent than the foliage: rays 20—30, not very narrow. Alpine summits of the Sierra. Aug., Sept.

19. **E. elegantulus**, Greene, Eryth. iii. 65 (1895). Stems densely cespitose, the short whitish branches thickly beset with narrowly linear strigose-pubescent leaves 1 in. long or more: peduncles slender, 2—3 in. high, leafy-bracted toward the base: heads 3 lines high: bracts of involucre in 2 series, the outer one-third shorter: rays 20—30, violet: pappus dull-white, simple, merely scabrous.—Dixey Valley, Lassen Co. *Baker & Nutting*.

20. **E. ursinus**, Eaton, Bot. King Exp. 148 (1871). Loosely cespitose: herbage dull-green, mostly glabrous, or the margins of the spatulate or oblanceolate leaves hirsute-ciliate: scapes 3—8 in. high, leafy-bracted below: involucre ¼ in. high, almost hemispherical, its bracts subequal, hirsute-pubescent: rays 40—50, rather broad, deep violet, or more

commonly pale, and even white: pappus simple. Frequent in alpine or subalpine situations above Donner Lake, and about the headwaters of the Truckee. July—Sept.

21. **E. Bloomeri**, Gray, Proc. Am. Acad. vi. 540 (1865). Densely tufted and low, green and glabrate, or sparsely strigose: leaves spatulate-linear, or the few bract-like ones of the scape filiform: scape 2–6 in. high: head almost ½ in., the involucre campanulate, its bracts equal, soft-villous or canescent: rays none: achenes glabrate; pappus simple.— Stony ground among the lower mountains and elevated plains northeastward east of the Sierra. June—Aug.

+ + *Leaves ternately divided; pappus simple.*

22. **E. compositus**, Pursh, Fl. ii. 535 (1814). More or less hirsute, 3–6 in. high: leaves with rather long and slender margined and setose-ciliate petioles, the blade 1–3-ternately divided into linear segments; scapes with a few linear bracts below the middle: involucre sparsely hirsute: rays 40–60, not very narrow, white or purplish. Var. **discoideus**, Gray. Mostly larger than the type, and the involucre narrower: rays none. Subalpine in the Sierra Nevada. July—Sept.

23. **E. trifidus**, Hook. Fl. ii. 17, t. 120 (1834). Much smaller than the last, and more hirsute; leaves usually only trifid and the lobes short and rounded, or the two lateral ones sometimes 2-lobed: involucre very hirsute: rays few, purplish.—Alpine on Mt. Whitney, *Coville*.

* * * *Taller, slender freely branching species; pappus double, the outer squamellate.*

24. **E. divergens**, T. & G. Fl. ii. 175 (1842). Finely and cinereously pubescent, diffusely branched biennial, often 1 ft. high: radical leaves spatulate-obovate, the cauline almost linear, all entire, none much exceeding an inch in length: heads 2–3 lines high; involucre hirsute; rays very slender and numerous (about 100), purple: inner pappus of scanty bristles, the outer of subulate squamellæ.—Plant of the Rocky Mountains and the Great Basin, reaching our borders in gravelly places along the Truckee River. June—Aug.

25. **E. Californicus**, Jepson, Bull. Torr. Club, xviii. 324 (1891). Allied to the preceding but perennial, cinereously hirsutulous, 6–10 in. high, leafy up to the base of the peduncles of the scattered heads: radical leaves small, pinnately parted into about 5 linear segments: cauline narrowly oblanceolate, entire: rays 60–75, not very narrow, purplish pappus of conspicuous squamellæ, and only 4 or 5 bristles.—Marysville Buttes, *Jepson*. A distinctively Californian and perhaps quite local species. May, June.

26. **E. multiceps,** Greene, Pittonia, ii. 167 (1891). Perennial, the thick tap-root bearing a stout multicipital densely leafy caudex: herbage cinereous with a fine appressed pubescence: lowest leaves obovate to oblanceolate, 1–2 in. long, the cauline nearly linear: slender ascending stems 6–10 in. long, loosely corymbose: involucre hirsutulous: rays about 75, purplish: pappus of 5—8 bristles and rather more numerous subulate squamellæ.—Mountains of Kern Co , *Palmer & Wright.*

* * * * *Tall, green, simple and leafy-stemmed biennial; pappus a crown of distinct or partly united squamellæ, with few or no bristles.*—Genus PHALACROLOMA, Cassini.

27. **E. strigosus,** Muhl. in Willd. Sp. iii. 1956 (1803). Roughish with a short appressed rather scanty pubescence; stem 2—4 ft. high: lowest leaves spatulate and sometimes toothed, the rest lanceolate, entire: heads small, loosely corymbed at summit of the stem: rays short, white: slender bristles of the pappus wanting in the ray-flowers, and very deciduous in those of the disk.—Plumas Co., *Lemmon,* thence northward; also in the Atlantic States.

* * * * * *Stems equably leafy up to the subcorymbose or monocephalous summit; leaves narrow, entire; bracts of the involucre mostly in several series and imbricated.*

← *Rays usually numerous and narrow.*

28. **E. foliosus,** Nutt. Trans. Am. Phil. Soc. vii. 309 (1840). Scabrous, more or less strigose-pubescent, 1–2½ ft. high: leaves narrowly oblanceolate, 1–2 in. long, those of the branches reduced: hemispherical heads ½ in. broad; rays about 30; achenes with a few coarse bristly short hairs. Dry hills from Sonoma and Contra Costa counties southward. June–Sept.

29. **E. tenuissimus,** Greene, Pittonia, iii. 25 (1896). Very slender, 2 ft. high or more, nearly glabrous: leaves filiform, 1—2 in. long: flowering branches at summit of stem somewhat racemosely disposed, divaricate or even slightly deflexed, each with 1 or 2 small heads; bracts of the broadly campanulate involucre in 2 or 3 series, not very unequal, the outer ones strigose-pubescent: rays 30 or more, bluish: achenes oblong-linear, nearly glabrous. Hills of Ventura Co.

30. **E. Blochmanæ,** Greene, l. c. Stout, 1–2 ft. high; the stem striate and, with the spatulate-linear leaves, canescently hispidulous; the largest leaves 2 in. long or more, each with a short leafy branchlet in its axil: terminal corymb of 12–20 rather short-peduncled heads, these ½ in. high; bracts of the involucre subulate-linear, unequal, in about 3

series, canescent with an appressed pubescence: rays 50–60; achenes glabrous. Sandy places near the sea in the northern part of Santa Barbara Co., *Mrs. Blochman.*

31. **E. Hartwegi,** Greene, Eryth. iii. 21 (1895). Tufted stems very erect, a foot high, less rigid than in the foregoing, ending in a simply corymbose cluster of 3–7 rather large heads; pubescence of the whole plant scanty and strigose, in no degree hispid: leaves linear, with a very narrow revolute margin: rays numerous, rather broad, of a pale bluish purple: involucre minutely scabrous and strigose, the bracts in 2 slightly unequal series.—Foothills of the Sierra Nevada east of the Sacramento.

32. **E. Breweri,** Gray, Proc. Am. Acad. vi. 541 (1865). Stems slender, erect or ascending from horizontal wiry rootstocks; herbage scabrous-cinereous with minute spreading pubescence: leaves barely 1 in. long, narrowly spatulate, or the uppermost linear, obtuse: heads solitary, or several in a corymb, 3–4 lines high; involucre glabrous or granulose-glandular, its bracts unequal, obtuse: rays few (12—20) and broad, 3 lines long, violet. In open woods at middle elevations of the Sierra, from Kern Co. northward.

33. **E. confinis,** Howell, Eryth. iii. 35 (Feb. 1895); *E. Blasdalei,* Greene, l. c. 124 (Aug. 1895). Stems tufted from a lignescent base, slender, decumbent, ½—1 ft. long; herbage minutely and sparsely strigose-pubescent: leaves narrowly linear, plane, 1½ in. long or more: heads 1–5, rather large; bracts of the involucre linear-acuminate, imbricated in about 3 series: rays very many, rather broad, violet: achenes linear, strigose-pubescent. Lower foothills of the Sierra, on the Stanislaus River, *Blasdale, Dary,* and again in the Siskiyou Mts., *Howell.*

34. **E. Elmeri.** *Aster Elmeri,* Greene, Pitt. ii. 170 (1891). Tufted stems very slender, wiry, decumbent, 5–8 in. long, leafy up to the solitary small head: leaves oblong-linear, acute, strigose-pubescent or hispidulous: bracts of the involucre in 2 or 3 series, scabrous-puberulent rays few, purplish: achenes sparsely setulose-hairy; pappus indistinctly double, the outer of a few unequal short bristles.—Grand Cañon of the Tuolumne, *Chesnut & Drew.*

+ + *Rays wholly wanting.*

35. **E. supplex,** Gray, Proc. Am. Acad. vii. 353 (1868). Stems 3–6 in. long, decumbent or assurgent, in a tuft, leafy except at the naked and pedunculiform summit, which bears a single head; herbage more or less villous-pubescent: leaves numerous, entire, spatulate-lanceolate to linear:

head nearly ½ in. broad; bracts linear-acuminate, loose, nearly equal, villous on the back: rays none: achenes 2-nerved, hispidulous.—Sonoma Co., *Bioletti*, and northward, toward the sea.

36. **E. inornatus**, Gray. Proc. Am. Acad. xvi. 88 (1880). Stems tufted, erect, 10–18 in. high; herbage light green, glabrous and glandless even to the involucres: leaves broadly linear, 1 in. long or more, with narrowly revolute margins: heads 20–40, in a rather close corymbose panicle: bracts of the involucre lanceolate-linear, rigid, erect, in few series and little imbricated.—Chiefly of the Sierra Nevada at rather higher than middle altitudes, and from Placer Co. northward. July-Sept.

37. **E. viscidulus**, Greene, Pittonia, i. 174 (1888). Only a few inches high, minutely and densely glandular-puberulent: leaves 1 in. long, spatulate-linear, acute: heads 1–3, on short bracted terminal peduncles; involucral bracts more herbaceous than in the last, imbricated in about 3 series: achenes sparsely setulose. -Humboldt Co.

38. **E. angustatus**, Greene, Bull. Calif. Acad. i. 88 (1885). Stems tufted, 2 ft. high, rigid and brittle; herbage glabrous except a few short incurved hairs on the margins and midvein of the leaves, and a somewhat granular minute indument on the much imbricated turbinate involucres: leaves narrowly spatulate-linear, entire: corymbose panicle ample: bracts of the involucre with reddish tips: corollas of a deep golden yellow: achenes setose-hirsute.—Low hills on either side of Napa Valley. July -Oct.

39. **E. Biolettii**, Greene, Man. 181 (1894). Size of the preceding: whole plant scabrous-puberulent: leaves oblanceolate, obtuse, with sparsely but rigidly hispid-ciliate margins: corymb with branches less divergent: achenes appressed-pubescent. -On Hood's Peak, *Bioletti*, and Howell Mountain, *Jepson*.

40. **E. petrophilus**, Greene, Pittonia, i. 218 (1888). Half the size of the last, more leafy, the whole plant except the somewhat glandular heads canescently hirsute; the corymb less ample; bracts of involucre not as numerous.—Rocky summits of the inner Coast Range, from Mt. St. Helena to Mt. Hamilton, and in Monterey Co. July Oct.

41. **E. miser**, Gray, Proc. Am. Acad. xiii. 372 (1878). Stems closely tufted on a short woody caudex, erect or ascending, 3-6 in. high, rather softly hirsute, densely leafy up to the corymb of small heads: leaves from oblong-spatulate to short-linear, ½—¾ in. long: involucre minutely glandular, short; its bracts lanceolate or linear, acute. Rocky summits in the region of Donner Lake. Aug.—Oct.

30. **BRACHYACTIS,** *Ledebour.* Low branching annuals, with entire slightly succulent leaves, and numerous racemose or panicled heads of very inconspicuous whitish flowers. Bracts of the involucre imbricate, with prominent green-herbaceous tips. Rays very small, in several series and suberect. Style-tips lanceolate. Achenes compressed. Pappus copious, fine and soft, accrescent after flowering (as in *Baccharis*).

B. **frondosa,** Gray, Proc. Am. Acad. viii. 647 (1873); Nutt. Trans. Am. Phil. Soc. vii. 296 (1840), under *Tripolium.* A foot high or less, upright, or branched from the base and the branches spreading: leaves 1 in. long, spatulate-linear, the uppermost narrower, passing into the rather broad and obtuse involucral bracts: heads hemispherical, about 4 lines high; rays a line long: achenes narrow, appressed-pubescent.—In subsaline wet soils in the Sierra Nevada, at Sonora Pass, *Bolander,* and northward.

31. **BACCHARIS,** *Linnæus.* Diœcious shrubs or herbs, with striate or angled branches, alternate simple, often glutinous leaves, and small clustered discoid heads of white unisexual flowers. Involucre of scale-like imbricated bracts. Fl. of staminate heads with tubular-funnelform 5-cleft corolla; of the pistillate slender-tubular, truncate or minutely toothed. Style-appendages ovate to lanceolate, rarely coalescent. Achenes 5—15-costate, glabrous or pubescent. Pappus of fertile flowers very fine and soft, becoming elongated in fruit.

* *Herbaceous perennial.*

1. B. **Douglasii,** DC. Prodr. v. 400 (1836). Erect, 3—4 ft. high, simple up to the terminal corymb; leaves very glutinous, ovate-lanceolate, nearly or quite entire, 3—6 in. long: bracts of involucre erose-ciliate: pappus of pistillate fl. short, soft; of staminate clavellate and barbellate at summit. In moist lowlands. Sept. Nov.

* * *Suffrutescent or shrubby.*

2. B. **Plummeræ,** Gray. Proc. Am. Acad. xv. 48 (1879). Stems tufted, rigidly herbaceous from a woody base, 2—3 ft. high, simple up to the cymose-paniculate summit; herbage sparingly and rather roughly pubescent: leaves linear-oblong, 1—3 in. long, 3-nerved, irregularly, closely and sharply serrate: heads numerous, 4 lines high: achenes somewhat flattened and puberulent: mature pappus fuscous, nearly $\frac{1}{2}$ in. long. Northern part of Santa Barbara Co., *Mrs. Blochman,* and southward.

3. B. **glutinosa,** Pers., Syn. ii. 425 (1807). Shrub 6—12 ft. high: leaves lanceolate, acute, entire, denticulate or repand-dentate, 2—3 in.

long; heads in ample cymose panicles at the ends of long willowy leafy branches.- On banks of streams, from Butte Co. southward.

4. **B. pilularis,** DC. Prodr. v. 407 (1836). Low, slender, the depressed or prostrate diffusely branching stems 1-2½ ft. long; branchlets angular: leaves seldom ½ in. long, cuneate-obovate, angular-toothed or subentire, heads mostly solitary in the leaf-axils and at the ends of the broom-like fastigiate branchlets: involucral bracts acutish, fringed toward the tips.—Sandy soils along the seaboard.

5. **B. consanguinea,** DC. l. c. Compactly branching evergreen 8-12 ft. high: branchlets green, angular from the leaf-bases: leaves subcoriaceous, glutinous, 1 in. long and less, cuneate-obovate, coarsely toothed: heads sessile singly or in pairs or threes in the leaf-axils: bracts of involucre oblong-linear, obtuse, with subscarious fringed margins. - Hillsides and banks of streams everywhere; occasional hybrids occur between this and the preceding at San Francisco. Oct. Dec.

Suborder 3. GNAPHALIACEÆ.

Plants mostly white with floccose wool, the herbage apt to be more or less pleasantly or unpleasantly scented. Heads discoid: bracts of involucre various, often scarious and white or yellowish. Anthers caudate. Style-branches of perfect flowers blunt, unappendaged, the stigmatic lines running almost to the summit, which is sometimes papillose or penicillate. Pappus finely capillary or none.

Hints of the Genera.

* *Involucral bracts many; receptacle not chaffy.*

Involucre herbaceous; achenes large, naked, glandular - - - - - - 42
Involucre scarious; achenes small;
    Plants unisexual - - - - - - - - - 33, 34
    Male and female fl. in the same head - - - - - - 35
Involucre dry, but hardly scarious - - - - - - - - - - 32

* * *Involucral bracts few or none: receptacle chaffy.*

Fructiferous chaff or bract quite enclosing its achene;
    Achenes gibbous - - - - - - - - - - - - 40
    Achenes straight or curved;
        Receptacle columnar - - - - - - - - - - 41
        Receptacle globular or ovoid - - - - - - - - 39
Fructiferous bract scarcely enclosing its achene;
    Receptacle columnar;
        Pappus none - - - - - - - - - - - 38
        Pappus to sterile flowers only - - - - - - - 37
    Receptacle convex; pappus present - - - - - - - 36

32. **PLUCHEA,** Cassini. Herbs or shrubs with alternate leaves, and terminal cymose clusters of smallish heads; these many-flowered, the flowers largely pistillate only, their corolla reduced to a slender truncate or 2–3-toothed tube, that of the hermaphrodite (but sterile) flowers regularly 5-cleft. Achenes small, 4—5-angled or sulcate. Pappus a series of capillary bristles.

1. **P. camphorata,** DC. Prodr. v. 452 (1836); Linn. Sp. Pl. 2 ed. 1212 (1763), under *Erigeron*. Annual, stoutish, leafy, 2 ft. high; minutely and somewhat viscidly pubescent: leaves oblong-ovate to oblong-lanceolate, acute at both ends, toothed or denticulate, the larger (3—5 in.) petioled: heads short-pedicelled, dull-purple, crowded in a corymbiform cluster: involucral bracts ovate to lanceolate, often tinged with the dull pale purple of the corollas. - Borders of brackish marshes. Aug.—Oct.

2. **P. borealis,** Gray, Proc. Am. Acad. xvii. 212 (1882); T. & G., Emory's Rep. 143 (1848), under *Tessaria*. Shrub 6 ft. high more or less, with suberect slender and flexible willowy branches, very leafy up to the cymose clusters of smallish heads: leaves silky-pubescent, 1—2 in. long, linear-lanceolate, entire, acute at both ends: involucre campanulate; outer bracts ovate, obtuse, tomentose; inner ones narrowly linear, deciduous: corollas whitish, with a tinge of purple or red: pappus copious, the bristles of the infertile flowers clavellate-dilated, of the fertile ones, more slender. Banks of streams, from San Luis Obispo Co. southward.

33. **ANTENNARIA,** Gaertner. Tufted or matted white-woolly perennials with alternate entire leaves, and solitary or corymbose small heads of white or pinkish flowers, the plants all unisexual. Pistillate plants with filiform truncate corollas shorter than the 2-cleft style; staminate with 5-cleft corollas and truncate undivided style. Pappus a single series of capillary bristles; those of fertile flowers accrescent, very slender, connate at base; those of the sterile flowers mostly thickened at apex. Achenes small, terete or flattish.

* *Radical leaves larger, those of the scape-like stems reduced in number and size.*

+ *Stems with stolon-like leafy sterile shoots from the base.*

1. **A. dioica,** Gaertn. Fruct. ii. 410 (1791); Linn. Sp. Pl. ii. 850 (1753), under *Gnaphalium*. Gregarious, by multiplication of radical shoots, and forming broad mats, with rosettes of spatulate or oblanceolate silvery-tomentose leaves: scapiform stems 2–10 in. high, bearing mostly linear leaves, and several heads in a close corymb: bracts of involucre with

obtuse pearly white or pinkish tips: bristles of pappus in sterile flowers dilated into a broad flat tip.—Subalpine in the Sierra Nevada from near Yosemite northward.

2. **A. alpina,** Gaertn. l. c.; Linn. l. c. 856, under *Gnaphalium*. Less gregarious: flowering stems 1–4 in. high, bearing a close cluster of few subsessile heads: bracts of involucre livid-brown and thin-scarious acutish in the fertile but obtuse in the sterile heads; bristles of pappus in the latter with narrower and less abrupt tips.—In the Sierra at alpine elevations.

← ← *Stems clustered or matted, but not stoloniferous.*

3. **A. dimorpha,** T. & G. Fl. ii. 431 (1843); Nutt. Trans. Am. Phil. Soc. vii. 405 (1841), under *Gnaphalium*. Low and matted, only 1–2 in. high: leaves spatulate, silky-woolly, crowded on the branches of the caudex: heads solitary, terminating short pedunculiform stems: bracts of the turbinate involucre brownish, those of the sterile heads ovate-lanceolate, of the fertile narrower and acuminate: achene with a minute pubescence of short bi-uncinate hairs; pappus of sterile flowers barbellate at summit but scarcely dilated.—Foothills of the Sierra along our eastern borders.

4. **A. argentea,** Benth. Pl. Hartw. 319 (1849). Closely silky-woolly, the slender stems 8–16 in. high: lowest leaves spatulate, 4–5 lines wide; those of the stem linear to linear-subulate: heads small, numerous in a compound cyme: inner bracts of pistillate-involucre obtuse: achenes glandular-papillose; male pappus with petaloid-dilated tips.—Subalpine from Yosemite northward.

5. **A. microcephala,** Gray, Proc. Am. Acad. x. 74 (1874). Slender, silvery-woolly, 4–10 in. high: lower leaves spatulate, upper small and linear: heads many, very small, loosely panicled: bracts of involucre with scarious tips inconspicuous: achenes glandular; tips of bristles of male pappus much dilated.—Eastern slope of the Sierra northward.

* * *Suffrutescent; stems equably leafy.*

6. **A. Geyeri,** Gray, Pl. Fendl. 107 (1849). Stoutish and rigid, 4–8 in. high, densely white-woolly, leafy to the summit, the leaves oblanceolate, 1 in. long or less: heads cymose or subspicate, rather large: bracts of involucre with conspicuous usually rose-red tips: bristles of the male pappus moderately clavate.—Dry hills, east of the Sierra northward.

34. **ANAPHALIS,** *De Candolle.* Perennial equably leafy simple-stemmed herbs, mostly dioicous and the flowers yellow. Heads in a cymose corymb. Bracts of the campanulate involucre with short ap-

pressed subcartilaginous base, and ample white scarious upper portion, this spreading in age. Style in the staminate flowers cleft at apex. Pappus a single series of slender bristles, in the sterile flowers slightly clavellate.

1. **A. margaritacea**, Benth. & Hook. Gen. ii. 303 (1873); Linn. Sp. Pl. ii. 850 (1753), under *Gnaphalium*. Stems erect, 1–3 ft. high, from slender creeping rootstocks: leaves lanceolate or linear-lanceolate, 2–4 in. long, ascending, tapering at both ends, thinnish, woolly on both faces, but less so above: scarious main portion of involucral bracts ovate-lanceolate. Var. **occidentalis**. Stem more leafy, the leaves sessile by a broad auriculate-clasping base, glabrous and shining above, spreading rather than ascending on the stem. –Type common in the higher mountains. The very marked variety as common among sand hills of the seaboard from at least middle California to Alaska.

35. **GNAPHALIUM**, *Linnæus*, partly. Floccose-woolly. Leaves sessile, entire. Heads cymosely clustered, white, yellowish or rose-tinted. Receptacle flat, naked. Bracts of involucre scarious, imbricated. At least the outer flowers (usually all of them) fertile. Achenes terete or flattish. Pappus a single series of scabrous capillary bristles.

* *Pappus-bristles not united at base, falling separately.*

+ *Involucre woolly only at base; heads paniculately or corymbosely glomerate at summit of stem and branches; leaves more or less adnate-decurrent.*

1. **G. ramosissimum**, Nutt. Pl. Gamb. 173 (1848). Biennial, erect. 2–6 ft. high, paniculately much branched above the middle, the panicle often rather narrow and virgate: herbage glandular and very sweet-scented, only the stem slightly arachnoid, the leaves green on both faces, distinctly decurrent: heads only two lines high, reddish; bracts oblong-lanceolate, acutish. –Wooded hills along the seaboard. July—Oct.

2. **G. Californicum**, DC. Prodr. vi. 224 (1837). Stoutish, 2–3 ft. high, biennial, the leaves diminishing in size towards the broad cymose terminal loose cluster of large rather dull white heads: leaves lanceolate, glabrate above, glandular and balsamic-scented, very obviously adnate-decurrent: outer bracts of the involucre ovate or oblong, the inner acute. Common on dry hills in places partly shaded. May–July.

3. **G. microcephalum**, Nutt. Biennial, slender, with several erect branches 2 ft. high or more, loosely corymbose-paniculate above, the whole herbage white with a persistent wool, not at all glandular or

heavy-scented: leaves linear, or the lower spatulate, with slenderly decurrent base: involucres small, ovate, bright white; bracts ovate or oblong, obtuse.—Hillsides of both Coast Range and Sierra. July—Sept.

4. **G. Chilense,** Spreng. Syst. iii. 480 (1826): *G. Sprengelii*, H. & A. Annual and biennial, stoutish, 1–2½ ft. high, cymose-corymbose at summit: leaves lanceolate, more thinly floccose than in the last, the short decumbent leaves rather broad: involucre hemispherical, with a greenish-yellowish tinge; bracts thin, oval or oblong, obtuse. Var. **confertifolium,** Greene. Very stout and low: leaves linear, densely clothing the stem up to the sessile dense cluster of heads.—Very common and variable; the variety biennial; both flowering at almost all seasons.

+ + *Involucre embedded in loose wool.*

5. **G. palustre,** Nutt. Trans. Am. Phil. Soc. vii. 403 (1841). Low branching annual, floccose with long wool: leaves spatulate to oblong and lanceolate: heads glomerate, leafy-bracted, a line high: tips of linear involucral bracts white, obtuse.—In low moist lands. May—Aug.

* * *Pappus-bristles united at base, deciduous in a ring.*

6. **G. purpureum,** Linn. Sp. Pl. ii. 854 (1753). Biennial, simple or branching, erect or decumbent, 6—10 in. high, canescent with a dense coating of close wool: leaves spatulate, obtuse, usually becoming glabrate and green above: heads crowded in an elongated more or less interrupted spiciform inflorescence: involucre brownish: achenes sparsely scabrous. In open grounds. March—May.

36. **FILAGO,** *Tournefort.* Erect rather slender floccose-woolly herbs, with alternate and entire leaves, and small heads in capitate lateral and terminal clusters. Rays 0. Receptacle plane, hemispherical or subconical; its naked summit bearing both sterile and fertile flowers having a pappus of capillary bristles. Base of receptacle bearing pistillate flowers, the achenes from these being destitute of pappus and enfolded by a concave bract. Achenes terete or slightly compressed, sometimes roughish-papillose.

1. **F. Californica,** Nutt. Trans. Am. Phil. Soc. vii. 405 (1841). A span high or more: heads ovate, slightly angular: convex: pistillate fl. 8—10, their bracts broadly ovate, deeply boat-shaped, incurved: inner bracts oblong, concave: achenes almost terete, obscurely papillose-granular.—Dry hills. May.

2. **F. gallica**, Linn. Mant. ii. 481 (1771). Receptacle nearly plane: heads pentagonal-conical: outer achenes completely enclosed in their conduplicate at length indurated bracts.—Introduced from Europe, but not rare with us.

37. **STYLOCLINE,** *Nuttall.* Low and diffuse white-woolly annuals. Heads terminal, subglobose, the broad thin bracts both of involucre and columnar receptacle deciduous with the mature fruit; those of the fertile flowers involute or saccate-conduplicate, embracing the obovate or oblong obcompressed achene (this without pappus); those of the few and central sterile flowers plane or concave; the rudimentary ovary of these flowers with a pappus of few caducous bristles.

1. **S. gnaphalioides,** Nutt. Trans. Am. Phil. Soc. vii. 338 (1840); Gray, Pac. R. Rep. iv. 101, t. 13. Stems 2–4 in. long; leaves linear or oblong: bracts hyaline, woolly on the back.—From Monterey and the San Joaquin southward.

38. **HESPEREVAX,** *Gray.* Low but rigid, leafy, with heads axillary and terminal. Bracts of the involucre and those of the receptacle subtending the pistillate flowers from oblong to obovate, becoming coriaceous, persistent, concave. Receptacle slender columnar from a broader base, sparsely villous, the pistillate flowers and their bracts crowded at its base; the summit bearing a whorl of 3–7 coriaceous obovate or rounded open bracts subtending a few sterile flowers; these with cleft style but no ovary. Achenes pyriform obovate, somewhat obcompressed, very smooth. Pappus none.

1. **H. caulescens,** Gray, Proc. Am. Acad. vii. 356 (1868); Benth. Pl. Hartw. 319 (1849), under *Psilocarphus*. *Evax involucrata,* Greene, Man. 185 (1894). Stout, strictly erect, simple, or rarely with one or more ascending long branches from the base, 8 in. high or more: heads only in a terminal hemispherical cluster $\frac{3}{4}$ in. broad, surrounded by a conspicuous whorl of 15 or 20 leaves, these of firm texture, with spatulate-obovate cuspidate blade 1½ in. long, only a third the length of the slender petiole, this abruptly dilated at base to half the width of the blade; cauline leaves shorter and narrower.—Plains of the lower Sacramento. May.

2. **H. humilis.** *Evax acaulis,* Greene, Bot. Gaz. vii. 256 (1882) excl. Syn.; Man. 184. Stout and low, the very short branches horizontal: leaves thin and flaccid, with short blade and greatly elongated petiole, this less dilated at base: heads glomerate at the ends of all the branches, none in the axils.—Moist plains near Antioch, and southward. April—June.

3. **H. sparsiflora.** *Evax caulescens*, var. *sparsiflora*, Gray. Slender but rigid, 2–5 in. high, with few or many ascending branches from the base, all leafy and floriferous throughout: leaves 1 in. long, the petiole little longer than the cuspidately acute blade: heads small, few in the axils of all the leaves and not more numerous at the summit, narrowly oblong.—Common on open hillsides in middle parts of the State toward the interior and on the seaboard. May.

4. **H. brevifolia.** *Evax caulescens*, var. *brevifolia*, Gray. Habit of the last, but only 1–2 in. high, more woolly, the small leaves very short-petiolate and the heads mostly clustered and terminal, few or none in the axils; all the parts small.—Humboldt and Mendocino counties and northward. April, May.

5. **H. acaulis.** *Stylocline acaulis*, Kell. Proc. Calif. Acad. vii. 112 (1877). *Evax caulescens*, var. *minima*, Gray, partly. Very dwarf, stemless, the whole plant only $\frac{1}{4}$–$\frac{1}{2}$ in. high, white-woolly: leaves spatulate-oblong, mucronately acute: heads glomerate and sessile among the leaves, very small, few-flowered: bracts subtending the one or more sterile flowers narrow, acute, woolly on the inner face, glabrous without.—Hills of Fresno Co. and northward.

39. **PSILOCARPHUS,** *Nuttall*. Small usually depressed much branched floccose annuals, with opposite leaves and globose heads sessile in the axils or at the forks. Fructiferous bracts numerous, on the globular or oval receptacle, cucullate-saccate, semiobovate or semiobcordate, rounded at top, herbaceo-membranaceous, apex introrse, the ovate or oblong hyaline appendage inflexed or erect. Achene loose within the bract, oblong or narrower, straight, slightly compressed. Pappus none.

1. **P. tenellus,** Nutt. Trans. Am. Phil. Soc. vii. 341 (1840). Prostrate, forming a dense mat 3–6 in. wide: heads very many: leaves spatulate, $\frac{1}{4}$–$\frac{1}{2}$ in. long: fructiferous bracts scarcely a line long: achene ovate-oblong.—In rather low or shaded grounds among the hills. May.

2. **P. brevissimus,** Nutt. l. c. 340. Dwarf, with very few and rather large woolly heads: leaves oblong or lanceolate, 2–5 lines long, seldom surpassing the heads: achene cylindrical or slightly clavate, 1 line long. Plains of the interior in low places. May.

3. **P. globiferus,** Nutt. l. c. Branched from the base and spreading or prostrate: leaves linear or narrowly spatulate, the uppermost little

surpassing the very woolly heads: achenes obovate-oblong, scarcely over ½ line long.—Santa Barbara and eastward.

40. **GNAPHALODES,** *Tournefort.* Low floccose-woolly annuals, with alternate entire leaves. Heads scattered, several-flowered. Pistillate flowers on a small receptacle, each enclosed in a conduplicate bract, the tip of which is scarious-appendiculate; the few hermaphrodite-sterile ones mostly naked. Involucre outside of the fruiting bracts scanty and scarious. Achene gibbous, obovate, enclosed in its bract and falling away with it. Pappus none.

1. **G. Californica,** Greene, Man. 183 (1894); F. & M. Ind. Sem. Petr. 42 (1835), under *Micropus.* Slender, erect, 6–12 in. high: leaves mostly linear: fructiferous bracts 5 or 6, firm-coriaceous, somewhat semiobcordate or semiobovate, straight anteriorly, the erect beak-like tip largely scarious. Open ground; very common. May.

2. **G. amphibola,** Greene, l. c.; Gray, Proc. Am. Acad. xvii. 214 (1892), under *Micropus.* Fructiferous bracts about 10, somewhat imbricated on an oblong receptacle, membranaceous or merely chartaceous at maturity, the beak, an ovate almost hyaline appendage, at maturity porrect.- Hills of Contra Costa and Alameda Counties. May.

41. **ANCISTROCARPHUS,** *A. Gray.* Low canescently flocculent annual, erect, branched from the base, with alternate entire leaves, and more or less glomerate heads. Fertile flowers 5–9, loosely disposed on a slender receptacle, their enclosing bracts cymbiform, firm except the narrow hyaline tip. Sterile flowers involucrate by 5 larger bracts, these ovate-lanceolate, tapering into a rigid incurved-uncinate cusp, persistent and at length stellate-spreading. Achene ovate-fusiform, obscurely obcompressed, the pericarp distinct from the seed and faintly nerved. Pappus none.

1. **A. filagineus,** Gray, Proc. Am. Acad. vii. 356 (1868). Leaves linear to spatulate: heads capitato-glomerate, the hooked empty bracts at maturity ¼ in. long.—In open grounds; not common.

42. **ADENOCAULON,** *Hooker.* Perennial, with alternate dilated leaves on long margined petioles; the slender stem naked and paniculate above, bearing small heads of whitish flowers; the peduncles beset with stalked glands. Involucre of few thin-herbaceous bracts. Receptacle flat, naked. Achenes ovoid-oblong or subclavate, far exceeding the involucre, the upper part beset with stout stipitate glands.

1. **A. bicolor**, Hook. Bot. Misc. i. 19, t. 15 (1829). Stem 2 ft. high: leaves ample, deltoid-cordate, coarsely sinuate-dentate or slightly lobed, green above, white-cottony beneath: involucral bracts 4 or 5, in one series, ovate, reflexed in fruit, small by the side of the 4—6 clavate achenes.—Redwood forests of the Coast Range; also in the Sierra.

Suborder 4. AMBROSIACEÆ.

Heads small, greenish, the fertile flowers without corolla, or this reduced to an obscure rudiment. Rays none. Staminate involucres mostly forming a raceme above the axillary and few pistillate ones. Anthers but slightly united or quite distinct. Pappus none.

Hints of the Genera.

Heads all alike, and only in the leaf-axils - - - - - - - - 43
Staminate heads racemose above the others - - - - - - - - 44, 45
Staminate heads glomerate; fertile head becoming a bur - - - - - 46

43. IVA, *Linnæus*. Perennial herb with simple mostly alternate leaves, and discoid heads nodding on short pedicels in their axils. Involucre of few scales in 1 series, commonly joined into a cup. Marginal fl. pistillate and with short tubular corolla; the other and more numerous fl. staminate, with funnelform 5-lobed corolla and undivided style: anthers nearly distinct. Receptacle with linear or spatulate scales subtending the sterile fl. Achenes thick, naked.

1. **I. axillaris**, Pursh, Fl. ii. 743 (1814). Branching sparingly, 1—1½ ft. high: leaves from obovate and spatulate to broadly linear, sessile, entire, 1 in. long or more: heads hemispherical: scales of involucre about 5, united at base, or beyond the middle. Var. **pubescens**, Gray. Villous with loose spreading hairs; the involucre turbinate, almost entire.—Solano Co. and southward, mostly on subsaline plains, or near the coast.

44. AMBROSIA, *Tournefort*. Weedy aromatic coarse perennials with mostly alternate and pinnately divided leaves. Flowers unisexual, the staminate heads several-flowered and arranged in erect spikes or racemes resembling aments. Pistillate heads mostly in the axils of the upper leaves, 1—4-flowered, their involucres closed and achene-like, in maturity bearing protuberances toward the summit. Achene ovoid or obovate, thick.

1. **A. psilostachya**, DC. Prodr. v. 526 (1836). Stems erect, from horizontal rootstocks, 2 ft. high or more, with strigose pubescence and

somewhat scabrous: leaves once or twice pinnatifid: fr. mostly solitary in the axils, turgid-obovoid, less than 2 lines long, obtusely short-pointed, rugose-reticulate, either unarmed, or with 4 short blunt or sharp tubercles. Borders of fields in uncultivated land near the Bay; plentiful on Point Isabel.

45. GERTNERA, *Medicus*. Genus in all respects like *Ambrosia*, except that the fertile involucre is 1—4-flowered and becomes in fruit a sharply prickly bur.

1. G. acanthocarpa, Britton, Mem. Torr. Club. v. 332 (1894); Hook. Fl. i. 309 (1833), under *Ambrosia*. *Franseria Hookeriana*, Nutt. (1840). Annual, diffuse, hirsute or hispid, the stems and branches 1—3 ft. long: leaves of ovate or roundish circumscription, 1—3 in. broad, bipinnatifid: sterile racemes numerous, short: fruiting involucre with flat lanceolate-subulate spines. - Sandy soils southward ; but migrating northward along railway embankments.

2. G. bipinnatifida, O. Ktze. Rev. Gen. i. 339 (1891); Nutt. Trans. Am. Phil. Soc. vii. 507 (1841), under *Franseria*. Perennial, very stout, procumbent, 2 -3 ft. long, somewhat hirsute: leaves ovate, 1 3 in. long, twice or thrice pinnately parted into oblong-linear divisions and small oblong lobes, canescent with a silky pubescence: sterile raceme dense, the heads large: fruit ovate-fusiform, armed with short thick flattish spines, their tips often incurving. -Sandy beaches; very common. June —Dec.

3. G. Chamissonis, O. Ktze. l. c. (1891); Less. in Linnæa, vi. 507 (1831), under *Franseria*. Size, habit, etc., of the last, but leaves cuneato-obovate, or oblong-ovate with cuneate base, obtusely serrate, only some of the lower laciniate-incised: fruiting involucre ovate, the spines broad and channeled.—Habitat of the preceding, but less common.

46. XANTHIUM, *Tournefort*. (COCKLE-BUR.) Coarse annuals, with branching stems, alternate lobed or toothed leaves, and clustered heads of greenish flowers; the staminate clusters uppermost, the pistillate in the leaf-axils. Involucre of staminate heads 1 or 2 series of narrow bracts. Stamens monadelphous but anthers merely connivent. Fertile head a closed ovoid bur-like 2-celled and 2-flowered involucre, 1 - 2-beaked at apex: each flower a single pistil, becoming a thick ovoid achene, the two enclosed in the hardened prickly involucre.

1. X. SPINOSUM, Linn. Sp. Pl. ii. 987 (1753). Widely branching from the base, 2 ft. high: leaves ovate-lanceolate, more or less lobed or

pinnatifid, glabrate and green above, white-tomentose beneath; burs ¾ in. long, armed with short weak prickles.--By waysides, common; native of tropical America.

2. X. Canadense, Mill. Dict. 8 ed. (1768). Stout, branching above only: leaves broad-ovoid, slightly lobed, scabrous: bur an inch long, densely beset with stoutish prickles, and at apex strongly 2-horned.—In low fields, where it may have been introduced; but also apparently native, in more slender form, on dry ground, amid salt marshes. Aug.-- Oct.

Suborder 5. HELIANTHACEÆ.

Plants commonly with balsamic-resinous juice, and coarse roughish or woolly foliage. Rays conspicuous; receptacle strongly chaffy; anthers not caudate; involucre not scarious; pappus never of capillary bristles.

Hints of the Genera.

Involucre of 1 or 2 series of similar bracts;
    Rays small, white - - - - - - - - - - - 47
    Rays large, yellow, or wanting - - - - - - - - - 48
Involucral bracts imbricated in several series;
    Pappus none; achenes oblong - - - - - - - - 49
    " of more or less united awns or paleæ - - - - - - 50
    " of two or more thin caducous paleæ - - - - - - 51, 52
    " none; achenes obovoid, compressed - - - - - - 52
Involucre double; outer series of bracts spreading, inner erect;
    Pappus not retrorsely barbed - - - - - - - - 54
    " retrorsely barbed or aculeolate - - - - - - - - 53

47. ECLIPTICA. *Rumphius.* Flaccid low riparian herbs with opposite leaves, and scattered small heads of whitish flowers. Involucre broad, of one or two series of herbaceous bracts. Bracts of flattish receptacle reduced to awn-shaped chaff or bristles. Rays short, fertile. Achenes thick, those of the ray triquetrous, of the disk compressed, margined. Pappus of 2—4 short teeth or awns or none.

1. E. alba, O. Ktze. Rev. Gen. i. 334 (1891); Linn. Sp. Pl. ii. 902 (1793), under *Verbesina.* Annual, 1-3 ft. high, decumbent, minutely strigose-pubescent: leaves lanceolate or oblong, sparingly serrate, sessile, or the lower short-petioled: peduncles from the upper axils long or short; rays about equaling the disk: disk-achenes corky-margined, truncate.—Banks of the lower Sacramento, *Jepson.* Sept., Oct.

48. **RUDBECKIA,** *Linnæus.* Tall perennials, with alternate leaves, and few long-peduncled terminal heads; ray-flowers neutral or none; those of the disk dark-brown, forming a conical head, the receptacle being nearly columnar. Achenes 4-angled, laterally compressed, crowned with a persistent chaff-like cup, or with 4 teeth more or less united.

1. **R. Californica,** Gray, Proc. Am. Acad. vii. 357 (1868). Stem 3 ft. high, mostly simple and with a long naked monocephalous peduncle: leaves ovate-lanceolate, coarsely and sparingly serrate or incised, some of the cauline lyrately 3-parted: cylindric disk $1\frac{1}{2}$ in. high, the broad yellow rays nearly as long.—At the Mariposa Big Trees. *Bolander.*

2. **R. occidentalis,** Nutt. Trans. Am. Phil. Soc. vii. 355 (1840). Size of the last: leaves ovate or ovate-lanceolate, 4–8 in. long, irregularly and sparingly toothed, the upper sessile by a rounded or subcordate base, the lower narrowed to a short winged petiole: heads as in the last, but rays none. Mountains of Butte Co., *Bidwell,* and at Summit Soda Springs, *Greene.*

49. **BALSAMORRHIZA,** *Nuttall.* Rather coarse but low mostly acaulescent vernally flowering perennials, with thick roots which exude a terebinthine balsam, and bear a tuft of long-petioled leaves and several monocephalous scapes. Involucre broad; bracts large, imbricated. Chaff of receptacle linear-lanceolate. Rays large, fertile. Achenes destitute of pappus, those of the ray oblong, of the disk quadrangular.

1. **B. Hookeri,** Nutt. Trans. Am. Phil. Soc. vii. 349 (1840). Canescent with a fine appressed pubescence: leaves a foot long, once or twice pinnately parted, lanceolate in outline: scape often 2-leaved near the base: involucral bracts linear or lanceolate, acuminate.—Hills of Sonoma and Alameda counties, where it seems out of place, being plentiful on dry elevated plains far northeastward.

2. **B. sagittata,** Nutt. l. c. 350. Silvery-canescent with a somewhat woolly pubescence: leaves entire, cordate-sagittate to deltoid-hastate, 4–9 in. long, erect on elongated petioles: scape mostly with 2 or 3 heads: involucre woolly.—Eastern slope of the Sierra mainly.

3. **B. deltoidea,** Nutt. l. c. 351. Green and somewhat pubescent, or glabrous: leaves deltoid-cordate, usually serrate, sometimes entire, 3–9 in. long, long-petioled: heads 1 or 2 to each scape: bracts of the involucre lanceolate or linear, obtuse.—Moist ground in the Coast Range.

4. **B. Bolanderi,** Gray, Proc. Am. Acad. vii. 356 (1868). Glabrous, glutinous, stout, low, leafy-stemmed, with mostly scales instead of

leaves from the rootstock: leaves about 3, cordate or ovate, entire, 3-5 in. long: outer bracts of involucre foliaceous, oval; inner narrow and villous.—Foothills of the Sierra, from Amador Co. to Placer.

50. WYETHIA, *Nuttall*. Vegetative characters of *Balsamorrhiza*, and like that genus, flowering in spring, but the stout stems in our species leafy. Achenes prismatic-quadrangular, crowned with a short pappus of united or nearly distinct rigid scales or awns.

1. W. helenioides, Nutt. Trans. Am. Phil. Soc. vii. 353 (1840); DC. Prodr. v. 537 (1836), under *Alarconia*. Soft-tomentose, becoming less so in age, 1-2 ft. high, very stout and leafy: leaves oblong, or oblong-ovate, the lowest a foot long or more: heads very large, the disk 2 in. broad, the rays also long and broad: outer bracts of the foliaceous involucre ovate and ovate-lanceolate, occasionally toothed: achenes somewhat pubescent at summit.—Hillsides of the Mt. Diablo Range; very early-flowering.

2. W. glabra, Gray, Proc. Am. Acad. vi. 543 (1865). Size and habit of the preceding, but glabrous and balsamic-viscid; leaves occasionally serrate: achenes glabrous.—Middle Californian, like the last, but of the Coast Range proper, and to the seaward.

3. W. longicaulis, Gray, Proc. Am. Acad. xix. 4 (1883). Glabrous and balsamic-viscid, tall and rather slender, 2 ft. high: leaves lanceolate: heads solitary or panicled, the disk scarcely 1 in. broad: outer bracts of the involucre oblong or spatulate, foliaceous, surpassing the disk; rays only 1 in. long: achenes with a short erose-denticulate crown. Prairies of eastern Humboldt Co., *Rattan*.

4. W. reticulata, Greene, Bull. Calif. Acad. i. 9 (1884). Tall and leafy, puberulent-hispidulous without woolliness, the heads several and corymbosely disposed: leaves shining, ovate or subcordate, 2—4 in. long, 3-5-nerved, and with veins and veinlets much reticulated: heads only ½ in. high; bracts of involucre short, obtuse: rays few and small: achenes glabrous; pappus a short erose-denticulate crown.—Banks of Sweetwater Creek, El Dorado Co., *Mrs. Curran*.

5. W. ovata, T. & G., in Emory's Rep. 143 (1848). Canescent with a soft but not woolly pubescence, 2-3 ft. high from running rootstocks, branching: leaves ovate, the upper subcordate, acute, 3-nerved, scarcely reticulate: bracts of involucre broadly lanceolate, not surpassing the disk: pappus a chaffy several-toothed crown.—Obscure species of southeasterly habitat, perhaps beyond our limits.

6. **W. mollis,** Gray, Proc. Am. Acad. vi. 544 (1865). White with floccose wool when young, 1-3 ft. high, with 1 or more heads: involucre rather narrow; bracts unequal, the outer larger: leaves oblong and ovate: achenes pubescent at summit: pappus a truncate chaffy crown, with also 1 or more subulate awns.—Eastern slope of the Sierra.

7. **W. angustifolia,** Nutt. Trans. Am. Phil. Soc. vii. 352 (1840). Stems scapiform, with a few reduced leaves toward the base, 1-2 ft. high, more or less hirsute: radical leaves 1-1½ ft. long, elongated-lanceolate, acuminate at both ends: head naked; bracts of involucre many, broadly linear or lanceolate, foliaceous, loose, ciliate with villous or hirsute hairs: achenes crowned with 1—4 stout hirsute awns, with some short intervening scales. Very common on dry plains and low hills. May, June.

51. **HELIANTHUS,** *Linnæus.* (SUNFLOWER.) Annuals and perennials, flowering in summer and autumn. Leaves simple, the lowest of them opposite. Heads peduncled. Rays conspicuous, yellow. Disk-corollas yellow or dark purple, with short tube and long cylindric throat. Chaffy bracts of receptacle partly embracing the compressed-quadrangular or 2-edged achenes. Pappus a pair of caducous thin scales, with occasionally a few smaller intervening ones.

* *Annuals 3—6 feet high.*

1. **H. lenticularis,** Dougl. in Bot. Reg. t. 1265 (1829): *H. ovatus,* Lehm? (1828): *H. annuus,* Linn.?? (1753). Robust, hispid or scabrous: stem often 1 in. thick at base, mottled or spotted with purple: leaves ovate, acute or acuminate, more or less regularly serrate, 4-10 in. long, petiolate: involucral bracts broadly ovate to oblong, aristiform-acuminate: dark-purple disk 1 in. or more in diameter: rays often 2 in. long.—Plains of the San Joaquin, but probably introduced from the Rocky Mountain region. The supposition that this is the parent of *H. annuus* seems to us too much a mere speculation. July—Oct.

2. **H. Bolanderi,** Gray, Proc. Am. Acad. vi. 544 (1865): *H. scaberrimus,* Benth. (1844), not of Elliott (1824). Not as stout, a yard high, scabrous-hispid: leaves ovate to oblong-lanceolate, entire or coarsely serrate, 2—5 in. long: disk 1 in. wide or less, brownish-yellow; rays about 1 in. long: chaff of receptacle subulate-aristiform, equaling the disk-flowers.—Sonoma Co., and northward and eastward. June—Sept.

3. **H. exilis,** Gray, l. c. 545. More slender, seldom a yard in height: leaves lanceolate to ovate-lanceolate, sparingly denticulate, tapering into a slender petiole: cusp of the chaff a slender awn surpassing the

disk-flowers. Lower Sacramento plains; thence northward. July—Sept.

* * *Perennials, the roots more or less tuberiform; stems 6 - 10 feet high.*

4. **H. Californicus,** DC. Prodr. v. 589 (1836). Stem very leafy throughout: leaves lanceolate, entire or serrate, 6—12 in. long, short-petioled: heads about ⅜ in. high in a terminal corymbose panicle: involucral bracts linear-subulate, often somewhat hirsute: rays over an in. long: disk-corollas canescently puberulent toward the base: achenes glabrous; paleæ of the pappus broadly lanceolate.—Plentiful along streams, and borders of marshes. Aug.—Nov.

5. **H. Douglasii,** T. & G., Fl. ii. 332 (1842). Stems branching, hispidulous; upper rhomboid-oblong to spatulate-lanceolate, tapering into winged petioles, obtuse, entire, 1 2 in. long: heads ½ in. high: involucral bracts mostly foliaceous, hispidulous; outer narrowly oblong, obtuse, reflexed or spreading, longer than the disk; innermost shorter, erect, acute or acuminate: rays ½ in. long; chaff entire.- Obscure and long lost species, collected only by *Douglas*, near Santa Clara, sixty years since.

52. HELIANTHELLA, *Torr. & Gray.* Low subacaulescent perennials, with habit of some eastern *Helianthi;* differing from that genus in the more compressed and thin-edged achenes, either with no pappus, or one of a pair of chaffy awns with some delicate squamellæ between them.

1. **H. Californica,** Gray, Pac. R. Rep. iv. 103 (1857). Minutely hirsute-pubescent, slender, 2 ft. high, sometimes branching: all save the radical leaves opposite, all tapering into petioles and of spatulate-lanceolate outline: heads foliaceous-bracted, the disk ¾ in. wide: rays ¾ in. long: achenes black, obovate-oblong, smooth and glabrous, obcordate at summit, narrowly margined.—Common at considerable elevations among the Coast Range hills of Marin and Contra Costa counties and northward. May—Aug.

2. **H. castanea,** Greene, Eryth. i. 127 (1893). Stouter, seldom a foot high, rough-pubescent with short spreading hairs: leaves scabrous, lanceolate, nearly equaling the stem; heads nearly 2 in. wide; rays 1 in. long: achenes cuneate-obovate, neither strongly compressed nor thin-edged, those of the ray thicker and triquetrous, all dull-black at base, chestnut-brown above the middle; apical notch short and deep. Summit of Mt. Diablo. June.

3. **H. Nevadensis**, Greene, Bull. Calif. Acad. i. 89 (1885). Rigidly erect, 2 ft. high, the stems bearing at summit 3 or more short-peduncled heads: leaves lanceolate, the upper mostly alternate; achenes narrower, less obcordate than in the foregoing, obovate-oblong in outline: pappus of 2 short persistent awns and several equally persistent intermediate squamellae.—Plentiful at subalpine elevations in the Sierra Nevada. July–Sept.

53. **BIDENS**, *Tournefort*. (BEGGAR TICKS.) Branching herbs with opposite leaves, the heads with double involucre, the outer series of bracts foliaceous and spreading, the inner membranaceous and erect. Achene bearing a pappus of 2 or more rigid retrorsely hispid or aculeate awns.

1. **B. frondosa**, Linn. Sp. Pl. ii. 832 (1753). Somewhat hairy, 2–6 ft. high: leaves pinnately 3—5-divided into lanceolate-serrate petiolulate leaflets: involucre often very leafy: rays inconspicuous: achenes obovate or oblong, 2-awned.—Fields of the lower Sacramento, and southward. Aug.–Oct.

2. **B. lævis**, B. S. P. Cat. N. Y. 29 (1888); Linn. under *Helianthus*. Glabrous, stout, more or less decumbent, 1–2 ft. high: involucre not leafy, surpassed by the oval inch-long yellow rays: achenes often with more than 2 awns.—In very wet grounds only, near lakes and rivers. Aug.–Nov.

54. **LEPTOSYNE**, *DeCandolle*. Glabrous annuals, perennials, and half-shrubby plants, with dissected foliage, and usually long scapiform erect peduncles bearing showy heads of yellow flowers. Involucre double; an outer series of narrow foliaceous spreading bracts, and an inner of broad membranaceous erect ones. Rays broad. Chaff of receptacle linear, thin, scarious, deciduous with the fruit. Achenes flat, or somewhat concavo-convex, margined. Pappus a minute callous cup, or a pair of paleæ.

* *Low annuals, mostly almost stemless.*
+— *Achenes callous-winged and meniscoid.*

1. **L. Douglasii**, DC. Prodr. v. 531 (1836). Leaf-divisions filiform: peduncles slender; head an inch wide: achenes sparsely beset with capitate rigid bristles, the margin at length corky; cup-like ring in place of the pappus entire.—Attributed to the vicinity of San Francisco; perhaps erroneously; common from Monterey southward.

2. **L. Stillmani,** Gray, Bot. Mex. Bound. 92 (1859). Stouter, more leafy below, and with manifest short branches: leaf-divisions linear, a line broad: achenes somewhat obovate, smooth and naked on the back, papillose or tuberculate on the inner face, at least along the slightly ridged centre, the corky wing somewhat rugose.—Valley of the Sacramento.

← ← *Ray-achenes oval, flat, glabrous; those of the disk marginless, villous on the edges, and with paleaceous pappus.*—Genus
PUGIOPAPPUS, Gray.

3. **L. Bigelovii,** Gray, Syn. Fl. 300 (1884); Pac. R. Rep. iv. 104 (1857), under *Pugiopappus.* Size and habit of *L. Douglasii:* leaves once or twice ternately parted into linear lobes: involucre ½ in. high, its outer bracts linear, the inner oblong-ovate: ray-achenes oblong, with narrow callous margin; disk achenes narrower, villous on the edges, sparsely so or naked on one or both faces; pappus of two linear triquetrous paleæ.—Southern, perhaps only beyond our limits.

4. **L. calliopsidea,** Gray, l. c.; DC. Prodr. v. 569 (1836), under *Agarista.* Taller, more branching, with shorter and less scapiform peduncles: bracts of outer involucre broadly ovate, of the inner narrowly ovate: ray-achenes oval, thin-winged, those of the disk cuneate-oblong, long-villous on the edges and inner face; pappus as in the last.—Plains of the interior, from the Sacramento southward.

* * *Perennials, with fleshy caudex or stem; heads large and showy; achenes plane, smooth, glabrous; pappus almost none.—*
Genus TUCKERMANNIA, Nutt.

5. **L. gigantea,** Kellogg, Proc. Calif. Acad. iv. 198 (1873). Stout simple or branched naked stems 2–8 ft. high, bearing at summit an ample tuft of leaves, and stout peduncles of corymbosely arranged heads: leaves tripinnately divided into filiform segments: achenes oblong or obovoid, obscurely 3–5-nerved, narrowly callous-winged; pappus a slight coroniform cup.—Remarkable tree-like plant of almost palm-like habit, plentiful on islands off Santa Barbara; but a strong colony of it was recently discovered on the mainland at Point Sal, Santa Barbara Co., by *Mrs. Blochman.*

Suborder 6. MADIACEÆ.

Herbs with watery juice, but herbage mostly viscid and glandular. Involucre of a single series of equal bracts. Ray-flowers fertile, the achene of each partly embraced by, or closely enfolded within its involucral bract. Chaff of receptacle often in a single row between ray and disk, and united into a cup. Disk-corollas usually hairy. Style-branches subulate, hispid.

Hints of the Genera.

* *Ray-achenes distinctly incurved.*

Ray-corollas ample and showy, yellow - - - - - - - - - 56
Ray-corollas minute, usually very few - - - - - - - - 57

* * *Ray-achenes commonly gibbous, but not incurred.*

Ray-achenes smooth, laterally compressed, nearly or quite enclosed by their bracts;
    Receptacle glabrous - - - - - - - - - - - 55
    Receptacle fimbrillate-hirsute - - - - - - - 58
Ray-achenes smooth, somewhat obcompressed, half enclosed by their bracts;
    Pappus none in either disk or ray - - - - - - - - 59
    Pappus present, at least in the disk;
        Pappus plumose - - - - - - - - 60, 70
        " paleaceous - - - - - - - - 61
Ray-achenes tuberculate or rugose, half enclosed by their bracts;
    Receptacle bracted only between ray and disk;
        Bracts of involucre 5 to 8, with short erect rigid tips - - - - 61, 62
        Bracts of involucre 15 or more, with longer spreading soft tips - - 63
    Receptacle chaffy throughout, *i. e.*, a bract to each flower;
        Involucral bracts gland-tipped - - - - - - - - 62, 64
        " " spinescent-tipped - - - - - - - 65

* * * *Ray-achenes obcompressed, or else clavate and not compressed, nearly or quite enfolded by their bracts.*

Rays 8 to 15, showy; pappus of bristles or sharp-pointed paleæ, the latter often very villous below - - - - - - - - - - - - 66
Rays short, inconspicuous; pappus of blunt silvery-white scales - - - - 67
Rays 5 to 8, large in proportion to the small heads;
    Pappus none; rays yellow - - - - - - - - 68
    Pappus of ray cupuliform; rays white - - - - - - - 69
    Pappus of disk and ray plumose - - - - - - - - 70

55. MADIA, *Molina.* (TARWEED.) Glandular and viscid heavy-scented herbs, with at least the upper leaves alternate, entire, or merely toothed. Heads axillary and terminal, the yellow flowers vespertine, closing in sunshine. Involucre angled by the salient carinate backs of the uniserial involucral bracts; these usually completely enfolding the laterally compressed smooth achene, and having free herbaceous tips. Receptacle flat or convex, bearing a single series of chaff united and forming a cup which separates between rays and disk-flowers, otherwise naked. Ligules 3-lobed. Bracts of involucre deciduous with the mature achenes (except in n. 2); these smooth, beakless.

1. **M. sativa**, Mol. Chile, 1 cd. 136 (1782). Stout, 1—4 ft. high, pubescent with slender hairs and beset with pedicellate very viscid glands, ill-scented: leaves lanceolate, nearly entire: heads ½ in. high, short peduncled or sessile in the upper axils and at the ends of some short branches: cup of receptacle broadly campanulate, enclosing many diskachenes, these cuneate oblong and 4-angled; ray-achenes falcate-obovate. By waysides and in cultivated lands; native of S. America, perhaps only naturalized in California. July—Sept.

2. **M. capitata**, Nutt. Trans. Am. Phil. Soc. vii. 386 (1841). Size of the last, but the more viscid herbage honey-scented, the loose hairs hispid: leaves linear, sessile by a broad base: heads longer and narrower, capitate congested at the ends of stout ascending short branches: involucre very hispid: cup of receptacle narrow and nearly closed, the achenes within it very few: ripe involucral bracts and achenes semipersistent.—Marin Co., and far northward. Very distinct; early-flowering. April June.

3. **M. dissitiflora**, T. & G. Fl. ii. 405 (1843); Nutt. l. c. 387 (1841), under *Madorella*. Slender, loosely branching, 2 ft. high, viscid: heads scattered, broad-ovate, 1/4 in. high: cup of receptacle ovoid but not closed: achenes thin, but none angular. Borders of thickets and along mountain roads; not in open plains or cultivated lands. May—July.

4. **M. anomala**, Greene, Bull. Calif. Acad. i. 91 (1885). Lower and stouter than the last, otherwise of the same aspect: chaff of receptacle not joined into a cup: achenes of ray 3—5, of disk 3 only, none either compressed or angled, all somewhat gibbously obovate. Mountain districts from Marin Co. northward; perhaps rare.

5. **M. citriodora**, Greene, Bull. Torr. Club. ix. 63 (1882); Gray, Syn. Fl. 307 (1884), under *Hemizonia*. Stem simple, 1—1½ ft. high, bearing at summit an almost corymbose cluster of peduncled heads; villous-hirsute, glandular and keenly lemon-scented: rays 8 or 9, greenish-yellow: marginal bracts of receptacle only very slightly united: achenes little more than half enclosed by the involucral bracts, rounded on the back, scarcely compressed.—Foothills of the Sierra east and north of the Sacramento Valley. June.

6. **M. glomerata**, Hook. Fl. ii. 24 (1834). A few inches to a foot high, slender, usually with only some virgate branches at summit; herbage hirsute, the inflorscena glandular: leaves narrowly linear, entire: heads glomerate, very narrow; rays 1—5, very short and inconspicuous;

disk-flowers not more numerous, enclosed in a narrow cup of united receptacular bracts: achenes 2 lines long or more, narrow, those of the disk 4—5-angled, of the ray slightly curved, 1-nerved on each face.— Mountain districts mostly or altogether east of the summit of the Sierra, thence far eastward and northward.

56. **ANISOCARPUS,** *Nuttall.* Rather sparingly branching hirsute and glandular herbs with undivided foliage. Ray-corollas rather numerous, large and showy, trifid, vespertine; their achenes more or less curved, usually completely enclosed within the involucral bracts and destitute of pappus. Receptacle flat, smooth, with a single circle of bracts between ray and disk. Disk-achenes straight, angular, usually crowned with a pappus of subplumose paleæ.

1. **A. madioides,** Nutt. Trans. Am. Phil. Soc. vii. 388 (1841). *Madia Nuttallii,* Gray. Perennial, slender, 2 ft. high: leaves opposite, linear-lanceolate, remotely serrate: heads loosely panicled, 4 lines high, slender-peduncled: bracts of involucre 8—12, with short tips: rays $\frac{1}{2}$ in. long: cup of receptacle deeply cleft, enclosing many sterile flowers, these with a pappus of small paleæ. Borders of redwood forests, and elsewhere in damp shades of the coast mountains. July—Oct.

2. **A. Bolanderi,** Gray, Proc. Am. Acad. vii. 360 (1868), and l. c. viii. 391 (1872), under *Madia.* Perennial by horizontal branching rootstocks: stems 2—4 ft. high: leaves opposite, linear, 3—10 in. long: heads few, $\frac{1}{2}$—$\frac{3}{4}$ in. high: involucral bracts and rays 12—16; bracts of receptacle linear, not connected: ray-achenes linear-falcate, commonly with rudimentary pappus; disk-achenes numerous, nearly straight, the outer ones fertile, all with a pappus of long narrow brownish paleæ.- At middle or higher elevations of the Sierra Nevada, from Mariposa Co., northward. July—Sept.

3. **A. Rammii.** *Madia Rammii,* Greene, Bull. Calif. Acad. i. 90 (1885). Annual, erect, slender, 1—2 ft. high, very leafy below, the leaves alternate, linear, villous-hirsute especially below the middle: heads numerous, long-peduncled, forming a loose corymbose panicle: rays about 10, deep golden-yellow, $\frac{1}{2}$ in. long; disk-flowers 18—30: bracts of the involucre minutely hispid, closely investing the lunate laterally apiculate achene, the apiculation with a rudimentary ciliolate pappus: pappus of disk-flowers of 5 slender soft barbellate awns. -Dry foothills of the Sierra Nevada, from Fresno Co. northward to the upper Sacramento. May—July.

4. **A. Yosemitanus.** *Madia Yosemitana*, Parry; Gray, Proc. Am. Acad. xvii. 219 (1882). Small and slender, sparingly branched, only 4—6 in. high, leaves few and the lowest opposite; hirsute pubescence only scanty: rays 5, light-yellow; disk-flowers about 3: ray-achenes semi-obovate (scarcely incurved); pappus as in the last. —Damp moss, at the foot of the Upper Yosemite Fall, *Dr. Parry.*

5. **A. radiatus.** *Madia radiata*, Kell. Proc. Calif. Acad. iv. 190 (1873). Annual, erect, 1—3 ft. high, hirsute and viscid: larger leaves broadly lanceolate, denticulate: bracts of involucre 10—20, with short tips; rays rather light-yellow, $\frac{3}{4}$ in. long; the receptacular chaff between rays and disk united: disk-achenes (all but the central ones fertile) somewhat clavate and 4-angled, nearly straight; those of the ray flat and obovate-falcate, tipped with a small reflexed beak, all destitute of pappus. —Lower part of the San Joaquin Valley, near the mouth of the river. A very peculiar species, local, and rarely collected; more vernal in its flowering than the other species.

57. **HARPÆCARPUS,** *Nuttall.* Small annuals, with glandular-viscid sweet-scented herbage, entire narrow leaves, and numerous pedicellate small few-flowered heads. Ray-flowers fertile, 4—8, the rays minute. Disk-flowers 1—4, enclosed in a tubular cup of an united circle of receptacular chaff. Achenes slender, compressed or obcompressed; pappus none.

* *Leaves alternate; achenes laterally compressed.—* HARPÆCARPUS proper.

1. **H. exiguus,** Gray, Bot. Mex. Bound, 101 (1859); Sm. in Rees' Cycl. (1816), under *Sclerocarpus*. *Harpæcarpus madarioides*, Nutt. Trans. Am. Phil. Soc. vii. 389 (1841). *Madia filipes*, Gray, Proc. Am. Acad. viii. 391 (1873). *Madia exigua*, Greene, Eryth. i. 90 (1893). Slender, 1 ft. high more or less, hirsute, glandular above, paniculately branched, the small heads on long filiform naked peduncles: leaves linear: involucral bracts 5—8, lunate, almost destitute of free tips, hispid-glandular: cup of receptacle prismatic and very narrow, enclosing a single straight obliquely obovate achene; ray-achenes obovate-lunate, pointed by a small disk.—Open woods and glades among the higher hills, especially northward. June—Aug.

* * *Leaves mostly opposite; achenes obcompressed.—*Genus
HEMIZONELLA, Gray.

2. **H. parvulus.** *Hemizonia parvula* & *Durandi*, Gray, Proc. Am. Acad. vi. 549 (1865). *Hemizonella parvula* & *Durandi*, Gray, l. c. ix. 189 (1874). Much branched from the base, 2- 3 in. high, hispidly hirsute

with whitish hairs: leaves narrowly linear, the uppermost involucrately subtending the small subsessile heads (or some earlier heads slender-peduncled): achenes narrowly oblong-obovate or somewhat fusiform, manifestly obcompressed with the inner face slightly angled, tipped with a short incurved beak.—In the mountains from the Yosemite region northward.

3. **H. minimus.** *Hemizonia minima*, Gray, l. c. (1865) and *Hemizonella minima*, l. c. (1874). Very dwarf, only about 1 in. high: lowest leaves oval or oblong, the others linear, but scarcely more than ½ in. long: achenes of the ray broadly obcompressed, rounded at summit, beakless. Eastern slope of the Sierra, from Mariposa Co. northward.

58. **MADARIA,** *DeCandolle.* Erect glandular-hairy pilose or somewhat hispid annuals with somewhat corymbosely panicled heads of showy yellow vespertine flowers; the foliage lanceolate, never divided, usually entire. Receptacle convex, densely fimbrillate-hirsute, usually with only a circle of bracts between ray and disk. Achenes of ray compressed, not incurved, of disk abortive, all destitute of pappus.

1. **M. elegans,** DC. Prodr. v. 692 (1836); Don, in Bot. Reg. t. 1458 (1831), under *Madia*. Stout, 3–5 ft. high, with scattered lanceolate leaves entire or serrate, sessile by a broad base; whole herbage viscid with abundant stalked glands or short gland-tipped hairs, the peduncles and involucres copiously hirsute with long white hairs: heads numerous in an ample corymbose panicle: rays 12–15, nearly 1 in. long, yellow, often with dark red base: achenes rather thin and flat, dark-brown or blackish.—Frequent in partly shaded ground among the hills and lower mountains of middle Calif. toward the coast. Aug.—Oct.

2. **M. densifolia.** *Madia densifolia*, Greene, Pitt. iii. 107 (1897). In size and in floral characters like the preceding, but leaves all very narrowly linear and entire, densely short-hirsute without glands, 4–8 in. long and crowded upon the basal portion of the tall stem, the much reduced and scattered cauline leaves and rameal bracts sparsely bristly-ciliate with hairs longer than the width of the leaf: the ample rays often red at base; achenes less flattened than in the last, being distinctly compressed-trigonous.—Same range with *M. elegans*, but in more open ground; extremely distinct from, and more common than *M. elegans*, with which it has long been confused, and for which I mistook it formerly; this being exactly the *Madia elegans* of the *Bay Region Manual*, but by no means that of Don. July—Oct.

3. **M. hispida,** DC. Prodr. v. 692 (1836); Greene, Pitt. ii. 217 (1891).

under *Madia*. Clothed with long almost hispid hairs, and a short little more than scabrous indument, a few gland-tipped hairs of intermediate length interspersed: stem 1—2 ft. high: lowest leaves in opposite rather remote pairs, oblanceolate, obtusish, not even the scattered reduced and linear upper ones acute: corymbose panicle ample: rays narrow, $\frac{1}{2}$—$\frac{3}{4}$ in. long, yellow: achenes lunate-clavate, brown dotted with black.—Mountain sides and summits of the Coast Range from Sonoma Co. to Monterey. An early-flowering species. May, June.

4. **M. corymbosa,** DC. l. c.; Greene, Pitt. ii. 218 (1891), under *Madia*. More slender and less branching than the last, seldom more than a foot high, and simple up to the corymbose summit; pubescence more scanty and villous, with a scabrous-hispid short indument and usually abundant small stalked glands: leaves all linear, the lowest in remote pairs and sometimes remotely serrate: rays rather few, light-yellow, deeply 3-cleft, the middle segment much narrower than the others: achenes compressed-trigonous, mottled with dark and light brown. -Very common at middle elevations of the Sierra on both slopes, thence northward; flowering and fruiting from April to July, according to altitude; thus a truly vernal species like the last.

5. **M. polycarpha.** *Madia polycarpha*, Greene, Pitt. iii. 167 (1897), under *Madia*. Slender, 2 ft. high, corymbose-panicled: herbage appressed-strigulose, minutely glandular and with some setose-hispid hairs: lowest leaves opposite, oblanceolate, obtuse, entire, hispid-ciliate below the middle, the reduced ones of the branches linear, ciliate throughout: bracts of the involucre not glandular, only sparsely hirsute: rays 8-10, yellow: chaff of receptacle almost wholly scarious, disconnected and in several series: ray-achenes compressed-trigonous, nearly enclosed by their bracts.—Foothills of the Sierra Nevada; apparently not common.

59. **HEMIZONIA,** *DeCandolle.* Annuals, with the glandular pubescence and vespertine rays of *Madia;* differing scarcely generically by the thicker and rather obcompressed than laterally compressed achenes, these more stipitate, smooth. Receptacle flat; chaff mostly as in *Madia*. Flowers in the typical species white; their season summer and autumn.

* *Heads corymbose or panicled.*

1. **H. congesta,** DC. Prodr. vi. 692 (1836). Soft-hirsute but not lanate, 2 ft. high, the inflorescence glandular: bracts of the involucre with lanceolate foliaceous tips little surpassed by the white rays: marginal bracts of the receptacle lightly connate or distinct: achenes with

conspicuous inflexed stipe. Marin Co., near Olema, and elsewhere in middle sections of the State, but not common.

2. **H. citrina,** Greene, Man. 194 (1894). Lowest leaves opposite, oblong-lanceolate, 3-nerved, obtusish, glandular-pubescent, not lanate: heads at first few, in a simple corymbose-panicle: flowers lemon-yellow: tips of involucral bracts short and broad; bracts of receptacle joined into a cup: achenes with inconspicuous stipe.—Northern part of Marin Co., a vernal species, flowering in April and May.

3. **H. luzulæfolia,** DC. Prodr. vi. 692 (1836). Lowest leaves opposite, narrowly linear-lanceolate, acute or acuminate, silvery-canescent with a fine appressed silky wool: inflorescence at length diffuse, very glandular and ill-scented: rays white or pinkish: involucral bracts, achenes, etc., as in the last. Var. **lutescens,** Greene. Flowers from rich cream-color to lemon-yellow, the branches more slender: leaves narrower.—The type abundant in fields and waste lands generally. The variety in Contra Costa Co., near San Pablo, and in Marin, about San Rafael. June— Dec.

* * *Heads racemosely arranged along simple branches.*

4. **H. Clevelandi,** Greene, Bull. Torr. Club, ix. 109 (1882). Lower leaves narrowly linear, 1-nerved, silky-lanate as in the last, racemose or spicate flowering branches villous with long spreading hairs: fl. white: achenes nearly as in the last.- Common from Marin and northern Napa Co. to Oregon.

60. **BLEPHARIZONIA,** *Greene.* Stout and rather coarse glandular-viscid and hirsute heavy-scented annuals, with linear entire lower, and oblong upper leaves. Ray-flowers 7 10, with 3-lobed white ligules. Disk-flowers 10—30, the outer ones subtended by linear chaff. Achenes silky-hirsute, 10-striate, those of the ray partly embraced by the involucral bracts and with scanty pappus; those of the disk surmounted by many densely plumose awns.

1. **B. plumosa,** Greene, Bull. Calif. Acad. i. 279 (1885); Kell. Proc. Calif. Acad. v. 49 (1873) under *Calycadenia*. Somewhat paniculately branching from the base, the branches bearing, racemosely, many heads. these 15 20-flowered: ray achenes with a minute crown of short scales. those of the disk 20 or more erect plumose bristles half as long as the achene. -Plains near Antioch. Aug.—Oct.

2. **B. laxa,** Greene, l. c. Larger, 3 6 ft. high, loosely paniculate above, the large heads borne singly at the ends of the branches, 20—25

flowered: pappus of ray- and disk-achenes alike, short and spreading, less plumose than in the preceding, only a fifth as long as the achene.— Habitat of the other species.

61. CALYCADENIA *DeCandolle*. Rigid strict virgate more or less hispid annuals. Lowest leaves opposite, the others alternate, all narrowly linear, entire, revolute; those of the axillary fascicles and about the heads subulate, but obtuse, commonly ending in a large saucer-shaped gland. Receptacle small, flat, the chaff herbaceous and only encircling the disk flowers. Rays 1–5, white or yellow, vespertine, palmately 3-lobed or -parted; the head as a whole narrow and small. Ray-achenes obovoid-triangular, the terminal areola low, nearly central. Disk-achenes turbinate-quadrangular, the outer fertile, all bearing a conspicuous chaffy pappus (except in n. 2).

* *Yellow-flowered species.*

1. **C. truncata**, DC. Prodr. v. 695 (1836); Gray, Proc. Am. Acad. ix. 192 (1874), under *Hemizonia*. Slender, 1–2 ft. high, glabrous, or some of the lower leaves sparsely hispid; herbage keenly benzine-scented, though rigid and dry: heads sessile and scattered along the virgate branches: short uppermost leaves and bracts truncate by a large sessile flattish gland: fl. yellow; rays 5 (rarely more); disk-fl. 10—24; receptacle-chaff distinct, or nearly so, truncate: pappus of disk-achenes of 7—10 oblong fimbriate-toothed pointless paleæ.—Foothills of Coast Range and Sierra, from Marin and Amador counties northward, in dry open ground. July—Oct.

2. **C. scabrella**. *Hemizonia scabrella*, Drew, Bull. Torr. Club. ix. 151 (1889). More slender than the preceding, with loose and spreading, not at all virgate, mode of branching: leaves minutely scabrous: rays 3—5: chaff of receptacles not united: achenes rugose, glabrous, the abortive ones of the disk, glabrous and devoid of pappus.—Humboldt and Trinity counties to southern Oregon. The northern analogue of *C. truncata*, and wholly distinct, the heads solitary at the ends of the slender branchlets.

3. **C. mollis**, Gray, Proc. Am. Acad. vii. 360 (1868); l. c. 191 (1874), under *Hemizonia*. Stem only puberulent, 1—2 ft. high: leaves cinereously puberulent, those of the fascicles and near the heads like the involucral bracts tipped with a short-stalked dark gland: ray-corollas 3—5, 3-parted: chaff of receptacle forming a 6—8-toothed cup: ray-achenes obpyramidal, glabrous: disk-flowers 5—10, with pappus of 5 or 6 subulate awned paleæ nearly twice the length of the achenes, and one

or two small pointless ones. —Foothills of the Sierra in Merced and Tuolumne counties.

4. **C. ciliosa.** Simple, or toward the summit somewhat virgately branching, 1—2 ft. high, strigose-pubescent and sparsely pilose: leaves 1–2 in. long, scabrous and scabrous-serrulate, the short fascicled and floral ones tipped with a subsessile gland and densely hirsute-ciliate; united bracts of the receptacle about 5, nearly truncate and mucronate, woolly-ciliate within at apex: rays 3, their black achenes smooth and glabrous; those of the disk hairy on the angles, their pappus of alternately short and blunt and aristate-acuminate paleæ.—Lake and Humboldt counties, *Pringle, Chestnut & Drew.*

5. **C. bicolor.** Seldom more than a foot high, erect, simple, scabrellous throughout: leaves very narrowly linear, 2—3 in. long, mostly with a few scattered setiform hairs along the margin near the base; the fascicled and floral short ones gland-tipped, and those next the involucre with small glands up and down the back, the margins sparsely setose-ciliate: cup of the receptacle as in the last, but less strongly woolly-ciliate: rays 2 or 3, cream-color (drying yellow), red at base: ray-achenes semiobovate, carinate-nerved on the back, more or less strigose-pubescent, especially about the summit; disk-achenes also with flattened ventral face, this pubescent, the back usually glabrous; pappus of about 10 unequal paleæ, 2 or 3 being short and acute, the others though unequal all much longer and aristate-acuminate.—Very common upon a narrow belt of the Sierra foothills from Butte Co. to Calaveras. Herbage with the offensive odor of *Hemizonia luzulæfolia.*

6. **C. hispida.** *Hemizonia hispida,* Greene, Bull. Torr. Club. ix. 63 (1882). Stout and rigid, 1½—3 ft. high, scabrous-pubescent and hispid, virgately short-branched and floriferous almost from the base: leaves narrowly linear, 7—3 in. long, hispid-ciliate toward the base, those of the axillary fascicles shorter and with a tack-shaped gland at apex, the floral bracts hirsute-ciliate throughout: cup of receptacle acutely about 10-toothed: rays 3—5, sulphur-yellow, large and showy; their achenes with a more or less conspicuous appressed pubescence: disk-achenes similarly pubescent throughout, and with a pappus of narrowly linear-acuminate subequal paleæ of twice their own length.—Sandy plains of the middle and lower San Joaquin. A gigantesque and very distinct species, apparently somewhat local. Heads mostly glomerate on short branches.

* * *White-flowered species.*

7. **C. campestris.** Near *C. hispida,* but rather slender and not virgate,

nor hispid, the loosely thyrsoid-panicled branchlets with scattered smallish white-rayed heads: floral leaves not crowded, very narrow, recurved, hispid-ciliate and gland tipped: teeth of the cup of the receptacle also gland-tipped: rays 3, their obovate-triquetrous smooth achenes hirsutulous on the angles and around the summit: disk-achenes also angular, strigose-pubescent throughout, with 5 linear abruptly acute paleæ of their own length, and as many linear-acuminate ones twice as long.—Plains about Stockton; collected only by *Dr. Parry*.

8. **C. villosa**, DC. Prodr. v. 695 (1836). Simple or sparingly branching, 1-2 ft. high, somewhat villous-pubescent, the margins of reduced leaves and bracts villous-ciliate: heads mostly solitary and subsessile in the axils of the scattered leaves: bracts or chaff of the involucre acute, not gland-tipped, only slightly united and that toward the base: rays 3—5, white; their achenes broad and truncate at summit, where they are also conspicuously villous; those of the disk-flowers villous-hairy on the angles and also at summit, bearing a pappus of 10 subequal linear-subulate, long paleæ.—Hills along and near the seaboard in Monterey and San Luis Obispo counties; in recent years collected by *Hickman, Norton, Palmer*.

9. **C. cephalotes**, DC. l. c. Seldom more than a foot high, simple or variously branched, pubescent, the leaves with hirsute margin, the floral more or less glandular: heads very narrow and densely glomerate terminally and in the axils of the upper leaves; the narrow cup of the receptacle acutely about 5-toothed and decked exteriorly with small stalked glands: rays 1 or 2, or in some heads none, their achenes tuberculate, prominently but obtusely angled, minutely hispidulous at summit and occasionally upon the angles: disk-achenes 5-angled, canescently strigose, most of the pappus-paleæ ovate-lanceolate and aristate-pointed, little longer than the achene, but 2 or 3 short and blunt.—Hills of Marin and Sonoma Counties. Pleasantly balsamic-scented. Species perhaps including the obscure and only partly described *C. multiglandulosa*, DC.

10. **C. spicata**. *Hemizonia spicata*, Greene, Bull. Torr. Club, ix. 16 (1882). Slender, 1—2 ft. high, floriferous throughout, the heads glomerate-spicate, subsessile in the axils of all the leaves: floral leaves terete and glabrous except at the dilated and ciliate base, all truncate and surmounted by a large stalked gland: ray-flowers 1 or 2, their achenes obovate and scarcely angled, canescently villous with appressed hairs: those of the disk also canescent, but less so, crowned with a pappus of 10 subulate awn-pointed paleæ.—Lower foothills of the mountains in Calaveras Co., about Milton, where it was collected, first by *Dr. Parry*, in 1881, and again by the author, in 1889; not otherwise known.

11. **C. oppositifolia.** *Hemizonia oppositifolia*, Greene, l. c. 110. Very slender, mostly less than a foot high, simple or branched, all the leaves opposite and with long internodes, the heads few and sessile in the leaf-axils, or terminal: stem pubescent: narrowly linear leaves ciliate at base, the floral similar except as to length, and with usually several slenderly stalked glands at and near the apex: teeth of the cup of the receptacle gland-tipped: rays 3, white, changing to rose-red, their achenes black, obovate-trigonous, nearly smooth, glabrous: paleæ of disk-pappus all short, the alternate ones truncate or obtuse. Foothills of Butte Co. *Parry, Mrs. Bidwell.*

12. **C. Fremonti,** Gray, Bot. Mex. Bound. 100 (1859). Erect, slender sparingly branched, seldom a foot high, more or less hirsute; leaves narrowly linear, scabrous, setose-hispid at base, the floral ending in a subclavate stalked gland: heads solitary, terminal and axillary, subsessile: cup of the receptacle with about 12 obtuse teeth: rays 5—7, their achenes smooth; disk-achenes about 20, hirsutulous, their 10 pappus-scales all of equal length and subulate-pointed, little longer than the achene.- Collected only by *Fremont*, a half-century ago, the special locality not recorded.

13. **C. elegans.** Slender, erect, about a foot high, somewhat fastigiately panicled above, with solitary heads sessile in the leaf-axils and terminal: leaves hispid-ciliate below the middle, the floral very few and short, adorned with one terminal and usually several lateral stalked glands: cup of receptacle elongated, acutely 3-5 toothed and the whole exterior thickly beset with stalked glands: ray solitary, its achenes glabrous, faintly tuberculate: disk-flowers 3—6, their achenes elongated, pubescent, crowned with a very short pappus of dark brown paleæ which are alternately obtuse and aristate pointed.—Dry open hills, at the northern base of Mt. St Helena, associated with *Hemizonia Clevelandi*, apparently collected only by the author, in 1888.

14. **C. pauciflora,** Gray, Bot. Mex. Bound. l. c. Stem 1 2 ft. high, parted at the middle into many divergent long loosely spicate slender branches: pubescence scanty, but main leaves bristly-ciliate below the middle, the few oblong gland-tipped floral ones more distinctly ciliate throughout: cup of the receptacle sparsely strigose and with some subsessile glands on the outside, the 4 or 5 teeth obtuse and mucronate: achenes of the single ray-flower smooth and glabrous, short, obovate-trigonous: those 3—5 disk-flowers hirsutulous and with a very short pappus of alternately pointed and pointless paleæ.—Lake Co. *Pringle.* Also collected by *Fremont*, but precise locality unknown.

15 **C. ramulosa.** Size and slender habit of the last, but diffusely

branched, the heads nearly all terminal upon slender branchlets: achene of the solitary ray very short: disk-flowers 3, each almost enclosed by its proper receptacular bract, and these 3 wholly distinct and with acute somewhat recurved tip; disk-achenes prismatic, hirsute, all their paleæ acuminate.—Near Lakeport, Lake Co. *C. G. Pringle*, Aug. 1882.

62. DEINANDRA. HARTMANNIA, DC. (1836), as to his type species, not of Spach (1835). Plants erect, rigid and brittle, balsamic-viscid, with mostly small few-flowered panicled heads. Leaves entire, or serrate. Tips of the few involucral bracts short, rigid, erect. Rays about 5, broad, 3-toothed, diurnal. Receptacle chaffy only next the rays. Ray-achenes gibbous, tuberculate-rugose, the terminal areola raised upon a distinct curved beak from the angle of the inner face of the achene. A paleaceous pappus crowning the mostly sterile disk-achenes, in some species.—A natural genus, well established by the illustrious De Candolle; though the name assigned by him was preoccupied.

* *Receptacle chaffy only next to the rays; disk-ovaries with a pappus.*

1. **D. fasciculata.** *Hartmannia fasciculata*, DC. Prodr. v. 693 (1836); T. & G. Fl. ii. 397 (1843), under *Hemizonia*. Hirsute or hispid below, glabrous and viscid-glandular above, 6—18 in. high: heads small subsessile, usually faciculate-clustered: involucral bracts glabrous or glandular-hispidulous, those of the receptacle slightly united: pappus of disk-ovaries of 6—10 linear paleæ. Hills of the Mt. Diablo Range, near Walnut Creek and Livermore, thence far southward. June—Aug.

2. **D. Wrightii.** *Hemizonia Wrightii*, Gray, Proc. Am. Acad. xix. 17 (1883). Slender, diffusely and widely branching; the filiform branchlets terminating in a single head; lower leaves laciniate-pinnatifid: disk-ovaries with pappus of 8 or 9 oblong firm paleæ.—Native of San Bernardino Co., but found on the Oakland mole in 1881; at that time a species still undescribed. It has not reappeared in this last named district.

3. **D. Kelloggii.** *Hemizonia Kelloggii*, Greene, Bull. Torr. Club. x. 41 (1883). Hirsute below, loosely paniculate above, 1—3 ft. high, the heads on slender pedicels: lower leaves pinnately parted: involucre ¼ in. high; bracts glandular on the back: ray-achenes with a slender curved beak; pappus of the sterile ones of the disk long, almost equaling the corolla, lacerately truncate and united into a tube from base almost to summit.—Abundant in fields of grain on the lower San Joaquin from Antioch southward. Aug., Sept.

* * *Virgately racemose species: all the flowers subtended by bracts; disk-achenes with no pappus* (except in n. 6).

4. **D. virgata.** *Hemizonia virgata*, Gray, Bot. Mex. Bound. 100 (1859). Nearly or quite glabrous, 2–4 ft. high: flowering branchlets very leafy: their leaves short-linear, a line long, glandular-truncate: bracts of oblong involucre also ending in a truncate gland, and stipitate-glandular on the back: disk-flowers 7–10.—Plains of the Sacramento and San Joaquin, a weed of fields and waysides. July—Sept.

5. **D. Heermanni.** *Hemizonia Heermanni*, Greene, Bull. Torr. Club, ix. 15 (1882). Viscid and pubescent, heavy-scented, 1–3 ft. high: minute leaves of the flowering branchlets scattered: bracts of hemispherical involucre viscid-pubescent and beset with stalked glands: terminal gland inconspicuous: disk-flowers 10–15.—Mt. Diablo and southeastward. Aug.—Oct.

6. **D. Lobbii.** *Hemizonia Lobbii*, Greene, l. c. 109 (1882). Habit and pubescence much as in the last, but smaller and more slender, the branches almost filiform: heads very narrow, and with only 3 ray- and 3 disk-flowers: achenes of the ray obovate-oblong; those of the disk with a pappus of 8–10 linear chaffy paleæ. Common in the interior of Monterey Co., about Jolon, etc.

63. **ZONANTHEMIS.** Branching annuals, soft-pubescent and viscid-glandular, with membranaceous pinnatifid or serrate foliage and rather showy heads of yellow flowers terminating corymbose or panicled branches. Bracts of involucre with long soft tips (as in *Madia* and *Hemizonia*) Receptacle chaffy only next the ray-flowers, where it forms a cup. Rays numerous (8–25), narrow, diurnal. Ray-achenes nearly as in *Hemizonia*, but with some unevenness of surface (as in *Deinandra*), and a short apiculation.

1. **Z. corymbosa.** *Hartmannia corymbosa*, DC. Prodr. v. 694 (1836): *Hemizonia corymbosa*, T. & G. Fl. ii. 398 (1843). Pubescent, viscid and glandular, 1 ft. high, corymbosely and widely branching: radical leaves pinnately divided into linear segments, the uppermost linear entire: heads 1½ in. high, 1 in. broad including the rays, these 15—25, oblong-cuneate: beak of ray-achenes short and stout; pappus of sterile disk ovaries of paleæ cut into chaffy bristles, or nearly obsolete.—Plains and hills about San Francisco Bay and southward. June—Oct.

2. **Z. H. angustifolia.** *Hemizonia angustifolia*, DC. Prodr. v. 692 (1836). Hirsute and viscid-glandular, widely branching from the base:

leaves mostly entire, linear, less than 1 in. long: rays 12—15; ray-achenes with prominent upturned beak; pappus of sterile disk-achenes none, or a row of minute bristles rather than scales.—Less frequent than the preceding, and of the same range.

64. **HOLOCARPHA.** Rigid upright branching annual, the serrate or entire leaves not pungent, the upper gland-tipped, the subglobose heads solitary or glomerate. Receptacle convex, the numerous flowers each subtended by a persistent bract, the outer and involucral ones beset on the back with prominent stout clavate and gland-tipped papillæ. Rays diurnal, numerous, narrow, their achenes half-enclosed by their bracts, obovate-triquetrous, nearly smooth, obtusely 4-angled on the back, the one ventral angle ending above in a short beak. Disk-flowers sterile. Pappus none.

1. **H. macradenia,** *Hemizonia macradenia*, DC. Prodr. v. 693 (1836). Stout, hirsute, viscid-glandular, 1—2 ft. high, leafy below, parted abruptly above the middle into few and widely diverging spicate branches: leaves linear, sharply laciniate-toothed or entire, the chaff of receptacle, floral bracts and uppermost leaves linear-subulate, abruptly gland-tipped and more or less beset with smaller gland-tipped hairs: heads often sessile and glomerate, $\frac{1}{2}$ in thick: ray-flowers very many, with short yellow ligules: achenes dull-black, scarcely rugose or glandular, with an angle on the ventral face and 5 dorsal nerves; the apiculation very short.—Rich open ground about San Francisco Bay and southward. Aug., Sept.

65. **CENTROMADIA,** *Greene.* Rigid corymbosely or diffusely branching annuals, with alternate pinnatifid or entire spinescent foliage and involucral bracts; the whole plant more or less resiniferous or glandular and scented. Receptacle convex, chaffy throughout and the bracts distinct, persistent. Bracts of involucre subulate, pungent, embracing the ray-achenes, persistent. Ray-flowers 30—40, small, open all day; their achenes destitute of pappus, triangular, the inner angle terminated by a short apiculation, the whole surface nearly smooth, or faintly rugose-tuberculate. Disk-achenes mostly sterile and with or without a paleaceous pappus.

* *Herbage yellowish-green, scentless, or with aromatic or sweet odor.*

+– *No pappus to disk-achenes.*

1. **C. pungens,** Greene, Man. 196 (1894); H. & A. Bot. Beech. 357 (1836), under *Hartmannia*. Erect, 2—4 ft. high, stout and with rigid

ascending branches; hirsute or hispid, scarcely viscid and nearly or quite scentless: lower leaves doubly, the upper simply pinnatifid, all the lobes pungent-tipped; chaff of receptacle rigid and pungent: ray-achenes nearly black, rather glossy, about a line long not strongly compressed, the ventral angle carinate, and with a short apiculation, the plane sides and rounded back faintly tuberculate-rugose. Plains of the lower San Joaquin, and towards Monterey. July—Oct.

2. **C. maritima**, Greene, l. c. Stout as the last, but only 1–2 ft. high, less rigid, darker green, more villous or hirsute, and with widely spreading and divaricate branches at summit of the erect stem; leaves nearly all pinnatifid and softer, but setose-pungent; chaff of receptacle sharply mucronate, scarcely pungent: ray-achenes dull greenish-brown, scarcely $\frac{3}{4}$ line long, not compressed, though with angled face more prominently and acutely apiculate, the summit and back quite prominently and sharply rugulose, the whole surface obscurely roughened. – Borders of salt marshes about San Francisco Bay. July–Nov.

+ + *Disk-achenes with 3 or more slender linear paleæ.*

3. **C. rudis**, Greene, l. c. 197. With the aspect of *C. pungens*, but only 1–2 ft. high, commonly branched from near the base, and the branches ascending, sparsely hispid-hairy and scabrous-pubescent, slightly resinous and distinctly honey-scented: earliest cauline leaves pinnatifid, all the others linear-subulate, entire hispid-ciliate, the margins in age revolute: achenes of ray black, about a line long, strongly compressed, semi-obcordate in outline, the surface nearly smooth, the apiculation infra-terminal and rather prominent, though short.– Sacramento Valley, near Vacaville, Jepson. Long supposed to be mere *C. pungens*, to which it bears a very close general likeness. But the disk-achenes have the pappus of the next species, and those of the ray are altogether peculiar. May--Aug.

4. **C. Parryi**, Greene, l. c., and Bull. Torr. Club. ix. 16 (1882), under *Hemizonia*. Widely branching, 1–2 ft. high, sparsely hirsute, minutely resinous-glandular, aromatic: lowest leaves pinnatifid, the cauline linear, entire, sharply pungent, spreading, the uppermost pilose-ciliate toward the base: heads scattered rather than glomerate: ray-achenes dull black, $\frac{3}{4}$ lines long, somewhat compressed, smooth on the sides, but with a few coarse tuberculations on the back: those of the disk with 3 or more paleæ exceeding the corollas: chaff of receptacle not pungent.— Plentiful about the warm springs at Calistoga; herbage with the fragrance of Wintergreen. June--Aug.

* * *Herbage dull and dark, ill-scented: disk-achenes with pappus.*

5. **C. Fitchii,** Greene, l. c.; Gray, Pac. R. Rep. iv, 109 (1857), under *Hemizonia*. Stout, widely branching, 1 2 ft. high, villous hirsute, somewhat viscid, more or less beset with stalked glands: leaves mostly entire, linear-acerose, the very lowest pinnately divided into about 3 pairs of linear-acerose segments: bracts of the involucre conspicuous, subulate; those of the receptacle soft, pointless, long-villous: ray-achenes obovate-triquetrous, light-brownish, obscurely if at all tuberculate, indistinctly angled; those of the disk with 8–12 linear pappus-paleæ.—Very common on plains and among the foothills of the interior, from the borders of Oregon southward.

66. **BLEPHARIPAPPUS,** *Hooker* (LAYIA). Vernal annuals, with alternate leaves, and mostly showy broad heads of white or yellow diurnal flowers. Bracts of involucre flattened on the back, with dilated and thin margins the whole more or less completely enfolding its obcompressed achene. Rays 8–20; their achenes obovate-oblong or narrower, without pappus. Disk-flowers with cylindraceous-funnelform 5-lobed corollas; their achenes linear-cuneiform, usually with a pappus of bristles or awns. Receptacle flat, bearing a series of chaffy bracts between ray- and disk-flowers, these, with the involucral bracts, deciduous when mature, leaving a naked receptacle.

* *Pappus of 10–20 bristles which are stout, and, below the middle,*

long-plumose.

+—*Hairs of pappus-bristles not interlaced.*

1. **B. heterotrichus,** Greene, Pittonia, ii. 245 (1892); DC. Prodr. v. 694. (1836), under *Madaroglossa*; H. & A. Bot. Beech. 358 (1841), under *Layia*. Erect, 1 ft. high or more branching from the base, rough-hirsute or hispid and glandular: lower leaves lanceolate, laciniate-pinnatifid or incised, the upper entire: rays large, white: long-villous hairs of the pappus-bristles all erect and straight.—Eastern base of the Mt. Diablo Range, on sandy plains. April, May.

2. **B. graveolens,** Greene, l. c. 246; also Bull. Calif. Acad. i. 92 (1885), under *Layia*. Stout, erect, 2 ft. high or more, sparingly branching, hirsute, and with numerous rigid gland-tipped hairs interspersed: leaves all entire: heads very large, rays of a creamy white: achenes slenderly clavate: pappus when mature deciduous in a ring, the villous wool of the bristles all straight and erect and two-thirds their length.—Said to occur on Mt. Tamalpais; but this may be doubted. It is of the interior of the State, in Kern Co and northward. Apr.—June.

3. **B. carnosus,** Greene, l. c.; Nutt. Trans. Am. Phil. Soc. vii. 393 (1841), under *Madaroglossa*. Dwarf, depressed, branched from the base, pubescent; leaves succulent, 1 in. long, linear-oblong or spatulate, entire, or the lowest sinuate-pinnatifid: heads small: rays white, reduced and inconspicuous: pappus-bristles sparsely plumose with straight villous hairs.—Sands of the sea beaches from Marin Co. to Monterey. April–June.

4. **B. hieracioides,** Greene, l. c ; DC. Prodr. v. 694 (1836), under *Madaroglossa*. Erect, rather strict, 2—3 ft. high, stoutish, hispid: leaves linear to oblong, laciniate-dentate: rays yellow, short, little exceeding the disk: hairs of the pappus all straight and erect. Var. **anomala**, Bioletti. Involucral bracts open-boat shaped, hardly enfolding the achenes, and persistent on the receptacle after the falling of the fruit. A coarse weedy species of wooded or bushy hills, in half shady places. May–July.

5. **B. gaillardioides,** Greene, l. c.; H. & A. Bot. Beech. 148 (1840), under *Tridax*. Freely branching below, 1 ft. high or more, hispid: leaves commonly laciniate-pinnatifid: rays orange-yellow, 1_2-3_4 in. long: pappus dull-white or sordid, the bristles about twice as long as their copious straight villous basal hairs.—Western sections of the State, from Mendocino Co. southward.

6. **B. nemorosus,** Greene, Man. 200 (1894). Rather slender, sparingly branched above, 1—2 ft. high, hispidulous: foliage and heads much as in the preceding, but rays pale yellow below the middle, white above it: pappus short, the bristles often scarcely surpassing their copious brownish villous hairs.—A beautiful species of shaded slopes on Mt. Tamalpias, Mt. Diablo, and the Berkeley Hills. May, June.

+ + *Villous hairs of pappus bristles more or less interlaced.*

7. **B. glandulosus,** Hook. Fl. i. 316 (1833); H. & A. Bot. Beech. 358 (1841), under *Layia*. Seldom more than 6 in. high, branched from the base, rough with short hispid hairs: lowest leaves sparingly pinnatifid, upper linear, entire: heads of middle size; rays 8—13 rather ample and showy, white: disk-achenes appressed silky-villous; pappus bright white, the copious villous wool much shorter than the stout bristles, the inner portion crisped and interlaced. Dry open ground along the eastern base of the Sierra; more common far to the eastward of California. Apr.–June.

8. **B. hispidus,** Greene, Pittonia, ii. 246 (1892), and l. c. 20 (1889),

under *Layia*. A foot high or less, diffusely branched from the base, rather densely hispidulous throughout, and with a few small dark stipitate glands on the involucre: leaves all narrow and entire: heads small, rays white, but short and inconspicuous: pappus bright white, of 10 aristiform bristles, with copious short villous hairs, the innermost of which are interlaced. — Near the summit of Mt. Diablo, and near Tehachapi, Kern Co., at considerable elevation. April — June.

9. **B. elegans**, Greene, l. c.; Nutt. Trans. Am. Phil. Soc. vii. 393 (1841) under *Madaroglossa*. Habit of the last but much larger, more or less stipate-glandular throughout: lower leaves pinnately toothed; upper entire: rays yellow, $\frac{1}{2}$ in. long: pappus white, its copious villous hairs much shorter than the aristiform bristles. — More widely diffused than the last, and at lower elevations. May, June.

* * *Pappus of naked aristiform bristles.*

10. **B. platyglossus**, Greene, l. c.; F. & M. Ind. Sem. Petr. ii. 31 (1835), under *Callichroa*. Sparingly branching, 1 ft. high more or less, hirsute and stipitate-glandular: lower leaves pinnatifid into linear lobes: rays $\frac{1}{2}$ in. long, yellow, with white tips: disk-achenes silky-hirsute; pappus of 15—20 upwardly scabrous stout awn-like bristles. — Common in open grounds. April — June.

11. **B. pentachaetus**, Greene, l. c.; Gray, Pac. R. Rep. iv. 108, t. 16 (1857) under *Layia*. Somewhat hirsute and viscid-pubescent, scarcely hispid, 1—2 ft. high, corymbosely branching, lowest leaves laciniately pinnatifid, the lobes narrowly linear: rays wholly orange-yellow: disk-achenes glabrous or minutely pubescent; pappus of about 5 rigid smooth bristles, or sometimes wholly wanting. — Foothills of the Sierra Nevada, in middle section of the State.

* * * *Pappus, when present of flattened awns or paleæ rather than bristles.*

12. **B. Fremonti**, Greene, l. c.; T. & G., Journ. Bost. Soc. v. 140 (1844), under *Calliachyris*. Strictly erect; branching above the base 1 ft. high, minutely pubescent, not glandular: leaves pinnately cut into short lobes: rays $\frac{1}{2}$—$\frac{3}{4}$ in. long, yellow at base, white above it: pappus-paleæ ovate to oblong-lanceolate, tapering into a subulate awn, entire at the margins and with a few long-villous hairs. — Plains of the lower Sacramento, etc. April, May.

13. **B. Douglasii**, Greene, l. c. 247; H. & A. Bot. Beech. 356 (1841), under *Calliglossa*. Habit and flowers of the last, but plant nearly or

quite glabrous: pappus of 10 18 very unequal rigid subulate awns, these slightly hirsute near the dilated base, Var. **oligochæta**, Greene. Pappus reduced to 2 marginal awns and some rudiments of intervening ones. Open plains and hills, chiefly in the region of San Francisco Bay. April June.

14. **B. Jonesii**, Greene, l. c.; Gray, Proc. Am. Acad. xix. 18 (1883), under *Layia*. Hispidulous and viscid, also with some sessile glands about the involucre: lower leaves pinnatifid, upper narrowly linear, 3-lobed or entire, hispidulous-ciliate: rays only $\frac{1}{4}$ in. long, yellow at base: paleæ of pappus ovate to oblong-ovate, acuminate, often erose-denticulate, not exceeding the corolla-tube; the achenes villous-hirsute. San Luis Obispo Co.

15. **B. nutans**, Greene, l. c ; also Pittonia, ii. 227 (1892), under *Callichroa*. Low, slender, with divergent branches above the base, 3 6 in. high: leaves all linear, entire, the lower pairs opposite, all hirsute-ciliolate: branches, peduncles and involucres glandular-pubescent: rays 5 7, yellow, $\frac{1}{4}$ in. long: achenes hispidulous: pappus of some 10 unequal linear-lanceolate acuminate white paleæ, their margins barbellate: heads small, nodding both in bud and in fruit.—Mountains of Sonoma Co. April June.

16. **B. chrysanthemoides**, Greene, l. c.; DC. Prodr. v. 693 (1836), under *Oxyura*. Aspect of n. 10 preceding; flowers the same: achenes destitute of pappus and all wholly glabrous. -Common in the vicinity of San Francisco Bay. April June.

67. **ACHYRACHÆNA**, *Schauer*. Soft-pubescent sparingly branching annual, with narrow leaves, and rather large oblong-campanulate seemingly rayless heads terminating pedunculiform branches. Rays 6—8 very short, erect: their achenes slightly obcompressed, enclosed, without pappus. Disk-achenes chiefly fertile; clavate, 10-costate, bearing a showy pappus of 10 elongated-oblong obtuse silvery-scarious paleæ. Chaff of receptacle in a single row, not united, between ray and disk.

1. **A. mollis**, Schauer, Del. Sem. Hort. Vratisl., 1837, p. 3; DC. Prodr. vii. 292. Erect, 1 ft. high or less; branches fastigiate: heads 1 in. long: rays very short and involute, light yellow, soon changing to scarlet: globose heads of mature achenes with expanded pappus very showy. May, June.

68. **LAGOPHYLLA**, *Nuttall*. Slender annuals, rigid and brittle, paniculately branching, with many small heads of pale salmon-colored or yellow vespertine flowers. Ray-achenes 5 only, obcompressed, en-

closed completely by their involucral bracts, their terminal areola not protuberant. Disk-achenes slender, abortive; no pappus to any, whether of ovary or disk. Receptacle bearing a circle of chaff between ray and disk, this and all the achenes and bracts deciduous at maturity.

* *Rays small and inconspicuous.*

1. **L. ramosissima,** Nutt. Trans. Am. Phil. Soc. vii. 391 (1841). Canescent with a loose silky pubescence, 1—4 ft. high, diffusely paniculate: lowest leaves spatulate-obovate, cauline lanceolate and linear, all entire: heads 1/4 in. high, 1/2 in. broad in expansion of rays: achenes 1 1/2 lines long.—Mountain districts north and south, and in both Coast Range and Sierra. June—Oct.

2. **L. congesta,** Greene, Bull. Torr. Club, x. 87 (1883). Tall as the preceding and robust, not paniculate, but the heads twice as large, densely glomerate on short branches; lowest leaves oblanceolate, remotely serrate: achenes 2 lines long.—Habitat of the last.

* * *Rays more showy, the expanded heads an inch broad.*

3. **L. dichotoma,** Benth. Pl. Hartw. 317 (1849). Stem 1-2 ft. high, dichotomously paniculate; branchlets puberulent: cauline leaves spatulate, the lower ones toothed, those of the branchlets short, hirsute-ciliate as are also the involucral bracts, sometimes bearing a few glands: achenes obovate, the ventral face with no keel or nerve.—Plains of the upper Sacramento; also in Lake Co. May—Sept.

4. **L. glandulosa,** Gray, Proc. Am. Acad. xvii. 219 (1882). Often 2 ft. high, virgately paniculate: leaves cinereous-puberulent, linear or spatulate-lanceolate, mostly entire, the upper ones beset with subsessile glands: bracts of the involucre and subtending small leaves little or not at all ciliate: achenes clavately obovate-oblong, with a keel or strong nerve down the ventral face.—Foothills of the Sierra from Butte Co. to Mariposa. June Sept.

5. **L. serrata,** Greene, Bull. Calif. Acad. i. 280 (1885). Stem 1—3 ft. high, diffusely paniculate above, the herbage puberulent, not at all glandular, except by a few minute stalked glands on the involucre: spatulate-oblong leaves 1 3 in. long, with remote but distinct serratures: heads showy, the rays 1/2 in. long. Middle foothills of Nevada and El Dorado counties. June—Sept.

69. **HOLOZONIA,** *Greene.* Perennial, spreading by creeping root-

stocks. Leaves opposite. Heads small, on slender pedicels, in an ample panicle. Flowers diurnal, white. Ray-achenes 6 -8, obcompressed, completely enclosed, smooth, surmounted by a short saucer shaped hyaline entire persistent pappus Disk-achenes with a pappus of 2 slender, deciduous paleae. Receptacle flat, with a circle of united chaff between disk and ray.

1. **H. filipes,** Greene, Bull. Torr. Club, ix. 122 (1882); H. & A. Bot. Beech. 356 (1841), under *Hemizonia*. Stems decumbent, 2 ft. high; slender branchlets and filiform peduncles glabrous or glandular: cauline leaves linear, minutely villous; those of the branches with some short-stipitate dark glands: involucre loosely villous: rays white or rose-tinted, deeply cleft into 3 linear lobes.—By streamlets in the hills east of Napa Valley, and in low fields along Napa River; also foothills of the Sierra in Calaveras Co. July - Oct.

70. PTILONELLA, *Nuttall.* Slender annuals with alternate narrow entire leaves and corymbose-panicled heads of white vespertine flowers. Heads with 3 - 6 rays, and nearly twice as many disk-flowers. Receptacle chaffy throughout, the paleae thin, subscarious, and like the involucral bracts partly embracing the achenes. Achenes turbinate, silky-villous, crowned with a pappus of pectinate-plumose narrow scales.

1. **P. scabra,** Nutt. Trans. Am. Phil. Soc. vii. 386 (1841); Hook. Fl. i. 316 (1833), under *Blepharipappus.* A few inches to a foot high, scabrous throughout, also somewhat hispid below, and glandular above; the narrowly linear leaves with revolute margins when mature or dry: heads many, terminating the slender and usually somewhat fastigiate branchlets; flowers vespertine. Var. **subcalva** (Gray), differs from the type in having the pappus very much reduced, or even obsolete. -Common along the eastern base of the Sierra, especially northward. June - Sept.

2. **P. laevis.** *Blepharipappus laevis,* Gray, Bot. Gaz. xiii. 73 (1888). Glabrous and mostly smooth up to the heads, these smaller than in the preceding, their fewer flowers diurnal: leaves all small and appressed, the uppermost reduced almost to mere scales. —Range of the preceding, but less common.

Suborder 7, HELENIOIDEÆ.

Herbs seldom viscid or balsamic. Receptacle naked. Bracts of involucre herbaceous mostly uniserial and equal, sometimes concave behind the ray-achenes, but never enfolding them. Style branches of perfect flowers with either truncate or appendiculate tips. Pappus mostly paleaceous or none.

Hints of the Genera.

* *Bracts of the involucre in 2 or 3 series and somewhat imbricated.*

Receptacle globose or hemispherical; achenes with paleaceous pappus - - 72
Receptacle flat; pappus 0;
    Large herb, with broad thin leaves - - - - - - - 71
    Low salt-marsh herb, with fleshy leaves - - - - - - 85
Receptacle flat; pappus present; achenes 2–4-angled, villous - - - - 82

* * *Bracts of the involucre in 1 or 2 series and nearly equal.*

Receptacle flat, or nearly so:
  Rays present, narrow;
    Leaves divided; pappus 0, - - - - - - 86
      " entire; pappus paleaceous;
        Achenes flat, villous-ciliate, - - - - - 76
        " terete, rugose, - - - - - 83
  Rays none, or not ligulate; pappus hyaline:
    Achenes slender, angled or striate;
      Outer flowers of the head larger, - - - - 80
      " " not enlarged, - - - 80, 81
    Achenes compressed, villous on the margin - - - - 77
Receptacle low-convex to conical or even subulate;
  Bracts of involucre joined into a cup;
    Opposite-leaved glabrous herbs, - - - - 73
    Leaves opposite or alternate, floccose-tomentose
      at least beneath, - - - - - 75, 79
  Bracts of involucre not united;
    Plants white-woolly; achenes 4-angled, - 78, 79
    Plants hirsute or glabrous, not viscid, - - 74
    " glabrous, viscid, fragrant, - - - - 84

71. VENEGASIA, *De Candolle.* Stout perennial, branching and leafy, with scattered large heads of yellow flowers. Involucre hemispherical, broad, the round-ovate bracts imbricated in several series, the outer somewhat foliaceous, the innermost narrow and scarious. Receptacle flat, naked. Rays many, long and narrow. Disk-corollas cylindrical. Style-tips of the disk-flowers very obtuse. Achenes oblong-linear, 5-angled, many-nerved, destitute of pappus.

1. **V. carpesioides,** DC. Prodr. vi. 43 (1837). Tall, widely branch-

ing, glabrous, leafy throughout: leaves thin, ovate-deltoid or ovate-cordate, acute, crenate, 3—4 in. long, petioled, resinous-dotted beneath.—Wooded cañons at Santa Barbara.

72. **HELENIASTRUM,** *Vaillant.* Perennial herbs, erect, with sessile mostly decurrent leaves, and long-peduncled heads; the herbage more or less resinous-dotted. Rays numerous, cuneate; disk-flowers very many. Involucre of 1 or 2 series of small herbaceous bracts. Receptacle globose, naked. Style branches with capitate-truncate tips. Achenes turbinate. Pappus in ours of awn-pointed paleæ.

1. **H. Hoopesii,** O. Ktze. Rev. Gen. i. 342 (1891); Gray, Proc. Philad. Acad. 1863, p. 65, under *Helenium*. Stout, 1—2 ft. high, leafy, bearing 1—6 large terminal heads on rather slender peduncles; young herbage somewhat woolly, in age glabrate: leaves large, oblong-lanceolate, or the lowest spatulate with a long tapering base: rays many, 1 in. long, cuneate-linear, 2—3-toothed, tardily reflexed: paleæ of the pappus lanceolate, tapering into an awn-like point.- Subalpine species, found long ago at Sonora Pass, by *Brewer* and *Bolander*, but not since observed in California.

2. **H. Bolanderi,** O. Ktze, l. c.; Gray, Proc. Am. Acad. vii. 358 (1868), under *Helenium.* Stout, 1—2 ft. high, simple, or with few pedunculiform branches, these long, naked, thick, enlarged at summit: leaves obovate to broadly oblanceolate, entire: heads large, the disk 1 in. in diameter; rays about 1 in. long, cuneate, 3-lobed, deflexed: paleæ of the pappus lanceolate or subulate, usually beset with 3 or 4 almost bristle-like teeth, and tapering into a slender awn.—Moist meadows near the sea, in Humboldt and Mendocino counties.

3. **H. Bigelovii,** O. Ktze, l. c.; Gray, Pac. R. Rep. iv. 107 (1857), under *Helenium.* Stout, 2—4 ft. high, erect, parted above into several stoutish very erect pedunculiform monocephalous branches: leaves lanceolate, 6 in. long or more, thickish and somewhat fleshy, almost gummy to the touch, and with some tomentose pubescence: rays showy, $\frac{3}{4}$ in. long: disk brownish-yellow: pappus-paleæ ovate-lanceolate, tapering into a long awn.—Common in brackish marshes at the upper end of San Francisco Bay, in Sonoma and Solano counties; plentiful near Suisun. This is my *Heleniastrum occidentale*, of the *Manual*, but is doubtless the type of Gray's *H. Bigelovii*. July—Oct.

4. **H. rivulare.** Commonly 2 ft. high, simple, or parted above the middle into several long pedunculiform monocephalous branches: leaves oblong-lanceolate or oblanceolate, thin and membranaceous, not in the

least gummy, quite glabrous: rays ⅜ in. long: paleæ of the pappus broadly ovate, abruptly ending in a long slender awn. — Very common along rivers and streamlets at middle elevations in both Coast Range and Sierra, especially northwards; forming by far the greater proportion of the "*Helenium Bigelovii*" of the herbaria, but very distinct from the type of that species, both in character, and as to distribution. June. —Sept.

5. **H. puberulum**, O. Ktze, l. c.; DC. Prodr. v. 667 (1836), under *Helenium*. Minutely cinereous pubescent, 2—4 ft. high, with slender widely spreading monocephalous branches: leaves lanceolate, entire, all but the radical strongly decurrent: involucre and reflexed rays very short and inconspicuous: globose disk of red-brown flowers ½ in. thick: paleæ of pappus ovate, short-awned.—Banks of streams, and in other moist places, on the plains, and along the seaboard. July—Dec.

73. **LASTHENIA**, *Cassini*. Mostly annuals and low, glabrous, slightly succulent. Leaves opposite. Heads middle-sized, on slender peduncles. Receptacle conical to subulate, muricate with projecting points on which the achenes are inserted. Involucres hemispherical, their uniserial bracts usually united and forming a toothed cup. Rays oval or oblong. Disk-corollas with slender tube and campanulate 5-toothed limb. Achenes linear, subclavate, or linear-cuneate, more or less flattened or angled, naked at summit, mostly destitute of pappus.

* *Achenes with a paleaceous pappus.*

1. **L. glaberrima**, DC. Prodr. v. 664 (1836). Stems weak, decumbent, 1 ft. long or less, very glabrous: leaves linear, entire: heads nodding in bud: involucre about 15-toothed: rays very short; all the corollas shorter than their broadly linear pubescent achenes: pappus of 5—10 firm chaffy scales, 2 or 3 of them subulate-pointed or short-awned, the others not so.—Subaquatic herb of shallow winter pools on low plains or in depressions among the hills; not very common.—May, June.

* * *Pappus wholly wanting.*

2. **L. glabrata**, DC. in Lindl. Bot. Reg. t. 1780 (1835). Stout, sparingly branching, 1—2 ft. high, peduncles few, elongated, erect: leaves, at least the upper pairs, ovate-lanceolate, coarsely but irregularly toothed, conspicuously connate at the dilated base and forming an open cup rather than sheath: heads very large, 1 in. wide: achenes dark, smooth. —Borders of salt marshes only; not common. June.

3. **L. Californica**, DC. in Lindl. l. c., also at t. 1823. More slender,

almost diffusely branching, the many and not elongated peduncles forming a corymbose top to the herb as a whole: leaves narrow, entire, divaricately spreading, not dilated or manifestly connate below: heads ⅓ in. wide.—Abundant on low plains, and even moist slopes of hills. Much more common than the last, and confounded with it for many years by Gray. May, June.

4. **L. chrysantha**, Greene, Man. 204 (1894), and Bull. Calif. Acad. i. 93 (1885), under *Crockeria*. Habit, foliage and flowers of the last, but plant smaller: achenes obovate-oval, much compressed; surrounded by a border of very short clavate closely packed hairs.—Plains of the San Joaquin in Tulare Co, *Greene*. April.

5. **L. conjugens**, Greene, Pittonia, i. 221 (1888). Only a few inches high; leaves narrowly linear, only the lowest entire, the others cleft into several pairs of long linear segments, these entire or toothed: involucral bracts united below the middle only; achenes very small (1 line long) olive-green and polished: pappus none.—Subsaline soil near Antioch; closely connecting this genus and the following. April, May.

74. **BAERIA**, *Fischer & Meyer*. Plants mostly annual, with opposite leaves, and middle-sized heads on slender peduncles: the herbage usually pubescent, but only sparingly so, never hoary or tomentose. Bracts of the campanulate involucre uniserial or nearly so, and distinct, usually carinate below. Rays few or many often short. Achenes clavate linear or linear-cuneiform. Pappus of few awns or paleæ, or both, or occasionally none.

* *Pubescence hirsutulous; leaves entire or merely toothed; receptacle conical; pappus aristiform or wanting.*—BAERIA proper.

+— *Root perennial; pappus none.*

1. **B. macrantha**, Gray, Proc. Am. Acad. xix. 21 (1883); Greene, Man. 205 (1894), under *Lasthenia*. Perennial, stout and nearly simple, decumbent at base, with peduncles 4—8 in. long: leaves somewhat 3-nerved and obtuse, linear, 4—8 in. long, hispid-ciliate toward the base: heads ½ in. high and as broad: involucre of about 12 hirsute-pubescent thickish herbaceous bracts: ligules ½—¾ in. long.—Moist lowlands of western Marin Co., and northward. June.

+—+— *Root annual; pappus present some.*

2. **B. chrysostoma**, F. & M. Ind. Sem. Petr. ii. 29 (1835); Greene, l. c.

under *Lasthenia*. Hirsutulous, 1 ft. high more or less, freely branching: leaves narrowly linear: heads 3-4 lines high: bracts of involucre, and the rays, 7-12, the latter 3-4 lines long: achenes glabrous; pappus wanting. -Rich fields and sunny slopes. April, May.

3. **B. hirsutula.** *Lasthenia hirsutula*, Greene, Man. 206 (1894). Stout and low, from a strictly annual root, mostly branching very freely: the whole herbage rather roughly short-hirsute: leaves broadly linear, often with saliently projecting scattered teeth, the lower conspicuously connate, sheathing the stem: involucral bracts obovoid, obtuse or acutish: rays oblong: achenes mostly very smooth, rounded at summit, manifestly compressed; pappus of 2 brownish very slender-subulate aristiform bristles — Plentiful on open rocky and grassy hills along the seacoast, from Marin Co. southward. May, June.

4. **B. gracilis,** Gray, Proc. Am. Acad. ix. 196 (1874); DC. Prodr. v. 664 (1836), under *Burrielia*. Whole habit and aspect of n. 2, but achenes linear-cuneate, with pappus of white lanceolate or ovate slender awned paleae, or the paleae sometimes almost obsolete.—Very plentiful and variable; often very small, slender and simple, not rarely as large as *B. chrysostoma*. April-June.

* * *Pubescence if any soft-hirsute or somewhat woolly; leaves often cleft or divided.*

+- *Pappus uniform, paleaceous.*—Genus BURRIELIA, De Candolle.

5. **B. microglossa.** *Burrielia microglossa*, DC. Prodr. v. 664 (1836); Greene, l. c. 205, under *Lasthenia*. Slender, only a few inches high, with a few narrow and almost cylindric involucres of apparently rayless heads, but the rays present, though very short and inconspicuous: receptacle subulate: achenes fusiform-linear; pappus of 2—4 attenuate-subulate paleae. -Hills and valleys of the Coast Range from Alameda and Santa Clara counties southward. April.

6. **B. leptalea,** Gray, Syn. Fl. 325 (1884), and Proc Am. Acad. vi. 546 (1865), under *Burrielia*. Glabrous, the filiform stems only a few inches high: leaves almost filiform, $\frac{1}{2}$ in. long: involucre 2 lines high, more campanulate, the rays obvious and numerous: pappus of 2 or 3 small paleae tapering to long awns. Interior of Monterey Co., on the Salinas and Nacimiento rivers. April.

7. **B. debilis,** Greene, in Gray, l. c. Pubescence minute, somewhat woolly; slender and weak stems 6—10 in. high, sparingly branched: leaves flaccid, linear, entire, 1 in. long or more: involucre campanulate.

2–3 times high: rays several, short and broad, little more than a line long, nearly as broad: pappus of 3 or 4 ovate-lanceolate awned paleæ, or sometimes wanting.—Foothills of the Sierra in Fresno and Kern counties, *Eisen, Greene*. April.

8. **B. platycarpha**, Gray, Proc. Am. Acad. ix. 196 (1874); and Bot. Mex. Bound. 97 (1859), under *Burrielia*; Greene, l. c. 205 (1894), under *Lasthenia*. Purplish-stemmed and very wiry, 5–8 in. high, with erect or ascending branches: leaves linear, or pinnatifid into filiform segments: involucral bracts 6 or 7, 3-nerved, the middle nerve carinately prominent: pappus-paleæ bright-white, ovate, slender-awned, the awn as long as the achene.—Subsaline plains of Solano and Contra Costa counties. April, May.

9. **B. carnosa**, Greene, Bull. Torr. Club, x. 86 (1883), and l. c. under *Lasthenia*. Leaves all filiform, entire: bracts of involucre with a single strongly carinate nerve: pappus of 4 or 5 subulate-awned ovate paleæ.— Border of salt marsh north of Vallejo: rare or local. April.

+ + *Pappus of several blunt paleæ and one or more slender awns, or sometimes wholly wanting.*—Genus DICHÆTA, Nuttall.

10. **B. tenella**. *Dichæta tenella*, Nutt. Trans. Am. Phil. Soc, vii. 383 (1841); Greene, l. c. 205, under *Lasthenia*. Erect, sparingly branching, 4–6 in. high, somewhat canescent with deciduous woolly hairs: leaves linear, entire, or some of the lower laciniate: rays oval or oblong, short: paleæ and awns each usually 2, but pappus not rarely wholly wanting.- Low plains of Contra Costa Co. April, May.

11. **B. uliginosa**, Gray, Proc. Am. Acad. ix. 197 (1874); Nutt l. c., under *Dichæta*; Greene, l. c., under *Lasthenia*. Stouter than the last, freely branching, often decumbent, somewhat flocculent: leaves linear-ligulate, the lower, if any, entire, the upper laciniate-pinnatifid into linear entire or cleft segments: involucral bracts and long-exserted rays 10–13: pappus of about 4 slender awns, and as many or twice as many broad truncate laciniate-fimbriate paleæ.—Common in low grounds. April, June.

12. **B. maritima**, Gray, Proc. Am. Acad. ix. 196 (1874), and l. c. vii. 358 (1868), under *Burrielia*. Diffuse, fleshy, the growing parts with a little woolliness: leaves oblong-linear, 1 in. long, entire, or sparingly laciniate-toothed: involucral bracts short and obtuse; rays 6 or 8 also, short and broad: pappus of 3–5 subulate awns, and as many narrow laciniate paleæ.—Farralloues Islands, this long the only known locality; but now recently detected on an islet off Vancouver Island.

13. **B. Fremonti,** Gray, l. c.; Benth, Pl. Hartw. 317 (1849), under *Burrielia;* Greene, l. c. under *Lasthenia.* Erect, slender, 8—10 in. high, only hirsute-pubescent: leaves mostly palmately parted into linear lobes: involucre broad; bracts 10 or 12; rays as many, the oval ligules not longer than the width of the disk: pappus of 4 slender awns and as many or more numerous small paleæ, or rarely none.—Moist plains from Napa and Solano counties southward. May.

14. **B. Burkei,** Greene, Bull. Calif. Acad. ii. 151 (1886); Man. 204 (1894), under *Lasthenia.* Much like the last, but taller, 1—2 ft. high: pappus of 8—10 minute entire acute paleæ and a single very long and slender awn.—Southern Mendocino Co.; to be expected in Sonoma.

75. **MONOLOPIA,** *De Candolle.* Annuals white with floccose wool. Leaves alternate, not linear, mostly toothed or entire. Ray-corollas with ample ligule, usually bearing at base, and opposite the ligule, a rounded denticulate appendage. Achenes black, without pappus. Genus otherwise like *Lasthenia,* and too near it; but also as easily referable to *Eriophyllum.*

1. **M. major,** DC. Prodr. vi. 74 (1837). Stout, nearly simple, or with several pedunculiform naked monocephalous branches, 2 ft. high, expanded heads more than 1 in. wide: bracts of involucre joined into a broad-campanulate toothed cup: achenes 2 lines long.- In rich fields, or on hillsides. May.

2. **M. gracilens,** Gray, Proc. Am. Acad. xix. 20 (1883). Slender, diffusely paniculate, bearing scattered short-peduncled heads less than 1 in. wide: bracts of involucre distinct to the base: achenes 1 line long.-- Plentiful on Mt. Diablo near the summit, thence southward to the Santa Cruz seaboard. May--July.

3. **M. minor,** DC. l. c. Loosely lanate and only a few inches high: cauline leaves cleft into 3—5 linear lobes: heads $\frac{1}{4}$ in. high: bracts of the involucre about 10, oblong, separate to below the middle.--An obscure and long lost middle Californian plant, collected only by Douglas more than sixty years since.

76. **EATONELLA,** *Gray,* Small woolly annual with rosulate undivided leaves, and small subsessile heads. Involucre of 8 oblong equal bracts. Receptacle plane, naked. Rays small. Disk-corollas short, 5-toothed. Achenes compressed, the callous margins densely villous-ciliate. Pappus of 2 broad nerveless paleæ.

1. **E. nivea,** Gray, Proc. Am. Acad. xix. 19 (1883); D. C. Eaton in

Bot. King. Exp. 174, t. 18 (1871), under *Burrielia;* Gray, Bot. Cal. i. 379 (1876), under *Actinolepis.* Dwarf, nearly stemless, loosely very woolly; the small heads nearly sessile in the midst of a rosulate tuft of spatulate entire leaves: bracts of the involucre 8, narrowly oblong; rays short, scarcely exserted; achenes linear-oblong, black and shining, the margins ciliate with snow-white soft hairs; pappus paleae 2, opaque, broad, subtruncate, erose-dentate, produced in the middle into a subulate awn.—Eastern slope of the middle Sierra Nevada.

77. **LEMBERTIA.** Leafy-stemmed and flocculent small annual, with habit of a small *Eriophyllum,* and similar involucre. Rays none. Disk-corollas 4-toothed. Achenes of oval outline, the outer ones triquetrous-compressed the others flatly compressed, the angles villous-ciliate. Pappus-paleae few, hyaline, erose-laciniate, scarcely longer than the hairs of the margins of the achene.

L. **Congdoni.** *Eatonella Congdoni,* Gray, Proc. Am. Acad. xix. 20 (1883). Stem simple, or sparingly branched, erect, 6 in. high, sparsely leafy: leaves oblong-linear, sinuate-toothed or repand: bracts of the involucre 5, broadly oblong: achenes pubescent on the flattened surfaces as well as villous-ciliate marginally.—Deer Creek, Tulare Co., *Congdon,* also in the valley of the San Joaquin at some unrecorded station, *Parry.*

78. **ACTINOLEPIS,** *De Candolle.* Small floccose-woolly freely branching annuals, with small heads of yellow flowers, the rays few, broad and usually short, sometimes white or reddish rather than yellow. Bracts of the obovate or oblong narrow heads few, thinnish, sometimes concave and partly embracing the achenes. Receptacle convex or subconical, or nearly flat. Achenes oblong or subclavate and 4-angled, crowned with a scarious or more opaque paleaceous pappus of several scales.—Plants sufficiently distinct from *Eriophyllum* in habit, texture of involucres, and character of pappus.

1. **A. multicaulis,** DC. Prodr. v. 655 (1836); Gray, Proc. Am. Acad. xix. 24 (1883), under *Eriophyllum.* Stems depressed, slender, 3–6 in. long, much branched, very fragile at the joints when mature, the wool then deciduous: leaves cuneate or spatulate, obtusely 3-toothed or 3-lobed at apex: heads many, sessile, leafy-bracted: scales of the pappus 10–15, slender, acuminate, unequal, in the disk-flowers sometimes wanting.—San Luis Obispo Co., and plains of Kern Co. and southward. April, May.

2. **A. Pringlei.** *Eriophyllum Pringlei,* Gray, l. c. 25. Only 1–2 in. high, or, the prostrate branches forming a mat several inches broad:

herbage copiously and permanently white-woolly: rays none: achenes villous; pappus of about 10 oblong-lanceolate pointless rather large paleæ. — Plains and valleys east of the Sierra Nevada, from Inyo Co. southward. May, June.

3. **A. nubigena.** *E. nubigenum*, Greene, in Gray, l. c. Stems erect, with loose and open mode of branching, the herbage densely white-woolly: leaves lanceolate-spatulate: heads narrow, short-peduncled: involucre of 5 oblong bracts: rays oval, little exceeding the disk-flowers; receptacle with small conical elevation in the centre: paleæ of the pappus about 10, oblong, obtuse, erose, nerveless, one-third the length of the achene.— On Cloud's Rest, above the Yosemite Valley, *Mrs. Curran.*

4. **A. Wallacei,** Gray, Proc. Am. Acad. ix. 198 (1874), and Pac. R. Rep. iv. 105 (1857), under *Bahia*. Diffusely branched and densely white-woolly: leaves alternate, obovate or spatulate, entire: heads short-peduncled: bracts of the involucre about 8, in maturity somewhat carinate-concave, with scarious margins embracing the ray-achenes: rays short and broad, yellow: achenes glabrous; paleæ of the pappus 10, very short, obtuse, nerveless.— Inyo Co., Shockley, and southward. The rays are said to be sometimes reddish instead of yellow.

5. **A. lanosa,** Gray, l. l. c. c., first as *Burrielia*. More slender than the last, the leaves almost linear, entire: achenes very slender; pappus of 10 dissimilar paleæ, 4 or 5 being aristate-pointed, the rest not so: rays either yellow or pinkish.— Range of the last.

79. **ERIOPHYLLUM,** *Lagasca*. Ours mostly perennial or suffruticose, floccose plants, usually with divided leaves. Involucres oblong or campanulate, the bracts of firm texture, permanently erect. Rays few, short and broad. Disk-corolla with distinct slender proper tube. Style-tips truncate, obtuse, or obscurely conical. Achenes clavate-linear to cuneate-oblong, mostly 4-angled. Pappus of nerveless and mostly pointless hyaline paleæ, or, in some small annual species, wanting.

* *Suffruticose; heads smallish, terminally clustered.*

1. **E. stæchadifolium,** Lag. Nov. Gen. et Sp. 28 (1816). Stem and lower face of leaves white with a close pannose tomentum; shrub much branched, 2—5 ft. high, very leafy throughout: leaves subcoriaceous, oblanceolate and entire, or pinnately parted into several narrow divisions: heads compactly corymbose-cymose; involucres oblong, angular, $\frac{1}{4}$ in. high or more, of linear bracts: receptacle convex: rays 6—8:

pappus-paleæ 8—12, the 4 over the angles of the achene somewhat longer. -Sandy hills and slopes near the sea; the typical state, with leaves crowded and almost entire, at Monterey only. May- Dec.

2. **E. confertiflorum**, Gray, Proc Am. Acad. xix. 25 (1883); DC. Prodr. v. 657 (1836), under *Bahia*. Smaller, 1-2 ft. high; leaves on the flowering branches reduced and scattered, membranaceous, hoary-tomentose on both faces, ternately 3-7 parted into linear divisions; heads 2 lines high, short-peduncled or sessile in a dense terminal cluster: involucre obovoid-oblong, of broadly oval bracts: rays 4 or 5: paleæ of the pappus 8—14. Var. **discoideum**, Greene. More condensed and leafy; heads broader, with more numerous flowers, but no rays. Very common on all hills; the variety in Sonoma Co. June—Dec.

3. **E. Jepsonii**, Greene, Pittonia, ii. 165 (1891). Suffruticose, 2 ft. high; stem white with pannose tomentum; leaves hoary on both faces, pinnately divided into 5-7 narrowly linear revolute segments: inflorescence loosely cymose-corymbose, the peduncled heads 3—4 lines high, and, with 6—8 oblong rays expanded, 1 in. broad: bracts of involucre 6-8, coriaceous, ovate: achenes with a few short hispidulous hairs, and 2 unequal sets of pappus-paleæ, those of the inner circle exceeding the others. Mountains of Alameda Co., south of Livermore, *Jepson*. May.

* * *Species more herbaceous, but all perennial.*
+— *Heads large, solitary or scattered.*

4. **E. speciosum**, Greene, Eryth. i. 149 (1893). Suffrutescent and leafy at base, erect, parted into many long and nearly naked mono-cephalous branches; these, and the lower face of the leaves, hoary with some arachnoid or more dense and floccose tomentum: leaves glabrate above, 2—3 in. long, lanceolate, acute, entire, or some with a few coarse teeth or small lobes: peduncles a foot long: involucre short-campanulate, of 12-15 distinct oblong-ovate acute bracts, their tips recurved: rays 12—15, more than $1\frac{1}{2}$ in. long: achenes sharply angled, appressed-pubescent; pappus of about 10 paleæ, alternately linear-oblong, and narrower and much longer, all toothed at summit.—Foothills of the Sierra from near Chico to Amador Co., *Sonne, Hansen, Mrs. Austin*.

5. **E. grandiflorum**. *E. cæspitosum*, var. *grandiflorum*, Gray. Much stouter than the last, 2-3 ft. high, more densely woolly, the stout elongated peduncles thickened under the very large heads: stem more leafy, the lowest leaves spatulate-oblanceolate, dentate, the others with short pinnate segments or coarse serratures: involucre about $\frac{3}{4}$ in. high, of 14—18 somewhat elliptic and overlapping bracts: rays as many, and

broad, the head with expanded rays 2 in. wide: achenes as in the last.—
Plains and low hills, along the western base of the Sierra, bordering the
lower Sacramento and San Joaquin. April—June.

6. **E. obovatum,** Greene, Eryth. iii. 123 (1895). Stems several,
decumbent, about a foot high, rather equably leafy, but ending in
a short naked monocephalous peduncle; herbage densely white-tomentose: leaves all entire, broadly obovate to obovate-spatulate, 1- 2 in.
long, only the lowest opposite: involucres broadly campanulate or
nearly hemispherical; bracts 10—12, oblong-lanceolate; rays lightyellow: tube of disk-corollas hispidulous: achenes glabrous; pappus
conspicuous, of about 8–10 very unequal paleæ, some very short and
obscure, others long, all lacerate-toothed.- From Kern Co. southward, at
middle elevations of the mountains.

7. **E. croceum,** Greene, l. c. 124. Slender, the several decumbent
stems 8 to 18 inches high: foliage silky-lanate beneath, glabrate above:
leaves of narrowly cuneate-obovate outline, of thin texture and coarsely
toothed or lobed above the middle: monocephalous peduncles several,
terminating the branches: heads hemispherical: bracts of involucre
thinnish, 10 or 12; saffron-colored showy rays about as many: corollatube short, densely hispid: achenes sharply 4-angled, the angles toward
the base white-callous: pappus of the ray none, of the disk of 4 very
short and blunt incurved callous points rather than paleæ.—Amador and
Calaveras County hills, *Hansen, Blasdale.*

8. **E. integrifolium,** *Trichophyllum integrifolium,* Hook. Fl. i. 316
(1833); *T. multiflorum,* Nutt. Journ. Philad. Acad. vii. 35 (1834). Dwarf
and matted, the short depressed very leafy branches only a few inches
long, the monocephalous naked and scape-like peduncles 3–6 in. high:
leaves from narrowly spatulate or oblanceolate and entire to more
dilated and 3-lobed at apex, the whole plant floccosely hoary: campanulate involucre of about 6 or 8 oblong-lanceolate carinate-nerved bracts:
rays as many, yellow: achenes rather slender, glabrous or with some
short bristly hairs about the summit: pappus of rather many narrow
erect paleæ, these mostly lacerate, some of them aristate-pointed.—
Summit of the Sierra, and down the eastern slope, thence far northeastward. July, Aug.

9. **E. lanatum,** Forbes, Hort. Woburn. 183 (1838); Pursh, Fl. ii. 560
(1814), under *Actinella;* Nutt. Gen. ii. 168 (1818), under *Trichophyllum;*
Spreng. Syst. iii. 574 (1826), under *Helenium;* DC. Prodr. v. 657 (1836),
under *Bahia.* Somewhat cæspitose, but the erect branches 8–18 in.
high, and clothed with many pairs of opposite or alternate pinnatifid

leaves, the uppermost linear, entire: peduncles often much elongated: involucres broadly campanulate, of 12 15 bracts: achenes glabrous or sparingly glandular: pappus very short and inconspicuous, the larger paleæ obtuse, the alternate ones minute. Both sides of the Sierra Nevada, but northward only, in Butte, Plumas and Sierra counties. Possibly distinct from the original plant of Pursh, Nuttall and others, but certainly far removed from the *E. cæspitosum* of Douglas, nothing like which is found in any part of California.

+ + *Heads middle-sized, usually somewhat corymbosely clustered at summit of leafy stems.*

10. **E. arachnoideum**, Greene, Man. 207 (1894); F. & L. Ind. Sem. Petr. ix. 63 (1842), under *Bahia*. Loosely branching from a decumbent base, 1—2 ft. high, clothed with long floccose wool: leaves broad, from rhombic or cuneate in outline to oblong-lanceolate, thinnish, 3—5-lobed or -incised, the lobes or coarse teeth triangular or oblong: involucre hemispherical, 3—4 lines high: rays 10 13, large; disk-corollas with very glandular-hirsute tube: achenes short, thickish: pappus-paleæ short.—In the redwood districts of the coast. June Oct.

11. **E. achillæoides**, Greene, l. c.; DC. Prodr. v. 657 (1836), under *Bahia*. Leaves mostly basal, opposite, pinnately parted into 3 5 divisions, these incised or pinnatifid: heads somewhat corymbosely collected and short-peduncled; involucres hemispherical, the bracts and rays 9—13; pappus as in the last.—Dry hills of the inner Coast Range. June- Sept.

12. **E. tanacetiflorum**, Greene, Pittonia, ii. 21 (1889). Erect, slender, 2 ft. high, leafy throughout: leaves opposite, thinnish, sparingly floccose beneath, from palmately trifid to somewhat pinnatifid: heads 3 7, nearly or quite sessile at the summit of the simple stem; involucre oval, 2 lines high, the bracts broad and nearly equal, acutish; rays none; achenes with a few closely appressed hairs and no resin-glands; pappus of about 8 unequal linear obtuse or retuse paleæ.—Wooded hills of Calaveras County, between Sheep Ranch and Murphy's, *Greene*. June, July.

* * * *Smaller species, all annual; pappus obsolete in some.*

13. **E. Heermanni**. *Monolopia Heermanni*, Durand. Pl. Pratt. 93 (1855). Rather diffusely branched from the base, somewhat decumbent, 6 10 in. high, deciduously flocculent: leaves small, alternate, mostly pinnately parted into 5 short linear segments, the uppermost entire: heads small, on slender peduncles terminating the branchlets: involucre

small, broadly campanulate, of about 8 bracts: rays yellow, conspicuous: achenes 4-angled, pubescent: pappus wanting, or represented by 1 or more minute paleæ.—Foothills east of the Sacramento and San Joaquin, near Brighton, *Mrs. Curran*, and in Amador Co., *Mr. Hansen*. Wholly an *Eriophyllum*, not only as to habit, but as to character of the involucre and achenes.

14. **E. bahiæfolium.** *Monolopia bahiæfolia*, Benth. Pl. Hartw. 317 (1849). Smaller, simple and monocephalous, only 2–3 in. high: leaves ½ in. high, spatulate to linear, entire, or at apex 3-lobed: involucral bracts united below: immature achenes sparsely pubescent.—An obscure and long lost plant, found in the valley of the Sacramento by *Hartweg*.

15. **E. ambiguum,** Gray, Proc. Am. Acad. xix. 26 (1883); l. c. vi. 547 (1865); Bot. Calif. i. 382 (1876), under *Bahia*. Loosely and somewhat deciduously floccose, 3–10 in. high, loosely branching above: leaves alternate, entire, or 3-toothed: involucre campanulate, of 6–9 oblong-lanceolate not very firm bracts: rays 5–9, oblong: achenes pubescent, with or without a pappus of small paleæ.—Near Ft. Tejon and southward. Plant with more the habit of *Monolopia* than any of the foregoing.

80. **CHÆNACTIS.** *De Candolle*. Compound-leaved herbs, often more or less woolly, with discoid heads mostly solitary and pedunculate. Involucre campanulate, the linear bracts equal, uniserial, herbaceous. Receptacle flat, naked. Corollas with short tube, long narrow throat, and short teeth; but those of the outer circle in some more ample, approaching the nature of rays. Achenes slender, smooth. Pappus of hyaline nerveless paleæ.

* *Corollas yellow, the marginal ones enlarged; all the species annual.*— CHÆNACTIS proper.

1. **C. lanosa,** DC. Prodr. v. 659 (1836). Stems short, branching, bearing many long naked peduncles, the earliest scapiform: herbage floccose-woolly when young: leaves thickish, simply pinnately parted into few and narrowly linear lobes, or the uppermost entire: heads ½ in. high: involucral bracts nearly linear: pappus of 4 equal long paleæ.— Plains and hills, from Monterey and the lower San Joaquin southward. April–June.

2. **C. glabriuscula,** DC. l. c. Taller, more caulescent, branching above, the herbage thinly floccose, at length glabrate; peduncles long, stout: heads ¾ in. high: bracts of involucre thickish, glabrate, obtuse:

marginal corollas ample, much exceeding the others; pappus of 4 equal narrowly oblong acutish paleæ.—Plains of the lower San Joaquin. April—June.

3. **C. gracilenta.** Erect, slender, 5–10 in. high, simple, the flexuous stem leafy below, parted at summit into several nearly naked slender peduncles; herbage puberulent, perhaps flocculent when young: leaves pinnately parted into numerous short and simple lobes: involucre about 4 lines high: marginal corollas moderately ampliate: pappus of the minutely and sparsely hispidulous slender black achenes short and very unequal, the shortest squamellate and as broad as long, the longest obovate-oblong, not a fourth the length of the achene, all obtuse and erose-denticulate.—Dry ridges above Napa Valley, *Jepson*. May.

4. **C. tanacetifolia,** Gray, Proc. Am. Acad. vi. 545 (1865). Low, branched from the base, the leaves mainly in a rosulate tuft and bipinnate, the lobes short; young herbage woolly: flowering branches long, nearly naked and pedunculiform: heads 4–5 lines high: achenes very slender, crowned with a pappus of 4 oblong obtuse paleæ and as many or fewer very short and rounded ones.—Plains and hills, from Lake Co. and the upper Sacramento southward to Kern Co. *Bolander, Palmer.*

5. **C. heterocarpha,** Gray, Pl. Fendl. 98 (1849). Hoary-tomentose, in age glabrate; stem 3–8 in. high, often simple and monocephalous, more commonly with several peduncled and subcorymbose heads: leaves pinnately parted into linear segments: heads 1½ in. high: pappus of 4 lanceolate or oblanceolate paleæ equaling the corolla, and 2 or more extremely short ones.—Plains and hills from Butte and Humboldt counties southward. April—June.

* * *Corollas whitish or flesh-color, the outer not enlarged.*—Genus MACROCARPHUS, Nutt.

6. **C. Xantiana,** Gray, Proc. Am. Acad. vi. 545 (1865). Annual, erect, 4–10 in. high, stoutish, nearly simple, or with few branches, each terminating in a long stout peduncle and large head: leaves pinnately parted into 3–7 linear and distant lobes: involucre ¾ in. high, its bracts obtuse, pointless: pappus of 4 lanceolate paleæ equaling the corolla, and 4 very much shorter outer ones.—Mountains of Kern Co. and southward. May–July.

7. **C. Douglasii,** H. & A. Bot. Beech. 354 (1841); Hook. Fl. i. 316 (1833), under *Hymenopappus*. Canescent with a fine tomentum, 1–2 ft. high, from a biennial or perennial root, the stems usually several and

often slightly decumbent: leaves of broad outline, the lobes many and crowded: heads subcorymbose at summit of stem, the largest ⅜ in. high: pappus of 8–14 large and nearly equal mostly narrowly oblong paleæ as long as the corolla.—Frequent at middle or higher elevations of the Sierra Nevada. June—Sept.

8. **C. Nevadensis,** Gray, Bot. Calif. i. 391 (1876); Kell. Proc. Calif. Acad. v. 46 (1873), under *Hymenopappus*. Perennial, low and tufted, the leafy and monocephalous branches only 2–4 in. high: white-woolly leaves of ovate outline, bipinnately parted into many and short lobes: heads ½ in. high, on scapiform peduncles little surpassing the crowded leaves: paleæ of the pappus about 12, equaling the corolla.- High summits of the Sierra, in dry volcanic soil, from above Donner Lake to Lassen's Peak, *Kellogg, Mrs. Austin.* July—Oct.

9. **C. santolinoides,** Greene, Bull. Torr. Club. ix. 17 (1882). Perennial, with crowded radical leaves, and very long scapiform usually monocephalous peduncles: leaves of linear or lanceolate outline, the broad rachis beset with crowded short few-lobed and crisped divisions: peduncles 4—8 in. high, simple or forked: heads ½ in. high: pappus of 8—10 linear-ligulate paleæ. Mountains of Kern Co. and southward, in dry pine woods. June—Aug.

10. **C. suffrutescens,** Gray, Proc. Am. Acad. xvi. 100 (1880). Stems numerous, erect and simple, from a branched and depressed woody base, leafy in the middle, ending in a naked monocephalous peduncle; herbage canescently tomentose : leaves pinnately parted into 5–7 narrowly linear toothed or entire segments: heads ⅜ in. high: pappus of 10 or more linear or narrowly oblong paleæ a little shorter than the corolla, or those of the outer flowers notably shorter.- Rocky banks and hills along the upper Sacramento, towards Mt. Shasta, *Lemmon, Pringle.*

81. **OREOCHÆNACTIS,** *Coville.* Low slender branching annual, with narrow entire leaves, and narrow glomerate rayless heads of whitish flowers. Bracts of the involucre 4, about equal; flowers about as many. Style branches linear, obtuse, hairy. Achenes obovate-clavate, striate. Pappus of several obtuse hyaline erose-fimbriate paleæ which cohere at the base and are deciduous.

1. **O. thysanocarpha,** Coville, Death Valley Exp. 134 (1893); Gray, Proc. Am. Acad. xix. 30 (1884), under *Chænactis*. Viscid-puberulent, with some early but deciduous villous hairiness: leaves narrowly linear: involucres ¼ in. high: paleæ of the pappus spatulate, erose-fimbriate

down to the somewhat unguiculate base.—Kern Co., at higher than middle elevations of the Sierra, in pine woods or openings.

82. **HULSEA**, *Torrey & Gray*. Herbs leafy, viscid and somewhat balsamic-scented, or floccose-woolly. Leaves alternate, sessile or nearly so, entire, toothed or pinnatifid. Heads large, solitary or scattered. Rays yellow or purplish. Bracts of the involucre thin-herbaceous, linear to oblong, plane, in 2 or three series. Receptacle flat. Disk-corollas with long narrow throat and 5 short lobes. Achenes linear-clavate, or cuneate-oblong, villous. Pappus of 4 or 5 hyaline paleæ either erose or lacerate at summit, or dissected into capillary bristles.

* *Stems low, leafy at base, the merely bracted peduncles scape-like.*

1. **H. vestita**, Gray, Proc. Am. Acad. vi. 547 (1865). Rosette of spatulate entire or dentate radical leaves pannosely white-tomentose: flowering stem 1 ft. high, sparsely leafy below the middle, above scapiform and monocephalous, or bearing 2 or 3 long-peduncled heads: involucre $\frac{1}{2}$ in. high, of broadly lanceolate viscid-pubescent bracts: rays little surpassing the disk-flowers, or shorter, or even wanting, yellow changing to reddish: pappus of quadrate erose paleæ, either nearly equal, or 2 longer than the others.—Volcanic hills of Mono Co., and southward.

2. **H. algida**, Gray, l. c. Smaller, and from a deep perennial rootstock; clothed with a villous or cottony wool which is caducous, and a permanent viscid pubescence: leaves linear-ligulate, irregularly toothed, the teeth sometimes large, the crowded lower ones 2–5 in. long: broad involucre 1 in. high; its bracts linear, attenuate-acute, loose, villous-lanate and viscid: rays very many, narrow, $\frac{1}{2}$ in. long, yellow: pappus short, not longer than the hairs of the achene, the paleæ deeply fimbriate-lacerate.—Alpine summits of the Sierra, from Mt. Dana southward.

3. **H. nana**, Gray, Pac. R. Rep. vi. 76, t. 13 (1857). Only a few inches high, from long branching rootstocks, rising through volcanic ashes, viscid-pubescent and villous-lanate: crowded leaves oblong-spatulate, incised or pinnatifid, tapering to a marginal petiole: involucre $\frac{3}{4}$ in. high, of lanceolate bracts: rays 30, yellow: paleæ of the pappus usually longer than the breadth of the achene, broad, or splitting into narrow ones, incisely or fimbriately lacerate.— Summits of Lassen and Shasta Peaks; in a form more woolly than the typical plant of the Cascades in Oregon.

* * *Taller, leafy nearly to the summit; no floccose wool; heads solitary or subracemose.*

4. **H. heterochroma,** Gray, Proc. Am. Acad. vii. 359 (1867). Stout, 2 ft. high or more, annual: leaves oblong, saliently dentate: involucre ³⁄₄ in. high, of linear-lanceolate attenuate-acute bracts: rays many, 3—4 lines long, rose-purple, but occasionally reduced or obsolete: paleæ of the pappus oblong, the two over the angles of the achene longer than the others, the shorter ones truncate-lacerate.—Yosemite Valley and southward.

5. **H. brevifolia,** Gray, l. c. Slender, 1 ft. high; character of root unknown: stem or simple branches clothed with rather small spatulate-oblong denticulate leaves, the largest only 1½ in. long: involucre ½ in. high, of loose linear bracts: rays 10—12, only 3—4 lines long: paleæ of the pappus nearly entire.—Foothills along the Merced River, and near the Yosemite.

83. **RIGIOPAPPUS,** *Gray.* Heads with short narrow ligulate rays; all the flowers yellow. Involucre turbinate-campanulate, its bracts nearly linear, equal, rather rigid, involute in age. Disk-corolla small, with short tube, long narrow throat, and 3—5 short erect teeth. Style branches with slender-subulate hispidulous appendage and short linear stigmatic part. Pappus alike in disk and ray, of 3—5 opaque paleaceous awns; the linear pubescent achene transversely rugose.

1. **R. leptocladus,** Gray, Proc. Am. Acad. vi. 548 (1865). Slender erect annual, 6—12 in. high, simple below, above with few slender corymbose branches: leaves alternate, narrowly linear, sessile, erect, entire, hirsutulous or glabrate, those of the filiform branchlets subulate: paleæ of pappus ½—⅔ as long as the long slender rugulose achene.— Open glades among sparsely wooded hills. May.

84. **AMBLYOPAPPUS,** *Hooker & Arnot.* Rigidly erect panicled small seaside annual, with narrow leaves, gummy and sweet-scented herbage, and numerous small rayless heads. Bracts of involucre 5 or 6, broadly obovate, their middle part in age somewhat carinate-concave. Receptacle small, conical. Corollas all short-tubular; those of the pistillate flowers minutely 2—3-toothed, of the perfect, 5-toothed, the teeth soon connivent. Achenes obpyramidal, pubescent. Pappus of 8—12 oblong obtuse paleæ about equaling the corollas.

1. **A. pusillus,** H. & A. in Hook. Journ. Bot. iii. 321 (1841) *Aromia tenuifolia,* Nutt. Trans. Am. Phil. Soc. vii. 396 (1841). Plant ½—1 foot

high, somewhat corymbosely panicled and with a profusion of small heads: the very lowest leaves pinnately 3–5 parted and opposite, the segments and the undivided leaves narrowly linear, entire; involucre 2 lines high: flowers yellowish.—Along the beaches from perhaps Santa Barbara southward.

85. **JAUMEA,** *Persoon.* Procumbent very succulent perennial herb with opposite subterete leaves and solitary terminal short-peduncled heads. Involucre campanulate, the outer bracts shorter. Corollas glabrous. Style-branches papillose or hairy. Achenes 10-nerved. Pappus none.

1. **J. carnosa,** Gray, Wilkes Exp. xvii. 360 (1874); Less. in Linnæa, vi. 521 (1831), under *Coinogyne.* Common denizen of sandy salt marshes, associated with *Salicornia;* heads middle-sized, the rays few and rather short.—Flowering throughout the summer and autumn.

86. **BLENNOSPERMA,** *Lessing.* Low annual, with pinnately parted leaves, and pedunculiform branches bearing solitary radiate yellow heads. Involucral bracts uniserial, equal, oblong, herbaceous but purplish or yellowish. Receptacle flattish, naked. Rays 5–12, linear. Disk-flowers about 20; their narrow tube abruptly expanded into a campanulate limb; their style undivided, with capitate apex; their ovaries abortive. Achenes (of the ray) pyriform, obscurely 8–10-ribbed, with small areola and no pappus; the surface bearing minute papillæ which develop mucilage when wet.

1. **B. Californicum,** T. & G. Fl. ii. 272 (1842); DC. Prodr. v. 531 (1836), under *Coniothele.* A span high, diffusely branching, flaccid, glabrous: leaves alternate, pinnately parted into narrowly linear entire lobes: expanded heads ½ in. broad: ligules pale yellow within, brownish without: disk-flowers shorter than the involucre: style-branches of fertile flowers broad.—Early vernal plant of the valleys of the Mt. Diablo Range, and plains beyond. March, April.

Suborder 8. **ANTHEMIDEÆ.**

Mostly aromatic-scented plants, with a very bitter juice. Leaves often much dissected. Bracts of involucre imbricated, more or less scarious. Receptacle either naked or chaffy. Anthers not caudate. Style-branches of perfect flowers truncate, sometimes penicillate. Achenes small and short, with no pappus, or a mere paleaceous crown.

Hints of the Genera.

Heads scattered or solitary, terminating leafy branches or peduncles;
  Leaves finely dissected;
    Rays 0; heads ovoid, - - - - - - - - - - 88
    Rays present; heads hemispherical, - - - - - - - 94
    " " ; disk convex, - - - - - - - - - 87
  Leaves merely toothed or lobed;
    Heads rayless, - - - - - - - - - - - - - 94
    " radiate; involucre hemispherical, - - - - - - 89
  Heads sessile in the forks, rayless, - - - - - - - - - 93
  Heads cymose-corymbose;
    Heads small; flowers white, - - - - - - - - - 90
    " larger; " yellow, - - - - - - - - - 91
  Heads panicled, small, rayless, usually nodding, - - - - - - 92

87. CHAMÆMELUM, *Tournefort.* Herbs with pinnately dissected leaves, and rather large heads on peduncles terminating leafy branches. Involucre hemispherical. Ray-flowers white; disk-flowers yellow. Chaff of receptacle bristly. Achenes not flattened, glabrous; the truncate summit with or without a short coroniform pappus.

1. C. COTULA, Allioni, Fl. Pedem. i. 186 (1785); Linn. Sp. Pl. ii. 894 (1753), under *Anthemis.* (MAYWEED.) Strong-scented annual weed 1—2 ft. high, somewhat freely branching: receptacle conical, destitute of chaff near the margin: achenes 10-ribbed, rugose or tuberculate.—In waste grounds. July—Oct.

88. MATRICARIA, *Tournefort.* Our species annual herbs, with finely dissected sweet-scented foliage and rayless heads of green flowers terminating the branches. Receptacle conical or ovoid, naked. Achenes glabrous, 3–5-nerved on the sides, rounded on the back, nearly destitute of pappus.

1. **M. discoidea,** DC. Prodr. vi. 50 (1837). *Tanacetum suaveolens,* Hook. Fl. i. 327, t. 110 (1833). Low often diffusely branching, mostly less than 1 ft. high, sweet-scented: heads short-peduncled: bracts of involucre broadly oval, scarious, with green centre, not half the length of the ovoid disk: achenes oblong, somewhat angled, with an obscure coroniform margin at summit.- By waysides everywhere, mostly in hard sterile soil. April, May.

2. **M. occidentalis,** Greene, Bull. Calif. Acad. ii. 150 (1886). Erect, very stout, 1½—2½ ft. high, corymbosely branched at summit, the herbage nearly scentless: heads more than twice as large as in the last and 6–8 lines high: receptacle somewhat fusiform: achenes sharply angled and with a broad coroniform margin below the summit.—Grain fields

and by roadsides, chiefly in the interior; only occasional near the seaboard. May, June.

89. **CHRYSANTHEMUM,** *Matthiolus.* Herbs of various habit and foliage. Heads large, with yellow or white rays. Receptacle flat or hemispherical, naked. Achenes glabrous, 5—10-ribbed all around.

1. C. LEUCANTHEMUM, Linn. Sp. Pl. ii. 888 (1753). (OX EYE DAISY.) Tufted perennial, the decumbent base of the branches very leafy, the long monocephalous peduncle only leafy-bracted: leaves spatulate-obovate, more or less deeply and incisely serrate: bracts of the broad low-hemispherical involucre with a narrow dark-red margin: rays many, showy, pure white; disk yellow.—Common weed of European and East American meadows, now established in some valleys of our middle Sierra region.

2. C. SEGETUM, Lobel, Obs. 298 (1570). Annual, erect, 1–2 ft. high, with few ascending leafy monocephalous branches: leaves somewhat fleshy, glabrous, glaucous, oblong, incisely serrate, the upper sessile by a clasping base: heads yellow, 2 in. wide including the rays: disk-achenes compressed, corky-winged.—Rather common in fields and by waysides above West Berkeley, and east of Oakland.

3. C. PARTHENIUM, Pers Syn. ii. 462 (1807); Linn. Sp. Pl. ii. 890 (1753), under *Matricaria.* (FEVERFEW.) Biennial, erect, 1 2 ft. high, stoutish, branching, puberulent: leaves thin, pinnately parted, the oval or oblong divisions incised or toothed: rays white, obovate, 2—3 lines long: pappus a minute crown.—Only occasionally spontaneous; an escape from the gardens.

4. C. INDICUM, Linn. Sp. Pl. ii. 889 (1753). Perennial, erect, leafy up to the somewhat corymbosely panicled summit, more or less cinereously pubescent: leaves petiolate, of ovate-outline, variously but usually deeply incise-pinnatifid and the lobes toothed: bracts of the hemispherical involucre obtuse and with a broad scarious margin: rays longer than the disk, the colors various.—The much admired garden Chrysanthemum, escaping from cultivation in several localities

90. **MILLEFOLIUM,** *Tournefort.* Erect perennial herb, with pinnately multifid lanceolate leaves, and small heads in a dense terminal cymose corymb. Chaff of receptacle membranaceous. Rays few, short and broad, white, or reddish. Achenes obcompressed, callous-margined, glabrous, destitute of pappus.

1. **M. vulgare,** Hill, Brit. Herbal. 458 (1756). (YARROW.) Stoutish, 1—2 ft. high, villous-lanate to nearly glabrous: heads crowded in a fastigiate flat-topped cyme: involucre oblong; rays 4 or 5, white. Very common in sandy soils toward the sea; also in the mountain districts.

91. TANACETUM, *Tournefort.* (TANSY.)

Aromatic perennials. Leaves ample, 2—3-pinnately dissected into very many divisions or lobes. Heads discoid, erect. Receptacle naked. Flowers yellow. Anther-tips broad, obtuse. Achenes truncate, and with a coroniform-dentate pappus, or epappose.

1. **T. camphoratum,** Less. in Linnæa, vi. 521 (1831); Syn. Comp. 260 (1832), under *Omalanthus.* Camphoric-aromatic, villous-tomentose, very stout but decumbent, the stems 1–3 ft. long, leafy to the summit: pinnæ and segments of the very large leaves much crowded: heads in a large corymbose cluster, short-peduncled, the low-convex disk ½ in. broad; achenes 4-angled. Abundant on sand dunes near the sea, from San Francisco northward. Aug.—Dec.

2. **T. potentilloides,** Gray, Proc. Am. Acad. ix. 204 (1874), and l. c. vi. 551 (1865), under *Artemisia.* Silvery-silky; the several stems mostly less than a foot high, leafy chiefly at base, the upper leaves much reduced and scattered; the lowest bipinnately divided and petioled; the upper simply pinnate, the lobes linear, entire: heads several, in a loose corymbose panicle: involucre about ¼ in. broad, the bracts about 10, broadly obovate, silky-tomentose: achenes 3—5-angled, thin and viscicular, truncate.—Sierra Valley, *Lemmon*, and northward, east of the Sierra.

92. ARTEMISIA, *Tournefort.*

Bitter-aromatic herbs and low shrubs. Leaves alternate. Heads discoid, small, paniculately disposed, usually nodding. Flowers whitish or yellow, often sprinkled with resinous globules. Anther-tips slender and pointed. Achenes obovate or oblong, usually with small summit and no pappus.

* *Shrubby species, with leaves filiformly dissected.*

1. **A. Californica,** Less. in Linnæa, vi. 523 (1831). Branches ascending, 2–4 ft. high; leaves cinereous with a minute appressed pubescence, terebinthine-scented, the lowest parted into a few linear filiform segments, the upper entire: heads many, in leafy panicles; involucre 2 lines broad: achenes broadish and truncate at summit, with a squamellate or coroniform-dentate pappus.—Southward and westward slopes of hills. Sept.—Dec.

COMPOSITÆ. 455

* * *Herbaceous species, all perennial except* n. 5.

2. **A. Norvegica,** Fries, Novit. 1 ed. 56 (1817). Stems several, erect, a few inches to more than a foot high, stoutish, loosely villous, or glabrous: leaves mostly basal, bipinnately parted into linear-lanceolate or broader acute lobes, the upper cauline reduced to trifid bracts: heads large, in a naked panicle or loose raceme: bracts of the involucre oblong, brownish: achenes oblong, 5-angled.—Alpine in the Sierra Nevada, but rare. Wood's Peak, *Brewer.*

3. **A. pycnocephala,** DC. Prodr. vi. 99 (1837). Stout, erect, simple, very leafy up to the dense close thyrsoid or virgate panicle, densely silky-villous even to the involucre: leaves 1– 3-pinnately parted into few and short linear or spatulate lobes: heads 2 lines broad, only the marginal fl. fertile; style of the disk fl. undivided and tufted at apex: achenes glabrous; pappus none.—Sandy hills and shores along the seaboard. Aug.—Dec.

4. **A. dracunculoides,** Pursh, Fl. ii. 742 (1814). Stems clustered, ascending, 2—4 ft. high, virgately branched: herbage glabrous, pungent-scented when bruised, but neither aromatic nor bitter: lowest leaves 3-cleft at summit, the others linear, entire: heads little more than a line broad.—At Black Point, San Francisco, and in Alameda Co., thence southward and eastward. Aug.—Nov.

5. **A. biennis,** Willd. Phytogr. 11 (1794), and Sp. iii. 1842. Annual, erect, virgate, 1—3 ft. high, leafy to the summit: herbage deep green, glabrous and tasteless, nearly: leaves 1—2-pinnately parted into lanceolate or broadly linear laciniate or toothed lobes, or the uppermost only pinnatifid: heads small, in close glomerules on the spiciform short branches and main stems: achenes with small epigynous disk and no pappus.—Mostly in or near cultivated ground; not very common. Aug.—Oct.

6. **A. heterophylla,** Nutt. Trans. Am. Phil. Soc. vii. 400 (1841). Strictly erect, 3—5 ft. high, simple and leafy up to the short naked dense panicle; herbage bitter and aromatic: leaves white beneath with cottony tomentum, green and glabrate above, 2—4 in. long, lanceolate or broader, acute, often laciniately toothed or cleft, as often entire, of firm texture: heads very numerous, 2 lines high, seldom as broad.—Very common in low rich land, along streams, and among the hills. July—Oct.

7. **A. incompta,** Nutt. l. c. Rather slender but disposed to branch, 1—2 ft. high: thinnish leaves green and glabrous above, pale and slightly tomentose beneath, once or twice pinnately parted into broadly or

narrowly linear lobes: heads small, narrowly spicate-clustered on the upper parts of the stem and branches; involucral bracts ovate, scarious-margined, nearly glabrous.—Eastern slope of the Sierra, thence eastward in Nevada. A very pleasantly fragrant herb.

8. **A. Ludoviciana**, Nutt. l. c. Completely and somewhat flocculently white-tomentose, 1—2 ft. high, simple, or with some virgate branches: leaves of firm texture, oblong to linear-lanceolate, mostly undivided and entire, sometimes 2- 3-cleft: heads very small, glomerately paniculate: involucre campanulate in flower, ovoid in fruit.—From Monterey southward. July—Oct.

* * * *Canescent or silvery-foliaged shrubs, the leaves usually toothed but not divided.*

9. **A. arbuscula**, Nutt. Trans. Am. Phil. Soc. vii. 398 (1841). Seldom a foot high, the many erect stems from a stout woody base: leaves short, cuneate-flabelliform, 3-lobed or parted, the lobes obovate to spatulate-linear, occasionally themselves 2-lobed, those nearest the heads mostly narrow and entire: panicle strict and simple, often spike-like and with few heads: involucre 5 -9-flowered.—Near the summit of the Sierra, and on its high easterly slopes. Aug.—Oct.

10. **E. tridentata**, Nutt. l. c. Larger shrub, often 4—8 ft. high, much branched: leaves cuneate, obtusely 3-toothed or lobed at the truncate summit, the uppermost cuneate-linear: heads small, very numerous and densely paniculate: involucre 5- 8-flowered, its smaller outer bracts short, ovate, tomentose-canescent.—Eastern slopes of the Sierra Nevada. Aug.—Oct.

11. **A. Bolanderi**, Gray, Proc. Am. Acad. xix. 50 (1883). Smaller, at most only 1—2 ft. high, the pubescence somewhat scurfy-tomentose, heads rather large: leaves narrowly linear, acutish, entire, or some with one or two slender lobes: heads very many, densely glomerate-paniculate, about 14-flowered, often quite surpassed by one or two linear-subulate herbaceous accessory bracts.—Rare or local species, at Mono Pass in the Sierra Nevada, *Bolander.*

12. **A. Rothrockii**, Gray, Bot. Calif. i. 618 (1876); Rothr. in Wheeler's Exp. 366. t.' 13. Less canescent than the foregoing, 1 ft. high or less: leaves from cuneate and obtusely 3-lobed to spatulate lanceolate, or the uppermost linear, sometimes all entire: heads middle-sized, glomerate-paniculate, 9—12-flowered: bracts of the involucre all ovate or oval, glabrate.—Southern part of the Sierra Nevada, in Tulare Co. and eastward.

93. **SOLIVA,** *Ruiz & Pavon.* Small depressed herb with rigid short branches, petioled pinnately dissected leaves, and heads of greenish flowers sessile in the forks. Involucre of 5–12 nearly equal bracts. Receptacle flat. Outer and only pistillate flowers apetalous, their achenes pointed with the hardened and persistent style, obcompressed, with rigid callous margins. Pappus none.

1. **S. sessilis,** R. & P. Prodr. 113, t. 24 (1794). Plant depressed, seldom 6 in. broad, villous or glabrate: primary divisions of the leaves 3–5, petiolulate, parted into 3–5 narrow lobes: heads depressed: achenes broadly obovate, thin-winged, the wings entire or panduriform-incised near the base, spinulose-pointed at summit: persistent style long and stout.—In moist open ground, or, less frequently, in shady places. May.

94. **COTULA,** *Linnæus.* Low herbs with alternate lobed or dissected leaves, and slender-peduncled discoid short-hemispherical heads. Outer series of flowers pistillate only, and apetalous, the style deciduous. Disk-flowers 4-toothed. Bracts of involucre greenish, in about 2 ranks. Mature achenes pedicellate, obcompressed, thick-margined or narrowly winged, in ours nearly or quite destitute of pappus. Both species of our flora supposed to have come from Australia, in recent times.

1. C. CORONOPIFOLIA, Linn. Sp. Pl. ii. 892 (1753). Somewhat succulent glabrous stoutish and decumbent usually subaquatic perennial: leaves ligulate-linear, laciniate-pinnatifid, or the upper entire, the base clasping or sheathing: head much depressed, ⅙–½ in. broad: pistillate fl. in a single series, their achenes with a thick spongy wing: disk-achenes with wing reduced.—Abundant in shallow pools and on muddy banks of tidal streams; occasionally on higher lands; flowering throughout the year.

2. C. AUSTRALIS, Hook. f. Fl. Nov. Zel. 128 (1853). Slender, not fleshy, very diffusely branched: leaves bipinnately dissected into linear lobes and somewhat pubescent: heads small; pistillate fl. in 2 rows, their achenes pedicelled, those of the disk less so.—Plentiful in gardens, and along some streets in San Francisco and elsewhere. Feb.—June.

Suborder 9. SENECIONIDEÆ.

Plants herbaceous or suffrutescent, mostly with watery juice, but pungent; some genera bitter and aromatic. Bracts of involucre herbaceous, in 1 or 2 series. Receptacle naked. Anthers not caudate; sometimes sagittate. Style-branches of perfect flowers obtuse or truncate; in some penicillate. Pappus of capillary bristles, these mostly fine and scabrous, sometimes coarser and barbellulate.

Hints of the Genera.

* *Herbs with ample palmatifid leaves.*

Plants subdioicous; flowers whitish, - - - - - - - - - - 95
Heads large, rayless; flowers yellow, - - - - - - - - 96

* * *Shrubby or suffrutescent; heads rayless.*

Involucre imbricated; leaves reduced and scale-like, - - - - - 99
  " of few and subequal bracts;
    Leaves small, linear or lanceolate, - - - - - - - 100
    " broader and rounded, - - - - - - - - 101

* * * *Herbaceous (a few suffrutescent); foliage various.*

Plants depressed and diffuse; rays none, - - - - - - - 97
  " upright, slender-stemmed;
    Involucre hemispherical; rays present, - - - - - - 98
    " campanulate;
      Bracts united; stems scapiform, - - - - - - 102
      " distinct; stems leafy;
        Pappus none, - - - - - - - - - 103
        " barbellulate, - - - - - - - 104
        " fine and capillary, - - - - - - 105
    Involucre cylindrical, - - - - - - - - - - 105

95. **PETASITES,** *Tournefort.* Perennial herbs, with creeping rootstocks sending up early scapiform leafy-bracted flowering stems, and later ample long-petioled radical leaves which are cottony-tomentose at least when young. Flowers whitish or purplish, in a racemose corymb, subdioecious. Achenes narrow, 5—10-costate. Pappus elongating in age, very soft and white.

1. **P. palmata,** Gray, Bot. Calif. i. 407 (1876); Ait. Kew. ii. 118, t. 2 (1789), under *Tussilago.* Leaves of round-reniform outline, 7—10 in. broad, palmately 7–11-cleft to beyond the middle, these appearing later than the very early flowering stems.- Wet mountain woods from near Santa Cruz northward, in the Coast Range. March.

96. **CACALIOPSIS,** *Gray.* Floccose-woolly stout perennial, with ample palmatifid leaves appearing before the flowers. Stem stout, with few large rayless heads of yellow flowers. Involucre campanulate, of many lanceolate-linear acuminate bracts. Corolla with cylindraceous throat longer than the slender tube. Style puberulent below the flattish branches. Achenes 10-striate. Pappus copious, soft, white.

1. **C. Nardosmia,** Gray, Proc. Am. Acad. xix. 50 (1883); l. c. vii. 361 (1868), under *Cacalia*, and l. c. viii. 631 (1873), under *Adenostyles*. Robust, 1 ft. high or more: leaves mostly radical or at the base of the stem: heads 1 in. high, peduncled, corymbosely or racemosely arranged near and at the summit of the stem: flowers honey-yellow, sweet-scented.

Mountain summits of northern Napa, or adjacent Sonoma Co., thence northward to Oregon, in the Coast Range. June.

97. **PSATHYROTES,** *Gray.* Low diffusely branched and rather rigid desert annuals, with herbage of heavy or balsamic odor, broad leaves and scattered heads of yellowish flowers. Involucre campanulate, of 1 or 2 series of subequal bracts. Receptacle flat or convex, naked. Flowers all tubular and perfect; the corolla-tube short, 5-toothed. Anthers minutely sagittate-auricled. Achenes turbinate or oblong, villous or hirsute. Pappus of copious unequal rigid capillary bristles usually brownish.

1. **P. annua,** Gray, Pl. Wright. ii. 100 (1853); Nutt. Pl. Gamb. 179 (1848), under *Bulbostylis*. Scurfy-cinereous; the small ovate dentate leaves short-petiolate, seldom cordate, sometimes broader than long: achenes oblong-turbinate, densely villous.—Low desert plains on the eastern borders of the State.

98. **CROCIDIUM,** *Hooker.* Small winter annual, floccose-tomentose when young, with numerous slender scape-like stems from a tuft of small leaves. Heads solitary, terminal on all the branches. Involucre hemispherical, of 9—12 nearly equal oblong-lanceolate thin bracts and no subtending bractlets. Rays about 12, oval or oblong. Disk-corollas with slender tube and campanulate throat; the 5 lobes spreading. Receptacle convex, tuberculate. Achenes fusiform-oblong, obscurely 3—5-costate, beset with hyaline oblong papillæ, these when wetted throwing out a pair of spiral threads. Pappus a single series of white barbellate deciduous bristles, in the ray commonly wanting.

1. **C. multicaule,** Hook. Fl. i. 335, t. 118 (1833). Proper and subradical leaves obovate or spatulate, ¾ in. long, few-toothed; those of the scape-like stems bract-like and linear or lanceolate: heads small but showy, the flowers yellow. Elevated plains and low foothills from about Mt. Shasta, southward to Mariposa Co. April.

99. **LEPIDOSPARTUM,** *Gray.* A rigid green but scaly-bracted and almost leafless shrub, branching somewhat fastigiately and bearing the heads of pale yellow flowers somewhat corymbosely or racemosely. Involucre 10—15-flowered, its longer inner bracts 8—12, in 2 or more

series, these subtended by many imbricate smaller ones. Receptacle naked. Rays none. Corollas with long tube, and lanceolate-linear spreading lobes which much exceed the open-campanulate throat. Achenes oblong, terete, 8—10-nerved, with large epigynous disk. Pappus very copious, of soft white capillary bristles.

1. **L. squamatum,** Gray, Proc. Am. Acad. xix. 50 (1883); l. c. viii. 290 (1870), under *Linosyris,* and ix. 207 (1874), under *Tetradymia.* Seedling plants and young shoots floccose-tomentose, and with spatulate entire leaves, these not found in the green and broom-like scaly-bracted mature shrub of 2—4 feet high: heads 3—5 lines high, terminal on the many branchlets.—From the Salinas Valley southward, in the Coast Range hills and valleys.

100. **TETRADYMIA,** *De Candolle.* Low rigid canescently tomentose shrubs, with alternate narrow entire leaves, and cymose-clustered rayless heads of yellow flowers. Involucre long and narrow, of 4 or 5 bracts. Corollas with long tube, the narrow spreading lobes longer than the campanulate throat. Achenes terete, short, 5-nerved, from long-villous to glabrous. Pappus of fine and soft long capillary white or whitish bristles.

1. **T. canescens,** DC. vi. 440 (1837). Silvery-tomentose, 1—2 ft. high: leaves narrowly linear or linear-lanceolate, $\frac{1}{2}$—1 in. long: heads terminal in a corymbose cluster.—Dry hills of the eastern slope of the Sierra both north and south.

2. **T. glabrata,** T. & G. Pac. R. Rep. ii. 122, t. 5 (1854). Size of the last, the tomentum more cottony and less permanent: leaves rather fleshy, glabrate in age, the primary ones linear-subulate and conspicuously mucronate: heads, etc. as in the preceding.—East of the Sierra; less common than the last.

3. **T. stenolepis,** Greene, Bull. Calif. Acad. i. 92 (1885). Very white-tomentose with appressed wool: longer leaves elongated, slender and spinescent, others narrowly spatulate or linear subulate and short: heads elongated, fully $\frac{1}{2}$ in. high, 5-flowered: bracts of involucre linear, rigid and thick: achenes merely pubescent: pappus copious, rather rigid.—Mountains of Kern Co. near Tehachapi, etc.

101. **LUINA,** *Bentham.* Cottony-woolly low suffrutescent plant with many erect simple stems and alternate sessile entire but not narrow leaves, and terminal corymbs of rayless heads of yellow flowers. Involucre campanulate, of 10 or 12 narrow rigid carinately 1-nerved bracts of

equal length. Corollas with slender tube and funnelform 5-lobed limb. Style-branches linear-semiterete, minutely papillose-puberulent externally, obtuse. Achenes terete, obscurely 10-striate, glabrous, or with scattered fine hairs. Pappus of copious soft white bristles.

1. **L. hypoleuca,** Benth. in Hook. Icon. t. 1139 (1873). Stems 1 ft. high, equably leafy to the summit: leaves ovate-oblong to elliptic, obtuse, 1 in. long, reticulate-veiny on the glabrate upper face, very white beneath: heads ½ in. high, on rather slender pedicels, forming an open cluster: corolla-lobes about half the length of the funnelform throat.— A somewhat rare sea-coast plant, found in the hills behind Santa Cruz, and on Chimney Rock, Mendocino Co.

102. **RAILLARDELLA,** *Benth. & Hook.* Perennial tufted or matted scapose herbs, with narrow entire mostly radical leaves, and slender often long monocephalous scapes. Campanulate or cylindraceous involucre of a single series of linear bracts lightly connate below the middle. Receptacle flat. Ray-flowers, when present, with cuneate deeply 3—4-cleft ligules. Disk-corollas with rather short proper tube, narrow-funnelform throat, and 5 ovate obtuse teeth. Achenes linear, subterete, obscurely nerved, pubescent. Pappus of 12—15 long aristiform but soft-barbellate bristles.

1. **R. argentea,** Gray, Bot. Calif. i. 417 (1876); Proc. Am. Acad. vi. 550 (1865), under *Raillardia.* Leaves spatulate-lanceolate, silvery-silky on both sides: involucre 7—15-flowered, terminating a glandular scape 2—4 in. high: rays none.—Alpine slopes of the high Sierra, from above Yosemite northward.

2. **R. scaposa,** Gray, Bot. Calif. l. c., and Proc. Am. Acad. l. c., under *Raillardia.* Larger, the leaves glandular and hirsutulous rather than silky: scape often 1 ft. high and with 1 or 2 bracts: involucre often 20—30-flowered, and occasionally developing a few imperfect rays.— Range of the last.

3. **R. Pringlei,** Greene, Bull. Torr. Club, ix. 17 (1882). Scapes 1—1½ ft. high, from a branching rootstock, or rather from prostrate short leafy branches: leaves almost linear, some of them remotely serrate-toothed, glabrous: scape glandular under the large campanulate about 4-flowered involucre; bracts of involucre only lightly united near the base: rays about 10, orange-yellow: pappus bristles fewer than in the other species, and less plumose.—Subalpine in the Mt. Shasta region.

103. **WHITNEYA,** *Gray.* Low canescent perennial, with creeping rootstocks, and upright mostly simple stems with a few pairs of reduced

leaves, and one or more peduncled heads of yellow showy flowers; the plant with wholly the aspect of an *Arnica*, from which genus it is hardly well distinguished by its more firm and persistent ray-corollas, and its want of a pappus. Involucre campanulate, of 9--12 thin-herbaceous lanceolate-oblong or ovate lanceolate equal bracts, these partly biserial. Receptacle conical, villous. Rays 10--16, becoming somewhat papery and persistent; their achenes somewhat obcompressed, several-nerved. Disk-achenes infertile. Pappus none.

1. **W. dealbata,** Gray, Proc. Am. Acad. vi. 549 (1865). Stems 1 ft. high; herbage hoary with a fine close tomentulose pubescence: lowest leaves obovate or spatulate, entire or denticulate; the cauline reduced, opposite, few: rays 1 in. long.—From Mariposa Co. northward, at middle elevations of the Sierra.

104. **ARNICA,** *Ruppius*. Perennial herbs, somewhat glandular or viscid and aromatic. Leaves opposite. Heads one or several and large, at summit of the stem. Involucre broadly campanulate, not bracteolate at base; the herbaceous bracts lanceolate, equal, in about 2 series. Receptacle flat, naked. Disk-corollas yellow (as also the rays when present), with distinct long tube and funnelform or cylindraceous 5-lobed limb. Achenes linear, angled. Pappus a single series of rather rigid brownish scabrous or barbellate bristles.

* *Heads destitute of rays.*

1. **A. discoidea,** Benth. Pl. Hartw. 319 (1849). Stoutish, very hairy, 2 ft. high or less: leaves ovate or oblong, 2—4 in. long, coarsely toothed, cordate, or truncate, or sometimes slightly cuneate at base; the upper smaller, sessile, often alternate: heads $\frac{3}{4}$ in. high, rayless: involucre villous and glandular: achenes sparsely pubescent, not glandular.—Northward slopes of the higher coast mountains, in shady places chiefly. June, July.

2. **A. parviflora,** Gray, Proc. Am. Acad. vii. 363 (1867). Slender, pubescent, 1 ft. high or more, only slightly glandular: lowest leaves long-petioled and small, the blade narrowly deltoid or oblong, truncate or abrupt at base, 1—2 in. long: involucre 4—5 lines high, its linear bracts sparsely pubescent; rays none: achenes minutely glandular, not pubescent.—Humboldt Co., in thickets. *Bolander.*

3. **A. viscosa,** Gray, Proc. Am. Acad. xiii. 374 (1878). *Raillardella paniculata,* Greene, Eryth. iii. 48 (1895). Very viscid and pubescent, 1—2 ft. high, fastigiately branched above and paniculate, the monoceph-

alous branches clothed with small scattered ovate-oblong entire sessile leaves: involucres short, only ½ in. high, much exceeded by the salmon-colored tubular corollas: achenes glandular-hirsute.—At middle elevations on Mt. Shasta. A most anomalous species; the probable type of a genus.

* * *Heads radiate and showy.*

4. **A. Sonnei,** Greene, Pittonia, iii. 104 (1896). Two feet high, leafy at base, the upper 2 or 3 pairs of leaves much reduced and remote; the lower and radical ones with broadly lanceolate evenly and saliently dentate blade 3–5 in. long, on a strongly villous-lanate petiole of 1—3 in.; herbage loosely and scantily villous; heads 3–7, pedunculate: involucre campanulate, ½ inch high or more; the linear-lanceolate bracts loosely hirsute: rays (occasionally wanting) numerous, golden-yellow: disk-corollas rather narrowly funnelform, the tube and throat pilose: achenes glabrous below, distinctly hirsute above the middle: pappus dull-white, barbellate.—Mountain slopes along the Truckee River, *Sonne.*

5. **A. cordifolia,** Hook. Fl. i. 331 (1833). Usually 1–2 ft. high, or smaller at alpine elevations, the stems mostly hirsute and peduncles villous: leaves rather few, the radical and lower cauline ovate or ovate-cordate, coarsely dentate, ample and on long petioles, the upper cauline small, sessile: head commonly solitary; involucre ⅔ in. high, pubescent or villous: rays 1 in. long: achenes more or less hirsute.—Eastern slope of the Sierra, and not common.

6. **A. Nevadensis,** Gray, Proc. Am. Acad. xix. 55 (1883). Rather smaller than the last, of the same habit and general aspect, but leaves more rounded and obtuse, not cordate, but oval or oblong rather, usually entire: involucre ½ in. high: achenes minutely pubescent and glabrate. —From Lassen's Peak to the heights above Donner Lake, *Mrs. Austin, Mr. Sonne.*

7. **A. latifolia,** Bong. Sitch. 147 (1832). More slender, the cauline leaves ovate, obtuse, serrate-toothed, closely sessile by a broad base: heads several, ¾ in. high: achenes glabrous; pappus almost subplumose. —In the Truckee Valley and northwards.

8. **A. longifolia,** D. C. Eaton, Bot. King Exp. 186 (1871). Stems tufted and tall, 2–3 ft. high; herbage puberulent: cauline leaves strongly developed, elongated-lanceolate, tapering to both ends, denticulate, 3—6 in. long, the lower with connate-sheathing base: heads rather numerous, small and corymbose : achenes minutely glandular, not

hairy.— In moist places along the eastern base of the Sierra from the Truckee Valley southward.

9. **A. foliosa,** Nutt. Trans. Am. Phil. Soc. vii. 407 (1841). Slender, erect and rather strict, very leafy, canescently tomentose, 1 ft. high more or less: leaves lanceolate, denticulate, nervose, the lower with tapering base and connate: heads short-peduncled, corymbose, though few, rarely solitary: rays rather short: achenes hirsute-pubescent or glabrate.— Wet meadows of the Sierra chiefly on the eastern slope.

10. **A. fulgens,** Pursh, Fl. ii. 527 (1814). *A. alpina* of American authors, not of Olin. Strict, simple and monocephalous, 1—2 ft. high, leafy at base chiefly, the few pairs of cauline leaves reduced and bract-like; herbage pubescent, the inflorescence somewhat hirsute or villous: leaves thickish, from narrowly oblong to lanceolate or oblong-spatulate, 3-nerved, entire or denticulate: rays elongated, golden-yellow: achenes hirsute-pubescent.—Meadows east of the Sierra, in Plumas Co., etc.

105. SENECIO, *Matthiolus*. Plants extremely diverse in habit, foliage, etc., but leaves always alternate. Heads clustered cymosely, or solitary; either radiate or discoid. Involucre usually cylindrical, of many equal bracts, and with calyx-like bracteoles at the base. Flowers yellow; those of the disk 5-toothed or -lobed. Receptacle flat or convex. Achenes commonly glabrous, terete, often somewhat ribbed. Pappus of very copious fine white capillary merely scabrous bristles.

* *Annual species; achenes not glabrous.*

1. **S. vulgaris,** Tragus, Stirp. Hist. 284 (1552). Hoary-lanate when young, in age nearly glabrous; slightly fleshy, 1 ft. high more or less, branching, leafy throughout: leaves clasping at base, pinnatifid; the lobes and sinuses sharply toothed: scales at base of the small cylindric involucre with black tips: rays none.—Abundant in rich shady cultivated grounds; flowering at all seasons of the year. From Europe.

2. **S. aphanactis,** Greene, Pittonia, i. 220 (1888). Slender, 2—6 in. high, slightly arachnoid when young, glabrate in age, scarcely viscid, scentless: leaves $\frac{1}{2}$—$2\frac{1}{4}$ in. long, slightly fleshy, erect or ascending, the lowest linear-spatulate, the cauline linear to oblong, coarsely toothed or slightly lobed: heads very small, 2 or 3 together at the ends of the few branches: bracts of involucre linear-acuminate, not black-tipped: rays about 5, minute, recurved: achenes appressed-silky.—Clayey slopes, or open hilltops of the Mt. Diablo Range, and southward. A somewhat rare indigenous plant, strangely mistaken, by some, for the Old World weed *S. silvaticus,* to which it is not very closely related. March, April.

3. **S. Californicus,** DC. Prodr. vi. 426 (1837). Glabrous, at least when mature, slender, 8—18 in. high: leaves of lanceolate or almost linear outline, varying from denticulate to pinnatifid, the lobes short and obtuse, all but the lowest sessile and auriculate-clasping, 1- 2 in. long: heads few, loosely cymose at the almost naked summit of the stem: involucre small, campanulate, 3—4 lines high, the bracts narrow: rays oblong, 3—4 lines long, light-yellow: achenes canescent. — Low sandy plains from near Santa Barbara southward, especially toward the sea. Feb.—April.

* * *Suffrutescent, or at least tufted perennials, with linear or dissected leaves, and canescent achenes.*

4. **S. Douglasii,** DC. Prodr. vi. 429 (1837). Usually 3 ft. high, branching from the base, stoutish, loosely leafy; growing parts and young leaves whitish-tomentose, later glabrate—at least the upper surface of the leaves; lower leaves pinnately divided into about 5 narrowly linear lobes, the upper linear, entire, all with revolute margins: heads few, large, corymbose; rays conspicuous, light-yellow: achenes hoary with a short pubescence.—Frequent on dry hills of the Mt. Diablo Range from Alameda Co. southward. July—Nov.

5. **S. Blochmanæ,** Greene, Eryth. i. 7 (1893). Tufted stout and rigid stems erect, 3—4 ft. high, simple and very leafy up to the fastigiately corymbose summit; herbage glabrous, heavy-scented: leaves filiform-linear, entire, 2—4 in. long, rather fleshy, recurved or deflexed on the stem: numerous heads ½ in. high, cylindric: rays 5—8, light-yellow: achenes hoary with a short strigulose pubescence; pappus very fine and soft.—Along the Santa Maria River, San Luis Obispo Co., *Mrs. Blochman.*

6. S. Cineraria, DC. Prodr. vi. 354 (1837). Stout, 2—4 ft. high, white-tomentose: leaves of firm texture, petiolate, pinnately parted, the segments oblong, obtuse, more or less distinctly 3-lobed: heads many, in a terminal corymb: rays 10—12, short-oval: achenes puberulent.— An ornamental species of southern Europe, not infrequent as an escape from the gardens.

* * * *Perennials of various habit; achenes glabrous.*
← *Equably leafy throughout.*

7. **S. Fremonti,** T. & G. Fl. ii. 445 (1843). Stems many, 1 ft. high or less; herbage rather fleshy, glabrous: leaves small, round-obovate or spatulate, 1—2 in. long, obtuse, dentate, the lowest narrowed to a winged

petiole; uppermost sessile by a broad base: heads rather few and long peduncled, ½ in. high or less, subtended by some loose bractlets: rays 8–10, yellow.—On Lassen's Peak and elsewhere in the Sierra at high elevations.

8. **S. Clarkianus**, Gray, Proc. Am. Acad. vii. 362 (1862). Glabrous, stout, 3–4 ft. high: lowest leaves ⅙—1 ft. long, petiolate, the upper half as large, scarcely petiolate, all of lanceolate outline and laciniately toothed or pinnatifid: heads ½ in. high or more and 5—20 in a subcorymbose terminal cluster; linear-acuminate bracts of the involucre subtended by many loose filiform bractlets: rays 10—15, yellow.—Mountain meadows in Mariposa Co., etc.

9. **S. Andinus**, Nutt. Trans. Am. Phil. Soc. vii. 409 (1841). Glabrous, tall, very leafy, the light-green foliage lanceolate, acute, subsessile, sharply toothed or serrate: heads small and numerous, in a short and subcorymbose panicle; bractlets beneath the cylindric and narrow involucre many and subulate: rays 6—8, about as long as the involucre. —By streams along the eastern base of the Sierra, barely within the State limits, but more common in Nevada, etc. July—Sept.

10. **S. triangularis**, Hook. Fl. ii. 332, t. 115 (1833). Size of the last, and with like yellow-green hue of herbage, but more flaccid; the leaves petioled, the blade deltoid or triangular-lanceolate, or even subhastate, coarsely toothed: heads rather few, corymbose: involucres cylindric, many, ½ in. high: rays 6—12.—Species of the Rocky Mountains and northern Cascades, reaching our borders northward; but not in the Sierra proper, being there replaced by the following.

11. **S. trigonophyllus**, Greene, Pittonia, iii. 106 (1896). Size of the last, and with similar triangular outline of foliage, but herbage deep-green and leaves membranaceous: heads far more numerous, only one-half as large, the involucres subcampanulate rather than cylindric: achenes correspondingly short, greenish at maturity.—Common at subalpine elevations of the middle Sierra Nevada, where it has passed for *S. triangularis;* but it is very distinct. July—Sept.

+ + *Leaves largely basal, those of the stem more scattered and reduced.*

++ *Herbage glabrous, mostly glaucescent or glaucous.*

12. **S. hydrophilus**, Nutt. Trans. Am. Phil. Soc. vii. 411 (1841). Stout, succulent, glabrous, glaucescent, the stems 2—4 ft. high, leafy mostly at base: leaves lanceolate, the lower 5—9 in. long, with stout petioles, the upper successively shorter and sessile, all more or less

denticulate, or some entire: heads numerous, in a rather dense panicle, the involucres cylindric, scarcely bracteolate: rays few, small, light yellow. Var. **Pacificus**, Greene. Stems coarser and fistulous: involucres larger and cymose-corymbose rather than panicled: rays none.—The type only along the eastern base of the Sierra, in subsaline wet meadows; the very marked variety in brackish marshes about San Francisco Bay and its tributaries.

13. **S. Clevelandi**, Greene, Bull. Torr. Club, x. 87 (1883). Stems tufted on short creeping rootstocks, slender, 1—2 ft. high; the mostly radical leaves somewhat succulent, ovate-oblong, entire, obtuse, nearly veinless, not as long as the petioles; even the upper and smaller cauline also distinctly petiolate: involucre 4—5 lines high, rather broad, its bracts subulate-linear: rays 6—8, short, deep yellow: achenes short, angular.—Springy places in the mountains of Lake Co., *Cleveland*, *Pringle*.

14. **S. Laynew**, Greene, l. c. Strict and simple, 2 ft. high: the mostly radical leaves linear-lanceolate, entire, 3—4 lines wide, 1—2 in. long, on petioles nearly as long; the few cauline similar, smaller, less obviously petiolate: heads 5—7, in a cyme, all but the central one on peduncles 2—3 in. long; involucre large, campanulate, ½ in. high, naked at base; rays 7—10, oblong-linear, ¾ in. long, orange-yellow: the merely convex style-tips with 3 or 4 conspicuous central bristles.—On Sweetwater Creek, El Dorado Co., *Mrs. Curran*.

++ ++ *Herbage more or less flocculent or hoary, at least when young.*

= *Leaves lanceolate or broader, entire or toothed, never lyrate.*

15. **S. aronicoides**, DC. Prodr. vi. 426 (1837). Growing parts loosely woolly, afterwards glabrate: stem stout, 2—3 ft. high, leafy chiefly at base, the many small heads in a compound terminal cyme: leaves ovate to oblong and lanceolate, 3—6 in. long, irregularly and coarsely toothed, much reduced on the stem, the uppermost only bract-like: involucres short, subcampanulate, its few bracts lanceolate, acuminate, not black-tipped: flowers 10—20; rays none, or rarely 1 or 2.—Chiefly among thickets, on northward slopes of hills, both in the Coast Range, and western foothills of the Sierra Nevada. April, May.

16. **S. Sonnei**. More distinctly and quite permanently woolly, the stout striate stem 1-3 ft. high, the oval-lanceolate petiolate leaves doubly dentate: heads few and large, 5—9, on very stout peduncles and forming usually a simple corymb: involucre campanulate, more than $\frac{1}{2}$ in. high and as broad as high, the lanceolate bracts acuminate, and with

a few calyculate ones at base: flowers yellow; rays short and broad: achenes linear, prismatic, finely striate between the angles, 2½ lines long, whitish, crowned with a pappus of decidedly short and firm scabrous bristles.—Dry open stony ground near the base of Castle Peak (otherwise called Mt. Stanford), north of Donner Lake. June, July.

17. **S. leptolepis.** Rather slender, 1—2 ft. high, distinctly but sparingly woolly-hairy in maturity and most so about the inflorescence: leaves oval and oblong-oval, obtuse, irregularly dentate, 2—3 in. long, abruptly tapering to petioles as long or longer; the middle canline broadly linear, 4 in. long, entire, sessile and amplexicaul: heads 5—7, in a simple corymbose cyme, all but the central one on slender elongated peduncles: involucre cylindrical, ½ in. high; bracts narrowly linear, thin, glabrous, the calyculate bractlets woolly: rays none.—Dry woods of Amador Co., *Hansen*.

18. **S. Rawsonianus,** Greene, Pittonia, ii. 166 (1891). Robust, 2—3 ft. high, somewhat canescent with scattered short woolly hairs: leaves 6—8 in. long, ovate to lanceolate, acute, sinuately or laciniately toothed, or the upper repand-dentate: heads very numerous, in a compound corymb; involucres nearly cylindrical, 4 lines high, the bracts oblong-linear, abruptly acuminate, the calyculate ones few and short: rays none: tubular corollas salmon-color, soon concealed by the accrescent and copious white pappus.—Forests of Fresno Co., at middle elevations in the Sierra; a handsome, apparently almost white-flowered species.

19. **S. Whippleanus,** Gray, Proc. Am. Acad. xix. 54 (1883) Decidedly hoary, stoutish, 2—3 ft. high: leaves 6—8 in. long, from sinuately to laciniately toothed, rarely almost pinnatifid: canline reduced and sessile: heads few, large and broad, almost campanulate, in an ample loose cyme: involucral bracts fleshy, oblong-linear, abruptly acuminate: rays ½ in. long.—Pine woods of Calaveras Co., *Bigelow*, *Greene*.

20. **S. Mendocinensis,** Gray, Proc. Am. Acad. vii. 362 (1868). Lightly arachnoid-floccose, early glabrate: stem robust, 2—3 ft. high, nearly naked above, and with a dense corymb of 5—8 rather large heads: leaves somewhat succulent, oval or oblong, obtuse, dentate, narrowed to a petiole, the middle canline reduced, lanceolate, sessile, the uppermost passing into subulate bracts: involucres rather large, with scarious-margined lanceolate acuminate proper bracts, and many loose calyculate ones: rays 12—15, oblong, short.—Open plains of Mendocino and Humboldt counties.

21. **S. Scorzonella,** Greene, Pittonia, iii. 90 (1896). Stems scapiform and leafy-bracted, 12—18 in. high, from stout horizontal or ascending

branched rootstocks: herbage flocculent, or somewhat arachnoid throughout, not glabrate in age: leaves in a radical tuft, suberect, 5 6 in. long, broadly oblanceolate, spatulately tapering to a narrow scarcely petiolate base, the margin coarsely, irregularly and retrorsely lacerate-toothed in the broad upper portion, the tapering lower part entire: heads 20 or more, in a somewhat paniculate terminal corymb, of the size of those of *A. aronicoides* and radiate; involucral bracts 10–14, obtusish, the involucre subtended by 2 or 3 somewhat remote more foliaceous ones: ovaries glabrous.—Plumas Co., *Mrs. Austin*.

22. **S. Covillei.** Tufted on branched rootstocks like the last, and with similarly scapiform stems, but slender, almost glabrous from the first, only the inflorescence a little flocculent: lowest leaves 3–5 in. long, spatulate-linear or -lanceolate, *i. e.*, linear ligulate and entire toward the base, the main portion broader and coarsely serrate-toothed, the apex acute or acuminate; the cauline few, reduced and lanceolate, entire or toothed, amplexicaul: heads few and small, in a rather dense cyme: involucre short-cylindric, the broad bracts black-tipped.—Near Whitney Meadows in the southern Sierra, *Coville & Funston* (n. 1651), also in a broader-leaved form in the White Mountains, Mono Co., *Shockley*.

23. **S. astephanus,** Greene, Pittonia, i. 174 (1888). Lightly floccose-pubescent when young, at length nearly glabrous: leaves ample, thin, undivided, the radical nearly a foot long including the short petiole, elliptic-oblong, acute at both ends, coarsely dentate, the teeth spreading, triangular, callous-tipped, the sinuses rounded and the larger of them denticulate: heads few, slender-peduncled, nearly an inch high and two-thirds as thick: involucre calyculate at base, its proper scales lanceolate, acuminate: rays none.—Mountains of San Luis Obispo Co., *Lemmon*. Species imperfectly known; perhaps of the leafy-stemmed group.

24. **S. canus,** Hook. Fl. i. 333 (1833)? Stems tufted, subligneous at base and leafy, 8–16 in. high; herbage permanently white-tomentose: leaves (in our plant) all ovate or oblong, entire, acutish, 1 in. long, on petioles of greater length: heads 4–5 lines high, in a lax more or less compound cyme: rays oblong-linear: achenes glabrous. Near the summit of the Sierra, and down the eastern slope. Scarcely identical with the plant figured by Hooker from the far north, but perhaps confluent with it.

25. **S. werneriæfolius,** Gray, Proc. Am. Acad. xix. 54 (1883). White-tomentose like the last, but low and cespitose: leaves all subradical, spatulate-linear or oblong-linear, entire, the margins revolute: scape

only a few inches high, naked, bearing a corymb of small radiate heads at summit.—Alpine on Mt. Conness, *Harford;* otherwise known only from the Colorado Rocky Mountains.

26. **S. petrophilus,** Greene, Pittonia, iii. 171 (1897). *S. petraeus,* Klatt (1881), not of Boissier & Reuter (1842). Low, the caudex multicipital and its branches leafy; herbage rather succulent and glabrous, or early glabrate: leaves round-obovate or oval, ¼—1½ in. long, or cuneate-oblong and larger, entire or crenately few-toothed at the broad summit, abruptly petioled: scape 2—5 in. high, bearing 1 or more large heads (4—5 lines high): rays 6—10, golden-yellow, 3 lines long.—Alpine on the highest summits of the Sierra.

— *Leaves mostly pinnate, and lyrately so, or the pinnæ reduced or obsolete, the one terminal leaflet then constituting the leaf.*

27. **S. lætiflorus,** Greene, Pittonia, iii. 88 (1896). Stems tufted, erect, 10—18 in. high, rather loosely cymose-corymbose at summit: leaves mostly radical, somewhat fleshy, light green, glabrous except a very distinct arachnoid white tomentum on the margins of the dilated bases of the long slender petioles, the blade from spatulate-obovate and oval to round ovate, mostly crenate or serrate-toothed, rarely entire; cauline leaves few, sessile and pinnatifid, often lyrately so: involucres campanulate, 4 or 5 lines high and nearly as broad; bracts lanceolate, acuminate, fleshy and carinate; rays showy, pale-yellow: chestnut-brown achenes 5-angled and with 5 alternating ribs.—Truckee Valley and northward, in meadows along streams.

28. **S. Lemberti,** Greene, l. c. 89. Stems clustered, very leafy at base and sparsely so above, erect, 1—2 ft. high; herbage glabrous throughout and the leaves thin; the radical from round-ovate and crenate to spatulate-obovate and coarsely dentate, 4 to 8 inches long including the slender petiole; cauline reduced in number and in size, deeply and somewhat lyrately pinnatifid, with a broad and clasping base: heads 6—8, in a terminal rather condensed umbel; involucre 4 lines high, broadly cylindrical, bracts about 20, lanceolate, thin. purple-tipped: flowers saffron-color; rays none.—Subalpine above the Yosemite, *Lembert,* and on Mt. Conness, *Harford.*

29. **S. indecorus.** Stems erect, stoutish, 2 ft. high, sparsely leafy throughout; herbage flaccid, with traces of arachnoid woolliness, but glabrate in age: lowest leaves not numerous, oval, obtuse, truncate at base, from crenately to deeply and incisely serrate, ¾—1½ in. long, on slender petioles of 2—3 inches, the lower and middle cauline on short

petioles, these stipulately appendaged at base, the uppermost sessile, of lanceolate outline and pinnatifid: heads about 6 or 8, in a dense umbellate corymb: involucre 4—5 lines high, of 18—20 thin oblong-lanceolate obtuse bracts: rays none.—In wet meadows east of the Sierra, in Plumas and Lassen counties and northward. May possibly be a discoid variety of the *S. cymbalarioides*, Nutt., of Oregon.

30. **S. Bolanderi**, Gray, Proc. Am. Acad. vii. 362 (1868). Stems slender, 6—15 in. high, from slender creeping rootstocks; herbage somewhat tomentose when young, glabrate in age: radical and lower cauline leaves petioled and pinnately divided, of thin texture; leaflets 3—7, rounded or cuneate, incisely and obtusely lobed, the terminal one largest and sometimes subcordate: heads few, subcorymbose; involucre of 12—15 linear bracts: rays 4—6, yellow, 4—5 lines long.—Sandstone bluffs in Mendocino and Humboldt counties.

31. **S. Breweri**, Davy, Eryth. iii. 116 (1895). Biennial or perennial stout, erect, 2—3 ft. high, the stems sulcate and herbage wholly glabrous: basal leaves petiolate, lyrately pinnate with cuneate or subreniform coarsely toothed leaflets, or deeply pinnatifid; upper cauline smaller, sessile, semiamplexicaul, laciniately once or twice pinnatifid: heads in an ample terminal corymbiform cluster; involucres 5—6 lines high, about 4 lines broad; bracts 10—15, broad, lightly 2-ribbed, scarious-margined: rays few, broad, light-yellow.—Gravelly hills and slopes, in shaded ground, from Alameda Co. southward.

32. **S. caulanthifolius**, Davy, l. c. 117. Stems several from the root, stout, flexuous, striate, 2—3 ft. high; herbage cinereously tomentose, the foliage glabrate in age: leaves thin, the lowest from oval to oblong-lanceolate, 4—8 in. long, cuneately tapering to a long petiole, the margin from incisely lobed to sinuately dentate and repand-denticulate; upper smaller and subsessile, laciniate-pinnatifid: heads loosely corymbose, large; involucre 4 lines high, the bracts about 15, oblong, acuminate, often 2-ribbed, scarious-margined: rays broad: achenes glabrous, linear, 5-angled and 10-striate.—Near the limestone cave at Murphy's, Calaveras Co., *Davy*.

33. **S. eurycephalus**, Gray, Pl. Fendl. 109 (1849). Hoary-tomentose, or in age only cinereous, the stout tufted stems 1—2 ft. high: lowest leaves long-petioled, some simple, obovate and coarsely serrate-toothed, others lyrate; the cauline sessile and lyrate-pinnatifid, the lobes or leaflets 7—15, cuneate and acutely incised or cleft: heads in a loose corymb; involucre campanulate, ½ in. high and almost as broad, the linear-lanceolate bracts 20 or more, rather thin, not ribbed: rays 10—12, clon-

gated: achenes short, 10-angled, 5 alternate angles less prominent.—
Rocky places along streams, from Sonoma and Contra Costa counties
northward. Possibly Mr. Davy's *S. Breweri* may be the true and original
S. eurycephalus, and supposing this to be the case, I long since named
the present plant *S. Tidestromii* in the herbarium.

34. **S. Greenei**, Gray, Proc. Am. Acad. x. 75 (1875). Somewhat floccose, less than 1 ft. high, leafy at base: heads 1—3, large, peduncled,
terminal: radical leaves roundish, with abrupt or slightly cuneate base
and long petioles, coarsely crenate-toothed: heads $\frac{2}{3}$ in. long, with no
bracteoles at base; rays deep orange, $\frac{1}{2}$ in. long: style tips of disk-
flowers penicillate and with a central cusp.—Under bushes, and on more
open and rocky spaces among the higher mountains of Napa and Sonoma
counties and northward. May, June.

35. **S. ionophyllus**, Greene, Pittonia, ii. 20 (1889). Tufted perennial,
1 ft. high or more, somewhat floccose, or nearly glabrous: radical leaves
crowded, somewhat fleshy, small ($\frac{3}{4}$ in. long), cordate-orbicular or
slightly reniform, coarsely crenate or dentate, on petioles 2 in. long;
cauline narrower, more sharply toothed or somewhat pinnatifid: heads
large ($\frac{2}{3}$ in. high, $\frac{1}{2}$ in. broad), solitary, or more frequently 3—5 in a lax
terminal corymb; bracts oblong-lanceolate, the calyculate few, imbedded
in white tomentum: rays showy, light yellow: style-tips hirsute:
achenes prismatic-fusiform.— Pine woods of Kern Co. near Tehachapi
and southward. June, July.

* * * * *Tall leafy climbing exotic.*

36. S. MIKANIOIDES, Otto, Allgem. Gartenz. xiii. 42 (1845). Glabrous,
trailing and twining to the height of 20 feet: leaves roundish-cordate-
hastate, sharply 5—7-angled, 2—5 in. long, nearly as broad, on petioles
as long or longer, these with a reniform stipulaceous lobe on either side
at base: heads small, in compound corymbs terminating axillary branchlets: rays none: disk-fl. 9—15.—Plentiful along banks of streams at the
base of the Oakland and Berkeley Hills. Native of S. Africa; flowering
profusely in January.

Suborder 10. CYNAROCEPHALÆ.

Herbs with watery juice, usually prickly leaves, and flowers in dense
ovoid or globose heads; involucre imbricate. Receptacle densely setose,
or paleaceous. Flowers all alike and perfect, or the marginal larger
than the others and neutral. Corollas regular or slightly irregular.

deeply cleft into 5 long and narrow lobes; the marginal in some enlarged and palmatifid but never ligulate. Stamens syngenesious and also rarely monadelphous below. Anthers caudate. Style slightly or not at all cleft, commonly with a pubescent node or ring below the stigmatic part. Achenes thick, hard, smooth, basally or obliquely inserted. Pappus setose, plumose, or rarely paleaceous, sometimes wanting.

Hints of the Genera.

* *Receptacle setose.*

Pappus of many setose bristles;
    Ray-flowers enlarged, palmate, - - - - - - 106
    " " not enlarged, - - - - - - - - - 107
Pappus double, of 2 different sets;
    Both sets bristly, - - - - - - - - - 108
    " " paleaceous, - - - - - - - - 109
Pappus of many series of narrow barbellate paleæ, - - - - - 110
Pappus united at base, deciduous in a ring;
    In many series of plumose bristles, - - - - - - - 111
    In one series of plumose bristles, - - - - - - - 113

* * *Receptacle paleaceous.*

Pappus of few unequal paleæ, - - - - - - - - - - 112

106. CYANUS, *Tournefort.* Annual flocculently hoary unarmed herbs. Heads naked, in slender peduncles. Bracts of the ovate involucre narrow, fringed with scarious teeth. Marginal corollas much larger than the others, palmatifid and raylike. Pappus of unequal bristles, the longest about as long as the achenes.

1. C. SEGETUM, Buchoz, Dict i. 197 (1770). *Centaurea Cyanus*, Linn. Sp. Pl. ii. 911 (1753). (CORNFLOWER.) Slender, 1—2 ft. high, not prickly, whitened, at least when young, with floccose wool: leaves linear, entire, or the lower toothed or pinnatifid: heads naked, on slender peduncles: involucral bracts narrow, fringed with short scarious teeth: marginal flowers asexual, much enlarged, irregularly and somewhat palmately cleft and ray-like, blue or pinkish or white: pappus of unequal bristles about equaling the achene.—Escaped from gardens to waysides; also occasional in grain fields; everywhere admired for the beauty of its flowers.

107. CENTAUREA, *Linnæus.* Annual or biennial herbs of various habit. Leaves unarmed and decurrent, or spinescent and merely sessile. Heads ovate; the imbricated bracts ending in a stout spine. Achenes compressed or somewhat tetragonal, the insertion somewhat oblique or sublateral. Pappus of many slender scabrous bristles, mostly in two sets, or wanting.—All our species naturalized from Europe.

1. C. CALCITRAPA, Linn. Sp. Pl. ii. 917 (1753). Stoutish, rigid, 1–2 ft. high, very widely branching, leaves narrow, laciniate-pinnatifid: the uppermost somewhat involucrate-crowded about the sessile head: principal bracts of the involucre armed with a widely spreading long and rigid subulate spine, which bears 2 or 3 spinules on each side at base: corollas red-purple: pappus none.—Common along roadsides at and near Vacaville.

2. C. MELITENSIS, Linn. l. c. Erect, 2–4 ft. high, cinereous-pubescent, or when young slightly woolly: radical leaves lyrate-pinnatifid; cauline lanceolate, mostly entire, narrowly decurrent: principal bracts of involucre with a slender spine of about their own length which is pectinate-spinulose at base; innermost with spinescent tips: flowers yellow: pappus of very unequal rather rigid bristles or squamellate.—A very common field and wayside weed.

3. C. SOLSTITIALIS, Linn. l. c. Erect, 1–2 ft. high, canescent with cottony wool: radical leaves lyrate-pinnatifid; cauline lanceolate and linear, mostly entire, decurrent on the branches in narrow wings: heads pedunculate: middle bracts of the involucre with a long rigid spreading spine several times their own length, and having 1 or 2 spinules at base; outermost bearing a few small palmate prickles; innermost only scarious-tipped: corollas yellow: pappus double; outer of short and squamellate, inner of longer bristles.—Very common in Napa Valley, and sandy fields along the San Joaquin.

108. CNICUS, *Vaillant.* Annual with sinuate-pinnatifid leaves thinnish and reticulate-veiny, only weakly prickly. Heads enclosed within large leafy accessory bracts. Proper bracts of the involucre thincoriaceous, in few ranks, many or all of them abruptly tipped with an aristiform or spinescent and pectinate-prickly appendage. Receptacle densely setose with long and soft bristles. Achenes terete, strongly many-striate, the corneous margin at summit 10-toothed. Pappus double; each set consisting of 10 aristiform bristles; the outer set longer and naked, the inner short and fimbriolate.

1. C. BENEDICTUS, Linn. Sp. Pl. ii. 826 (1753). (BLESSED THISTLE.) A less rigidly thistle-like annual, with pale yellow flowers in large leafy-involucrate heads; occurring rarely as a ballast waif at San Francisco, but plentiful in fields along the Sacramento River above Sacramento.

109. CENTROPHYLLUM, *Necker.* Annual or biennial. Leaves amplexicaul, reticulate-veiny, rigid, pinnatifid, and spinescent. Outer bracts of involucre foliaceous, with few spinescent lobes; the inner firmer, appressed, not cleft, but with a dilated and spinescent tip. Receptacle densely setose-paleaceous. Flowers yellow. Achenes obpyramidal; those of the marginal row more plump, somewhat convex and gibbous on one side, more angular on the other; those of the disk more regularly and sharply rhombic-tetragonal; all truncate at the broad summit and surrounded by a crenulate margin. Pappus of the outer row of achenes reduced or wanting; of the others double, namely with an outer set of elongated but unequal ciliate paleae in several series, and an uniserial and much shorter inner set.

1. C. LANATUM, DC. in Duby, Bot. Gall. 2 ed. 293 (1828); Linn. Sp. Pl. 830 (1753), under *Carthamus*. A very rigid thistle-like yellow-flowered annual, naturalized about South San Francisco, *Bioletti*.

110. SILYBUM, *Vaillant*. Coarse and stout annual, with very large sinuate-pinnatifid prickly leaves veined and blotched with white. Heads large, at the ends of long pedunculiform branches. Involucre broadly ovoid; bracts foliaceous, spinose along the margins, tapering into a rigid prickle. Receptacle densely soft-bristly. Corollas all alike, red-purple. Stamens monadelphous below. Achenes compressed and smooth. Pappus multiserial, of narrow almost setiform barbellate paleae which are united slightly at base, and deciduous in a ring.

1. S. MARIANUM, Gaertn. Fruct ii. 378 (1802); Linn. Sp. Pl. ii. 823 (1753), under *Carduus*.- One of the most common and troublesome of rank thistles, in all waste lands in California.

111. CYNARA, *Vaillant*. Very stout perennials, with pinnatifid or bipinnatifid sessile though not decurrent leaves; the lobes spinescent-tipped. Heads very large; involucral bracts coriaceous. Receptacle fleshy, fimbrillate. Achenes obovate, compressed and somewhat 4-angled. Pappus of many series of plumose bristles.

1. C. SCOLYMUS, Linn. Sp. Pl. ii. 827 (1753). (ARTICHOKE.) Stout and low, with very ample hoary-tomentulose bipinnatifid leaves, the segments spine-tipped: involucral bracts ovate, obtuse or emarginate.— Escaped from gardens near Benicia, West Berkeley, Alameda, etc.

112. SCOLYMUS, *Tournefort*. Rigid erect herbs, with coarse decurrent pinnately lobed and spinescent leaves, and rather large heads solitary or glomerate at the ends of the branches. Involucre ovate, leafy-bracted at base, the proper bracts scarious-margined and spinescent-tipped. Receptacle paleaceous, the paleae more or less embracing the achenes. Pappus a crown of scarious paleae.

1. **S. Hispanicus**, Linn. Sp. Pl. ii. 813 (1753). Stem several feet high, pubescent, interruptedly winged by the decurrent leaf-bases: heads somewhat spicately crowded in the axils of the upper leaves: flowers yellow: achene beakless; pappus of few unequal scales.—Plentiful near Los Gatos, *Miss Cannon;* apparently of recent introduction.

113. **CARDUUS**, *Tournefort.* (THISTLE.) Stout herbs, mostly biennial. Leaves mostly sessile or decurrent, and with sharply spinose lobes or teeth. Heads large, ovoid or subglobose; the pluriserial and imbricated involucral bracts usually prickly-tipped. Receptacle densely villous-setose. Flowers all alike, crimson, purple or white, the segments of the corolla long and linear-filiform. Achenes obovate or oblong, compressed, smooth, not striate; pappus a single series of long and barbellate or plumose slender bristles concreted at base and deciduous in a ring, often clavellate-dilated at the naked tip.

* *Bracts of involucre with dilated and squarrose-spreading tips; heads short-peduncled.*

1. **C. fontinalis**, Greene, Proc. Philad. Acad., 1892, p. 363 (1893); Bull. Calif. Acad. ii. 151 (1886), under *Cnicus*. Stout, 2 ft. high, the widely spreading branches ending in middle-sized nodding heads: stem and upper surface of leaves glandular-pubescent: bracts of the involucre herbaceous, broad, squarrose-spreading or recurved, abruptly acute, with a short spinose tip and no glandular spot: flowers dull white: anther tips acute.—Along streamlets and in springy places about Crystal Springs, San Mateo Co.; a very peculiar seemingly local species. July.

* * *Only the outer involucral bracts squarrose; terminal spine in all erect.*

← *Heads not conspicuously long-peduncled.*

2. **C. remotifolius**, Hook. Fl. i. 302 (1833). Loosely arachnoid when young, glabrate in age, 3—5 ft. high; leaves from sinuately to deeply pinnatifid: heads on long rather slender peduncles, 1½ in. high, ovate; bracts rather loose and thin, linear-lanceolate, the outer narrowed to a small weak prickle, the middle ones more or less cartilaginous and lacerate as well as somewhat dilated toward the tip; the inner linear-subulate with narrow fimbrillate scarious margin: corollas ochroleucous, the limb only a third as long as the throat.—Marin Co.

3. **C. callilepis**, Greene, l. c. 358. Stem and leaves much as in the last: heads pedunculate, depressed-globose, barely 1 in. high: bracts of the involucre in many series and closely imbricated, the outer broadly obovate, all except the innermost exposing round-ovoid tips with deeply lacerate scarious or sub-cartilaginous margins, and an abrupt short rigid

erect terminal spine; the very innermost with lanceolate scarious-margined and fimbrillate tips: flowers small, ochroleucous, the limb of the corolla only a third as long as the throat.—Mt. Tamalpais and northward, along the coast.

4. **C. amplifolius,** Greene, l. c. 363; Pittonia, i. 70 (1887), under *Cnicus*. Somewhat fleshy, green and glabrous except a sparse arachnoid tomentum on the lower face of the leaves: stem stout, 3 4 ft. high: leaves very broad and ample, conspicuously decurrent, the ample lobes crowded and overlapping each other, trifid and spinose-ciliate: heads short-peduncled and clustered, 1½ in. high, leafy-bracted at base; the outer bracts loosely spreading and arachnoid, the inner more appressed and glabrous, their spinose tips often reflexed: corollas of a rich bright purple, their linear obtuse lobes much shorter than the throat.—On stream banks back of Point San Pedro.

5. **C. crassicaulis,** Greene, l. c. 357. Very stout and tall, 4–7 ft. high: stem strongly striate, simple to near the summit, there becoming rather close-paniculate, with 3–7 short-peduncled heads 1½–2 in. high: herbage permanently hoary-lanate: leaves small, pinnately parted, the segments spine-tipped and the whole margin spinulose-ciliate: involucral bracts rather loose, linear-lanceolate to lanceolate-acuminate, all tipped with a slender spine, the outer and middle ones with pectinate-spinescent margins: segments of the whitish or pinkish corolla about as long as the throat.—Abundant in moist grassy bottoms of the lower San Joaquin.

6. **C. edulis,** Greene, l. c. 362; Nutt. Trans. Am. Phil. Soc. vii. 420 (1841), under *Cirsium*. Robust, 3–6 ft. high, pubescent, leafy up to the short panicle: leaves oblong or narrower, sinuate-pinnatifid, weakly prickly: heads 1½ in. high, depressed-globose, leafy-bracted at base: involucre arachnoid when young: corollas deep but dull purple; segments shorter than the throat.—Along streams in the Coast Range.

++ *Heads solitary on stout peduncles.*

7. **C. Californicus,** Greene, l. c. 359; Gray, Pac. R. Rep. iv. 112 (1857), under *Cirsium*. Rather slender, 2–4 ft. high, canescently woolly: leaves sinuate-pinnatifid, moderately prickly, heads middle-sized, the lower bracts coriaceous-acerose, spreading and incurved, the others straight, all subulate-spinescent at the tip: corollas lilac-purplish or reddish, lobes shorter than the throat.—Mt. Diablo Range, and foothills of the Sierra in Stanislaus Co., thence southward. June, July.

8. **C. Andersonii,** Greene, l. c.; Gray, Proc. Am. Acad. x. 44 (1874), under *Cnicus*. Scarcely woolly except as to the lower face of the thin-

nish scarcely pinnatifid, often merely toothed leaves: stem 2—3 ft. high, slender, only sparingly leafy above the base: heads 1–3, long-peduncled: bracts of the almost turbinate involucre narrowly lanceolate, gradually attenuate, merely spinescent at the erect tips, the inner ones very long and thin: lobes of the crimson corolla about equaling the throat: style with long filiform appendage and indistinct node.—Subalpine woods and open places in the Sierra Nevada. July—Oct.

9. **C. Andrewsii.** *Cnicus Andrewsii*, Gray, Proc. Am. Acad. x. 45 (1874). Probably tall and paniculate, the loose wool deciduous except from the heads: cauline leaves lanceolate, laciniate-pinnatifid: involucres 1 in. high or more, their bracts with coriaceous oblong-ovate base, abruptly contracted into an aristiform spinescent appendage, the whole involucre cobwebby: corolla whitish, its lobes twice the length of the throat: anther tips deltoid, acute.—Known in but a single imperfect specimen supposed to have been obtained somewhere in middle California.

10. **C. candidissimus**, Greene, l. c. 359. Stout, erect, 2—3 ft. high, densely and permanently white arachnoid-tomentose throughout; leaf-outline as in the last; heads few, on shorter and stouter peduncles, 2 in. high; outer bracts of the involucre with dilated and closely appressed base and squarrose rigid linear-acerose spinescent tip, all densely arachnoid-tomentose; flowers crimson; pappus an inch long, plumose almost throughout.—Dry hills along our northeastern borders; also along the seaboard at Santa Barbara. June—Aug.

11. **C. venustus**, Greene, l. c. Stoutish, 3 ft. high, the foliage permanently arachnoid-tomentose: heads large (2 in. high), terminating long and almost naked pedunculiform branches: involucre glabrate, the many subcoriaceous bracts with closely appressed base, and long lanceolate-subulate abruptly short-spinescent tips: corollas bright crimson, the segments longer than the throat: pappus-bristles barbellate above the plumose part, the tips scarcely dilated.—Foothills of the Coast and Mt. Diablo Ranges. May, June.

12. **C. occidentalis**, Nutt. Trans. Am. Phil. Soc. vii. 418 (1841). Stout, 2—3 ft. high; peduncles stout, rather short: leaves deeply pinnatifid, glabrate above, canescently tomentose beneath: involucre subglobose; bracts straight, subulate-lanceolate, with short spines, the whole mass densely festooned with remarkably distinct cobwebby hairs: corollas red-purple: anthers distinctly bisetose and lacerate at base: pappus somewhat scanty.—Sandy hills along the seaboard only. May—Aug.

* * * *Involucral bracts appressed, the slender spine at their tips more or less abruptly spreading.*

13. **C. hydrophilus,** Greene, l. c. 358. Rather slender, freely branching above, 3—5 ft. high; when young pale with a fine thin arachnoid pubescence, in maturity green and glabrate: leaves not large, deeply cut into uniform 3-lobed segments: heads 1 in. high, somewhat clustered at the ends of the branches; involucre ovate, the appressed-imbricate bracts with a green and glutinous ridge toward the summit, and ending in a short slender somewhat spreading spine.- Brackish marshes about Suisun Bay. July– Sept.

14. **C. inamœnus.** Habit of the last, but smaller, 2 -3 ft. high, hoary-tomentose even in age: heads as small, similarly glomerate in twos or threes at the ends of the branches, but bracts of the broader and more spherical involucre more numerous and more closely imbricated, ovate, destitute of the glutinous ridge, and tipped with a more distinct and prominent slender spine: segments of the dull-white corolla scarcely half the length of the throat.—Common on the eastern foothills of the Sierra Nevada, from Plumas Co. southward to the Truckee Valley. It forms a part of the *Cnicus Breweri Vaseyi* of Gray, but is not the plant of Vasey's collecting. It is also my *C. undulatus Nevadensis.* July—Sept.

15. **C. Drummondii.** *Cirsium Drummondii,* T. & G. Fl. ii. 459 (1843). Dwarf, subacaulescent, the stout stem only a few inches high, very leafy, terminating in a few clustered subsessile small heads with pale purplish flowers: lanceolate pinnatifid leaves green above, slightly arachnoid-tomentose beneath, ciliate-spinulose: bracts of the involucre ovate and ovate-lanceolate, ending in a short often suberect stout spine: segments of the corolla shorter than the throat.—Moist subalpine meadows on the eastern slope of the Sierra. July—Sept.

16. **C. validus.** Very robust, 1—2 ft. high, from a biennial root, leafy throughout, pale-green and apparently glabrous, a lens revealing a short and scanty arachnoid pubescence especially on the lower face of the ample deeply laciniate-pinnatifid leaves, the lobes of these deeply cleft and rigidly spinescent: large subsessile or stout-peduncled heads 2 in. high, ovate; bracts numerous, closely imbricated, ovate and oblong-lanceolate, coriaceous, with rigid subulate and spinescent spreading tips: corollas white, their segments little shorter than the throat: pappus-bristles very fine, more than 1 in. long, naked at the slender tip, otherwise sparsely soft-plumose.- Hills of the interior, in Kern Co., etc. Has been confused with *C. quercetorum.* June, July.

17. **C. cymosus.** Stout and rigid, the strongly striate-angled and deciduously floccose-woolly stem 3—4 ft. high, simple below, shortly and cymosely branched at summit, with many short-peduncled heads forming a somewhat flat-topped general inflorescence: leaves rather small, pinnatifid and spinescent, white-floccose on both faces: heads about 1½ in. high; bracts of the involucre not very numerous, the outer ovate and lanceolate, closely appressed except at the stoutly spinescent tip, flocculent-tomentose, destitute of glutinous spot, the inner with linear-elongated thin tips: corolla dull-white, the segments little shorter than the throat: pappus with dilated and barbellate tips above the plumose part.—Dry hills of Alameda and Contra Costa counties. June, July.

18. **C. Breweri**, Greene, l. c. 363; Gray, Proc. Am. Acad. x. 43 (1874), under *Cnicus*. Permanently and densely white-woolly, 4—10 ft. high: leaves numerous and rather ample, pinnatifid: heads many, panicled, and often rather densely so, at summit of stem; involucre round-ovate, only 1 in. high, arachnoid when young, the outer bracts short and broad, with a glutinous or glandular spot at the tip, the prickle weak: lobes of the purple or whitish corolla shorter than the throat: tips of the anthers almost obtuse.—In the Sierra Nevada at middle elevations; also near the level of the sea in Mendocino and Monterey counties. June—Aug.

19. **C. quercetorum**, Greene, l. c. 362; Gray, Proc. Am. Acad. x. 40 (1874), under *Cnicus*. Perennial by branching horizontal rootstocks: sparingly villous-arachnoid when young, soon glabrate: stem stout, 1 ft. high or less, with few rather large heads: leaves mostly petiolate, the larger 1 ft. long, pinnately parted, the oblong divisions often 3—5-cleft, prickly: involucral bracts coriaceous, closely imbricated in numerous ranks, the outer with short prickles, the inner obscurely scarious at tip: corollas either dull-purple or white: anther-tips narrow, very acute.—Open grassy summits and higher slopes of hills, chiefly, if not exclusively, in the region of San Francisco Bay. June, July.

20. C. LANCEOLATUS, Linn. Sp. Pl. ii. 821 (1753). More or less villous or hirsute, seldom cottony; 2—4 ft. high, stem and branches interruptedly winged by the decurrent leaves, both leaves and wings prickly: heads nearly 2 in. high, arachnoid-woolly at first, the bracts lanceolate, attenuate into slender and rigid prickle-pointed spreading tips: fl. rose-purple.—A most troublesome Old World weed, already abundant on this coast far northward, only lately established in central California.

INDEX OF GENERA.

	PAGE.		PAGE.
Achyrachæna	431	Conyzella	385
Actinolepis	441	Corethrogyne	378
Adenocaulon	403	Cotula	457
Amblyopappus	450	Crocidium	459
Ambrosia	404	Cyanus	473
Anaphalis	398	Cynara	475
Ancistrocarphus	403	Deinandra	424
Anisocarpus	415	Dipsacus	355
Antennaria	397	Eastwoodia	361
Arnica	462	Eatonella	440
Artemisia	454	Ecliptica	406
Aster	380	Erigeron	386
Baccharis	395	Eriophyllum	442
Baeria	437	Eupatorium	357
Balsamorrhiza	407	Euthamia	372
Bellis	385	Filago	400
Bidens	411	Gaertnera	405
Blennosperma	451	Gnaphalium	399
Blepharipappus	428	Gnaphalodes	403
Blepharizonia	419	Grindelia	361
Brachyactis	395	Gutierrezia	360
Cacaliopsis	458	Harpæcarpus	416
Calycadenia	420	Hazardia	374
Carduus	476	Heleniastrum	435
Centaurea	473	Helianthella	410
Centromadia	426	Helianthus	409
Centrophyllum	474	Hemizonia	418
Chænactis	446	Hesperevax	401
Chamæmelum	452	Heterotheca	364
Chrysanthemum	453	Holocarpha	426
Chrysoma	370	Holozonia	432
Chrysopsis	364	Hulsea	449
Chrysothamnus	368	Isocoma	371
Cnicus	474	Iva	404
Coleosanthus	357	Jaumea	451

INDEX OF GENERA.

	PAGE.		PAGE.
Lagophylla	431	Pyrrocoma	366
Lasthenia	436	Raillardella	461
Lembertia	441	Rigiopappus	450
Lepidospartum	459	Rudbeckia	407
Leptosyne	411	Scabiosa	355
Lessingia	375	Scolymus	475
Leucelene	384	Senecio	464
Leucosyris	384	Sericocarpus	380
Luina	460	Silybum	475
Machæranthera	379	Solidago	373
Macronema	367	Soliva	457
Madaria	417	Stenotus	366
Madia	413	Stylocline	401
Matricaria	452	Tanacetum	454
Millefolium	453	Tetradymia	460
Monolopia	440	Trichocoronis	357
Oreastrum	384	Valeriana	353
Oreochænactis	448	Valerianella	353
Pentachæta	363	Venegasia	434
Petasites	458	Whitneya	461
Pluchea	397	Wyethia	408
Psathyrotes	459	Xanthium	405
Psilocarphus	402	Zonanthemis	425
Ptilonella	433